S0-DLQ-835

Fixed-Mobile Wireless Networks Convergence

Technologies, Solutions, Services

Do you need to understand the technical solutions and associated services that allow multimedia communications between established mobile cellular networks and any form of fixed wireless communications? If so, this practical book, presenting the fundamentals of individual fixed and mobile wireless technologies in terms of architectures, standards, management capabilities, and quality of service issues, is essential reading.

Adopting the term Fixed-Mobile Convergence (FMC), an analysis of the interworking between cellular networks and a variety of wireless technologies such as WLAN, WiMAX, RFID and UWB is provided. An in-depth study of the convergent solutions offered by UMA and IMS is also given, in addition to the commercial realities of implementing convergent solutions. Up-to-date information about technical solutions, products, vendors, and current service offerings is included. You'll also find criteria for analyzing and evaluating fixed-mobile convergent products and services, and numerous diagrams and feature/component tables. This practical text is ideal for engineers and practitioners in the field of telecommunications and wireless communications, as well as for graduate students of electrical and computer engineering.

Joseph Ghetie is a Network and Systems Engineer Consultant and Instructor for his own technical consulting and training business TCOM & NET. He previously worked for Telcordia Technologies (Bell Communications Research) for over 13 years, where he was responsible for developing architectures, requirements, and designing solutions for network management integration, providing consulting, and supporting management standards development. He has taught over 35 tutorials at major international conferences and symposia, and has authored 27 advanced technical courses in data communications, telecommunications, Internet networking, network and service management, and wireless communications.

Fixed-Mobile Wireless Networks Convergence

Technologies, Solutions, Services

JOSEPH GHETIE

CAMBRIDGE UNIVERSITY PRESS
Cambridge, New York, Melbourne, Madrid, Cape Town, Singapore, São Paulo, Delhi

Cambridge University Press
The Edinburgh Building, Cambridge CB2 8RU, UK

Published in the United States of America by Cambridge University Press, New York

www.cambridge.org
Information on this title: www.cambridge.org/9780521513562

First published 2008

Printed in the United Kingdom at the University Press, Cambridge

A catalog record for this publication is available from the British Library

Library of Congress Cataloging in Publication data

Ghetie, Iosif G.
Fixed-mobile wireless networks convergence : technologies, solutions, services / Joseph Ghetie.
p. cm.
Includes bibliographical references and index.
ISBN 978-0-521-51356-2 (hardback)
1. Internet telephony. 2. Mobile communication systems. 3. Cellular telephone systems. 4. Telephone. 5. Convergence (Telecommunication) 6. Bluetooth technology. 7. Radio frequency identification systems. I. Title.
TK5105.8865.G52 2008
004.6′2 – dc22 2008015446

ISBN 978-0-521-51356-2 hardback

I dedicate this book to my granddaughters,
Nadia and Talia,
with the wish they discover that science is the key to understanding our universe and that art is the key to understanding human beings and the human spirit.

Contents

Part III Fixed Wireless Technologies: Networking and Management

Disclaimer

Product and service information contained in this book are primarily based on technical reports and documentation, including publicly available information received from sources believed to be reliable. However, neither the author nor the publisher guarantees the total accuracy and completeness of information published herein. Neither the author nor the publisher shall be responsible for any errors, omissions, or damages arising out of use of this information. Any mention of products, suppliers, and service providers in this book is done where necessary for the sake of scientific accuracy or for background information to provide an example of a technology for illustrative or clarification purposes. No information provided in this book is intended to be or shall be construed to be as either positive or negative endorsement, certification, approval, recommendation or rejection of any product, supplier, application, or service.

How the Book is Organized

The book consists of six major parts organized into 20 chapters. The first part describes the fundamental concepts of wireless communication and networking along with the concepts of network and service management. The second part is focused on one of the entities of convergence, cellular mobile radio networks with their network and service management capabilities. The third part presents the other side of the convergence equation, the fixed wireless technologies that span from local area networks, to personal area networks, to near-field sensor networks, and metropolitan access networks. The fourth part introduces all the elements of fixed-mobile convergence analyzing the particular architectural solutions, products, and services that result from the integration between each form of fixed wireless network and mobile cellular network. Several convergent implementations case studies are analyzed. The fifth part is dedicated to an in-depth analysis of two major standardized sets of solutions and specifications that provide a total approach to convergence, namely, the Unlicensed Mobile Access (UMA) and IP-based Multimedia Subsystem (IMS) with particular references to signaling using the Session Initiation Protocol (SIP). The sixth part provides an overall analysis of convergent services, quality of service, service providers, industry trends, economics of convergence, and evaluation criteria for fixed-mobile solutions/products as well as issues regarding design, development, and implementation of fixed-mobile convergence.

The organization of this book is simple. First, we introduce the palette of individual wireless technologies with a magnifying glass to show what is important and relevant to convergence, i.e., architectures, standards, management capabilities, and quality of services. Then we look at the individual approaches taken for converging pairs of fixed wireless and cellular mobile networks. Next, we provide a more in-depth look at the global convergent solutions offered by UMA and IMS. Finally, we analyze the marketing projection of convergent solutions, keeping an eye on the real issues when implementing these solutions.

Part I. Wireless Communications: Networking and Management

Chapter 1 is an overview of the world of wireless communications as supported through various architectures with a diverse set of components and spectrum allocations.

Chapter 2 presents the overall concept of open network management systems, management platforms, layered communication architectures, management protocols, and specific management requirements for wireless communications along with network management products.

Chapter 3 presents the service management concept and its components, classes of services, quality of service, and service level agreements/specifications along with specific service management products.

Part II. Cellular Mobile Radio Networking and Management

Chapter 4 provides an overview of cellular mobile radio networking architectures, and radio link access methods. Identification of spectrum allocation for specific technologies, standards organizations, and overall generational evolution steps are all intended to provide a common view of the primary entity of the fixed-mobile convergence equation.

Chapter 5 investigates specific aspects of cellular mobile networks and service management with focus on GSM/GPRS classes of services and GPRS service profiles.

Part III. Fixed Wireless Technologies: Networking and Management

Chapter 6 analyzes the wireless local area network environment from architecture to specific standards implementation with a special focus on challenges raised by transmitting voice over wireless LANs, a key requirement for convergence.

Chapter 7 analyzes the wireless personal area network environment from architectures to major standards implementations (Bluetooth and ZigBee) including analysis of QOS in wireless PAN environment.

Chapter 8 analyzes the evolution of wireless metropolitan access solutions with focus on WiMAX architecture, solutions, standards, and products.

Chapter 9 is dedicated to the analysis of near-field sensor networks from architectures to major standard implementations (RF Identification, Near Field Communications and Ultra Wide Band) along with network and service management capabilities.

Part IV. Fixed Wireless Cellular Mobile Networks Convergence and Integration

Chapter 10 provides an overview of the fixed-mobile concept that includes terminology, architectural components, interfaces, and protocols, including the overall requirements, solutions, and technical forums created to advance fixed-mobile convergent concepts.

Chapter 11 presents the specific architectural solutions, applications, products, and services created as part of the technical effort to provide convergence between wireless LANs (IEEE 802.11 a/b/g/n and WLAN Mesh) and cellular mobile networks. Several convergent implementations case studies are analyzed.

Chapter 12 presents the specific architectural solutions, applications, products, and services born out of the convergence between wireless PANs (Bluetooth and ZigBee) and cellular mobile networks. Several convergent implementations case studies are analyzed.

Chapter 13 presents the specific architectural solutions, applications, products, and services aimed at incorporating wireless metropolitan access, represented by WiMAX technology, into wide area cellular mobile networks. Several convergent implementations case studies are analyzed.

Chapter 14 presents the specific architectural solutions, applications, products, and services created as part of the technical effort to extend fixed-mobile convergence to incorporate the convergence between wireless near-field sensor networks (RFID, NFC, UWB) and cellular mobile networks. Several convergent implementations case studies are analyzed.

Part V. Fixed Wireless Cellular Mobile Convergence: Standardized Networking Solutions

Chapter 15 is a comprehensive presentation of the Unlicensed Mobile Access (UMA) set of standard specifications. It includes an overview of UMA evolution, driving forces, and specific architectures to incorporate the convergence aspects in a unique approach that spans wireless LANs, PANs, WiMAX, and sensor networks. Details about UMA network design, the UMA Network Controller, signaling protocols for voice and data communications, and interoperability parameters in UMA operations are supplemented by UMA-based fixed-mobile convergence solutions and products.

Chapter 16 is a comprehensive analysis of the signaling protocol of choice for IMS-based networks, namely, the Session Initiation Protocol (SIP). Details about SIP development, acceptance, applicability and actual standards include SIP message formats, fields, optional parameters and specific implementation of this protocol in voice over IP operations and IMS-based network design and operations.

Chapter 17 is a comprehensive presentation of the IP-based Multimedia Subsystem (IMS) set of standard specifications. It includes an overview of IMS evolution, driving forces, high-level conceptual architecture, specific architectural components, sub modules, functions, and the all-important stepping-stones for convergence not only within the wireless world but also between wired and wireless networks. Details about standards, proposed amendments, signaling specifications, and interoperability parameters in IMS operations are supplemented by analysis of partial implementations of IMS-based fixed-mobile convergence products and services.

Part VI. Fixed-Mobile Convergence Services, Industry Trends, and Implementation Issues

Chapter 18 looks into Quality of Service (QOS) aspects of fixed wireless cellular mobile networking solutions from the perspective of both voice and data parameters, and associated QOS metrics. Integrated/Differentiated Services, Multi-Protocol Label Switching, and Policy-based Management are also analyzed. Specific services and service providers of convergent solutions in both UMA-based and IMS-based networks are identified and analyzed.

Chapter 19 provides the economic perspective of fixed-mobile convergence starting with the driving forces and opponents of convergence, This is followed by diagrams depicting projected growth of various convergent applications in the domain of wireless LANs, PANs, WiMAX, and sensor networks along with the growth of the telecommunications industry in general, and the mobile communications sector in particular. A comprehensive set of evaluation criteria was developed to analyze and qualify fixed-mobile convergent solutions and products.

Chapter 20 is a summary of benefits of fixed-mobile convergence followed by an analysis of the state of implementations and trends in fixed-mobile convergence. A detailed list of issues confronting fixed-mobile convergence implementation in both UMA-based and IMS-based versions concludes this chapter.

List of Figures

Chapter 1

Chapter 2

Chapter 3

Chapter 4

Chapter 5

Chapter 6

Chapter 7

Chapter 8

Chapter 9

Chapter 10

Chapter 11

Chapter 12

Chapter 13

Chapter 14

Chapter 15

Chapter 16

Chapter 17

Chapter 18

Chapter 20

List of Tables

Chapter 1

Chapter 2

Chapter 3

Chapter 4

Chapter 5

Chapter 6

Chapter 7

Chapter 8

Chapter 9

Chapter 13

Chapter 14

Chapter 16

Chapter 17

Chapter 18

Preface

The past two decades have marked an unprecedented growth in size and sophistication of almost every aspect of telecommunications. Two major achievements stand out in this development. First, the **Internet** network, the quintessence of data communications and public/private access to information. Second, **Cellular Mobile Radio** wireless communication, the untethered jewel of voice communication beyond borders that opened another door to worldwide conversation. As a result, today, there are hundreds of millions of users with continuous access to the Internet and over three billion users of cell phones.

Cellular mobile communications has moved in a decade through three generations. The first generation was marked by the need to build a large physical/geographical presence. The second generation has moved from analog formats to digital communications and added features that allow decent data services running along with voice communications. More recent advancements have led to the third generation of cellular mobile transmission with a focus on providing data services in the class of broadband communications.

In the same span of time, a similar growth has taken place in the wireless expansion of well-established networking technologies, commonly known as "**fixed wireless**". The most popular customer premises technology, the Local Area Network, has become **Wireless LAN**, matching its data rate with wired frontrunners. This metamorphosis has continued by adopting voice communication over a traditional data dedicated technology. Identical phenomena took place in the **Wireless Personal Area Network** thanks to the Bluetooth, ZigBee, and other technologies. On the same token, the **Near-Field Sensor Networ**ks, once liberated from the limitations of infrared and laser barcode-like technologies, found their own voice in the form of **Radio Frequency ID, Near Field Communications,** and **Ultra Wide Band USB** technologies. After some shaky attempts to provide wireless Metropolitan Area Network solutions, **WiMAX** technology has been accepted as the standard technology that brings wireless broadband access to the home and small businesses. The story of fixed wireless technologies does not end here because various levels of mobility have been added making the initial name "fixed wireless" somehow a misnomer.

Consequently, it was just a matter of time in the chase of ideal communication "**anytime, anywhere, any technology**", to realize the need of transparent communications for any form of media, be that data, voice, or video. Hence, the need for interoperability between cellular mobile radio networks of latest generations and any form of fixed wireless networks, i.e., **Fixed-Mobile Convergence**, the very subject of this practical book. Similarly, the obvious outcome of convergent solutions has become the use of **one handset/cell phone and one portable number** that works across any wireless network capable of transmitting digital information at the highest possible/available data rate.

Acknowledgments

First and foremost, I would like to thank my friend and colleague Gene Geer. His help in reviewing and editing the entire book, his comments and questions, and his utmost grace and diligence in improving the material made his contributions invaluable. Thank you Gene.

And, I thank three field colleagues who reviewed my initial proposal and who provided valuable feedback and guidance in writing this book: Dr. Petre Dini, Cisco Systems, a tireless promoter of technical conferences that always provided excellent sources and opportunities to exchange information; Dr. Michael Kincaid, MSK and Associates, who introduced me to the world of wireless communications almost two decades ago; and Victor Arabagian, Lucent/Alcatel, an engineer, who is confronting the reality of business constraints in the wired and wireless worlds.

I would like to express my gratitude and thanks to the management, editors, and contributors to technical magazines such as IEEE Communications, IEEE Network, Business Communications Review, Wireless Week, Wireless Design and Development, Network World, Telecommunications, and Telephony for the interesting and informative articles they have published on the theory, practice, product features, and real case studies related to fixed-mobile convergence.

And I thank, without naming, the staff of all companies that provided technical information and documentation of the many products and networking solutions I describe in the book. I thank the organizers of CEBIT exhibitions in Germany, an excellent source of products, information, and demonstrations. My heart goes to my friends Dieter and Marianne Eisenburger as well as Carmen Soare and Dan Pupeza for their utmost hospitality over several years while I attended the CEBIT exhibitions. I feel fortunate and I wish everyone could have friends like them.

Exactly one year ago, a gentleman from Cambridge University Press, Mr. Phil Meyers, approached me with a proposal to write a book that would capture the amazing developments taking place in the world of wireless communications in general, and in the realm of fixed cellular mobile networks convergence in particular. The goal of this convergence is straightforward: Transparent communications across all wireless technologies having as the centerpiece multi-mode cell phones or handsets. This goal is no longer just a dream; it has captured the imaginations of the technical community and the prospective customers who are discovering the true meaning of communications "anytime, anywhere" regardless of the wireless technology used.

Here I am after one year, writing, drawing, thinking, reading, correcting material, teaching and consulting, working almost every day to build the content of this book. And, as in any of my other achievements, I could not have succeeded without the direct support, understanding, and encouragement from my wife Veronica. Thank you very much Vera.

Joseph Ghetie
Fort Lee, December 30, 2007

Acronyms

A

Abstract Syntax Notation 1 (ASN.1)
Access Modules (AM)
Access Point (AP)
Access Tandem (AT)
Accounting Management (AM)
Action Request System (ARS)
Adaptive Differential Pulse Code Modulation (ADPCM)
Advanced Encryption Standard (AES)
Advanced Mobile Phone Systems (AMPS)
Advances to IMS (A-IMS)
Amplitude Modulation (AM)
AOL Instant Messaging (AIM)
Application Header (AH)
Application PDU (A-PDU)
Applications Programming Interfaces (API)
Applications Server (AS)
Association Control Service Element (ACSE)
Assured Forwarding (AF)
Asynchronous Transmission Mode (ATM)
Authentication, Authorization, Access (AAA)
Automatic Call Distribution (ACD)
Automatic Number Identification (ANI)
Average Return Per Unit (ARPU)

B

Base Station (BS)
Base Station Controller (BSC)
Base Station Subsystem (BSS)
Base Station Subsystem Management Application Part (BSSMAP)
Base Station Subsystem Protocol (BSSP)

Base Transceiver Station (BTS)
Basic Encoding Rules (BER)
Basic Trading Areas (BTA)
Best Effort (BE)
Beyond Third Generation (B3G)
Binary Phase Shift Keying (BPSK)
Bit Error Rate (BER)
Bluetooth Access Points (BAP)
Bluetooth Network Encapsulation Protocol (BNEP)
Border Gateway Control Function (BGCF)
Border Gateway Protocol (BGP)
Breakout Gateway Control Function (BGCF)
Broadband Access Server (BAS)
Broadband Networks Access Group (BRAN)
Broadband Radio Access Network (BRAN)
Business Management Layer (BML)

C

Cable Television (CATV)
Call Control (CC)
Call Session Control Function (CSCF)
CAMEL Applications Part (CAP)
Capital Expenditures (CAPEX)
Carrier Sense Multiple Access/Collision Avoidance (CSMA/CA)
CDMA 1xRTT One Carrier Radio Transmission Technology
Cellular Data Packet Data (CDPD)
Central Office (CO)
Centro de Tecnologia de la Comunicationes (CETECOM)
Channel Service Unit/Data Service Unit (CSU/DSU)
Class Based Queuing (CBQ)
Class 1 Generation 2 (C1G2)
Class of Service (COS)
Cluster Controllers (CC)
Code Division Multiple Access (CDMA)
Coding Scheme (CS)
Command Line Interface (CLI)
Commercial off-the-Shelf (COTS)
Common Channel Signaling (CCS)
Common Management Information Protocol (CMIP)
Common Management Information Service Element (CMISE)
Common Object Request Broker Architecture (CORBA)
Common Open Policy Service (COPS)

Complementary Code Keying (CKK)
Complex Instruction Set Computers (CISC)
Compound Annual Growth Rate (CAGR)
Computer Telephony Integration (CTI)
Configuration Management (CM)
Connectionless Network Protocol (CLNP)
Control and Provisioning of Wireless Access Points (CAPWAP)
Convolutional Coding (CC)
Core Network (CN)
Custom Local Area Signaling Services (CLASS)
Customer Network Management (CNM)
Customer Premises Equipment (CPE)
Customer Premises Network (CPN)
Customer Service Management (CSM)
Customized Applications for Mobile (Networks) Enhanced Logic (CAMEL)

D

Data Communications Network (DCN)
Data Exchange Format (DEF)
Data Link Connection Identifier (DLCI)
Denial of Service (DoS)
Dense Wavelength Division Multiplexing (DWDM)
Destination Address (DA)
Destination Service Access Point (DSAP)
Detect and Avoid (DAA)
Device Wire Adapter (DWA)
Device Wireless Bridge (DWB)
Differential Quadrature Phase Shift Keying (DQPSK)
Differentiated Services (Diff Services)
Differentiated Services Code Point (DSCP)
Digest Authentication (DA)
Digital Advanced Mobile Phone Systems (D-AMPS)
Digital Encryption System (DES)
Digital Enhanced Cordless Telecommunication (DECT)
Digital Subscriber Line (DSL)
Direct Sequence Spread Spectrum (DSSS)
Directory Name Services (DNS)
Discreet Multitone (DMT)
Domain Name System (DNS)
Downlink (DL)
Dual-Mode Headset (DMH)
Dynamic Host Control Protocol (DHCP)

E

Electronic Number Mapping (ENUM)
Electronic Product Code (EPC)
Electronic Product Code Information System (EPCIS)
Element Management Layer (EML)
Element Management System (EMS)
Enhanced Data for GSM Evolution (EDGE)
Enhanced Data Rate (EDR)
Enhanced UTRAN (E-UTRAN)
Enhanced Version, Data Only, Data and Voice (EV-DO, EV-DV)
Enhanced Wireless Consortium (EWC)
Enterprise Resource Planning (ERP)
Erasable Programmable Read-Only Memory (EPROM)
Ethernet Passive Optical Network (EPON)
European Computer Manufacturers Association (ECMA)
European Telecommunications Standard Institute (ETSI)
Expedite Forwarding (EF)
Extended Markup Language (XML)
Extensible Authentication Protocol (EAP)
Extensible Authentication Protocol over LAN (EAPOL)

F

Fault Management (FM)
Federal Communications Commission (FCC)
Fiber Distributed Data Interface (FDDI)
Fiber to the Curve (FTTC)
Fiber to the Home (FTTH)
Field Programmable Gate Array (FPGA)
First In First Out (FIFO)
Fixed-Mobile Convergence (FMC)
Fixed-Mobile Convergence Alliance (FMC Alliance)
Forwarding Equivalence Class (FEC)
Frame Relay (FR)
Free Space Optics (FSO)
Frequency Division Duplex (FDD)
Frequency Division Multiple Access (FDMA)
Frequency-Hopping Spread Spectrum (FHSS)
Frequency Modulation (FM)
Front End Processors (FEP)
Functional Modules (FM)

G

Gateway (GW)
Gateway GPRS Support Nodes (GGSN)
Gateway Mobile Switching Centers (GMSC)
Gaussian Minimum Shift Keying (GMSK)
General Packet Radio Service (GPRS)
Generic Access Network (GAN)
Generic Access Network Controller (GANC)
Geostationary Earth Orbit (GEO)
Global Information Infrastructure (GII)
Global Multi-Service Interoperability (GMI)
Global Positioning Satellite (GPS)
Global Systems for Mobile (GSM) Telecommunications
GPRS Support Network (GSN)
GPRS Tunneling Protocols (GTP)
Graphical User Interface (GUI)
GSM Applications (GSMA)
GSM Interworking Profile (GIP)
GSM/GPRS EDGE Radio Access Network (GERAN)
Guaranteed Time Slot (GTS)
Guideline for the Definition of Managed Objects (GDMO)

H

High Definition Television (HDTV)
High Frequency (HF)
High Speed Circuit Switched Data (HSCSD)
High Speed Downlink Packet Access (HSDPA)
High Speed Uplink Packet Access (HSUPA)
Home Location Register (HLR)
Home Subscriber Server (HSS)
Host Wire Adapter (HWA)
Hub Earth Station (HES)
Hybrid Fiber Coax (HFC)
Hypertext Transfer Protocol (HTTP)

I

Improved Mobile Telephone System (IMTS)
IMS Media Gateways (IMS MGW)

Industrial Scientific Medical (ISM)
Initial User Assignment (IUA)
Instant Messaging (IM)
Institute of Electrical and Electronics Engineers (IEEE)
Integrated Access Device (IAD)
Integrated and Differentiated Services (Int/Diff Services)
Integrated Go-to-Market Network IP Telephony Experience (IGNITE)
Integrated Optical Network (ION)
Integrated Services (IntServices)
Integrated Services Digital Network (ISDN)
Intelligent Network Application Part (INAP)
Interactive Voice Response (IVR)
International Telecommunication Union Radio Communication Sector (ITU-R)
International Telecommunication Union Telecommunications Sector (ITU-T)
Internet Assigned Number Authority (IANA)
Internet Engineering Task Force (IETF)
Internet Header Length (IHL)
Internet Nodal Processors (INP)
Internet Protocol (IP)
Internet Protocol over ATM (IP over ATM)
Internet Service Provider (ISP)
International Electrotechnical Commission (IEC)
International Mobile Equipment Identity (IMEI)
International Mobile Subscriber Identity (IMSI)
International Mobile Telecommunications 2000 (IMT-2000)
International Organization for Standardization (ISO)
Interrogating Call Session Control Function (I-CSCF)
Inter-System Handover (ISH)
Interworking Profile (IP)
IP-based Multimedia Subsystems (IMS)
IP-based Multimedia Subsystem Alliance (IMS Alliance)
IP Multimedia Service Control (ISC)
IP Multimedia Service Switching Function (IM-SSF)
ISDN User Part (ISUP)

J

Japanese Total Access Communications System (JTACS)

K

Key-Value Pair (KVP)

L

Label Distribution Protocol (LDP)
Label Edge Routers (LER)
Label Switching Paths (LSP)
Label Switching Routers (LSR)
Light Emitting Diode (LED)
Lightweight Directory Protocol (LDAP)
Lightweight Extensible Authentication Protocol (LEAP)
Line of Sight (LOS)
Link Header (LH)
Link Trailer (LT)
Liquid Crystal Display (LCD)
Local Area Network (LAN)
Local Multipoint Distributed Service (LMDS)
Logical Link Control (LLC)
Logical Link Control and Adaptation Protocol (LLCAP)
Long-Term Evolution (LTE)
Low Earth Orbit (LEO)
Low Frequency (LF)

M

Machine-to-Machine (M2M)
Major Trading Areas (MTA)
Management Application Platform (MAP)
Management Applications Programming Interface (MAPI)
Management Information Base (MIB)
Management Information Library (MIL)
Management Platform External Interface (MPEI)
Manager of Managers (MOM)
Master Station (MS)
Maximum Transmission Unit (MTU)
Mean Opinion Score (MOS)
Mean Time Between Failures (MTBF)
Mean Time To Provision (MTTP)
Mean Time To Repair (MTTR)
Mean Time To Report (MTTR)
Media Access Control (MAC)
Media Gateway (MGW, MG)
Media Gateway Controller (MGC)
Media Gateway Control Function (MGCF)

Media Gateway Control Protocol (MGCP)
Media Independent Handover (MIH)
Media Resource Control Function (MRCF)
Message Transfer Part (MTP)
Microprocessor Control Units (MCU)
Middle Earth Orbit (MEO)
Mobile Ad-hoc Networks (MANET)
Mobile Application Part (MAP)
Mobile Commerce (M-Commerce)
Mobile Core Network (MCN)
Mobile Internet Protocol (Mobile IP)
Mobile Multi-hop Relay (MMR)
Mobile Satellite Services (MSS)
Mobile Station (MS)
Mobile Switching Center (MSC)
Mobile Telephone System (MTS)
Mobile-to-Mobile (Mo2Mo)
Mobile User (MU)
Mobile Virtual Network Operator (MVNO)
Mobility Management (MM)
Motion Picture Expert Group (MPEG)
Multi-Band OFDM Alliance (MBOA)
Multi-Band Orthogonal Frequency Division Multiplexing (MB-OFDM)
Multi-channel Multipoint Distributed Service (MMDS)
Multimedia Message Service (MMS)
Multi-Mode Headset (MMH)
Multiparty Multimedia Session Control (MMUSIC)
Multiple Input Multiple Output (MIMO)
Multipoint Control Unit (MCU)
Multi-Protocol Label Switching (MPLS)

N

Naming Authority Pointer (NAPTR)
Narrowband AMPS (N-AMPS)
National Television Standards Committee (NTSC)
Near Field Communications (NFC)
Near Field Communications Forum (NFC Forum)
Near-Field Sensor Networks (NFSN)
Network Access Point (NAP)
Network Address Translation (NAT)
Network Element (NE)
Network Element Layer (NEL)

Network Header (NH)
Network Interface (NI)
Network Management Center (NMC)
Network Management Forum (NMF)
Network Management Layer (NML)
Network Monitoring and Analysis (NMA)
Network Operation Center (NOC)
Network Protocol Data Unit (N-PDU)
Network Provider (NP)
Network Subsystem (NS)
Next Generation Infrastructure (NGI)
Next Generation Mobile Network (NGMN)
Next Generation Network (NGN)
New Generation Operation Systems and Software (NGOSS)
New Generation of Wireless Networks (NGWN)
Next Generation of Wireless Networks (NGWN)
Non-Line-of-Sight (NLOS)

O

Object Exchange (OBEX)
Object Name Service (ONS)
Object Request Broker (ORB)
Open Data Base Connectivity (ODBC)
Open Handset Alliance (OHA)
Open Mobile Alliance (OMA)
Open Service Access (OSA)
Open Service Access Service Capability Server (OSA-SCS)
Open Shortest Path First (OSPF)
Open Software Foundation (OSF)
Open Systems Interconnection (OSI)
Operating Expenditures (OPEX)
Operating System (OS)
Operations Management Center (OMC)
Operations Systems Support (OSS)
Operators' Harmonization Group (OHG)
Optical Line Terminal (OLT)
Optical Network Unit (ONU)
Optical Transceiver (OT)
Optical Transport Network (OTN)
Orthogonal Frequency Division Multiplexing (OFDM)

P

Packet Binary Convolutional Coding (PBCC)
Packet Data Gateway (PDG)
Packet Encoding Rules (PER)
Packet Switched Data Network (PSDN)
Packet Switched Network (PSN)
Packet over SONET (PoS)
Passive Optical Network (PON)
Peer-to Peer (P2P)
Perceptual Evaluation of Speech Quality (PESQ)
Performance Management (PM)
Per-Hop-Behavior (PHB)
Peripheral Component Interconnect (PCI)
Personal Area Network (PAN)
Personal Communications Network (PCN)
Personal Communications Service (PCS)
Personal Computer (PC)
Personal Computer Memory Card International Association (PCMCIA)
Personal Digital Assistant (PDA)
Personal Earth Station (PES)
Personal Handyphone Service (PHS)
Phase Modulation (PM)
Phase Shift Modulation (PSM)
Plain Old Telephone Service (POTS)
Point of Presence (POP)
Point-of-Sale (PoS)
Point to Multipoint (PMP)
Point-to-Point (PTP)
Policy-based Management (PBM)
Policy Decision Function (PDF)
Policy Enforcement Point (PEP)
Policy Information Base (PIB)
Policy Management Tool (PMT)
Policy Repository (PR)
Power Line Communications (PLC)
Power Line Networking (PLN)
Power-over-Ethernet (PoE)
Presentation Layer (P-PDU)
Presentation Modules (PM)
Private Branch Exchanges (PBX)
Programmable Logic Controller (PLC)
Protocol Data Unit (PDU)
Proxy Call Session Control Function (P-CSCF)

Public Land Mobile Network (PLMN)
Public Switched Telephone Network (PSTN)
Push-to-Talk over Cellular (PoC)

Q

Quadrature Amplitude Modulation (QAM)
Quality of Service (QOS)
Quaternary Phase Shift Modulation (QPSM)

R

Radio Access Network (RAN)
Radio Frequency (RF)
Radio Frequency Identification (RFID)
Radio Link Control (RLC)
Radio Network Controller (RNC)
Radio Network Subsystem (RNS)
Random Early Detection/Discard (RED)
Real-Time (Transport) Protocol (RTP)
Real-Time Control Protocol (RTCP)
Real-Time Streaming Protocol (RTSP)
Record Type Definition (RTD)
Reduced Instruction Set Computers (RISC)
Regional Bell Operating Company (RBOC)
Registration Administration and Status (RAS)
Relational Data Base Management System (RDBMS)
Remote Access Dial-In User Service (RADIUS)
Remote Access Servers (RAS)
Remote Monitoring (RMON)
Remote Operations Service Element (ROSE)
Request for Comments (RFC)
Request for Information (RFI)
Request for Proposal (RFP)
Research in Motion (RIM)
Reservation Protocol (RSVP)
Residential Gateway (RG)
Resilient Packet Ring (RPR)
Resource Reservation Protocol (RSVP)
Response Time Reporter (RTR)
Return on Investment (ROI)

S

Satellite Broadband Access (SBA)
Scalable-Orthogonal Frequency Division Multiplexing (S-OFDM)
Seamless Converged Communications Across Networks (SCCAN)
Secure Trade Lane (STL)
Security Management (SM)
Service Access Point (SAP)
Service Assurance (SA)
Service Billing (SB)
Service Capability Interaction Manager (SCIM)
Service Capability Features (SCF)
Service Capability Server (SCS)
Service Control Point (SCP)
Service Delivery Platform (SDP)
Service Discovery Protocol (SDP)
Service Fulfillment (SF)
Service Level Agreements (SLA)
Service Level Management (SLM)
Service Level Specifications (SLS)
Service Management Layer (SML)
Service Management System (SMS)
Service Oriented Architecture (SOA)
Service Provider (SP)
Serving Call Session Control Function (S-CSCF)
Serving GPRS Support Nodes (SGSN)
Session Announcement Protocol (SAP)
Session Border Controller (SBC)
Session Description Protocol (SDP)
Session Initiation Protocol (SIP)
Session Initiation Protocol Forum (SIP Forum)
Session Layer (S-PDU)
Set-Top-Box (STB)
Short Message Service (SMS)
Signaling Connection and Control Part (SCCP)
Signaling Engineering Administration System (SEAS)
Signaling Gateway (SGW)
Signaling Switching Point (SSP)
Signaling System Number 7 (SS7 or SS#7)
Signal to Noise Ratio (SNR)
Signal Transfer Points (STP)
Simple Mail Transfer Protocol (SMTP)
Simple Network Management Protocol (SNMP)

Simple Object Access Protocol (SOAP)
SIP for Instant Messaging and Presence Leveraging Extensions (SIMPLE)
SIP Proxies (SIP-P)
SIP Registrar (SIP-R)
SIP User Agent (SIP-UA),
Small Office Home Office (SOHO)
Software Defined Radio (SDR)
Software Defined Networking (SDN)
Source Address (SA)
Source Service Access Point (SSAP)
Special Interest Group (SIG)
Spectral Voice Priority (SVP)
Static Random Access Memory (SRAM)
Stream Control Transmission Protocol (SCTP)
Structure of Management Information (SMI)
Subscriber Module (SM)
Subscriber Station (SS)
Synchronous Digital Hierarchy (SDH)
Synchronous Optical Network (SONET)
System Architecture Evolution (SAE)
System Management Application Service Elements (SMASE)
System on Chip (SoC)

T

Tamper Resistant Embedded Controller (TREC)
Task Group (TG)
Telecommunications Industry Association (TIA)
Telecommunications and Internet Converged Services and Protocols for Advanced Networking (TISPAN)
Telecommunications Management Information Platform (TeMIP)
Telecommunications Management Network (TMN)
Telecommunications Operations Map (TOM)
Temporal Key Integrity Protocol (TKIP)
Temporary Mobile Subscriber Identity (TMSI)
Third Generation Partnership Project (3GPP)
Third Generation Partnership Project 2 (3GPP2)
Time Division Duplex (TDD)
Time Division Multiple Access (TDMA)
Time Division Synchronous CDMA (TD-SCDMA)
Time-To-Live (TTL)
Total Access Communications System (TACS)
Traffic Conditioning Agreements (TCAs)

Transaction Capabilities User Part (TCAP)
Transaction Language 1 (TL1)
Transmission Control Protocol (TCP)
Transport Layer Protocol Data Unit (T-PDU)
Transport Layer Security (TLS)
Triple Digital Encryption System (Triple DES)
Trunk Information Record Keeping System (TIRKS)
Type-of-Service (TOS)

U

Ultra Dense Wavelength Division Multiplexing (UDWDM)
Ultra-High Frequency (UHF)
Ultra Mobile Broadband (UMB)
Ultra Wide Band (UWB)
Unlicensed Mobile Access (UMA)
Unlicensed Mobile Access Consortium (UMAC)
Unlicensed Mobile Access/Generic Access Network Consortium (UMAC)
Unlicensed Mobile Access Network (UMAN)
UMA Network Controller (UNC)
UMA Radio Link Control (URLC)
Unlicensed National Information Infrastructure (UNII)
Unified Protocol (UP)
Universal Mobile Telecommunications Systems (UMTS)
Universal Product Code (UPC)
Universal Resource Identifier (URI)
Universal/Uniform Resource Locator (URL)
Universal Serial Bus (USB)
Universal Terrestrial Radio Access Network (UTRAN)
Uplink (UL)
User Agent (UA)
User Datagram Protocol (UDP)
User Equipment (UE)

V

Very High Frequency (VHF)
Very Small Aperture Terminals (VSAT)
Video over Internet Protocol (IPTV)
Virtual Circuit Identifier (VCI)
Virtual Path Identifier (VPI)
Virtual Private Network (VPN)

Visiting Location Register (VLR)
Voice Call Continuity (VCC)
Voice over Internet Protocol (VoIP)
Voice over Wireless Local Area Network (VoWLAN)

W

Weighted Fair Queuing (WFQ)
Weighted Random Early Detection/Discard (WRED)
Wide Area Network (WAN)
Wideband CDMA (W-CDMA)
Wi-Fi Multimedia (WMM)
Wi-Fi Protected Access (WPA)
Wired Equivalent Privacy (WEP)
Wireless Access Gateway (WAG)
Wireless Application Protocol (WAP)
Wireless Ethernet Compatibility Alliance (WECA)
Wireless Fidelity (Wi-Fi)
Wireless Local Area Network (WLAN)
Wireless Local Loop (WLL)
Wireless Mesh Network (WMN)
Wireless Metropolitan Area Exchange (WiMAX)
Wireless Metropolitan Area Network (WMAN)
Wireless Personal Area Network (WPAN)
Wireless Terminal (WT)
Wireless Universal Serial Bus (WUSB)
Wireless USB (WUSB)
Wireless Wide Area Networks (WWAN)
Working Group (WG)
Work Station (WS)

Z

ZigBee Applications Objects (ZAO)
ZigBee Coordinator (ZC)
ZigBee Device Objects (ZDO)
ZigBee End Nodes (ZN)
ZigBee Gateway (ZGW)
ZigBee Network Management System (ZNMS)
ZigBee Routers (ZR)
Zero Generation (0G)
First Generation (1G)

Second Generation (2G)
Third Generation (3G)
3G Partnership Project (3GPP)
3G Partnership Project 2 (3GPP2)

Part I

Wireless Communications: Networking and Management

1 Wireless Communications and Networking

1.1 Communications Networks

Communications by voice and physical signaling are common means of interaction between human beings. In the simplest forms, there is an emitting entity of information and a receiving entity of information. As the sources move apart the need for telecommunications appears self-evident. This simple model of communications becomes more complex when the information transmitted is not just sound, speech, or music, but full motion video images or various forms of data such as text, shared files, facsimile, graphics, still images, computer animation or instrumentation measurements. This information can be transmitted using electrical or optical signals, the native analog information undergoing numerous conversions and switching to accommodate various communications technologies. Telecommunications can take place over various media be that twisted copper pairs, coaxial cable, fiber optic, or wireless radio, microwave, satellite, and infrared links. Communication can be limited to a small group of people or extended to departments, compounds, or campuses and covering whole metropolitan areas, regions, countries, or continents. Hence, a shared infrastructure is needed, i.e., a communications network.

Communications networks can be classified in many ways, as there are distinct technologies and network equipment needed for voice communications, data/computer communications, and video communications. Among these types of communications, data communications, by the virtue of digitization of any type of information, has become the convergent system. Classification can also be done corresponding to the kind of media used, i.e. wired and wireless communications or by the underlying technologies such as circuit-switched networks (traditional voice communications), packet-switched networks (essentially data communications), broadcast networks (direct video broadcast, cable TV), or message switched networks (store-and-forward electronic mail, voice mail, web serving). Networks classification can go even further, looking at geographical coverage, from wide area networks, to metropolitan area, local area, and to personal area networks.

A typical telecommunications network, historically focused on voice processing and voice communications, is presented in Figure 1.1. The network, a symmetric infrastructure that connects Customer Premises Equipment (CPE) as the source and destination of information (telephones), consists of four major systems: **access**, **transmission**, **switching**, and **signaling**. The access systems are part of the local loop and consist of

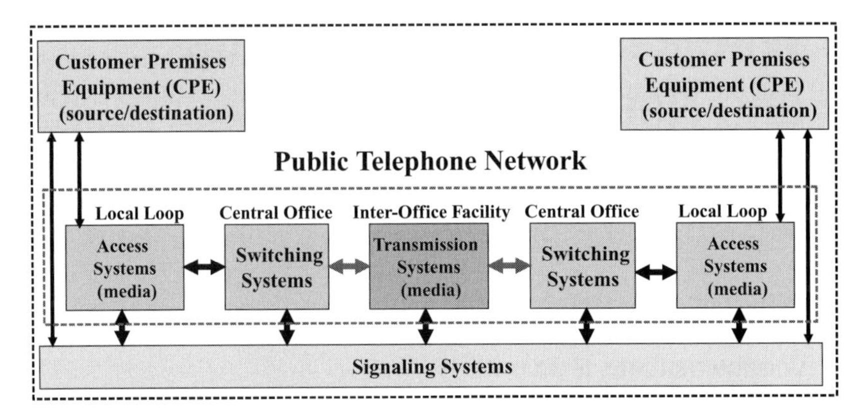

Figure 1.1 Telecommunications Network Model

copper-based twisted pair media or a combination of fiber optic and twisted pairs. The switching systems are part of the Central Office equipment connected on one end to the local loop lines and on the other end to the trunks that are part of the transmission systems. Last, but not least, signaling systems provide call control facilities for the whole network through a specialized overlay network. Currently, the same infrastructure is used to support not only voice but also data communications and other multimedia services.

A typical data communications network with clear roots in computer communications networks is presented in Figure 1.2.

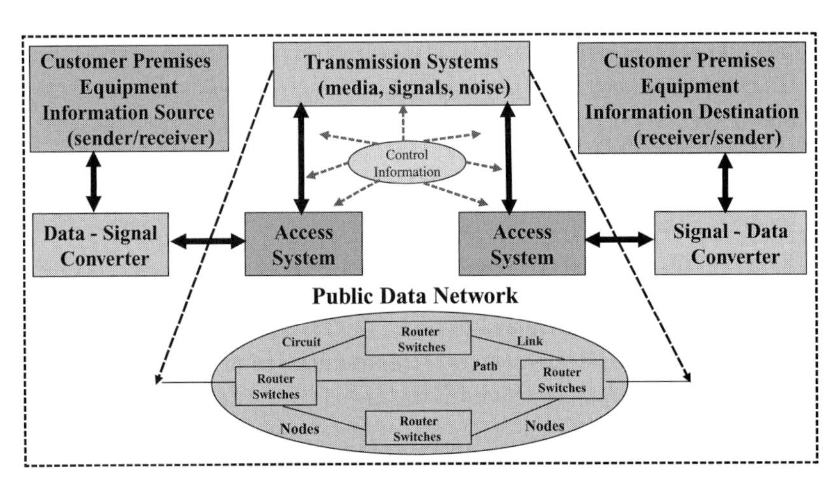

Figure 1.2 Data Communications Network Model

The data communications network model is a symmetric infrastructure that connects Customer Premises Equipment (CPE), as source and destination of information (desktop computer, laptop, server, fax machine, or instrumentation). It consists of four major systems: **conversion**, **access**, **transmission**, and **control systems**. The conversion systems (modems, coders/decoders, multiplexers) provide data-signal and signal-data conversion to accommodate the existent local loop infrastructure that was initially designed for voice communication. The access systems are mostly part of the same local loop used in telecommunications and consist of copper-based twisted pair media or a combination of fiber optic and twisted pairs. Security functions, service access, and billing components might be built into access systems. In many instances the access and conversion systems components and functions are bundled in real networking products. There are no circuit-switching systems involved since the information is packetized and the packets are routed as part of the transmission systems, a vast network of interconnected nodes, essentially packet routers. Depending on the transmission procedures adopted, a dedicated system may establish, control, and terminate the connection between the sender and receiver of information. Currently, to minimize cost, the same infrastructure is used to support not only data but also digitized and packetized voice and other multimedia services.

Both voice and data communications networks have evolved as part of the overall technical progress in telecommunications with the invention of alternative technologies, implementation of various architectural designs, and multiple service offerings. The ultimate goal of this evolution is the integration of voice and data, a multidimensional concept that combines IT infrastructures, networks, applications, user interfaces, and management aimed at supporting all forms of information media on all forms of networks. A narrower view of this integration is aimed at supporting voice and video over packet-based Internet networks.

1.2 Communications Architectures and Protocols

Communication between various network components requires a common understanding between sender and receiver regarding the transmission procedures, the types of signals carrying the useful information, and the format of the data transmitted. There are four major aspects involved in communications services: communication interfaces, communication protocols, layered communication stacks, and information models. **Communication interfaces** are connection points between network components designed according to standard or proprietary specifications. An interface is defined by mechanical, electrical, and functional characteristics that allow communication to take place between adjacent network devices. **Communication protocols** are formal descriptions of data unit formats and transmission rules for message exchange between network entities. Communication protocols are organized in multiple **layered communication stacks** where each layer provides services to the layer above. The **information models** are collections of abstracted managed object definitions and attributes for devices having common characteristics. Currently, there are many communications architectures and protocols in use.

The **transmission procedures** used to provide connectivity between network devices may include three distinct phases, as indicated in Figure 1.3. The first is the **establishment** phase which arises as a result of an inquiry about the availability and ability of the corresponding party to exchange information. Affirmative responses open the next phase of **data transfer** that takes place by continuous acknowledgement of data received and

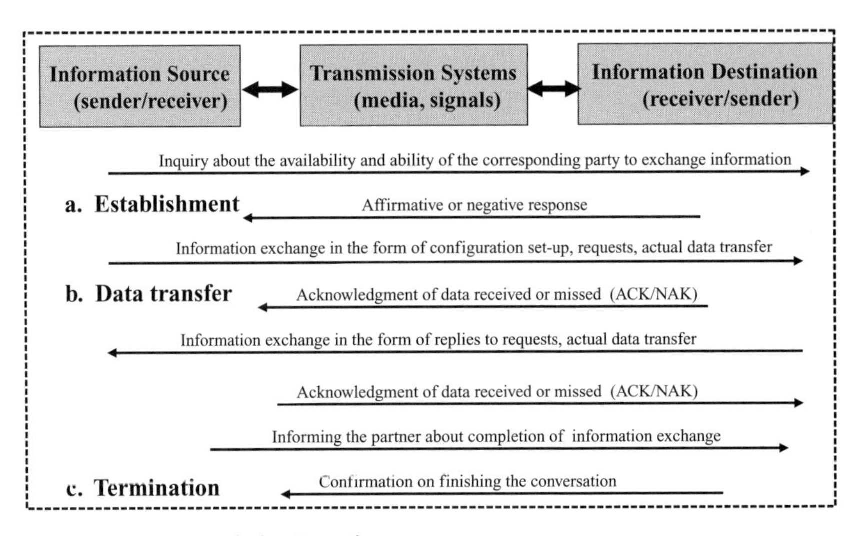

Figure 1.3 Data Transmission Procedures

its accuracy. At the completion of information exchange, the **termination** phase of connection is initiated.

Depending on the procedure adopted for connectivity, we may have two types of communication services, as indicated in Figure 1.4.

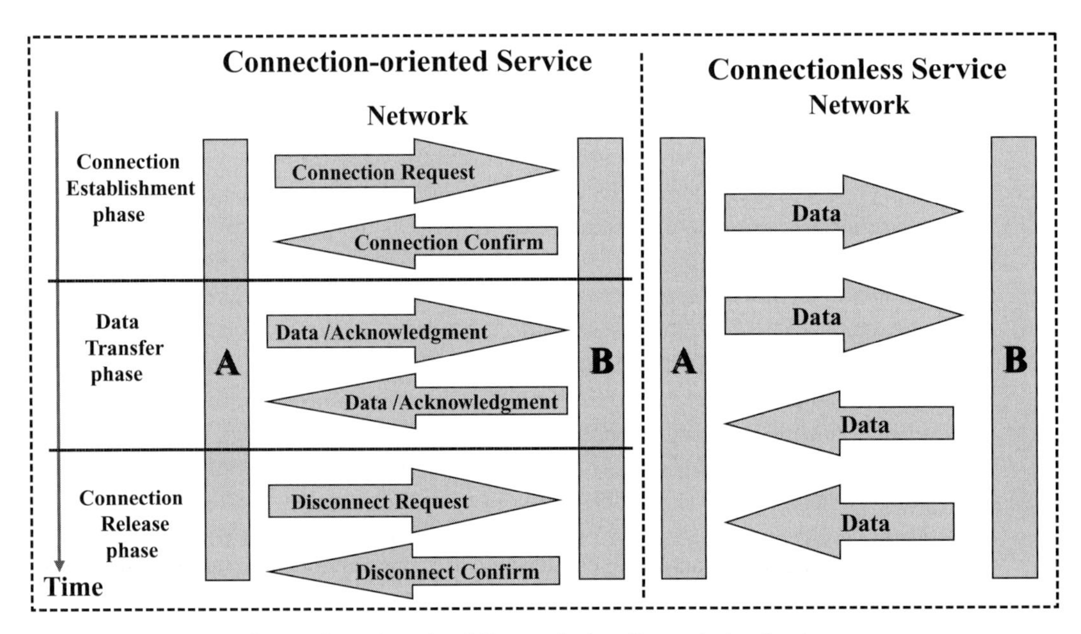

Figure 1.4 Connection-oriented and Connectionless Transmission Services

The first, **connection-oriented layer service** implies the requirement to establish a layer connection between the parties prior to the information exchange. At the end of data transfer, the layer connection will be terminated. This type of transmission gives full assurance about the ability of both communicating parties to exchange information with full control of every aspect of data transfer. Examples: phone calls, X.25 packet networks, Internet Transmission Control Protocol (TCP). The second, **connectionless layer service** assumes that no association between the transmitting parties takes place prior to actual data transfer, i.e. no connection establishment or connection termination takes place. This type of transmission does not give either party a guarantee that they are able to exchange information or, as a matter of fact, the data transfer has been successfully completed. Examples are the Open Systems Interconnection (OSI) Connectionless Network Protocol (CLNP) and Internet User Datagram Protocol (UDP).

Multiple layer protocols are aggregated into communication stacks. The communication between two applications across recognized interfaces takes place through identical peer protocol stacks as indicated in Figure 1.5. There is no direct communication between layers, except the physical layer. Peer relationships are established between corresponding layers via dedicated protocols/services. In the most common layered communication architecture, the International Organization for Standardization (ISO) Open Systems Interconnection (OSI) standards, the Physical, Data Link, and Network Layers are grouped and known as lower layers; the Transport, Session, Presentation, and Application Layers are also grouped and known as upper layers [1].

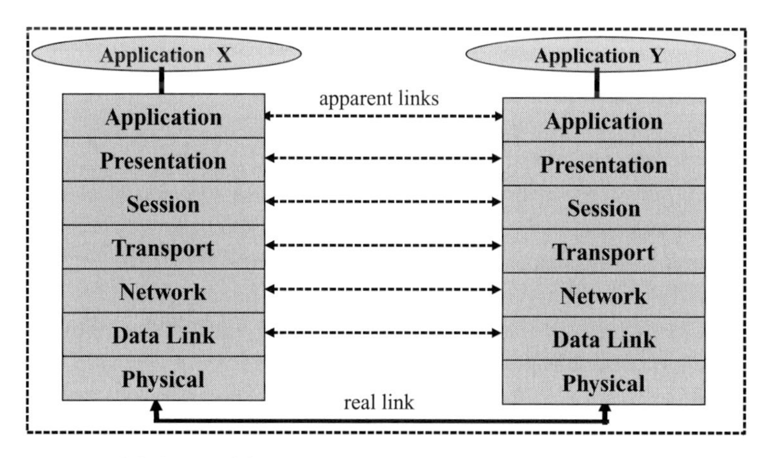

Figure 1.5 OSI Layered Communications Stack

The OSI Reference Model of seven layered communications architecture is an abstract concept of data communication that divides the functions required to exchange information into well-defined layers, services, access points, and standard protocols. There are numerous partial implementations and variations of the OSI architectural model. Some of these configurations pose serious interconnectivity and interoperability problems. A simplified view of OSI layer functions and services is given in Table 1.1. In most instances, the functions performed at the Data Link Layer are implemented in the hardware while the Network Layer functions and those associated with upper layers are implemented in software.

The basic format of information exchanged, which allows the communicating parties to understand and interpret the transmitted information, is the **Protocol Data Unit** (PDU).

Table 1.1 OSI Layers Main Functions and Services

OSI Layer	Main Functions and Services
Application	Provides means of application processes to access the OSI layered environment.
Presentation	Provides data format structure and representation according to an agreed syntax.
Session	Provides establishment and control of the dialog between applications in end systems.
Transport	End-to-end reliable mechanism for information exchange between processes or systems.
Network	Addressing, routing, and relaying of information transmitted across the networks.
Data Link	Data transfer flow control, error detection and correction mechanism, data recovery.
Physical	Functional, electrical, mechanical, and procedural interfaces to the communications media.

The seed of any PDU is the actual user/business/application/service information. This seed is augmented with a succession of distinct data fields specific to each layer to perform specific functions and services such as synchronization, addressing, sequencing, flow control, and error detection. More than that, the PDU can be changed at the same layer as the message passes different network interfaces. Each OSI layer is defined by a peer-to-peer protocol that operates across an apparent link (distinct PDUs for each layer) and by the services provided to the next higher layer. A header is appended to the user data or to the next lower layer PDU by encapsulation. The construct of OSI Protocol Data Units layer by layer is given in Figure 1.6.

In the message exchange between two end-systems represented by Application X (sender) and Application Y (receiver), User Data is encapsulated by the Application Header (AH) to create an Application PDU (A-PDU). Similarly, operations take place at the Presentation Layer (P-PDU), Session Layer (S-PDU), and Transport Layer (T-PDU). The next encapsulation, at the Network Layer, results in a Network PDU (N-PDU) that contains in the Network Header (NH) critical information that allows routing of packets between network nodes. The last encapsulation takes place at the Data Link Layer where a Link Header (LH) and a Link Trailer (LT) are added before the Data Link PDU (DL-PDU) is delivered to the Physical Layer. This is the PDU that is carried across the physical link connecting the two nodes using various types of physical signals.

At the other end of this link, the reverse process of deconstructing the message takes place as the message is passed through successive layers. Each layer removes the corresponding header and analyzes the remaining content before passing the resulting PDU to the layer above. This continues until the PDU reaches the Application Layer. Once the header of the A-PDU is removed, the pure user information is delivered to Application Y. A similar operation of constructing and deconstructing takes place when Application Y becomes the sender and Application X the receiver. In reality, multiple nodes might be used as the message crosses the network and this process of partial deconstructing and constructing the PDU might be repeated in many nodes. Therefore, each end-system or

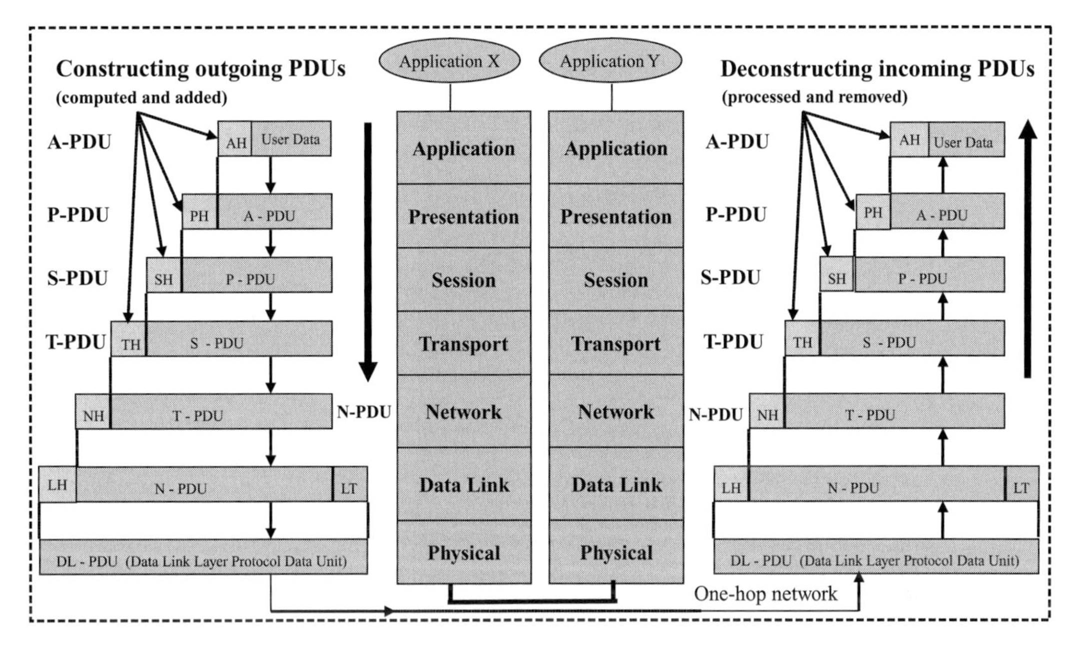

Figure 1.6 OSI Protocol Data Units Layer by Layer

intermediary node should have the same understanding of this process and the functions and information embedded in PDU headers.

1.3 Wireless Communications

Wireless communications, as the name indicates, is communication using wireless connectivity means between network components (as opposed to wired) to transmit information. Wireless networks can provide network access to phones, computers, applications, databases, and the Internet, within buildings, between buildings, within campuses, and between remote locations, giving the users the ability to move, roam, or work virtually from any location.

There are four main means used for wireless communications: **radio**-based, **microwave**-based, **satellite**-based, and free space optics **infrared**-based. Each of these means is based on distinct communications technologies, network elements, and distinct frequency spectrums for communications. The oldest wireless applications are those using radio waves for broadcasting and individual reception. **Radio Frequency** (RF)-based technologies have evolved from analog systems to digital systems and from voice to data communications. Two major areas of development took place. The first was the highly popular **cellular mobile** radio networking and services, initially geared to support voice communications. The second was radio-based **fixed wireless** technologies, such as Wireless Local/Personal/Metropolitan Area Networks WLAN, WPAN, WMAN, and near-field sensor networks, initially geared to support data communications. The

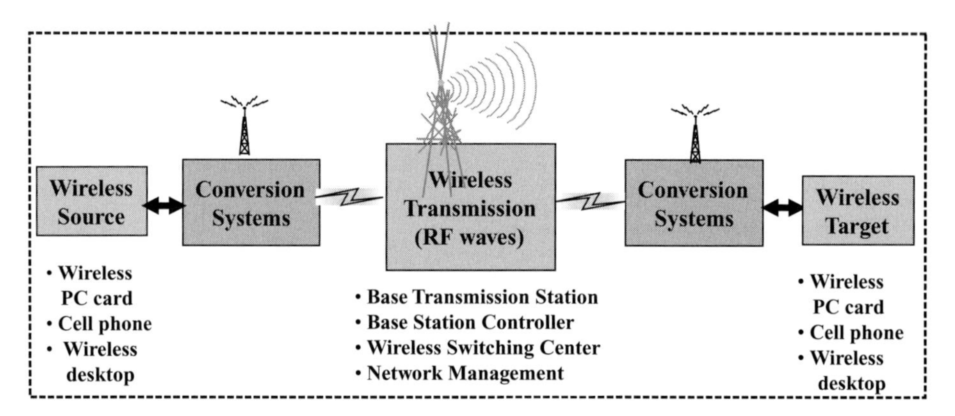

Figure 1.7 Wireless Communications Model

relationship between cellular mobile radio and fixed wireless networks will be examined as part of **Fixed Mobile Convergence (FMC)**, the very subject of this book. However, there is a plethora of other radio-based communications technologies and networks such as paging, push to talk, two-way radio communications, air-to-ground, Global Positioning Satellite (GPS), and cordless telephony that support niche applications born out of specific needs. These will be less examined in this book.

A high level model of wireless communications is very similar with the general model of communications, as indicated in Figure 1.7. This kind of communications can be limited to point-to-point exchange of information or just one instance of point-to-multipoint communications. It also can be a one-way communications (broadcast) or a two-way communications when the roles of **transmitter** and **receiver** are alternated, hence the name of **transceivers**. Specialized metallic devices, **antennas**, with the corresponding electronics, are used for transmitter and receiver. Conversion systems are needed to take the native information, be that voice, data, or image, and convert it into electromagnetic signals that spread out from antennas and propagate through space at the speed of light. A receiving antenna, similar to the transmitting antenna, intercepts the electromagnetic waves and converts them back into electrical signals that are decoded so the native information is delivered to the receiving party, a radio set, a cellular phone, or a television set.

Critical to the success of wireless communications are the RF antennas. There are many types, shapes, and sizes of antennas depending on the application provided, physical design, and ability to enhance the performance by concentration of RF wave energy. Further, there are two basic types: **omnidirectional antennas**, radiating signals in all directions, and **directional antennas**, radiating signals in a narrow direction or controlled pattern. Depending on their ability to concentrate the energy, there are **active** antennas and **passive** antennas. There are antennas with **diversity gain** that control radio signal fading. Antenna shapes include simple vertical antennas, folded dipole, parabolic, collinear, log periodic, or Yagi, panel or horn. There are striking differences between **source antennas**, truly towers in the sky, or hub earth stations, and **target antennas**, simple vertical polarizing sticks or plate-size satellite dishes. Last but not least, there are **Multiple Input Multiple Output** (MIMO) antennas for truly user dense wireless communications.

Key to wireless communications are the conversion systems; they use various techniques to convert the native information into electromagnetic signals. These techniques are based on modulation processes, where the characteristics of the native signal, originating from say a microphone or a video camera are changed to a suitable carrier signal, i.e., amplitude, frequency, and phase. The frequency of the carrier signal is higher than that of the native signal. Consequently, three fundamental types of modulation techniques are used: **Amplitude Modulation** (AM), **Frequency Modulation** (FM), and **Phase Modulation** (PM). However, there are many other fine-grained modulation techniques used which are variations and combinations of these basic modulation techniques.

In amplitude modulation, the amplitude of the carrier is changed according to the variations in amplitude of the native signal, e.g. voice, while the frequency of the carrier remains constant. In frequency modulation, the amplitude of the carrier signal remains constant while its frequency is changed according to the rhythm of the change in amplitude of the modulating signal. Special electronic components generate the carrier signal, a constant amplitude and frequency succession of sine waves, and control the process of modulation. On the receiving side, the radio waves are detected and demodulated to extract the original signal. Given the nature of the shared transmission environment, the earth atmosphere, we can expect natural attenuation of the transmitted signal's strengths and interference in the form of noise from other transmission sources or just from the very existence of cosmic and solar emitted rays. It is not the intent of this book to go into the details of antenna design, modulation techniques, RF equipment components and characteristics, physics of propagation, path loss, or radio link budget although they will be mentioned whenever it is necessary for the sake of clarity.

1.4 Wireless Communications Classification

The large variety of wireless networks introduced in the previous section can be classified using three major criteria: geographical coverage, level of mobility, and spectrum allocation. Figure 1.8 depicts the variety of wireless communications systems. Depending on the area of coverage, there are five main types of wireless networks as follows:

- **Wireless Wide Area Networks (WWAN)**: These networks provide direct network connectivity to large areas including disparate, remote sites. Top on the list of WWANs is the large family of mobile radio cellular networks spanning three solid generations (1G, 2G, 3G) with various mobile technology and service implementations (AMPS, GSM, CDMA). They support voice and data services that have reached data rates up to 2 Mbps. Cellular mobile systems use ground-based equipment with direct horizontal RF-based wireless connectivity. Satellite networks are in the same class of wide area, global coverage, but they use ground-to-satellite-and-back type transmission between ground-based Hub/Personal Earth Station (HES/PES) and Geostationary Earth Orbit, Middle Earth Orbit, or Low Earth Orbit (GEO/MEO/LEO) satellites. They are capable

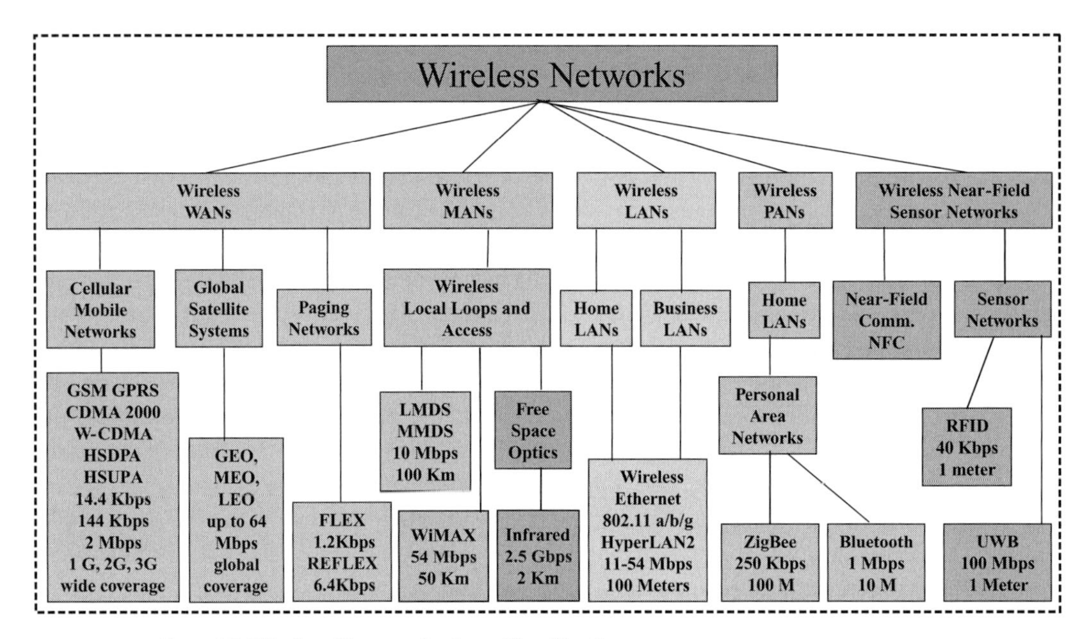

Figure 1.8 Wireless Communications Classification

of providing video broadcast, voice and data at aggregate transmission rates up to 64 Mbps. Given the high altitude positioning of geosynchronous satellites (22,300 miles or 35,900 km), this type of satellite transmission is characterized by significant propagation delays in communications. New generations of satellite communications technologies allow direct phone-to-phone global area communication.

Other well spread wireless networks are paging systems using a combination of satellite-based communication and radio terrestrial paging towers. Typically, in one-way wireless communication, the pager is a receiver that can be accessed via phone, computer, or operator displaying short alphanumeric messages such as phone numbers and names. The newer generation of pagers includes Short Message Services (SMS) with the ability to acknowledge messages received (two-way pagers). Data rates are up to 6.4 Kbps, as is the case for the REFLEX protocol. Some mobile cell phones incorporate pager capabilities.

- **Wireless Metropolitan Area Networks (WMAN)**: These networks provide direct broadband access and fixed wireless network connectivity within metropolitan areas. The first WMAN technologies were **Local Multipoint Distributed Service** (LMDS) and **Multi-channel Multipoint Distributed Service** (MMDS). LMDS is a **fixed broadband** wireless technology using directional antennas working in the microwave spectrum of 28–40 GHz. It requires line-of-sight between transmitters and receivers. It is sensitive to heavy moisture/rainfall. MMDS is a fixed wireless access technology alternative for broadband local loop services. It has less stringent line-of-sight requirements between transmitters and receivers since it works in the 2.1–2.7 GHz band. Top on the list of WMAN technologies is the **Wireless Metropolitan Area Exchange** (WiMAX). It is a direct line-of-sight wireless access technology alternative for broadband local loop services for residential and business customers. It operates in the 2.5, 3.5, 5.3, 5.8 GHz spectrum.

- **Wireless Local Area Networks (WLAN)**: These networks provide wireless **local data** and **voice access** to shared resources such as servers, printers, routers and PBXs within a limited area of an organization or business. WLAN builds on the success and resilience of LAN technologies and brings the users closer to the ideal of transparent communications across multiple networks. WLAN has already a strong presence in the retail, manufacturing, education, and health care sectors, and in hot areas such as airports, small/home offices, Internet cafes, and even in residential homes. WLAN is based on **short-range RF communications**, standardized in the USA by one of the Institute of Electrical and Electronics Engineers (IEEE) working groups. The European counterpart of WLAN is HyperLAN standardized by the European Telecommunications Standard Institute (ETSI). There are three main IEEE WLAN specifications currently implemented: **802.11a, 802.11b, and 802.11g**. The most popular of these is 802.11b, also known as Wireless Fidelity **(Wi-Fi)** or Wireless Ethernet. WLAN operates in one of the unlicensed spectrum bands at 2.4 GHz, 5 GHz, 1.9 GHz, or 900 MHz (exactly 902–928 MHz) at a distance up to 100m.
- **Wireless Personal Area Networks (WPAN)**: These networks provide wireless connectivity to various devices and appliances within the limited **area of a residence**. Also known as home networking, WPAN is aimed at integration and standardization of use and interaction of home end-devices and appliances. It facilitates use of multiple PCs and peripherals, data interconnection (printer, scanner, digital camera, video camera), voice and video communications, music distribution, and use of surveillance devices to command and control appliances (meter reading, temperature, light regulation). It may include small-scale WLANs. Several standards address the development of WPAN, i.e., IEEE 802.15 WPAN Bluetooth/ZigBee and of course IEEE 802.11a/b/g WLANs. **Bluetooth** wireless technology is a set of specifications for a low-cost, low-powered radio and associated protocol stack that provides a **short-range wireless link** between notebook/laptop computers, mobile phones, PDAs, and other portable devices. It operates in the **2.4 GHz Industrial, Scientific and Medical** (ISM) band (from 2.4 to 2.4835 GHz) with a maximum data rate of 3 Mbps. **ZigBee** is the IEEE **802.15.4 standard** set for wireless networking, control, and monitoring of devices using **low data rate radio communications** and **ultra low power firmware resources**. ZigBee operates in the range from 1 to 100m at peak data rates of 128 Kbps.
- **Near-field, Sensor-based Networks**: These networks provide communications in near-field environments using radio frequency identification and ultra-wide band transmission. Three technologies dominate this area of wireless communication: First, **Radio Frequency Identification (RFID)**, a wireless RF, tag-based, low cost technology with no direct power consumption. It requires a special reader and corresponding business processes. Depending on whether the tags are powered or not, we have two major types of RFID systems: passive and active. RFID is considered a replacement of line-of-sight laser-based bar code systems. RFID can also be used to provide a network of radio-based sensors and actuators in industrial automation and control networks. One of the RF spectrums for RFID is in the Ultra-High Frequency (UHF) range of

868–928 MHz with data rates in the range of 40 Kbps. Second, the **Ultra Wide Band** (UWB) radio technology promises data rates up to 480 Mbps at even lower power consumption, thus becoming a very competitive solution for WPAN Bluetooth/ZigBee technologies. Currently, UWB is approved in USA to work in the 3.1–10.6 GHz unlicensed spectrum at only the one microwatt power level; thus minimizing possible interference with Wi-Fi, GPS, and cell phones. Currently, the primary UWB application is the **Wireless USB** (WUSB) that resembles the 480 Mbps interfaces linking PCs to peripherals such as monitors or printers. It will have an immediate use in presence applications (location and tracking). Third, **Near Field Communications** (NFC) provides very short range wireless communications within few centimeters.

Given the relative position of network components, wireless networks can also be divided into: fixed wireless and mobile wireless.

- In **fixed wireless networks** the network components maintain a fixed position. Most of the technologies associated with WLAN, WMAN, WPAN, and near-field sensor networks fall in this category;
- In **mobile wireless networks** the network terminals are mobile. These networks are used for all cellular mobile technologies and some instances of WLAN and WMAN (e.g., nomadic WLAN and mobile WiMAX).

Details about classification of wireless technologies based on spectrum allocation will be given in one of the subsequent sections.

1.5 Wireless Communications Architecture

Having the picture of the range of wireless technologies we can develop an architectural model that comprises the major families, their interdependence, and the relationship with the wired world represented by the well-established voice-oriented wired telecommunications network and the data-oriented Internet network. The high level architecture of the wireless network is presented in Figure 1.9. In this symmetric layout of the wireless world, the **Wired Core Backbone Networks** dominate the very center. The wired networks facilitate all the connectivity needed between the wired and the wireless world (e.g., from a regular phone to a mobile phone), between mobile switching centers to assure roaming across large geographical areas, and between various mobile operators providing cellular mobile telephony services.

Next in this architecture are the **Core Cell-based Radio Mobile Networks**, the infrastructure, applications, and services provided by hundreds of mobile operators that make possible mobile communications to hundreds of millions of users. It is the gamut of three solid generations of networks developed in the past two decades under the flag of Personal Communications Services/Networks (PCS/PCN), International Mobile Telecommunications 2000 (IMT-2000), and Universal Mobile Telecommunications Systems (UMTS), to name a few. It is the set of networks that have advanced from analog mobile communications to digital mobile communications in the pursuit of providing more encompassing data services both in variety and available bandwidth.

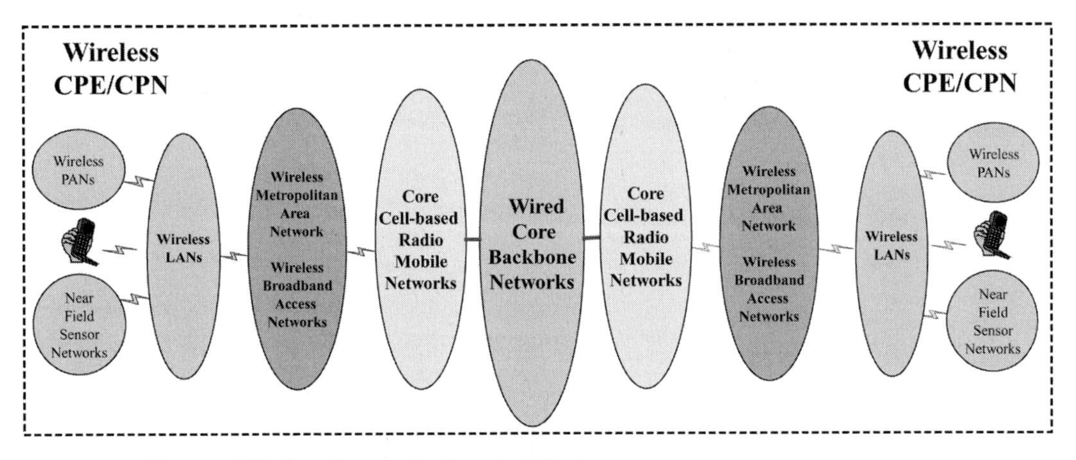

Figure 1.9 Wireless Communications Architecture

The architecture contains the **Wireless Metropolitan Area Networks** focused on providing Wireless Broadband Access that competes with terrestrial services such as Digital Subscriber Line (DSL) and Cable TV. The representative technology for WMAN is WiMAX that was introduced in the previous section as one of the fixed wireless technologies.

The last component set of wireless communications architecture is grouped as Customer Premises Equipment/Network (CPE/CPN). It includes the fixed wireless networking technologies also introduced in the previous section: **Wireless Local Area Networks** (WLAN), **Wireless Personal Area Networks** (WPAN), and **Near-Field Wireless Sensor Networks**. Paramount for the convergence of all wireless networks is the highly mobile handset, truly customer equipment, that communicates transparently across voice, data, and video wireless networks.

1.6 Wireless Communications Architectural Components

Each of the wireless network architectural sets, CPE/CPN, metropolitan broadband wireless access, core cellular mobile networks, and the wired core backbone networks are supported by a variety of competing networking technologies. Each of these technologies is supported by a set of communications protocols and standard specifications. This short chapter gives an indication of the actual technological components that stand behind the architectural boxes. An asymmetric layout of wireless architectural components is presented in Figure 1.10. This will be an opportunity to introduce many of the acronyms used in the book.

Starting from the right, the wired core backbone networks, the realm of major telecom carriers, is realized as a mixture of various technologies. Most of these networking facilities use fiber optics as the media for transmission systems such as Synchronous Optical Network (SONET) or Synchronous Digital Hierarchy (SDH), Dense Wavelength Division Multiplexing (DWDM) or Ultra Dense Wavelength Division Multiplexing (UDWDM), Packet over SONET (PoS), Asynchronous Transmission Mode (ATM), or Internet Protocol over ATM (IP over ATM). The development of this infrastructure is

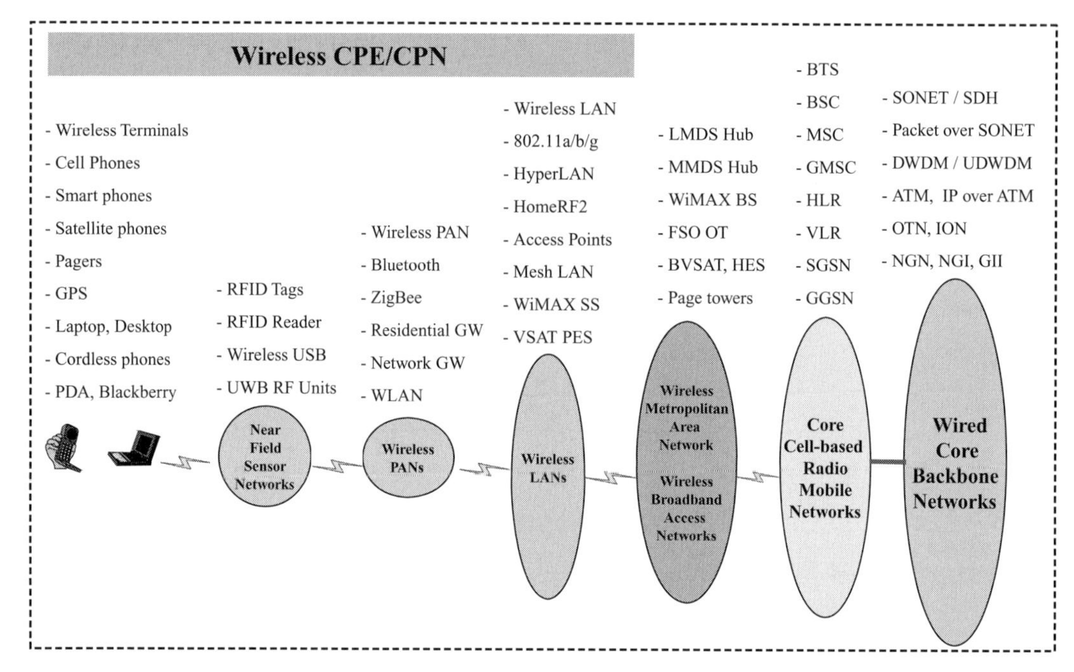

Figure 1.10 Wireless Communications Architectural Components

based on major concepts known as Optical Transport Network (OTN), Integrated Optical Network (ION), Next Generation Network (NGN), Next Generation Infrastructure (NGI), or Global Information Infrastructure (GII).

The core cell-based radio mobile network is an infrastructure that consists of the following major architectural components: Base Transceiver Station (BTS), Base Station Controller (BSC), Mobile Switching Centers (MSC), Home Location Register (HLR) and Visiting Location Registers (VLR) databases, and Gateway Mobile Switching Centers (GMSC). These components support primarily voice services. Support for data services requires additional architectural components. For example, in Global Systems for Mobile (GSM) Telecommunications, the General Packet Radio Service (GPRS) requires two additional architectural components: Serving GPRS Support Nodes (SGSN) and Gateway GPRS Support Nodes (GGSN). Both are part of the GPRS Support Network (GSN).

Wireless metropolitan area networks and wireless broadband access are supported by networking technologies such as Local Multipoint Distributed Service (LMDS), Multi-channel Multipoint Distributed Service (MMDS), and Wireless Metropolitan Area Exchange (WiMAX). The focal point for access is a multi-point LMDS or MMDS hub or a WiMAX Base Station (BS). In the same metropolitan class, two other technologies can be considered: Free Space Optics (FSO) infrared-based solutions using for connectivity Optical Transceivers (OT) and Satellite Broadband Access (SBA) solutions achieved through broadband Very Small Aperture Terminals (VSAT). VSAT pairs Hub Earth Stations (HES) with CPE Personal Earth Stations (PES). Paging systems using page towers as multipoint distribution centers can also be considered.

The wireless local area networks, one of the most dynamic environments, have an array of standardized IEEE technologies such as 802.11a, 802.11b, and 802.11g. In Europe, ETSI has developed standards such as Hyper LAN1, Hyper LAN2, and Broadband Radio Access Network (BRAN). Other WLAN technologies are 5 Unified Protocol (5UP) and Home RF2, the latter targeting Small Office Home Office (SOHO) environments. Keys in all these WLAN technologies are the Access Points (APs). Some support handover capabilities in the build-up of Mesh WLANs to cover a metropolitan area, provided that hundreds of Access Points are used. In many instances, WLANs have connectivity to WiMAX Subscriber Station (SS) or to satellite VSAT Personal Earth Stations (PES).

The wireless personal area network, a relatively new environment, has support from WLAN technologies as mentioned above (IEEE 802.11a/b/g, Hyper LAN1/2, BRAN) and from two specialized technologies, Bluetooth and ZigBee, both incorporated as IEEE standards. Other technologies pertinent to WPAN are based on Digital Enhanced Cordless Telecommunication (DECT) cordless telephony standards and Wireless Application Protocol (WAP) carrier-independent transaction-oriented protocol standards. Key in WPAN technologies is the Residential Gateway (GW) and Network Gateway.

Last in the list of architectural components are those technologies that belong to Near-Field Sensor Networks (NFSN), namely, the Radio Frequency Identification (RFID), Ultra-Wide Band (UWB), and Near Field Communications (NFC) technologies. The main components of RFID technology are the RFID tags, both passive and active, and RFID Reader. The main components of UWB are the Wireless USB devices (e.g., memory sticks) and the broadband RF units on both transmitting and receiving segments.

On the list of wireless architectural components, a special place is reserved to the actual wireless terminals, devices that use a range of wireless communication technologies and have varying degrees of mobility. An incomplete list includes mobile terminals such as cell or mobile phones, smart phones (dual or multi-mode handsets), satellite phones, pagers (receivers), GPS devices, any type of cordless phones. And, ultimately, we include here all the computing devices with built-in wireless capabilities that have various levels of mobility such as PDAs, BlackBerrys, iPhones, laptops, and even desktops.

1.7 Wired and Wireless Communications Networks

To understand the rapid expansion of wireless technologies and their positions relative to other technologies we have to put this growth in the context of the overall telecommunications industry. An attempt to capture the complexity of the overall data world is presented in Figure 1.11.

At high level, we can describe the data communications networks as an overlay of three major classes of networks. The set of **access networks** provides access facilities to the actual services and applications using specialized **access servers**. At the center there is a **core backbone network** that provides high-speed, long haul, data transmission interconnectivity across various networks. A **connectivity network** provides the link between access networks and the core backbone network. This is the layer where services are provided to the users through specialized **data communications servers**.

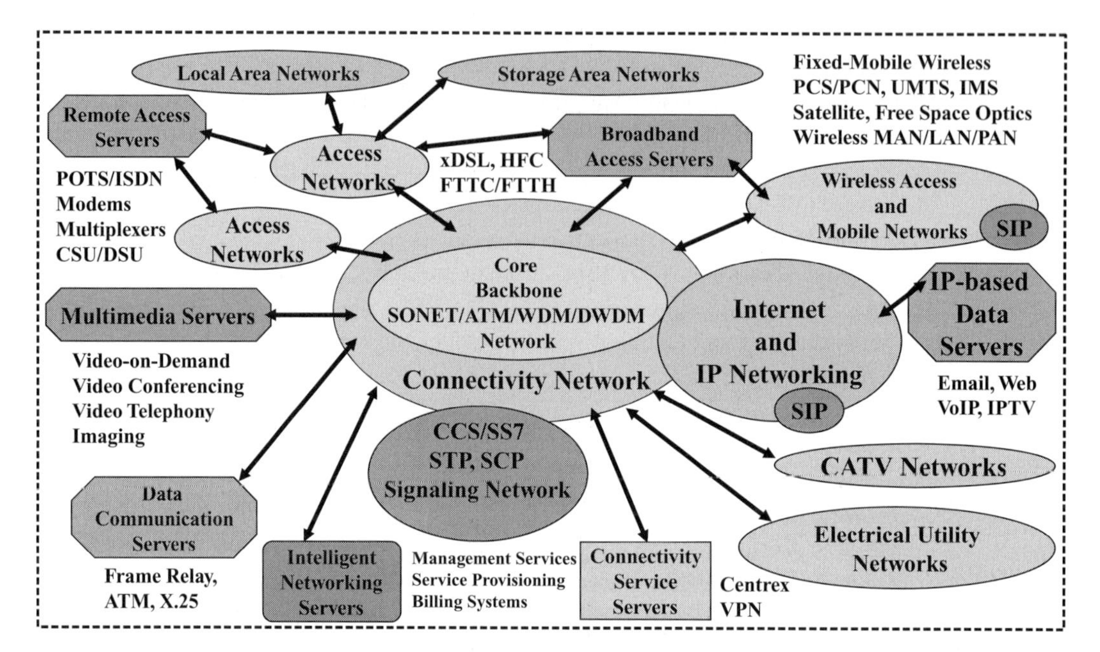

Figure 1.11 An Overall View of Data Communications Networks

This traditional model of data communications, represented by the evolution of the telecommunications network, is challenged by the development of full blown data networks such as Internet IP-based networks, wireless and cellular mobile networks, cable TV (CATV) networks, and even networks built on existent power line electrical utility facilities. All these challengers, represented in the right side of the diagram, are continuously encroaching into the realm of traditional data networks. The diagram also contains two examples of customer premises data networks, namely, the Local Area and Storage Area Networks (SAN).

A more detailed analysis of this diagram would include access technologies such as modems and multiplexers using Plain Old Telephone Service (POTS) and Integrated Services Digital Networks (ISDN) facilities through specialized Channel Service Units (CSU) and Data Service Units (DSU). Another group of access networks provide wired broadband access facilities using variations of twisted pair-based Digital Subscriber Line (xDSL), coaxial cable-based Hybrid Fiber Coax (HFC), or optical fiber-based Fiber-To-The-Curve (FTTC) and Fiber-To-The-Home (FTTH) technologies. A distinct group of access networks are those that are part of the wireless world technologies promoted along WMAN, WLAN, WPAN, NFSN, satellite, and free space optics-based networks and cellular mobile data support as reflected in the PCS/PCN, UMTS, and IP-based Multimedia Subsystems (IMS) architectures.

The plethora of supporting data communications components include Remote Access Servers (RAS), Broadband Access Servers (BAS), pure data communications servers in support of X25, Asynchronous Transfer Mode (ATM), Frame Relay (FR) and multimedia servers supporting video-on-demand, videoconferencing, video telephony, and imaging services. In their own class are the IP-based data servers in support of email, web,

Voice over IP (VoIP), and video over IP (IPTV) applications, to name a few. Finally, are the connectivity service servers in support of Centrex and Virtual Private Networks (VPN), and intelligent networking servers in support of management services, including provisioning and billing.

One of the layers in this diagram is the **signaling network**, a separate out-of-band data communication network. This network controls not only voice calls set up but also controls data transmission by assigning and setting up the proper resources. The dominant signaling network is based on the Common Channel Signaling architecture and the Signaling System #7 set of protocols. Major architectural components are the Signal Transfer Points (STP) and Service Control Points (SCP). The corresponding signaling protocol in Internet, developed by the Internet Engineering Task Force (IETF), is the Session Initiation Protocol (SIP), also embraced as the signaling protocol of choice for the wireless IP-based Multimedia Subsystem architecture.

1.8 Spectrum Designation in Wireless Communications

Given the multitude of wireless technologies, the need to avoid interference between groups of users (military, commercial, amateurs) and to impose the discipline of using common standards, strict allocation of spectrum to different applications is required. The governing body controlling the spectrum in the USA is the **Federal Communications Commission** (FCC). FCC was chartered in 1934 with the regulation of interstate and international communications by radio, satellite, and cable, both wired and wireless. Within FCC, the **Wireless Bureau** oversees all the regulations regarding wireless services, i.e., radio, mobile communications, and pagers. A separate Mass Media Bureau regulates television and radio broadcasting.

A **spectrum** is a range of allocated frequencies within which wireless applications with common characteristics can operate without interfering with each other or with other applications operating outside the given range. Monitoring efficient use of the allocated spectrum is an important activity of **spectrum management**. Before we analyze the actual spectrum allocation for various wireless technologies, we need to have a general understanding of the correlation between transmission media and the frequency of the signals used for communications, as well as transmission capacity of various technologies. This view is presented in Figure 1.12.

The media is either wired, such as copper based twisted pair, coaxial cable, optical fiber or it is wireless. Transmission rates on twisted pairs go up to hundreds of MHz, while on coaxial cable rates reach into GHz. The full exploitation of media capacity comes with fiber optic, where the infrared, visible light and ultraviolet spectrum can support rates into the THz (note the logarithmic scale adopted in the figure to accommodate the broad range of frequencies that are represented). The wireless domain has overlapping rates with both twisted pair and coaxial cable, pushing the limits of radio-based and microwave-based technologies close to 100 GHz. From a different angle, one can notice the modest transmission rates needed in basic telephony, followed by the technologies that support computer networking and mass media broadcasting, and then by the bands that have the capability of supporting true-broadband networking.

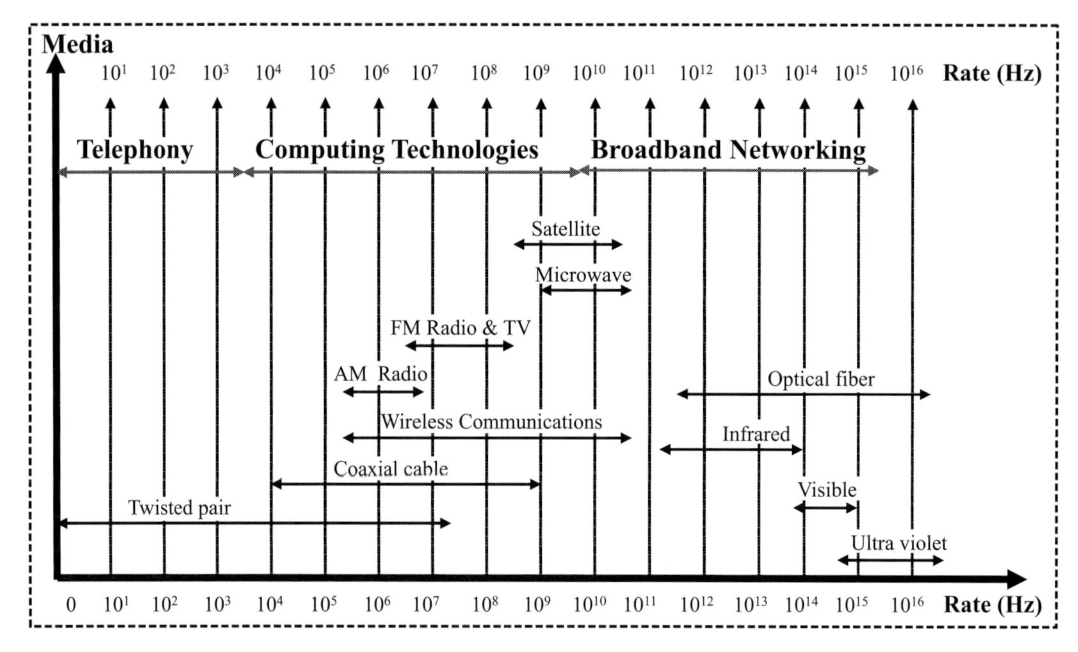

Figure 1.12 Communications Media and Transmission Rates

The correspondence between transmission frequencies and wavelength along with the commonly used term references associated with different wave types, are given in Table 1.2. One of the extremes of this spectrum, extremely low frequency, below 30 Hz, is the Ultrasounds. The other extremes, in Band 12 and 13, are X Ray and Infrared, with wavelengths below one millimeter.

Table 1.2 Relationships between Frequencies and Wavelengths

Wave Types	Wavelength	American Common Reference	Frequency	ITU Band designator
X Rays	0.1 mm to 1 mm	THF-Tremendously High Frequency	3–30 THz	13
Infrared			300–3000 GHz	12
Microwaves	10 mm to 100 mm	EHF-Extremely High Frequency	30–300 GHz	11
	to 1 m	SHF-Super High Frequency	3–30 GHz	10
		UHF-Ultra High Frequency	300–3000 MHz	9
FM Radio	10 m to 100 m	VHF-Very High Frequency	30–300 MHz	8
Radar waves		HF-High Frequency	3–30 MHz	7
AM Radio	1 km to 10 km	MF-Medium Frequency	300–3000 KHz	6
		LF-Low Frequency	30–300 KHz	5
Audio	100 km	VLF-Very Low Frequency	3–30 KHz	4
Range		ULF-Ultra Low Frequency	300–3000 Hz	3
Ultrasounds		ELF-Extremely Low Frequency	30–300 Hz	2

Details about spectrum partitions in those commercial bands opened to auctions will be given in later chapters dedicated to those technologies. Spectrum allocation may also vary from region to region (e.g., USA/Canada, Europe) or from country to country (e.g., Japan, South Korea). The most common spectrum allocation for wireless communications technologies introduced earlier follows [2]. All acronyms will be explained in subsequent sections.

Cellular mobile technologies

- **USA** initial **cellular** operating in the 850 MHz band (824–894 MHz). This band covers AMPS, D-AMPS (IS-136), and CDMAone (IS-95);
- **USA PCS** architecture-based operating in the 1900 MHz band (1850–1910 MHz receive and 1930–1990 MHz transmit). This band covers D-AMPS (IS-136), CDMA 2000 1xRTT (IS-95-C), and GSM 1900;
- **GSM (worldwide)** operates in the 400 MHz band (initially 450 MHz then 480 MHz), 850 MHz, 900 MHz (primary, extended, railway), 1800 MHz (DCS-1800), and 1900 MHz (PCS);
- **Worldwide 3G**, 1900 MHz and new spectrum allocations to support UMTS, W-CDMA, 1xEV-DO, HSDPA, and HSUPA technologies;
- **USA 3G UMTS-IMT2000** newly opened spectrum for auction (1432–1435 MHz, 1710–1755 MHz, 2110–2155 MHz, and 2500–2690 MHz);
- USA **iDEN** operates in the Specialized Mobile Radio (SMR) 800 MHz band (806–821 and 851–866 MHz) adjacent to the cellular frequency band;
- **CDMA** operates in the 832–925 MHz domain **(Japan)**, 1750–1870 MHz domain **(South Korea)**, and 2300–2400 MHz domain **(China)**.

WMAN technologies

- **LMDS** operates in several bands in the 27.5–31.3 GHz domains;
- **MMDS** operates in the 2.5–2.686 GHz domains;
- **WiMAX** operates in the 2–11 GHz domains (2.5 GHz, 3.5 GHz, and 5.8 GHz domains in Europe);
- **FSO** operates in the optical domains with wavelengths between 750 nanometers and 1 millimeter. The commonly used infrared domains have several bands with wavelengths between 1260 and 1675 nanometers.

WLAN technologies

- **IEEE 802.11b** and **IEEE 802.11g** operate in the 2.4–2.4835 GHz ISM band;
- **IEEE 802.11g** uses the 5 GHz ISM band (5.15–5.35 GHz, 5.47–5.825 GHz, 5.725–5.875 GHz);
- **HyperLAN** (Europe) operates in the 5 GHz band.

WPAN technologies

- **Bluetooth** operates in the 2.4–2.485 GHz unlicensed ISM band;
- **ZigBee**, according to the IEEE **tri-band WPAN** 802.15.4 standard, operates in the 2.4 GHz ISM band, the 915 MHz ISM band, and in the European 868 MHz band;
- **DECT** operates in the 1.9 GHz, 915 MHz, and 2.4 GHz bands.

Near-field Wireless Sensor Networks

- **UWB** with applications such as radar, imaging or positioning systems operates in the 3.1 GHz-10.6 GHz domains in a band that covers roughly 25% of the center frequency allocation;
- **RFID** operating in the **low-frequency band** (125–134.2 kHz and 140–148.5 kHz) and **high frequency band** (13.56 MHz) globally unlicensed bands and **ultra-high-frequency** (868 MHz-928 MHz) not globally licensed band (in North America can use the unlicensed 902–928 MHz band);
- **NFC** operating in the unlicensed band of 13.56 MHz, shared with RFID.

Satellite technologies

- **VSAT** satellite communications based on **GEO satellites** operates in the microwave domain of **C band** (Uplink 5.9–6.4 GHz, Downlink 3.7–4.2 GHz), **Ku band** (Uplink 14–14.5 GHz, Downlink 11.7–12.2 GHz), and **Ka band** (Uplink 28–28.9 GHz, Downlink 17–17.9 GHz).

MEO and **LEO** satellites-based communications use specific microwave frequency allocations according to international agreements.

Pager technologies

- **Pager receivers** mostly operate in the 900 MHz spectrum (910–930 MHz) unlicensed band.

In 1993, the FCC allocated three 1 MHz narrow bands (data) in the 900 MHz region (901–902, 930–931, 940–941 MHz) in support of enhanced paging and messaging (Narrowband PCS). A license is not needed by users to operate in these bands. The demand for wireless communications has induced FCC to open up other parts of the radio spectrum. WLANs, WiMAX, and Bluetooth users might share these bands as well. Business and consumer applications such as cordless phones, microwaves, and garage openers can also use these bands. These unlicensed bands include the following:

- ISM Spread Spectrum (Industrial, Scientific, Medical) opened in 1985:
  - 902 MHz–928 MHz, 2.4 GHz–2.483 GHz, 5.725 GHz–5.825 GHz;
- Unlicensed PCS (Personal Communications Services) opened in 1993:
  - 1.910 GHz–1.930 GHz, 2.39 GHz–2.4 GHz;
- Millimeter Wave, opened in 1995:
  - 59 GHz–64 GHz;
- U-NII (Unlicensed National Information Infrastructure), opened in 1998:
  - 5.15 GHz–5.35 GHz, 5.725 GHz–5.825 GHz;
- Millimeter Wave Expansion opened in 2001:
  - 57 GHz–59 GHz.

1.9 Wireless Communications at a Glance

To fully understand the complexity of wireless communications we have to look beyond just the communications facilities. This section looks, at a high-level, at the elements that constitute the ultimate goal of wireless technologies, wireless applications and services for the benefit of users and businesses. A first glance at the complexity of the wireless world and the knowledge involved is given in Figure 1.13. A more detailed analysis will follow in subsequent chapters.

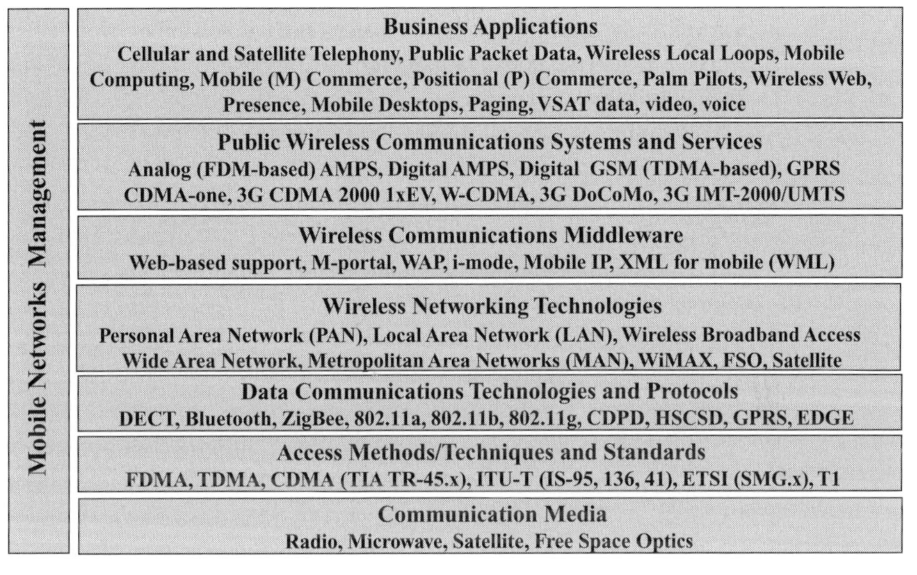

Figure 1.13 Wireless Communications at a Glance

This knowledge construct starts with the communication media, radio, microwave, satellite, and free space optics that represent the air interface and defines the very notion of wireless communications. The wireless media represent the physical layer of the layered communications model. The next level of knowledge is the use of one of the three major access methods, Frequency Division Multiple Access (FDMA), Time Division Multiple Access (TDMA), and Code Division Multiple Access (CDMA). There are numerous standards supporting these communication techniques and these standards or technical specifications are primarily defined by the International Telecommunication Union Telecommunications Sector (ITU-T), European Telecommunications Standard Institute (ETSI), and Telecommunications Industry Association (TIA).

Implementation of new types of data communications technologies and protocols further differentiates wireless communications from the older telecommunications model. This has led to the creation of Digital European Cordless Telecommunications (DECT), Bluetooth, ZigBee, 802.11a/b/g WLANs, specific data services such as Cellular Data Packet Data (CDPD), High Speed Circuit Switched Data (HSCSD) (a technology that allowed GSM users to use multiple TDMA channels and increasing the maximum data rate 8 fold compared with regular GSM that offered 9.6 Kbps), General Packet Radio Service (GPRS), and Enhanced Data for GSM Evolution (EDGE). The search for data transmission technologies that better fit a particular environment led to the consolidation of wireless communication under the banner of WLAN, WMAN, WPAN, UWB, WWAN or cellular mobile networks, satellite, and free space optics solutions. Critical in achieving this goal are Wireless Communication Middleware in form of web support, special mobility portal (M-Portal), Wireless Application Protocol (WAP), i-mode (Japan), Mobile IP, and Extended Markup Language (XML) for the mobile environment.

On the road of continuous development and progress in wireless communications, numerous wireless public systems and services were created. The most significant of these are in the cellular mobile world. Names such as Advanced Mobile Phone Systems (AMPS) the first cellular standard in the US, Digital Advanced Mobile Phone Systems (D-AMPS) the digital expansion of AMPS, Global Systems for Mobile Communications (GSM) and its General Packet Radio Service (GPRS), Code Division Multiple Access (CDMA) variations from CDMA-one to CDMA-2000 to Wideband CDMA, IMT-2000/UMTS conceptual architectures, all are part of the cellular radio wireless world. These technologies support a great variety of business applications such as Cellular and Satellite Telephony, Public Packet Data, Wireless Local Loops, Mobile Computing, Mobile (M) Commerce, Positional (P) Commerce, Palm Pilots, Wireless Web, Presence, Mobile Desktops, Paging, VSAT data, VoIP, and IPTV.

All the physical components, from communications media, wireless hardware, data link layer protocols, wireless technologies and networks to wireless middleware, applications and services should be managed to have mature wireless products and services. System management starts from managing the spectrum, the air interface and ends by providing complex management applications supporting fault, configuration, performance, accounting, and security management. The fundamentals of network and systems management as applicable to wireless communications are presented in the next chapter.

2 Network Management

2.1 Networks and Systems Management Concepts

To administer the enormous number of resources involved in communications, either wired or wireless, from customer premises to backbone networks and across all geographical and administrative boundaries, there is a need for specialized systems designed specifically for management. These systems comprise hardware and software components, applications and corresponding operations used together to monitor, control, operate, coordinate, provision, administer, diagnose and report faults, and account for the network and computing resources that allow communications to take place. As communications systems became an important part of any business, management systems evolved to support whole enterprises, i.e., to provide management of multivendor, multiprotocol and multitechnology network and systems environments. Management systems are more than just simple tools used to manage network and systems resources. They include standardized procedures and sophisticated communications protocols to collect and process the management information.

The high level paradigm of management consists of two entities: the **Managing Entity** and the **Managed Entity**. The relationship between managing entities and managed entities can be modeled as manager-agent, client-server, mainframe-terminal, master-slave, or peer-to-peer. In the manager-agent model, the management entity, also called the manager, represents the managing process while the managed entity, also called agent, represents the managed process. These models are depicted in Figure 2.1. Both manager and agent processes are software applications. The manager application provides management functions and services while the agent application provides access to the management information related to managed resources or managed objects.

The manager-agent is a hierarchical model in which the manager starts the management process by sending requests to the agent and receiving replies from the managed entity. Generally, a connection request to pass management information is initiated by the manager. However, the agent can also initiate passage of management information from the manager if the circumstances require doing so. The agent can have direct access to managed resources or it can use an intermediary entity, a sub-agent. When the manager does not have access to the agent because the agent does not understand the communication protocols used by the manager, a proxy-agent might be used. Communication is not limited to a manager-agent exchange of information. In a build-up of large management systems multiple managers may be needed and communication between them

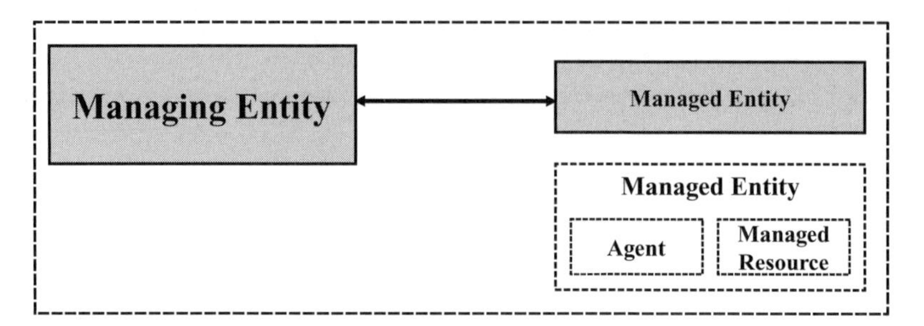

Figure 2.1 Basic Manager-Agent Management Model

could be required. There are three major topological frameworks used in analyzing the relationships between managers and agents: single managers, manager of managers, and cooperative managers. This is indicated in Figure 2.2.

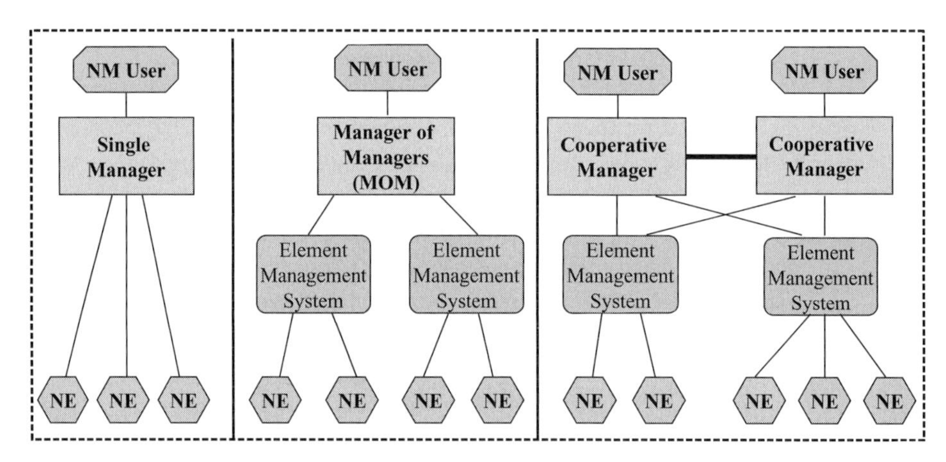

Figure 2.2 Network Management Topology Frameworks

The **Single Manager** framework assumes a single manager and multiple agents that all communicate with that manager. This framework is a fully centralized approach where the manager concentrates and processes all the information collected from **Network Elements** (NE) through agents. Network elements are various network components such as modems, multiplexers, switches, and routers. The concentration of management functions in one point, and practical limitations of managed resources presents both availability (single point of failure) and scalability problems. These limitations make this management topological framework applicable to relatively small networks and systems.

The **Manager of Managers** (MOM) framework assumes multiple layers of managers, each layer responsible for management of a certain class and number of managed resources. A MOM acts as the manager of multiple Element Management Systems (EMS). It acts as a logically centralized focal point. Detailed monitoring and control is performed by each individual element management system in a distributed fashion and only certain critical management information is processed by the manager of managers.

The **Cooperative Managers** framework, or network of managers, is a fully distributed topology where management information is exchanged between multiple peer managers. Although each element management system is responsible for its own domain, cooperative links are established between higher level managers to allow exchange of certain types of information. Each cooperative manager can take over the functions of other managers. This framework is suitable for very large interconnected networks.

A class apart is the management system that qualifies as an open management system. An open system, as opposed to a proprietary system, is one that is based on commonly accepted standards. Standardization is paramount for communication between managers and agents since the managers and agents can be the products of designs by different manufacturers. The openness of a management system can also come from the actual software design of managers and agents. The separation of management applications from the underlying hardware and operating system allows third parties and independent software vendors to create applications running on top of well-known hardware and operating system platforms.

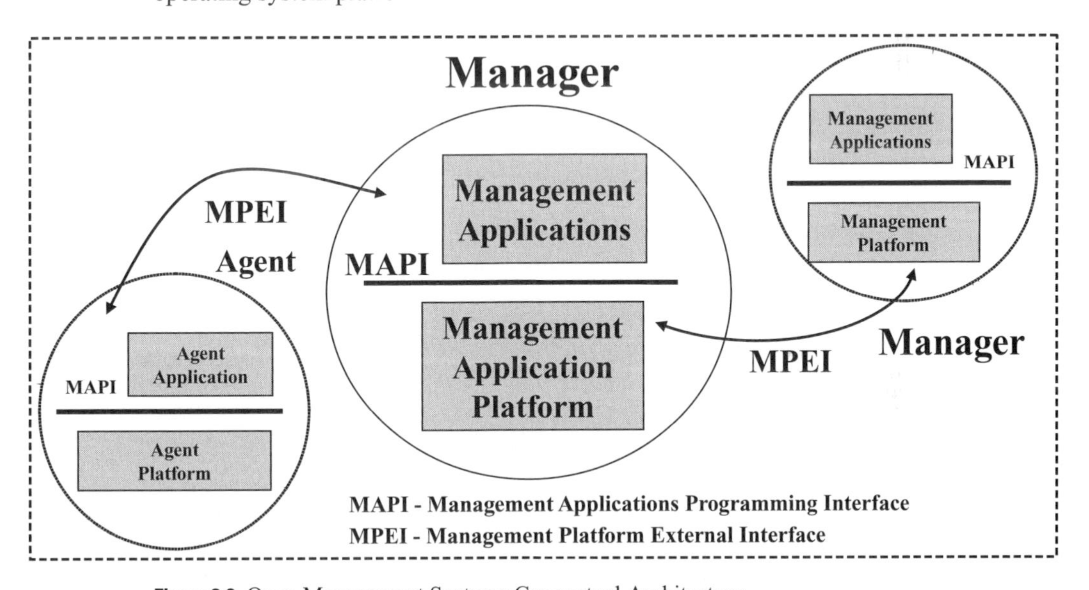

Figure 2.3 Open Management Systems Conceptual Architecture

The open management system conceptual architecture can be described as a collection of four major components: Management Application Platform (MAP), Management Applications, Management Applications Programming Interface (MAPI), and Management Platform External Interface (MPEI). The architectural model, given in Figure 2.3, shows both manager-to-agent and manager-to-manager external interfaces as similar types of interfaces. Open management systems assume that both the platform external interface and application programming interfaces are based on open published standard specifications. The characteristics and the details of the framework for management platforms will be given in subsequent sections.

Descriptions of each of the open management system components are given in Table 2.1.

Table 2.1 Open Management Systems Component Descriptions

Component	Description
Applications	Set of software programs and associated data that can run directly on an application platform.
Applications Platform	Set of software and hardware resources able to support running applications and providing services to applications.
Applications Program Interfaces (API)	Collection of service interfaces that make the characteristics of the application platform services transparent to the application developers.
Platform External Interfaces (PEI)	Interfaces which support information exchange between two open systems and the subject of Open Systems Interconnection (OSI) architectures and standards.

In the open management conceptual model we assumed that truly open managers and agents should be designed as management platforms. More detailed explanations about management platforms will be given in subsequent sections.

2.2 Network and Systems Management Models

The relationships between manager and agent, or between managers, are not limited to the communication interface. As indicated in Figure 2.4, several other aspects should be incorporated to describe the complexity of the management paradigm.

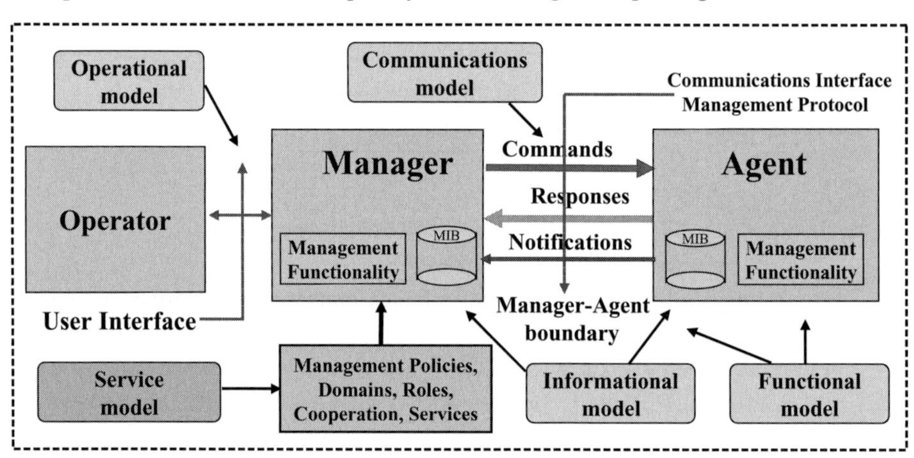

Figure 2.4 Models of the Wireless Manager-Agent Paradigm

The **communications model** is supported by an interface that includes a layered communication stack and dedicated management protocol. Most communications models rely on the commands initiated and sent by the manager to the agents attached to the managed resources. The agents, in turn, send replies or responses to the manager containing the requested information. The manager commands can be sent infrequently, or in a continuous polling fashion, searching all the agents and every managed resource for new management information. The communication model also allows agents to initiate communications when sending urgent or scheduled notifications containing critical or updated management information.

The **functional model** describes that functionality built into the management applications which is supported by the managers and agents. This functionality will determine the type and the class of management services. The most common model divides the management functions into five major categories: fault, configuration, performance, accounting, and security. **Fault Management** (FM) addresses processes that allow monitoring, detection, isolation, reporting, and correction of abnormal operation of network resources and the associated applications. **Configuration Management** (CM) includes network resource identification and collection of that data needed from the managed resources to initialize, start, provide, maintain, and terminate a service. **Performance Management** (PM) addresses processes that ensure operation of the managed resources and associated applications at levels needed to meet expected performance and service requirements. **Accounting Management** (AM) functions process and manipulate service and resource utilization to measure the usage and to determine the associated cost for users of services. **Security Management** (SM) addresses processes that control access and provide protection of network resources and associated applications from intentional or accidental abuse and unauthorized access.

The **information model** is the representation of the management information that may be transferred by management protocols across management interfaces. Managers and agents should have a common communications language and a common understanding of how to represent and describe managed resources and their characteristics. An instance of the management information model is called a **Management Information Base** (MIB) and is a conceptual repository of management information defined for a particular managed system. A **Management Information Library** (MIL) is a collection of various MIBs required to manage a network or system environment that contains very different managed resources.

The **operational model** describes the user interface that allows operators to perform management operations. Most management systems are operated by people who interact through an interface that can be a simple Command Line Interface (CLI) or, in most cases, a sophisticated Graphical User Interface (GUI). Using various symbols in the form of topological layouts and maps of networks, GUIs provides iconographic representation of the managed resources. The GUIs also provide display capabilities to depict errors, alarms, and general reporting information. They also accept user inputs in the form of commands of mouse clicks and keyboard inputs.

The **service model** describes operation of management systems in real environments as enshrined in overall management policies. The model contains descriptions of the services performed by management systems and the operators of management systems. For large management systems, with multiple operators, it describes, in the form of profiles, the role of each operator, the domain of operation, the level of cooperation between operators, and the hierarchical position of each operator.

2.3 Management Systems Classifications

Management systems can be viewed from different perspectives. By introducing the concepts of management in previous sections of this chapter we have already provided classifications of management based on topological frameworks, hierarchical relationships established between management systems components, and the major functionality provided during management operations. However, the most recognizable classification of management systems is along management domains. This is indicated in Figure 2.5.

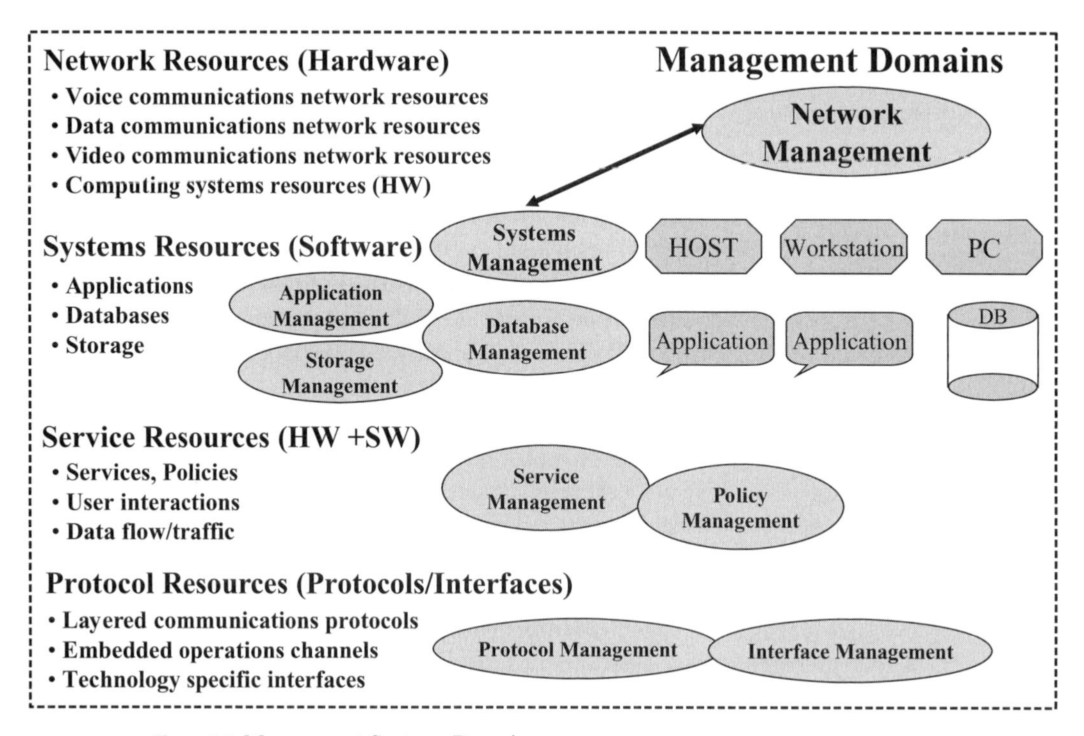

Figure 2.5 Management Systems Domains

There are four grand management domains: the first, the **network** domain, covers physical resources, essentially hardware-managed resources. The second, the **systems** domain, covers systems resources, essentially software-related managed resources. The third, **service** domain covers both physical and logical resources that provide services. The fourth, the **protocol** domain covers those communication resources used to transfer management information across management interfaces, essentially protocols and technology-specific management interfaces.

The **network management domain** includes all the various types of network components that support voice, data, video, and multimedia communications. Therefore, within this grand domain, we have management systems that are dedicated to **voice communications management** (telecommunications network resources), **data communications management** (Internet, mainframe-based, wide, local and personal area networks with all the physical components of computing networks), and **video communications**

management (all the video broadcast hardware components, including video terminals). In this domain belong hardware computing resources such as mainframes, front-end processors, cluster controllers, host computers, servers, desktop computers, workstations, PCs, laptops, and notebooks. Within the network domain the management systems can be classified as those serving **wired networks** and those serving **wireless networks**.

The **systems management domain** includes all those software resources that support communications and networking. Three distinct classes of systems resource management head the systems management list: **application management**, **database management**, and **storage management**. Applications management encompasses management of any collection of executable code, data files, processes, and operating systems that relate to identification, configuration, and performance management. Database management includes management of external relational and object-oriented databases or embedded databases that support access, manipulation, and data storage services. Storage management includes management of external data storage devices and the transmission channels and protocols used to store and retrieve mass information.

The **service management domain** is an umbrella domain of the hardware and software elements that constitute the network and system components providing services to the customer. **Service management** includes service provisioning, service monitoring to ensure that services are delivered at expected performance levels, and reporting use for billing. One of the important aspects of service management is **policy management**, monitoring those policies that apply to service management, service planning, and correlation with operational plans. Other aspects loosely belonging to the service domain include traffic and data flow measurements, and user interactions with service management tools. Service management will be analyzed in greater detail in the next chapter.

The **protocol management domain** includes management of logical resources such as layered communication stacks and individual layer protocols with all the attributes, including the application layer management protocols. This domain also includes management of embedded operations channels, dedicated in-band or out-of-band management channels, special management related fields, or management data streams.

2.4 Management Systems Evolution

Initially, management systems were created out of necessity to deal with the ever-increasing size and complexity of networks and computing systems. The need for monitoring, controlling, preventing, and ultimately solving network problems appears self-evident. Otherwise, the consequences can be severe if networks cannot be kept running and problems cannot be quickly found and resolved in a timely manner. Historically, management system developments have been afterthoughts in the world of proprietary designs. This explains the difficulty in achieving standardization in the areas of network, systems, and service management. A depiction of management systems evolution is presented in Figure 2.6.

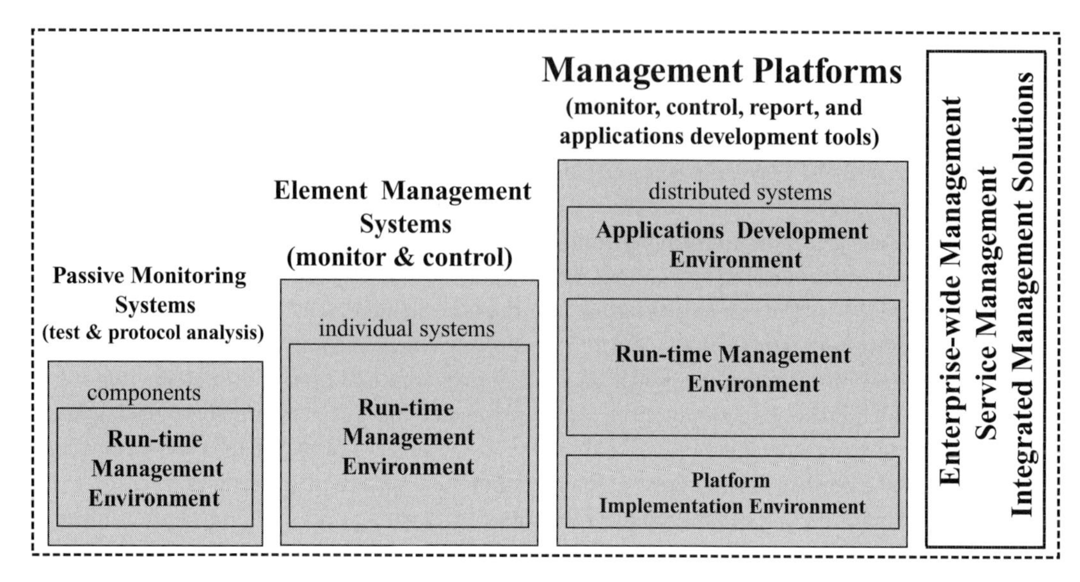

Figure 2.6 Management Systems Evolution

In the evolution of management systems, in the 1970s and early 1980s, **passive monitoring systems** were first. The core, run-time, and operational environment of these systems had limited functionality built around testing network components for their strict availability and performing tedious protocol analyses. The second generation of systems augmented basic monitoring capabilities by adding controlling features. The scope of management moved from simple components to individual networks. This was the generation of **proprietary Element Management Systems**, dedicated to managing a certain class of equipment (e.g., modems or multiplexers) manufactured by a particular vendor. The third generation of management systems, developed in the late 1980s to the present, introduced the idea of **management platforms**. These are open management systems capable of monitoring, controlling, and reporting the management information of large distributed networks and systems using standardized communication interfaces. These management platforms are not just the core, run-time, or operational management platforms; they also provide an application development environment that allows the platform designers and independent software vendors to write management applications that run on top of management platforms. The success of management platforms is assured by careful implementation through policies that are incorporated in the platform implementation environment. This work began in the early 1990s, and has gone through a decade of consolidation and maturation. Currently, development of a new generation of management systems is taking place. Management platforms now have the functionality to provide enterprise-wide management solutions and integrated management across multi-vendor and multi-technology management systems. Network and system management has been elevated to the rank of service management.

The evolution of management systems can be better understood by marking some chronological events [3]:

- IBM NetView, an integrated mainframe-based proprietary network and systems management set of tools (1986);
- AT&T Unified Network Management Architecture (UNMA) concept (1987);
- Element Management Systems from Timeplex, Paradyne, GDC, Codex (1987–1989);
- Open System Interconnection Network Management international standardization work starts in 1989;
- Network management platforms concept is advanced by several vendors and organizations such as DEC, HP, IBM, Sun Microsystems, OSF (1990);
- Internet Simple Network Management Protocol version 1 is adopted as recommended standard for TCP/IP-based networks (1991);
- ITU-T, ANSI T1M1, and ETSI start work on Telecommunications Management Network (TMN) concept (1992);
- Commercial management platforms on the market: HP OpenView, DEC DECmcc, Tivoli TME (1993);
- Network Management Forum (NMF), later the TeleManagement Forum international consortium is formed to promote OSI management standards implementation (1994);
- Home grown management platforms from Cabletron, NetLabs, Computer Associates, IBM, Sun Microsystems (1996–1997);
- TeleManagement Forum Telecommunications Operations Map (TOM) simplified service management concept and its extensions (1999–2002);
- Service management products from Micromuse Inc., Concord Communications, Visual Networks (2000–2002).

Today, we have a solid set of surviving network and systems management platforms although the new generations of service management platforms are far from the ideals of open, standardized, and distributed management platforms. OSI's CMIP-based management standards have been relegated to a few island implementations in both wired and wireless telecommunications. Although it has many shortcomings, the dominant protocol remains SNMP since it runs on the highly popular Internet.

2.5 Network and Systems Management Platforms

At the beginning of this chapter, when we provided an open management system architectural model, and in the previous section dealing with management systems evolution, we introduced the term "management platform" without giving any details. In this section, we explain the term in some detail.

Management platforms do not represent new topological frameworks or management paradigms. The fundamental difference between management platforms and other management systems is the fact that management platforms include a run-time operational environment and a set of applications development tools. These are collectively known as application development environments. The run-time environment allows normal management operations while the application development environment allows portability of management applications and integration of applications with platform services. A high level framework of management platforms is given in Figure 2.7.

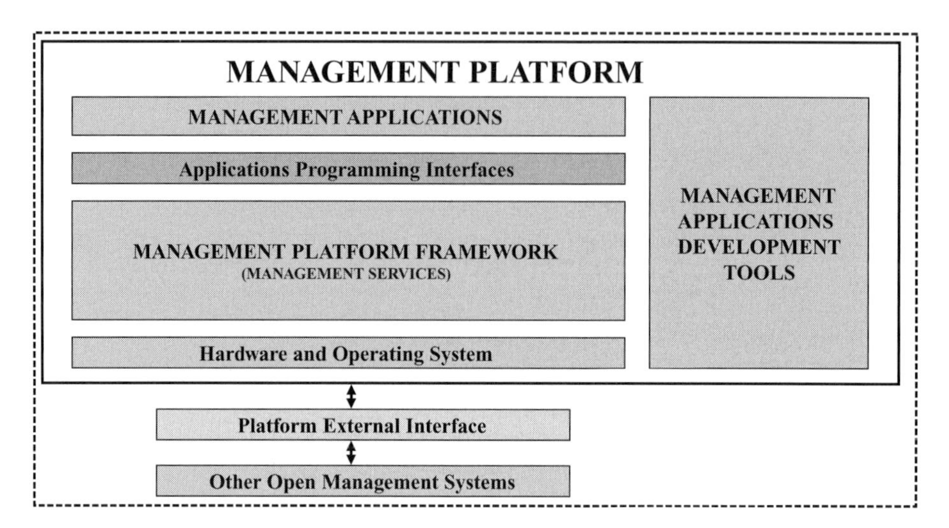

Figure 2.7 Management Platforms Frameworks

Three environments characterize a management platform: run-time, applications development, and implementation environment. The run-time environment consists of platform core management services and distributed applications running independently from the platform services. The application development environment consists of the run-time environment, software development tools, and a set of Applications Programming Interfaces. The implementation environment provides specifications of acceptance testing, conformance testing, interoperability testing and tools for installation, troubleshooting, and maintenance.

In the light of this framework, we consider the management platform as an example of an open system with interoperability across standard interfaces and with the portability of applications, due to the use of standardized Applications Programming Interfaces (APIs). The run-time operational platform should provide several basic management functions. These would include communication services to collect and exchange management information, event services to sort through the myriad of messages exchanged between manager and agents, user interface services to provide support for a graphical user interface, database services to store management information, and, ultimately, object manipulation services. We use the term object manipulation because the management information is modeled and manipulated using object modeling techniques. All these services run on top of one of the hardware and operating system combinations. The management platform architectural model is presented in Figure 2.8 [3].

Management operations support services provide common services to all platform core services. This is done by managing the background processes that are associated with the platform hardware, operating system, and the management of applications running on top of the platform. The architectural model, in addition to these management services, contains several distributed services such as time (synchronization), directory (naming), and security services. Additional service applications are needed for software

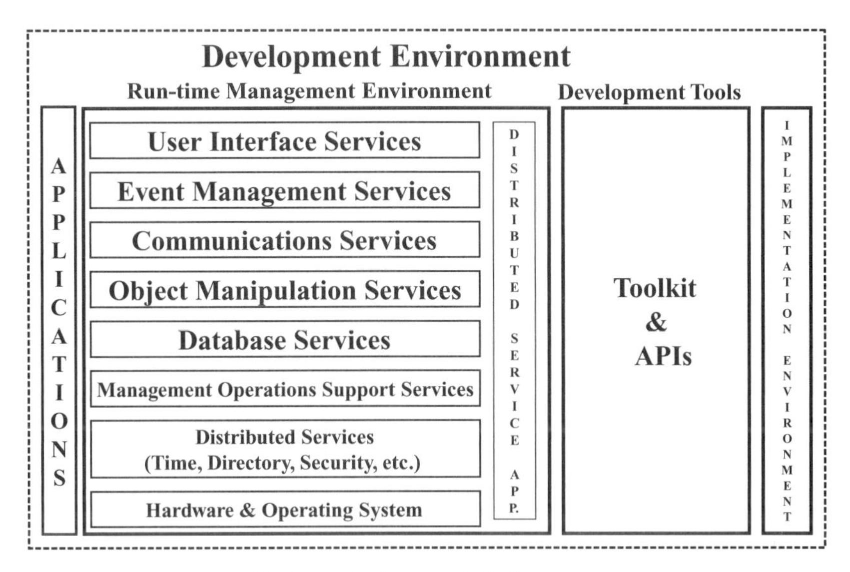

Figure 2.8 Management Platforms Architectural Model

distribution regarding the agents. Explanation of operations performed by each platform core management service is given in Figure 2.9.

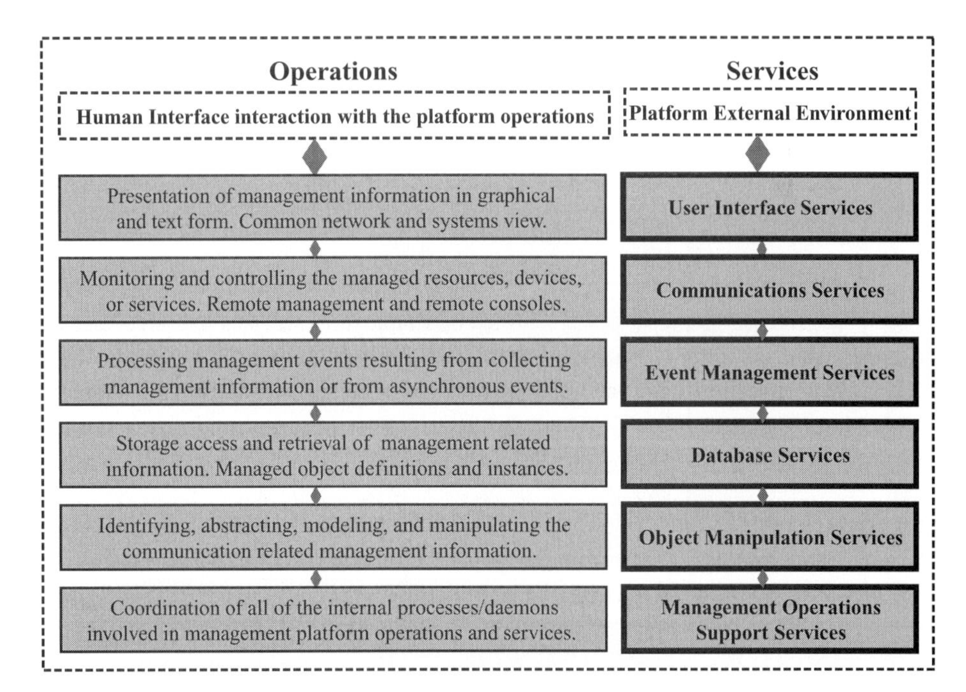

Figure 2.9 Management Platform Core Operations and Services

There are features that clearly differentiate management platforms from other management systems, mainly the proprietary management systems such as those listed below.

- Separation of management applications from the underlying platform services;
- Management applications development environment;
- Open, standardized Applications Programming Interfaces;
- Modularity of management platform core services;
- Support for integration of management applications;
- Open, standardized platform external interfaces;
- Common graphical user interface;
- Common application services (time, directory, security);
- Common management data repository;
- Software distribution and licensing service;
- Cost effective software development environment.

The human interface interaction with the platform operations is considered outside the management platform core services.

2.6 Internet SNMP-based Management

In the first chapter we explained the meaning of layered communication stack and we used as an example the first internationally standardized communication stack and set of protocols, collectively known as the International Organization for Standardization **Open Systems Interconnection** (OSI) stack. Another recognized, and later internationally standardized layered communication stack, is the one used in the Internet, also known as TCP/IP stack, so named because of its two main protocols, the Transport Layer **Transmission Control Protocol** (TCP) and Network Layer **Internet Protocol** (IP). The Internet communication stack is presented in Figure 2.10.

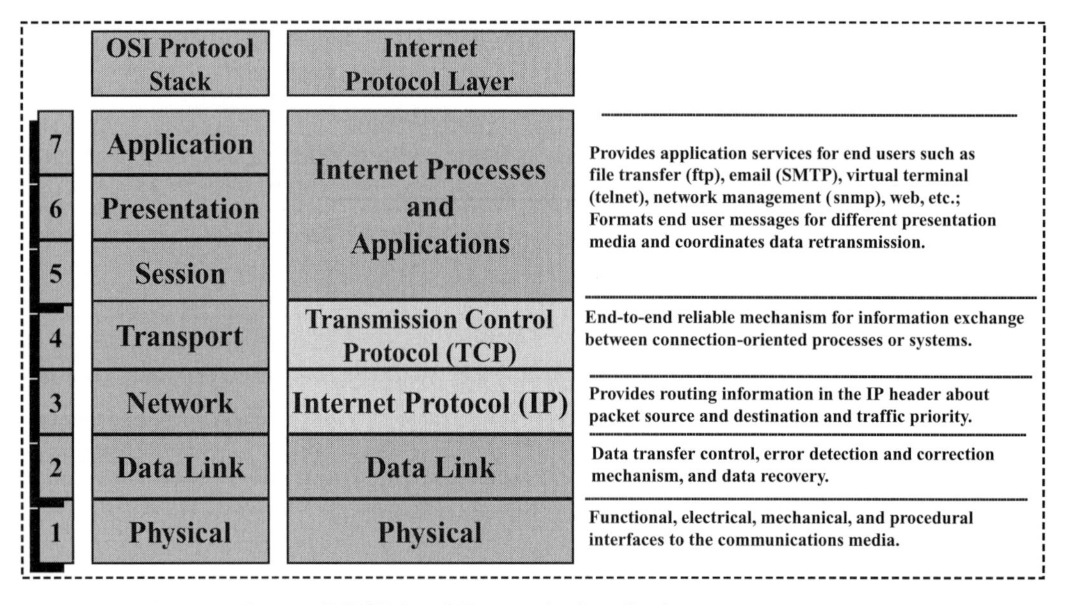

Figure 2.10 Internet TCP/IP-based Communications Stack

By comparison with OSI's seven-layer protocol stack, the Internet recognizes only five layers. The first two layers' functions and practical implementations, Physical and Data Link Layer, are identical in OSI and Internet models. The first major difference comes at the Network Layer where the Internet has adopted a connectionless protocol, IP. IP is the only Network Layer protocol used throughout all Internet-based global and local networks. The main function of IP is to support routing of data packets between source and destination based on the use of unique addresses. There are two types of addressing schemes: one that is based on a four-byte-length address, known as the IP Version 4; and the other is based on a twelve-byte-length address scheme, known as IP Version 6.

The transmission layer protocol can be either a connection-oriented protocol such as TCP or one of the connectionless protocols such as User Datagram Protocol (UDP). The TCP protocol provides a reliable mechanism for information exchange between processes and systems, while the reliability of UDP-based information exchange is based solely on the reliability of the intervening data communications network. The upper layer of the Internet communication stack, layer 5, is called the **Application Layer** because it provides support for a myriad of application services for end-users such as telnet, web, file transfer, and network management based on **Simple Network Management Protocol** (SNMP). The SNMP-based protocol stack in a manager-agent communication paradigm is shown in Figure 2.11.

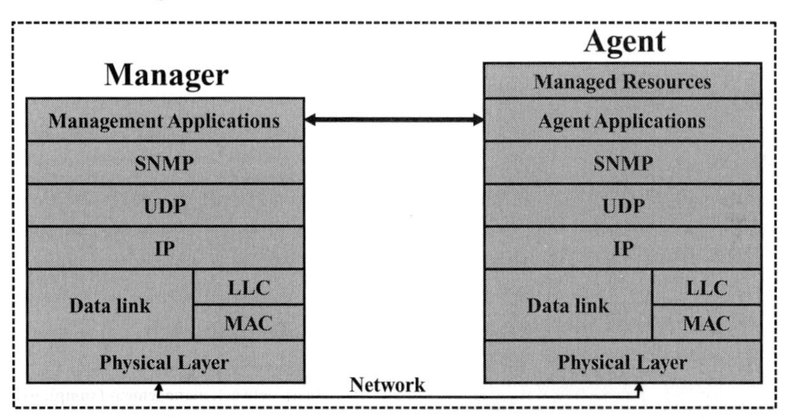

Figure 2.11 SNMP-based Protocol Communication Stack

SNMP is an application layer protocol running on top of the User Datagram Protocol (UDP). SNMP, standardized as the Version 1 in 1991, was initially designed solely for the management of Internet TCP/IP-based network environments. However, given the simplicity and versatility of this protocol, as opposed to the OSI-based management protocols, it quickly gained a wide acceptance. Therefore, it has become the protocol of choice for manufacturers of network devices and software systems in all management environments requiring basic monitoring.

Four components are part of the SNMP-based manager-agent paradigm: **Manager** or Management Station, **Agent**, **Management Information Base** (MIB), and the **SNMP protocol**. This is shown in Figure 2.12. The manager exchanges management information

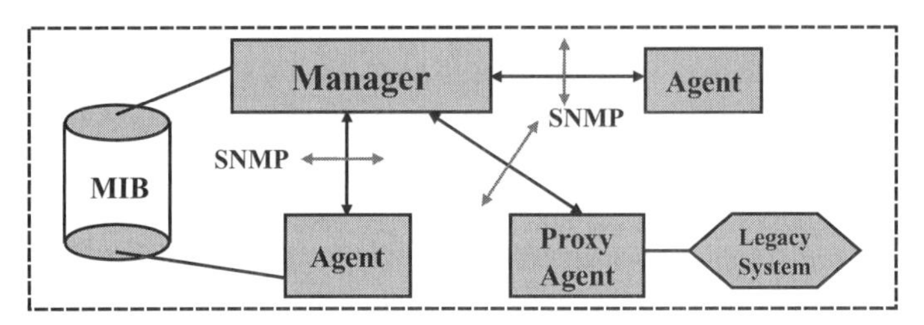

Figure 2.12 SNMP-based Manager-Agent Paradigm

with multiple agents using the SNMP protocol stack. In cases of legacy systems, with no IP-based communication support, a proxy agent is added that performs adaptation/mediation between the IP environment and the non-standard proprietary environment.

The Management Information Base (MIB) is a generic term for the repository of management information that is used to characterize a network/systems/service or computing system. The contents of the MIB represent, in abstract terms (i.e., objects), the real managed resources (hardware, software, services, protocols components) in accordance with well-defined information models (IETF RFC 1156).

The model used to create the MIB is termed the Structure of Management Information (SMI), and it is one of the IETF recommended standards that accompany the use of SNMP. This is indicated in Figure 2.13. It is not the intent of this book to analyze management information modeling techniques and the implementation of these models in specific management protocols (IETF RFC 1155).

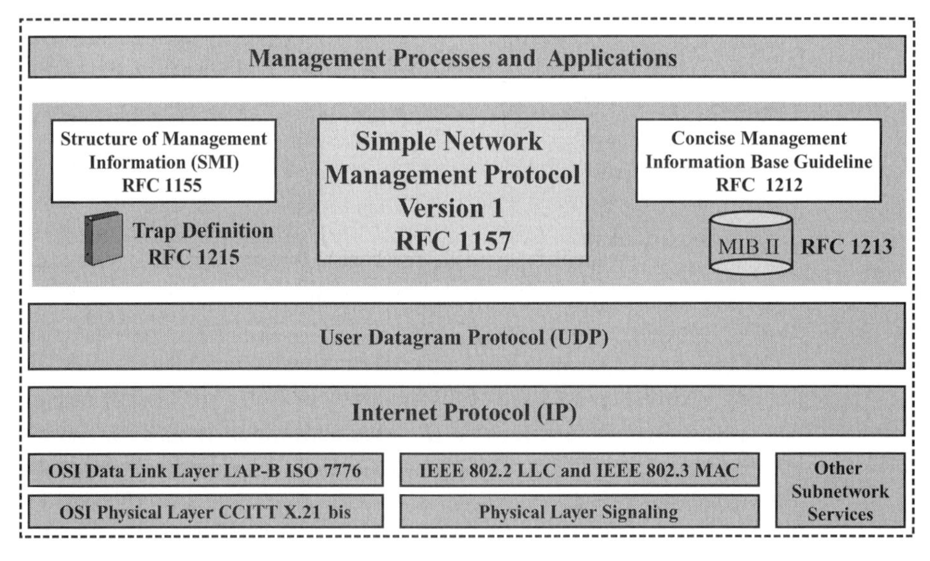

Figure 2.13 Internet Management Layered Architecture and Standards

As indicated in Figure 2.14, SNMP-based management uses four **types of operations**: Get, Get Next, Set, and Traps [4].

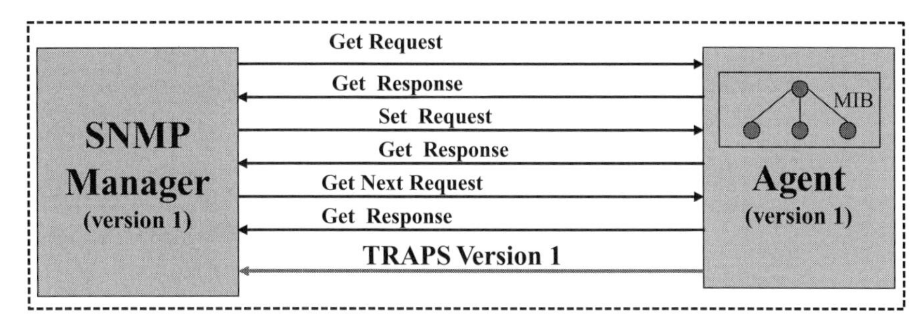

Figure 2.14 SNMP-based Manager-Agent Services

Three types of **Protocol Data Units** (PDUs) are associated with manager-agent messages. The first incorporates all requests initiated by the manager: SNMP Get Request, Get-Next Request and Set Request. Using a unique PDU format, the second type covers responses: SNMP Get Response. The third format is dedicated to the asynchronous messages sent by the agent to the manager, and is called SNMP Traps. An evaluation of strengths and weaknesses of SNMP-based management is given in Table 2.2.

Table 2.2 SNMP-based Management Evaluation

	Evaluation/Description
Strengths	– Simplicity of information model and agent implementation. – Allows off-line extensions of manager/agent domain specific MIBs. – Allows addition of enterprise specific traps. – Universal market acceptance by both vendors and users. – Compatible with proxy-agent implementations. – Portability across multiple operating systems platforms. – Built on top of the Internet TCP/IP distributed network architecture.
Weaknesses	– Limited network scalability because of polling mechanism. – Use of unreliable transport mechanism, the User Datagram Protocol (UDP). – Limited set of traps events and lack of of traps confirmation. – Lack of adequate security mechanism (unencrypted community password). – Inefficient retrieval mechanism for large amounts of data. – No built-in filtering, scoping or MIB instance creation mechanism. – Lack of standardized conformance testing procedures.

From this list of weaknesses, particularly the limited scalability and lack of distributed management, it is clear that building large-scale management networks based solely on SNMP will be difficult. Many of the concerns noted in Table 2.2 were addressed in SNMP Versions 2 and 3. However, implementation of the newer versions has met resistance because of their inherent complexity and saturation of the market after a decade of widespread implementation of Version 1. An example of SNMP-based management in the Internet environment is given in Figure 2.15.

The enterprise network can take various forms as indicated in this diagram. All these customer premises networks are connected to one or two of the Internet Nodal

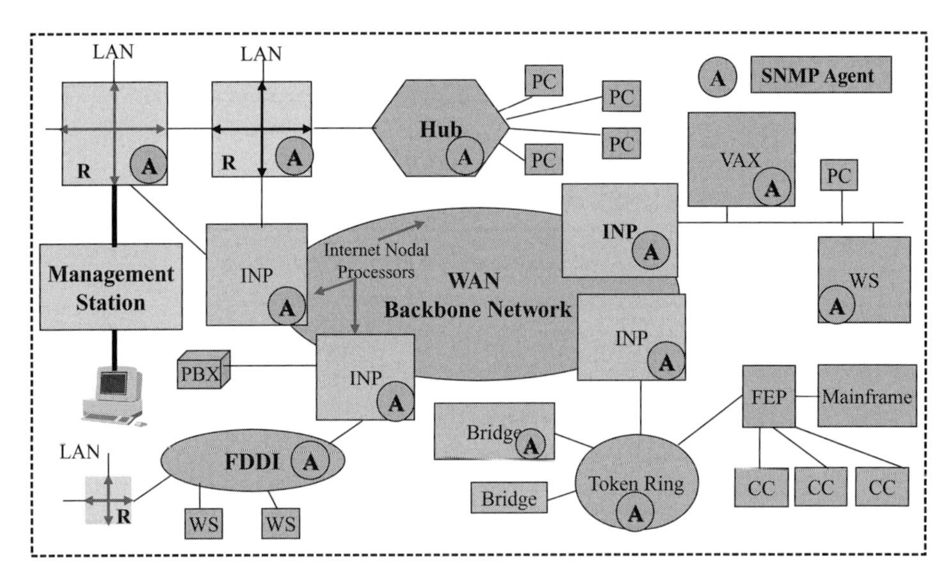

Figure 2.15 Enterprise Network SNMP-based Management Architecture

Processors (INP) that constitute the connectivity nodes of the Internet core backbone network, a Wide Area Network (WAN). These networks, each of them with their own network resources, can be Ethernet, Token Ring, or Fiber Distributed Data Interface (FDDI) types of LANs, customer premises Router (R)-based Internets, or digital Private Branch Exchanges (PBXs). Enterprise networks can also be based on proprietary IBM networks with mainframes, Front End Processors (FEP), Cluster Controllers (CC) or formerly Digital Equipment Corporation (now HP/Compaq) VAX minicomputer-based networks.

Internet SNMP-based management is an in-band type of management. This means that the network used for normal connectivity to exchange user or business information will also be used for exchanging management information. This information is exchanged between a Management Station that acts as a manager, and a multitude of agents collocated with managed resources.

The management arrangement has two aspects. First is the discovery of all the network and system resources. That is possible if all the resources are at least IP addressable components. Secondly, to manage those resources, the resources must have attached a SNMP Agent (A). For those resources that are not IP addressable, or do not support a SNMP agent, proxy agents should be provided. These agents convert/translate proprietary protocols stacks and management protocols into SNMP-based communication compatible stacks and protocols.

2.7 ISO OSI CMIP-based Management

In the first chapter we introduced the ISO standardized communication stack and set of protocols. OSI management is performed through this interface having two components:

a connection-oriented application layer protocol called **Common Management Information Protocol** (CMIP) and a Common Management Information Service Element (CMISE). CMISE works with two other application layer service elements, Association Control Service Element (ACSE) and Remote Operations Service Element (ROSE). A group of System Management Application Service Elements (SMASE) provide support for specific systems management functions such as fault, configuration, performance, accounting, and security. This OSI management layered architecture is presented in Figure 2.16.

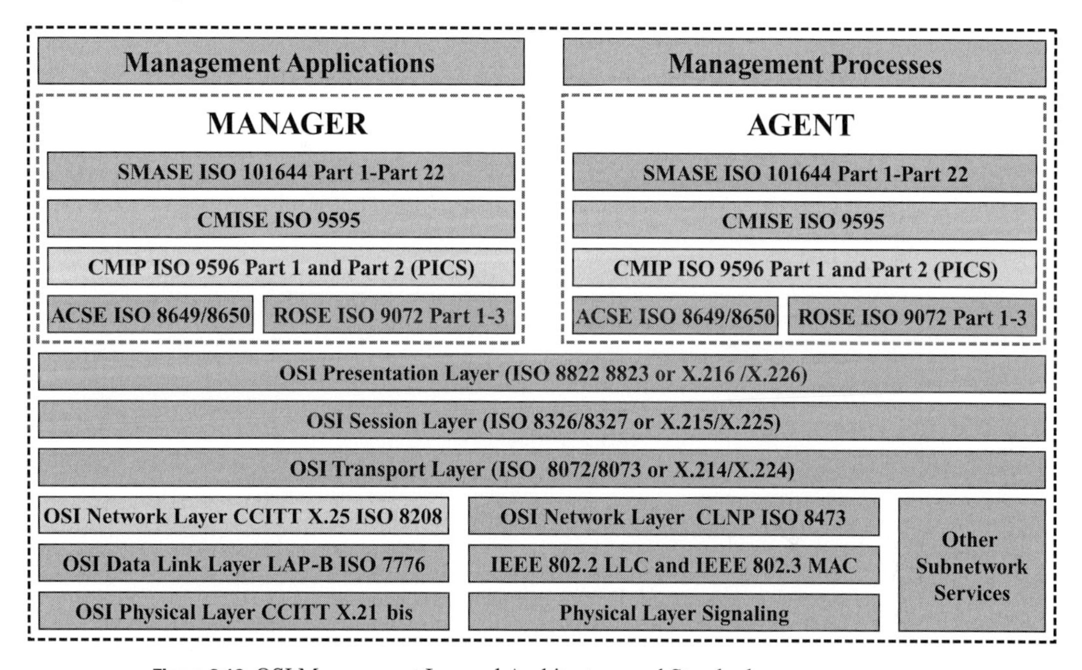

Figure 2.16 OSI Management Layered Architecture and Standards

The diagram lists ISO standards that contain specifications for each of these modules and for the protocols supported at Presentation, Session, and Transport Layer. These are all connection-oriented protocols. The Network Layer has two alternatives: the packet layer (X.25) and a Connectionless Layer Protocol (CLNP). The Data Link Layer and Physical Layers have any one of the numerous implementations that are common across various data communications technologies. One, the System Management Application Service Elements (SMASE), is used in wide area communications. The other data link layer example is the sub-layer-based Logical Link Control (LLC) and Media Access Control (MAC), specific to local area network environments. At the physical layer, CCITT X-21bis is mentioned as an example of a standardized alternative to the popular RS 232C.

Four components are part of the ISO OSI CMIP-based manager-agent paradigm: **Manager**, **Agent**, **Management Information Base** (MIB), **CMIP protocol** along with **CMISE services**. As shown in Figure 2.17, the manager exchanges information with multiple agents using the CMIP-based protocol stack. With those legacy systems having

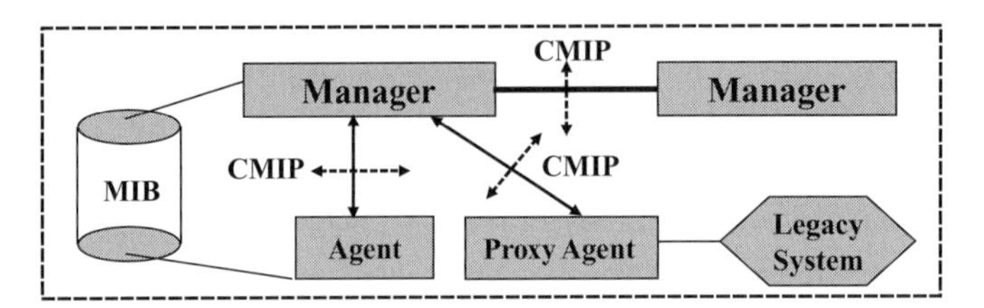

Figure 2.17 OSI-based Manager-Agent Relationship

no OSI-based communication support, there is a need to add a proxy agent. This agent performs adaptation/mediation between the standard OSI-based environment and the non-standard environment.

CMISE/CMIP-based management uses **six types of operations** and **one notification**: M-CREATE, M-DELETE, M-GET, M-SET, M-ACTION, M-CANCEL-GET, and M-EVENT REPORT. These exchanges are indicated in Figure 2.18 [5].

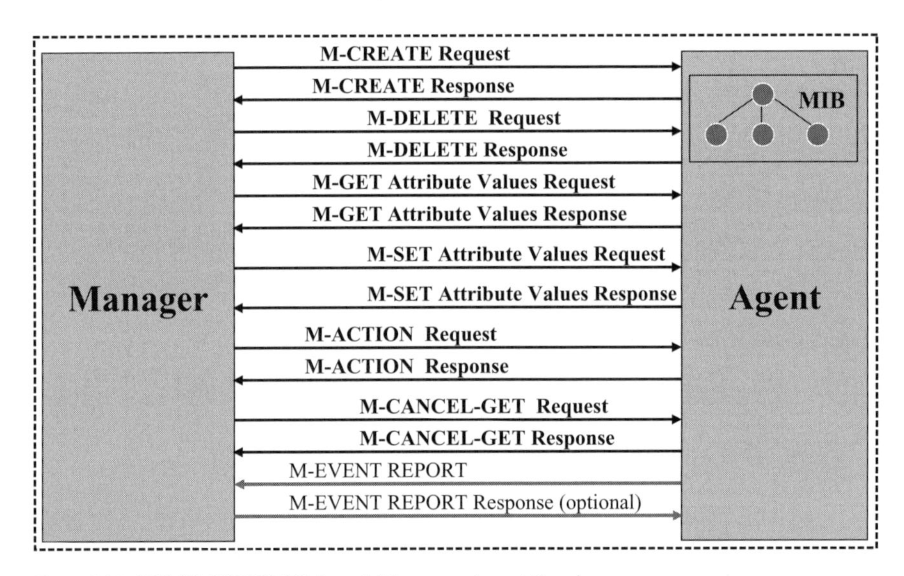

Figure 2.18 OSI CMISE/CMIP-based Manager-Agent Services

M-CREATE is a request to create a new instance of a managed object. M-DELETE is a request to delete managed objects. M-GET is used to retrieve managed object attributes. M-SET is used to change attribute values. M-ACTION is a request to perform an action on managed objects. M-CANCEL-GET is used to cancel a previously invoked M-GET service. The agent to report an event to the manager or to another open system uses M-EVENT REPORT. All requests are confirmed. Abstract Syntax Notation 1 (ASN1) and Basic Encoding Rules (BER) are used for PDU representation and encoding.

An evaluation of the strengths and weaknesses of OSI CMIP/CMISE-based management is given in Table 2.3. OSI management was internationally standardized in 1995, four years after SNMP-based management. This explains the near-dominance of SNMP-based management.

Table 2.3 OSI-based Management Evaluation

	Evaluation/Description
Strengths	– Rich set of protocol management service primitives. – Reduced management traffic due to the use of event reports. – Assumes a reliable connection-oriented transport mechanism. – The information model is based on object-oriented paradigm. – Rich set of standardized systems management functions. – Standardized security features as part of association establishment. – Support for multiple functional units (kernel, scoping, filtering, multiple object selection, linked reply, and cancel-get). – ISPs available for protocol and systems management functions. – Standardized conformance testing procedures.
Weaknesses	– Slow standardization process and lack of prior implementations. – Very complex protocol, system functions, and information model. – Overhead of the full seven layer OSI reference model architecture. – Sizable resources are required to implement the CMIP standards. – Limited market support for CMIP/CMISE-based products. – Lack of publicly available agent implementations and training.

The burden of complexity and slow standardization, combined with the opposition to OSI standardization by major companies having their own proprietary solutions, has hindered the implementation of OSI Management specifications. The few installations are limited mainly to islands in the telecommunications industry.

The best example of an OSI-based management implementation is the Telecommunications Management Network (TMN). TMN is a dedicated standardized network architecture, specifically designed for the management of telecommunication networks and services. It includes operations, administration, maintenance, and provisioning of heterogeneous telecom networks and services. Large telecom network infrastructures, wireless networks, cable networks, and even Internet driven data services/networks can be managed by using interfaces based on OSI Management standard protocols and services and by deploying OSI management agents in TMN NEs, as indicated in Figure 2.19 [6].

The TMN consists of Operations Systems (OSs), truly implementations of OSI-based managers, Network Elements (NEs) having deployed OSI-based Agents (A), Workstations (WSs) for operator and user interactions, and Data Communications Networks (DCNs). The management information is exchanged through Q3 and Qx interfaces based on OSI CMIP/CMISE standard specifications. Interoperability between telecom jurisdictions is specified as the X interface, also an OSI CMIP-based management interface.

Three basic components are defined within the TMN generic network model: physical architecture, functional model, and information architecture (ITU-T Recommendation M.3010). The physical architecture describes the layout of the physical components and their connections to transport, process, and store the information related to the management of telecommunication networks. The TMN architecture presented above is a schematic simplified view of physical aspects. Behind each link, we may have full networks with various networking technologies and standard specifications. Mediation

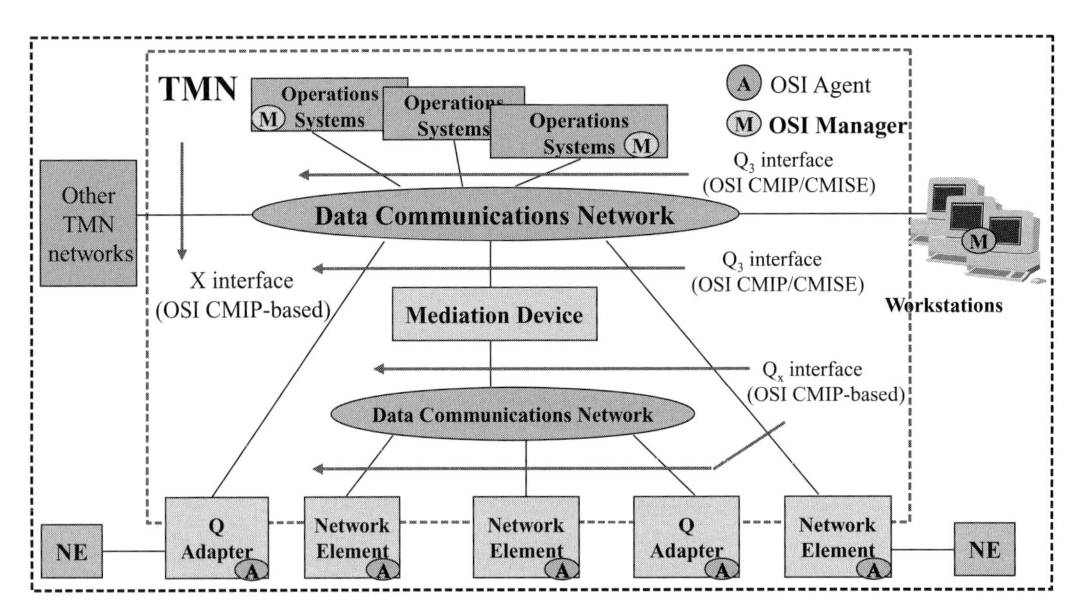

Figure 2.19 TMN OSI-based Network Management Architecture

devices or networks might be necessary to provide exchange of information across heterogeneous and legacy environments.

The basic description of the TMN functional model that will be used to analyze functionality built into wireless network applications and services is presented in Figure 2.20. The information architecture describes the management information model used to represent the managed resources and their characteristics and the collection of managed objects that describe the managed systems in order to be monitored, modified, and controlled by the managers.

	Fault Management	Configuration Management	Performance Management	Accounting Management	Security Management
BML					
SML					
NML					
EML					
NEL					

Figure 2.20 Network Management Functions Decomposition

The rows display the OSI-adopted functional areas mentioned earlier: fault, configuration, performance, accounting, and security management. The columns display functions specified in the TMN model. The Business Management Layer (BML) includes strategic planning, management policies, and agreements between service operators. Functions in the Service Management Layer (SML) include managing services, quality of service, and service level agreements. The Network Management Layer (NML) functions include provisioning and managing networks to support OS-NE interactions. Element Management Layer (EML) functions include managing collections of network elements and providing gateways to the network. Network Element Layer (NEL) includes management of individual network elements.

2.8 Network and Systems Management Requirements

Following the principles developed so far in this chapter, a set of high-level network and systems management requirements, applicable to any wired and wireless telecommunications network, can be summarized as follows:

- Provide remote access and configuration of physically distributed managed resources;
- Ability to monitor and control end-to-end network and systems components;
- Ability to report management-related information, and to backup and restore lost information;
- Secure management operations, user access, and management data transfer;
- Real-time management and automation of routine operations;
- Flexibility for system expansion and the ability to accommodate various technologies.

Support for open management systems adds additional requirements as follows:

- Open, standard communication interfaces;
- Modular management architecture with separation of management modules from signaling, transmission, and access modules;
- Interoperability with other TMN environments;
- Support for common graphical user interfaces;
- Support for various standardized layered protocol environments;
- Portability of management applications;
- Reduced cost of management network and operations.

Specific functional requirements can be associated with those management platforms used for managing large telecommunications networks.

- Separation of management applications from the underlying platform services;
- Modularity of underlying management platform framework services;
- Support for the applications development environment and development tools;
- Support for object-oriented information modeling and technologies;
- Support for heterogeneous communications stacks and protocols;
- Support for building enterprise-wide management systems and service management.

2.9 Network and Systems Management Products

In the span of two decades, there has been a multitude of designs and implementations for management systems. If we take TMN as a reference point for the sophistication required for management systems, we have three groups of products that have evolved:

- Management Platforms and Management Applications Development Tools;
- CORBA-based Applications and TMN Development Tools;
- OSI-based Applications, Development Tools, and Conformance Testing.

In the group of management platforms and management applications development tools:

- **Bull Inc.**: Open/Master platform for telecom and enterprise management;
- **HP**: TeMIP (prior DEC/Compaq) network and service management platform;
- **HP**: OpenView DM management platform and telecom applications;
- **IBM/Tivoli**: Workbench for TMN platform and TMN development tools;
- **Sobha Renaissance** (prior Agilent Technologies and Objective Systems Integrators Inc.): NetExpert management platform.

In the group of CORBA-based applications and TMN development tools:

- **IONA**: Orbix, a CORBA enterprise platform and associated tools;
- **Lumos Technologies** (now Micromuse Inc.): TL1 and CORBA gateway and development tools;
- **OpenCon Systems**: TMN inter-layer/intra-layer gateway, TL1/CMISE/CMIP mapping;
- **Platinum Technology** (now Computer Associates): Paradigm-Plus OO tools, UML mapping to CORBA/GDMO;
- **UH Communications**: TMN CMIP Q3ADE development tools, CORBA/GDMO translation, CORBA to CMIP gateway.

In the group of OSI-based applications, development tools, and conformance testing:

- **National Computing Center** (NCC) UK: TMN test tools and emulators;
- **Spirent Communications**: Testing across all seven layers of the OSI model;
- **Vertel Corporation**: Standard middleware and Enterprise Applications Integration, TMN TeleCore Designer for Q3-based interfaces.

3 Service Management

3.1 Service Management Conceptual Model

In general terms, the notion of **service** is an abstraction. However, in an actual context such as **communication service**, it becomes more meaningful. If we add service qualifiers such as availability, speed, security, integrity, and responsiveness, we have something palpable to the user. The ultimate goal of any service is to satisfy the customer. That means services that assure the good functioning of a network, system, processes, and overall business in a cost-effective way. Service Management is an area of management that was designed specifically to address these issues. **Service Level Management** (SLM) is that part of management that encompasses networks and systems, binding together service providers with users of services. It does this by identifying, defining, tracking, and even proactively monitoring the services.

Communication services are provided on a contractual basis between **customers** (organizations, businesses, or individual subscribers) and one or multiple **service providers**. These services are to be delivered with the assurances that the exchange of information on data communication networks will be done under well-defined conditions and with guaranteed performance. A simplified communications service conceptual model is presented in Figure 3.1. Service providers with adequate applications and network infrastructure provide communication service to customer systems. Multiple service providers are usually involved in providing communications services because of the particular media used (wireless, fiber optic, cable, twisted pair), the type of network (access, connectivity, backbone), the type of information (voice, data, video, multimedia), and the communications technologies adopted. In this context, the need for service management systems that are incorporated into the service provider's infrastructure is self-evident.

There are different perceptions of a communications service from the service provider's and customer's points of view. Three main components are used to characterize communication services: **Class of Service** (COS), **Quality of Service** (QOS), and **Service Level Specifications/Agreements** (SLS/SLA). These components are illustrated in Figure 3.2.

The first priority of the service provider is to identify the service needs of the users and to design and provision those services. Service parameters should be identified and measured to assess the quality of service provided. These parameters should be readily understood and perceived by both the service providers and the customers. From the

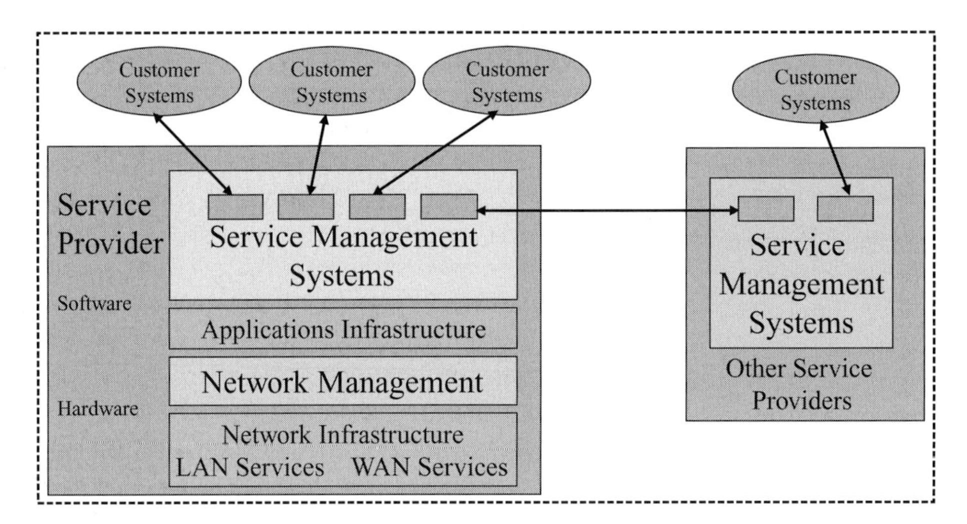

Figure 3.1 Communication Service Conceptual Model

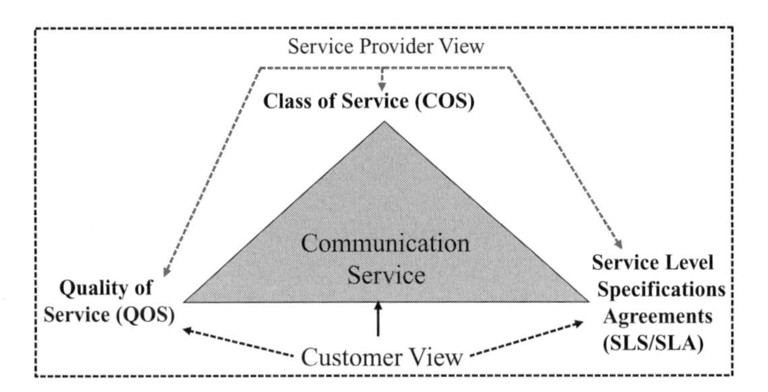

Figure 3.2 Communication Service Components

customer's view, quality of service is most obvious. To make explicit and to enforce the expected quality of service, a contract should be agreed upon between service provider and customers. This is called a Service Level Agreement and is based on well-defined service level specifications.

3.2 Classes of Services

Class of Service (COS) is a business-based grouping of common applications and users with similar service requirements into one of several broader services or priority classes. Classes of Service are used by service providers to **differentiate services** along performance, cost, and other offered criteria. Class of service assignment has different meaning according to the type of network environment: public switched telephone network, customer premises Private Branch Exchange (PBX)-based switched network, or a packet switched network, like Internet.

In public switched networks, different service classes can be established for individual users or for businesses. For instance, classes can be different for individual or party line users, between local and long distance callers, or between standard phone services and value added services such as call forwarding, call waiting, distinct rings, automatic number identification, etc. In a PBX-based environment, classes of services can be based on access to long distance, international calls, 800 or 900 calls, and other special privileges.

In a packet switched public network, such as the Internet, the initial designs were based on a "best effort" delivery premise. Packets were given equal treatment even when the network became congested. With the advent of time-sensitive and bandwidth-sensitive applications, such as VoIP, videoconferencing, IPTV, and others, the need to handle some packets differently from others, for example when traffic congestion occurs, became clear. Unequal treatment of packets is one aspect of differentiated services. Standardization of special fields within data packets is the best known way to allow prioritization. One example is the 802.1p class of service that allows LAN devices such as switches to recognize high-priority packets and deliver them accordingly. Another example is the Multi-Protocol Label Switching (MPLS) technology with corresponding protocols and applications. The following COS functions and characteristics are associated with packet networks:

- Classification of traffic according to coding of fields in the packets and the frame/packet headers;
- Providing preferential treatment (minimum delay of certain traffic in congestion points);
  - First priority: Interactive voice;
  - Second priority: Financial transactions; and
  - Third priority: E-mail, Web surfing, common applications.
- Bypassing queues on transmit interfaces;
- Providing extra buffer space in network elements;
- Providing extra bandwidth in network connections.

3.3 Quality of Service Parameters

A service is a function provided by a service provider to the subscriber of that service. Quality of Service (QOS) is a measure of service quality to the customer. The measurement of quality is based on **service parameters** or **QOS metrics**. Service parameters are values with significance for both the service provider and the subscriber. The difficulty of using service qualifiers, however, comes from the fact that the same parameters can have different significance to the user than to the provider.

The service parameters can be categorized as quantitative, qualitative and relative. The **quantitative parameters** are measurable in absolute terms. However, parameters that can be easily measured might not be relevant to the user. The reverse is also true. Parameters significant to the subscriber might not be easily measured. **Qualitative parameters** are subjective perceptions regarding the service intelligibility, manageability, security,

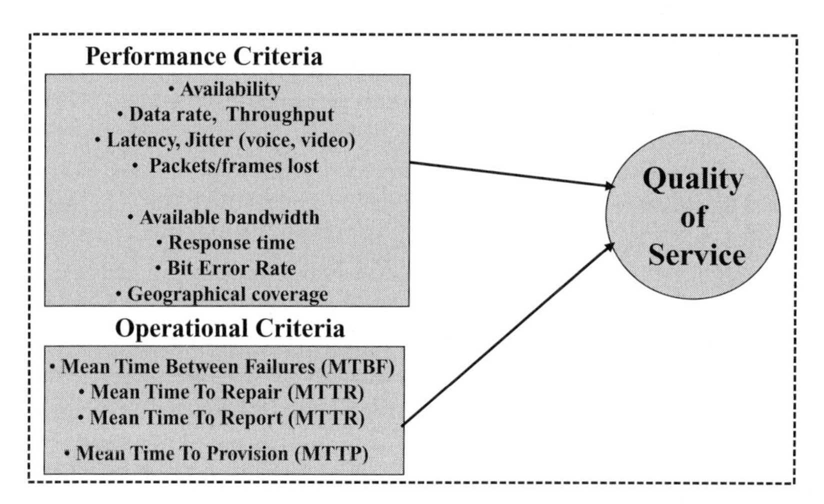

Figure 3.3 Communications Quality of Service Parameters

or signal strength in broader terms or categories such as strong, moderate, or weak. **Relative parameters** are used by comparisons to measurable service parameters in terms of higher, equal, or lower than a known parameter.

Quantitative service parameters can be classified in two major groups: performance and operational. This is indicated in Figure 3.3.

The most used QOS performance criteria are: availability, data rate, throughput, latency, jitter, and packet loss. Another group of performance parameters are: available bandwidth, response time, bit error rate, and geographical coverage. All the operational criteria are target parameters such as guaranteed minimum Mean Time Between Failures (MTBF), Mean Time To Repair (MTTR), Mean Time To Report (MTTR), and Mean Time To Provision (MTTP). These parameters are all valuable indications of the ability of the service provider to restore the service in case of outages. Outages can be caused by physical failures of network components, software errors/bugs, or natural disasters. QOS parameters are collected by managing functional areas covering physical layer, bandwidth, traffic, policy, and applications. A detailed description of the main QOS parameters follows:

Availability is the percentage of time the network, applications, or services are available when the user needs them. It is a ratio between the total time the system is used in normal conditions and a given time interval (month, year). The traditional benchmark, inspired by PSTN known availability, is 99.999% (five-nines). This equates to outage time of 5.25 minutes per year. Accordingly, 99.99% availability assumes almost 53 minutes, 99.9% 8 and a half hours, 99.5% almost 44 hours, and 99% 3 days and 16 hours downtime a year.

Data rate is the measure of how quickly data can be transmitted on a copper line, fiber optic cable, or in space. It is one of the factors that determine how much information can be sent per unit of time. It is measured in Kbps, Mbps, Gbps, or Tbps. A loose "interchangeable" term for data rate is available bandwidth. **Throughput** is a measurement of the actual amount of user data/information transmitted per unit of time. Throughput depends on amount of accompanying redundant information carried along with the transmitted data, traffic queuing/aggregation mechanisms, congestion conditions, and priority handling policies applied to particular data flows. It is related to data rate (always less) and it can be a fraction of the maximum available bandwidth.

Delay or Latency is the average transit time of packets and cells from the ingress to egress points of the network. There are the end-to-end delays and the individual delays along portions of the network. End-to-end delay depends on the propagation rate of data in a particular communication medium (satellite or terrestrial), the distance, the number and type of network elements (design, processing, switching, and buffering capabilities), routing schemas (dynamic, static, queuing, and forwarding mechanisms), bit error rate in transmission (hence the number of lost or retransmitted packets), and policies regarding priority treatment of data flows.

Jitter is one form of delay variation caused by the difference in delay exhibited by different packets that are part of the same data flow. Jitter is caused primarily by differences in queuing delays for consecutive packets, and by alternate paths taken by packets because of routing decisions. Severe jitter is sometimes considered a more important voice transmission impairment than high latency.

Packet loss is typically measured as a percentage of the **ingress** and **egress traffic**. Packets can be lost because they are dropped at congestion points, traffic violations (synchronization, signaling, unrecoverable errors), excessive load, or natural loss included in compression/decompression mechanisms. Voice and video communications are more tolerant to the loss of packets than in the usual data communications (often not noticed by users or by brief flickering on the screen) if they do not exceed 5%. The Internet TCP and ISO's TP4 Transport Layer protocols can handle dropped packets because they allow retransmission of information (absolutely necessary for pure data but unnecessary for voice or video).

From the description of QOS parameters it is clear that **QOS Management is** a multidimensional development. A breakdown of the functions that must be performed in QOS management, according to the corresponding protocol layers in which they are usually implemented, is given in Figure 3.4. Service parameters such as data rate and bandwidth availability are extracted from layers 1 and 2 of the communications stack but most of the other parameters are extracted from the upper layers, including the application layer. At the base of the pyramid, Physical and Data Link Layer monitoring and control are performed regarding link continuity, signal conversation, and bandwidth management.

At the top of the pyramid, Policy Management decisions are made regarding which operational parameters will be adopted and how they will be used in the QOS measurement procedures.

3.4 Service Level Specifications/Agreements

Service Level Specifications (SLS) are quantifiable service parameters or QOS metrics that support the assessment of service performance. SLSs represent the core components

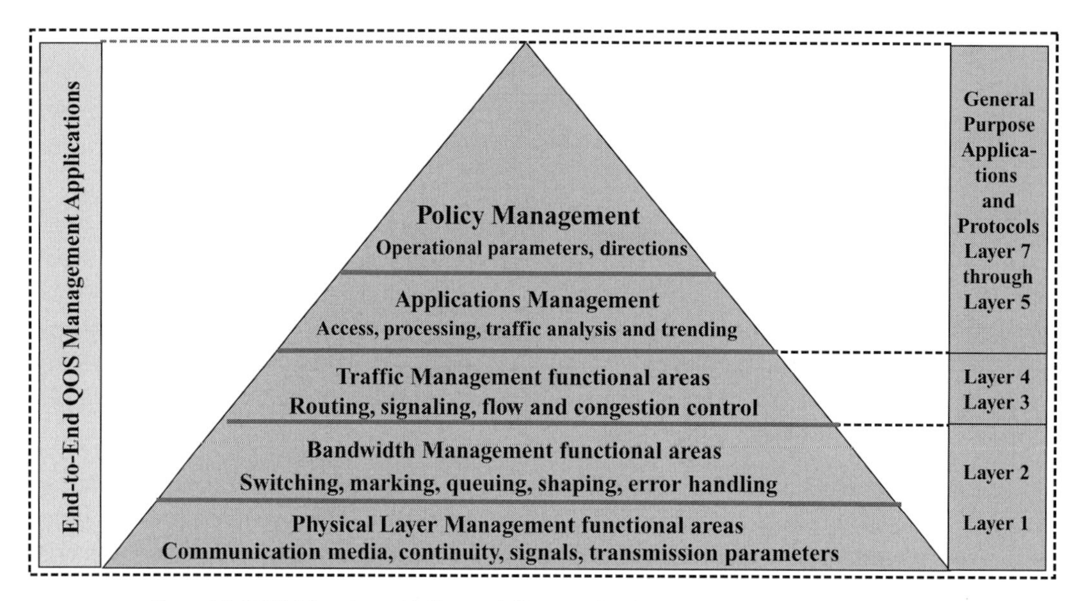

Figure 3.4 QOS Mapping with Layered Communications

of **Service Level Agreements** (SLA). SLAs are contracts between service providers and customers that specify the performance parameter bounds within which the services are provided. SLA management is a loop process targeting service aspects such as monitoring input and output QOS parameters, tuning network components, and, if necessary service redesign. These concepts are shown in Figure 3.5.

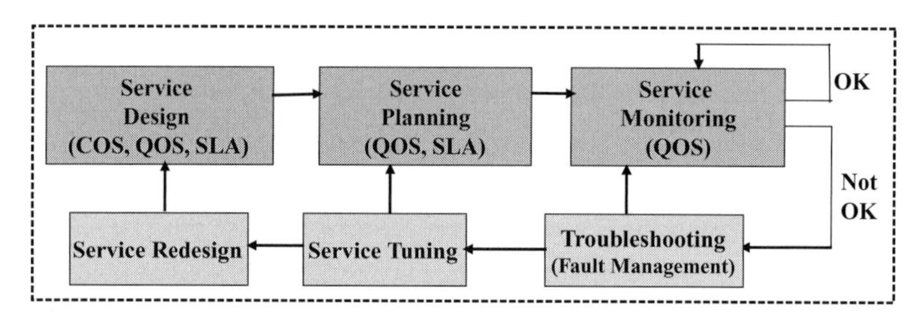

Figure 3.5 SLA Management Model

In normal service monitoring, QOS parameters are collected and compared with SLSs. This cycle continues if the results are satisfactory (OK). Any discrepancies caused by failures or performance degradation will launch a troubleshooting operation to identify and fix the problem. If the degradation persists, service tuning will be required. This will result in changing the overall service plans. Persistence in degradation, or repeated failures, will require major redesign of the service. This may affect not only the QOS and SLA but also the class of service. Services have a life cycle, as indicated in Figure 3.6.

Any service can be modeled along its natural **life cycle**. The cycle starts with service design and ends with service termination once the service becomes obsolete or no longer

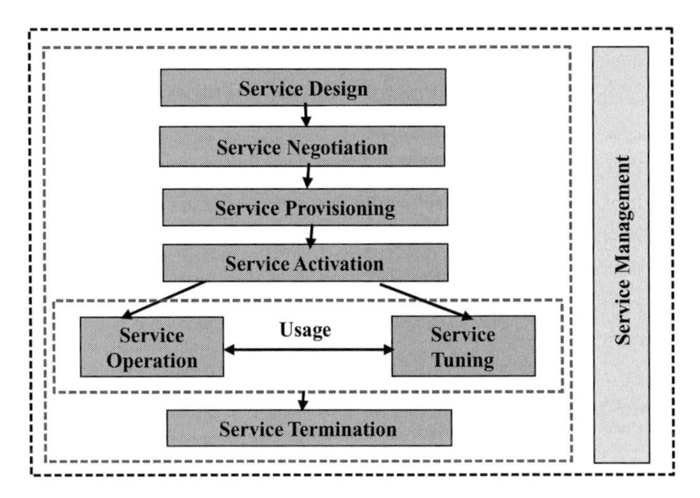

Figure 3.6 Service Life Cycles

justified economically. Each of these phases can be the subject of dedicated service management.

3.5 Guidelines for Establishing SLAs

Currently, many SLAs are focused less on SLSs, as captured through QOS parameters, and more on assigning contractual responsibilities. These SLAs may also include specification of overall commitments and goals, reporting of operations, escalating procedures, and help desk availability to fix problems. A typical SLA, as recommended in technical circles, should include the following:

1. Definitions of the service provided and the parties involved, and the effective date of the agreement;
2. Specifications of the hours and days when the service is offered, including when testing, upgrades, and maintenance will be scheduled;
3. Specifications of the numbers of users, locations, hardware for which the service will be offered (customer data);
4. Explanation of the problem-reporting procedures, escalation to the next level, and expected time to report a problem;
5. Explanation of change-request procedures. This portion may include expected times for scheduled, routine updates, and periodic maintenance that may affect normal functioning;
6. Service Level Specifications (SLS) of the target level of service quality, including explanations of how the QOS metrics are collected, calculated, and reported. These metrics include:
   - Average availability per service period (month, year) and lowest availability;
   - Average response time and percentage of time when the average response time is met;

- Average throughput;
- Guaranteed bandwidth or committed information rate;
- End-to-end latency and maximum jitter for real-time interactive voice/video communication;
- Maximum packet loss.

7. Specification of charges associated with provision of services, levels of usage, type of rating (flat, usage-based, tied to different levels of service quality), and compensation/credit given if SLAs are not met;
8. Specification of the user/customer responsibilities (training, maintenance, correct use of applications);
9. Description of procedures for resolving service-related disagreements;
10. Process for amending the SLS/SLA.

3.6 QOS Measurement Mechanisms

Implementation of the QOS/SLA specifications requires, in addition to clearly defined QOS metrics, a set of cost effective **QOS measurement mechanisms**:

- **Polling mechanisms** of counter-specific managed objects and associated MIBs. This is achieved using SNMP polling messages. Typical MIB information collected this way is usually related to traffic, error conditions, and status of interfaces, but not delay or jitter values;
- **RMON1 and RMON2** proactive remote monitoring using special Remote Monitoring (RMON) messages sent via SNMP. RMON provides a local collection of statistics, periodic off-load operations, management bandwidth allocation, and communication to multiple managers;
- The **Response Time Reporter** (RTR), used by Cisco routers, an IOS operating system extension, enables monitoring of a network's performance and its resources. It does this by tracing route commands to determine active ports and measuring round trip delay between hops;
- **Dedicated WAN Probes** in **CSU/DSU** network components. Embedded service analysis elements collect Frame Relay, ATM, and IP level traffic statistics for individual applications such as telnet, ftp, ICMP, and DNS;
- The **Advanced Network** and **Service Surveyor** measures transmission delays within 50 μs accuracy. Based on the Internet Performance Measurement Project specifications uses 50+ customized PCs and dedicated surveyors, each synchronized using Global Positioning Satellite (GPS) systems.

3.7 Service Management and COS/QOS/SLA

Successful execution of a comprehensive service and service management plan requires complex interactions between the service provider's domain and the customer's domain.

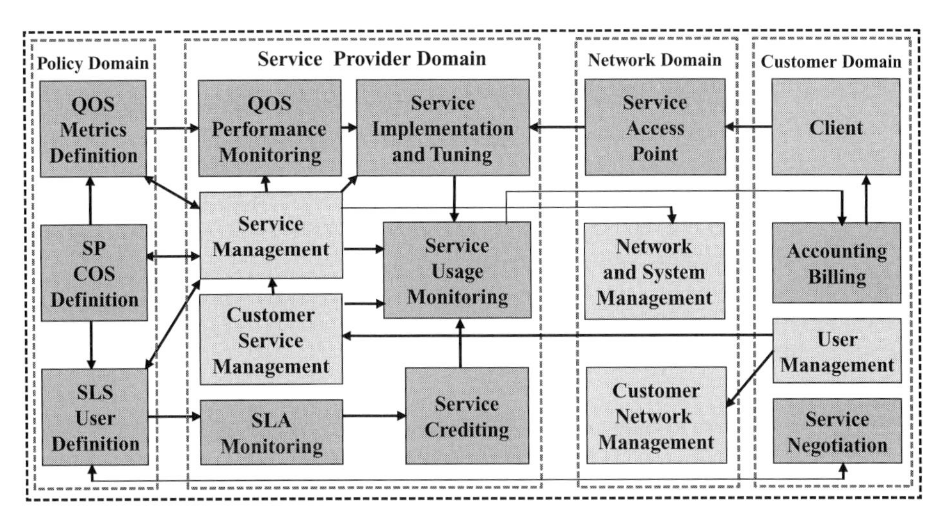

Figure 3.7 Service Management Processes

This interaction involves all the service management entities representing Quality of Service (QOS) metrics, Class of Service (COS), Service Level Specifications/Agreements (SLS/SLA), and Customer Network/Service Management (CNM/CSM). An overall depiction of this is presented in Figure 3.7.

This complex relationship starts in the policy domain where the Class of Service, QOS metrics are adopted, and the Service Level Specifications are defined. A service negotiation process may take place as initiated by the customer. Actual service management takes place in the service provider domain with processes supporting service implementation and monitoring, and tuning, if necessary. A separate process allows automatic service crediting in accordance with agreed compensation in cases of failures or performance degradation. Other processes are involved in monitoring service usage and reporting results to the accounting/billing management applications. The subscriber has the ability to interact with network management and service management systems through dedicated customer management processes. These interactions take place through the network domain that controls access to the service.

3.8 High-level Service Management Requirements

Service Management represents collection points of services to customers built on top of network services in accordance with business-based management policies. It is highly desirable to be network technology and topology independent. Regardless of the type of service management some **high-level functional requirements** can be considered:

- Identification of customers' access to network and services;
- Delivering services as contracted;
- Interfaces with other service providers;
- Support for customer network management;

- Reporting service usage for billing;
- Resolving network failures and performance degradation;
- Maintaining and reporting Quality of Service (QOS) data;
- Managing Service Level Agreements (SLAs).

A set of more granular requirements can be developed along all systems management functions following the ITU-T TMN architectural model and functional decomposition as presented in the next section.

3.9 Service Level Management

In the TMN model, service management has become an entity by itself, providing the linkage between the business management and network management layers, as indicated in Figure 3.8.

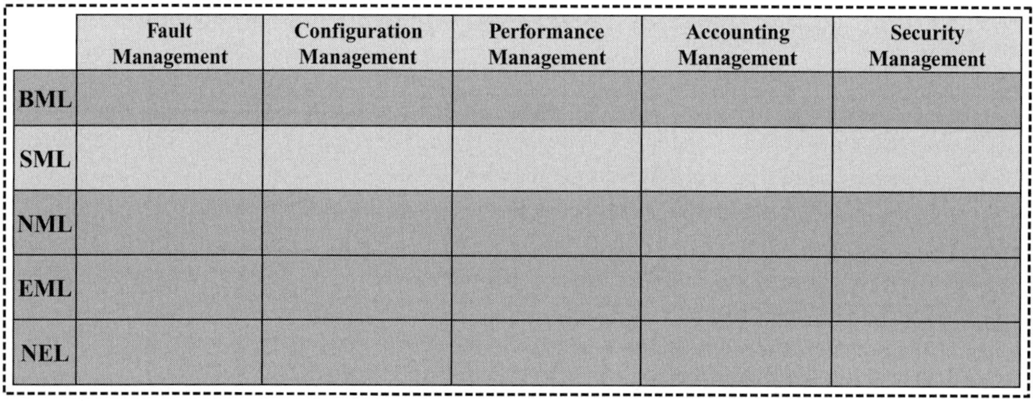

	Fault Management	Configuration Management	Performance Management	Accounting Management	Security Management
BML					
SML					
NML					
EML					
NEL					

Figure 3.8 Service Management and TMN Layered Architecture

Business management includes those processes that allow financial support of the enterprise and the elaboration of policies that govern all aspects of running the business and services provided to the customer. Service management is a complex set of processes that are dedicated to the management of services offered to subscribers. Thus, they provide access by the customer to the service, maintain the agreed upon quality of service, develop new services, and solve all the problems related to the services. Another service provider, or an operator providing only transmission services, delivers the services through a network infrastructure that might be owned and operated by a telecom carrier.

A simplified version of the TMN layered architecture, with focus on service management, is provided by the TeleManagement Forum (TM Forum) as the Telecommunications Operations Map (TOM) [7]. This is shown in Figure 3.9.

TOM is focused on those key business processes that provide **Service Fulfillment** (SF), **Service Assurance** (SA), and **Service Billing** (SB). TOM is a starting point for developing sub-processes, relationships between processes and detailed requirements, information models, and business agreements for both wired and wireless

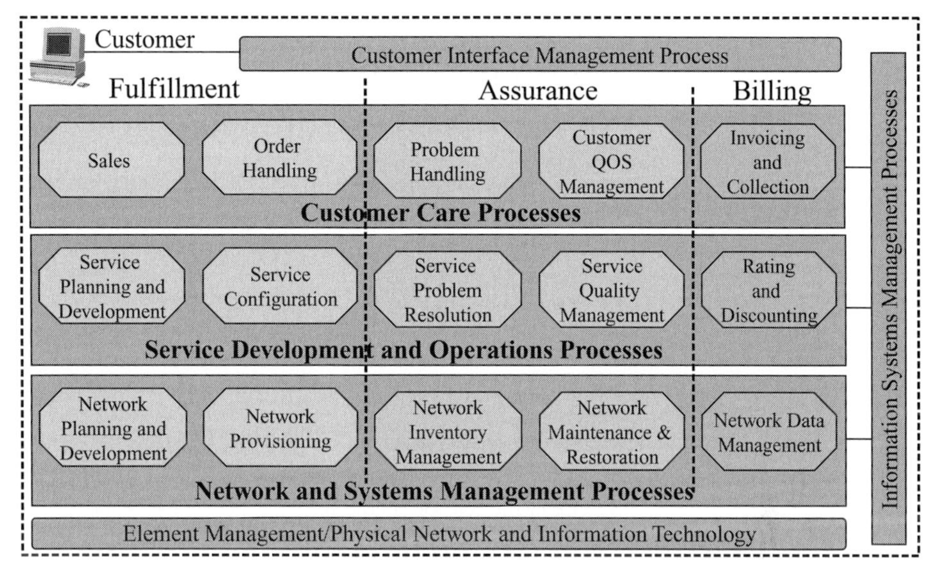

Figure 3.9 Telecommunications Operation Map

communications. All service management processes are divided in two major categories: service development and operation and customer care processes.

Service development and operation processes include all those activities that are related to service design, planning, development, configuration, activation, delivery, problem handling, QOS monitoring, and usage data collection. Customer care processes include all the interactions with customers from sales, service order handling, problem handling, help desk support, customer access to service management, and billing aspects such as invoicing and fee collection. A more detailed enumeration of service management activities is presented in Table 3.1.

Service level management is highly dependent on the collection of data related to the QOS parameters. One view of this process is shown in Figure 3.10.

QOS management means the collection of QOS metrics related data. This data must be interpreted and the results reported to the service providers and customers. QOS management can be organized to look at the performance of the media used (the data pipes) and the data streams itself. QOS of the pipe is mainly the concern of the network providers. They are responsible in providing the correct service class, bandwidth, Maximum Transmission Unit (MTU), and packet size. Subscribers' view is mainly on the QOS of the data stream itself, and they focus on throughput, delay, jitter, and packet loss as the relevant parameters.

3.10 Service Management Products

Service Management products can be grouped into three major categories: standalone application suites, management platforms with specialized service management

Table 3.1 Service Management Layer Functions and Activities

Functions	Service Management Activities
Service Design (Fulfillment)	Service design, planning, and negotiation (to insure basic/value-added services). Service feature definition (to meet specific customer needs, class of services). Network planning and engineering (to insure overall robustness and performance). Customer identification (records, profiles, market segments, prospective customers). Marketing (advertising, campaign planning, telemarketing, customer needs, surveys). Service provisioning and activation (policies, installation, access, directories, status). Service assurance and evaluation (benchmarking against relevant IT and BU metrics).
Service Operation (Delivery)	Service status and control (access state, network state, priority service restoration). Service quality (QOS assessment, performance summaries, outage reports, compensation). Service billing (usage collection/aggregation/distribution/storage, pricing, ratings, tariffs). Secure service operations (secure exchange, customer data protection, customer audit). Capacity planning and engineering (bandwidth, configuration/change management). Network management (proactive monitoring, trouble reporting, asset management).
Customer Management	Customer relationship (identification, screening for affordable services, service feedback). Service performance (progress reports, customer performance summaries, help desk, CSR).
Link to other Service Providers	Other service providers (identification, options, performance, cost, SLAs, forecasting). Vendor or supplier relationship (procurement, contracting, dispatch, installation).

packages, and carrier-based sets of products/policies supporting service providers and customers.

Standalone Application Suites for Service Management

- **Micromuse Inc**. (now IBM/Tivoli): NetCool/OMNIbus, PROVISO Service Level Management (former Quallaby/Micromuse), (www.ibm.com);
- **Concord Communications** (now Computer Associates): eHealth, (www.ca.com);
- **InfoVista**: VistaWatch, VistaTroubleshooter, VistaAPI, (www.infovista.com);
- **Cisco Systems**: NetSys Service Level Management Suite, Cisco QOS Policy Manager, (www.cisco.com);

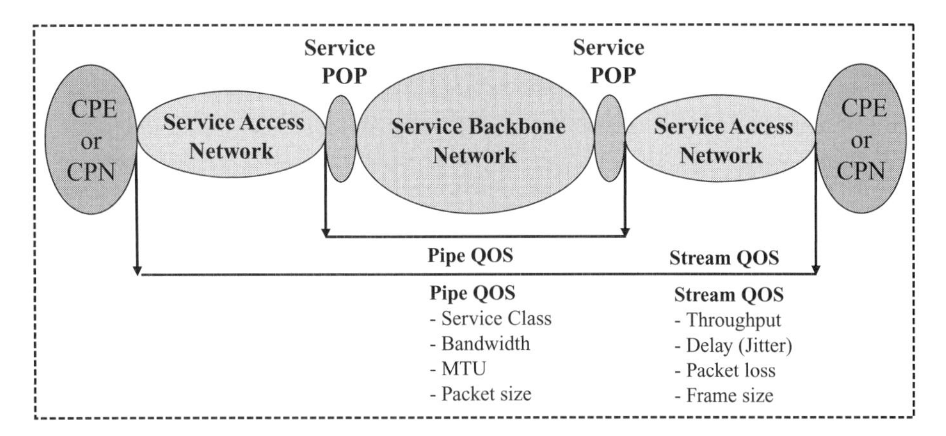

Figure 3.10 End-to-End QOS-based Service Level Management

- **NetScout Systems**: nGenius Performance Manager, nGenius Application Fabric Monitor, Sniffer (Network General), (www.netscout.com);
- **Network Associates** (now McAfee): Intranet Management Suite with Service Level Manager components, (www.mcafee.com);
- **Nortel**: Presidio Portfolio of Service Management (service providers), Optivity Suite of Products for Unified Management (enterprises), (www.nortel.com);
- **Opticom**: iView Service Management, (www.opticom.com);
- **Visual Networks** (now Fluke Corporation): Visual UpTime Select (WAN SML), Visual Trinity (E-business service management), (www.visualnetworks.com).

Management Platform Packages for Service Management

- **Aprisma Management Technologies** (now Computer Associates, formerly Cabletron, Enterasys, Concord): SPECTRUM Solutions for Service Providers, Response Time Management, Application Management, (www.ca.com);
- **BMC Software**; BMC Performance Manager (PATROL), (www.bmc.com);
- **Hewlett Packard**: HP OpenView, IT Service Manager, Internet Service Management, Application Management, (www.hp.com);
- **Tivoli Systems**: Performance, Availability, and Service Delivery Management, (www.tivoli.com).

Carrier Packages for Data Service Management

- **AT&T**: Managed Networks Solutions, (www.att.com);
- **Sprint**: Insite Network Management, (www.sprint.com/data).

Part II

Cellular Mobile Radio Networking and Management

4 Cellular Mobile Radio Networking

4.1 Cellular Mobile Radio Communications Concepts

In the first chapter we provided an extensive introduction to wireless communications. Major aspects such as general models for wireless communications, architectural components, and networks classification were introduced. We have also discussed, mainly at the level of acronyms, many of the elements that constitute the complex world of wireless communications. This was done in the context of the even larger legacy wired world. Key to wireless communications is the wireless link established between transmitter and receiver. The type of this link will determine the kinds of wireless communications. In this section we will limit analysis to terrestrial mobile cellular radio networks that, in simple terms, are the radio links established between Mobile Stations (MS) such as handsets and the Basic Transmission Stations (BTS).

Initially, the approach to mobile radio was the same as that in radio or television broadcasting, where a BTS was placed at the highest point of the desired area to be covered. As the number of mobile users increased, congestion eventually occurred because of the limited available spectrum. To assure that frequencies can be reused across geographical regions, mobile communication uses the concept of individual micro cellular radio systems. The cells can be created by earth-based radio tower transmitters/receivers or by satellite footprints. Clusters of 7 terrestrial cells provide an area coverage and separation of commonly used groups of frequencies. The number of total channels supported, hence the network capacity, will be determined by the number of clusters that are implemented. This is shown in Figure 4.1, resembling, at a certain scale, the United States map. Cell size will determine geographical coverage and the particular technique used for the radio link will determine the maximum density of users roaming within each cell that can be supported.

Cell boundaries are determined by the power and height of transmitting antennas. The reuse of frequencies is determined by the interference caused by signals that are close to each other. Signal power is inversely proportional to the square of the distance between the transmitting and reception points. Microcells or picocells using BTS equipment that fits into closets, can be used inside buildings, tunneled passages, pedestrian areas, or in areas with weak signal coverage.

Bell Laboratories developed the concept of cellular radio-based communication with the first design called Mobile Telephone System (MTS). This design, later evolved to the Improved Mobile Telephone System (IMTS). Both systems used analog

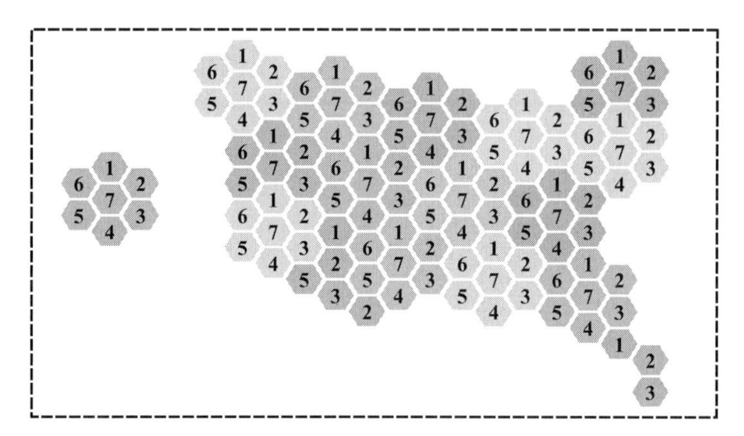

Figure 4.1 Cell-based Mobile Radio Communications Concept

technology. The first commercial mobile cellular service was provided in Chicago, Illinois, in 1983.

Transmission between MS and BTS is not straight line-of-sight communications. Obstacles like hills or buildings may block the radio waves. Reflection may occur on certain surfaces creating multi-path interferences. Sometimes, these reflections can have favorable effects when they penetrate hidden corners and improve coverage in high-density metropolitan areas. Excessive moisture in the air can weaken radio wave strength or cause undesired refraction. Interference may occur between adjacent channels or multiple operators. Most of these physical aspects of radio transmission plus modulation techniques, polarization, antenna design, and methods to increase spectrum efficiency have already been described in the literature. It is not the intent of this book to describe that material again here.

4.2 Mobile Radio Link Access Methods

There are three basic access techniques or methods to allocate a fixed spectrum to different mobile users: **Frequency Division Multiple Access** (FDMA), **Time Division Multiple Access** (TDMA), and **Code Division Multiple Access** (CDMA). These methods are depicted in Figure 4.2. There are fundamental differences between these access methods reflecting the technical evolution from FDMA to TDMA to CDMA. Each step has brought a 3–4 times increase in the number of users sharing the same spectrum.

In FDMA radio links, each user is assigned a discreet frequency bandwidth, one for each direction of transmission, i.e., MS-BTS and BTS-MS. In cellular telephony each user occupies 25 kHz of frequency spectrum. Within the limits of the spectrum many simultaneous communications can take place. Guard bands are required between adjacent frequencies to minimize the cross talk between channels. This technique was used for the first generation of analog mobile systems.

In TDMA radio links multiple users are assigned individual multiplexed time slots on common frequencies for both directions. Multiple time slots allocated to a user form a

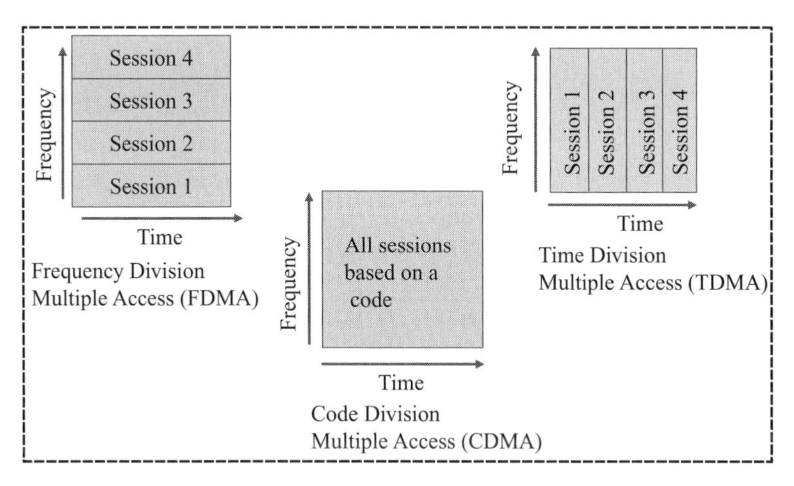

Figure 4.2 Cellular Mobile Communications Access Methods

frame. A preamble uniquely identifies various users. Multiple time slots, i.e., bandwidth, can be assigned on-demand. In digital cellular mobile communications using TDMA, 10 timeslots are allocated for each channel. Several packets of information are sent during each timeslot. All the packets are reassembled on the receiving site to reconstitute the original voice transmission. TDMA is typically used for GSM cellular and PCS 1900 systems, the most highly implemented mobile systems in the world, and for the digital enhancement of Advanced Mobile Phone Service (D-AMPS). One example of a TDMA variation is the extended version of TDMA. It uses Digital Speech Interpolation to take advantage of the quiet times within a conversation, to allow about 15 times as many calls.

In CDMA-based radio links, the original signal is combined with a code and spread over the available frequency bandwidth in a process called **spread spectrum**. Multiple users spread their signal/energy over a common wide frequency band by using individual codes. The carrier frequency, fc, is changed rapidly over a wide range of frequencies in a pseudo-random sequence that is known in advance by the receiver. The spread signal is picked up from the background signal by the user who was assigned the unique code.

Spread spectrum can be implemented using any of the three techniques: with Direct Sequence Spread Spectrum (DSSS) fc is modulated by a digital code; in the Frequency-Hopping Spread Spectrum (FHSS) technique, fc is shifted in a pattern generated by a code sequence; and with Time-Hopped Spread Spectrum (TH) technique, the transmitted signal is divided into frames and slots and within each frame only one slot is modulated with a message.

CDMA allows the reuse of frequencies in every cell, offering up to 20 times more handling capacity than the FDMA techniques. The power required for transmission is only one tenth that of FDMA and TDMA handsets. Transparent soft handoff, and overall quality of service make CDMA systems, originally developed for digital cellular mobile communications by Qualcomm, very attractive. Thus, this technique was adopted as standard in North America and in various forms.

The story of TDMA and CDMA does not end here. Numerous improvements have been made to these access methods, all intended to alleviate the handover issues, to make more efficient use of the spectrum, to provide larger data bandwidth, and to offer more sophisticated services. Today, in many countries, the number of cell phone users has surpassed the number of fixed telephone wired users. Mobile cell phone penetration in some areas has surpassed the number of people in those areas. It is estimated that every second four cell phones are sold somewhere in the world. Countries that have bare-bones wired infrastructure have jumped ahead in providing quick and cost-effective mass communications. And, this comes with all the broadband Internet connectivity and services provided by the third generation of mobile communications.

4.3 PCS/PCN Communications Architecture and Components

The best description of modern cellular mobile communications is provided by the architectural construct offered as part of the second generation of mobile systems, known as **Personal Communications Services (PCS)** [8] and, by extension, **Personal Communications Networks (PCN)**. PCS is an architectural concept, network, and set of standardized interfaces encompassing a broad range of wireless communication services. These services provide terminal (handset) and personal mobility on the basis of a unique personal number. PCS architecture and components are presented in Figure 4.3.

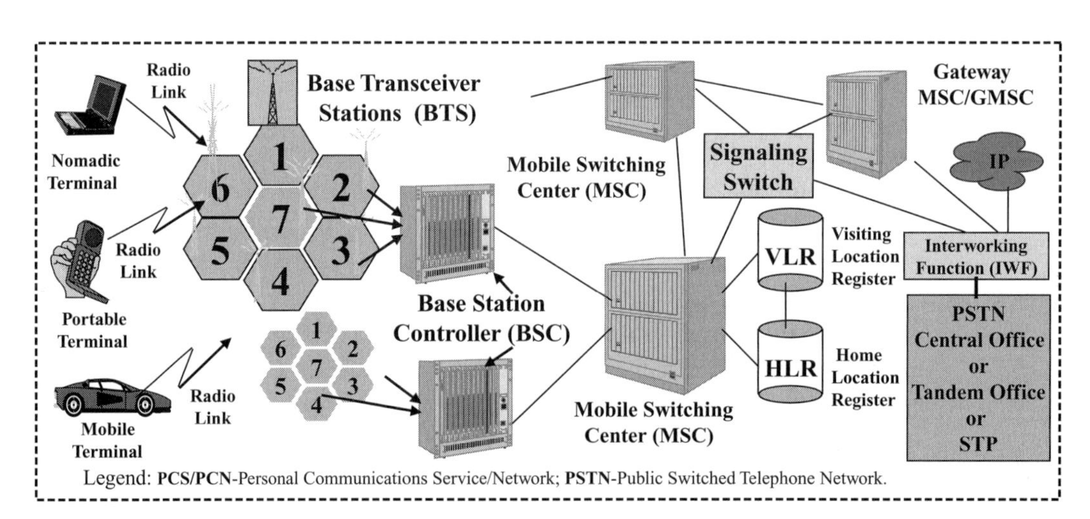

Figure 4.3 PCS/PCN Network Architecture and Components

There are seven major functional components in any PCS network:

- **Radio Terminals** or **Mobile Stations (MS)** (portable handsets or cell phones, nomadic laptops, or mobile terminals);
- **Radio Links** to the actual radio ports;

- **Base Transceiver Stations (BTS)** performing the transmission and reception of radio signals;
- **Base Station Controllers (BSC)** that control the send and receive processes from the BTS and pass instructions and acknowledgments from the MSCs;
- **Mobile Switching Centers (MSC)** that provide channel assignments, protocol conversion, and signal quality monitoring as well as sending, receiving, and processing of call information from external Public Switched Telephone Networks (PSTN);
- **Home Location Register (HLR)** and **Visiting Location Register (VLR)** PCS databases; and;
- **The external PSTN** represented by central switching offices, tandem switching offices, signaling systems, and PSTN Operations Systems Support (OSS) not represented in this picture.

In the simplest terms, the mobile communications process starts when the customer, using a handset/cellphone or a mobile terminal, activates the mobile unit to make a call. In the background, once the handset is switched "on" but idles, it monitors signals from neighborhood BTSs and selects the strongest set up channel that is very likely the closest cell site. Once the "send" button is "pressed", a call request is sent via radio link with the dialed phone number. This request is passed through the BSC to the MSC where the caller information is checked in the HLR/VLR databases against valid subscribers. The mobile terminal is identified based on two sets of data: the cell phone number and the **International Mobile Equipment Identity (IMEI)** number, unique to every GSM and mobile phone. IMEI is used to identify valid devices and can be used to stop a stolen phone from accessing the network regardless of the SIM card used. IMEI has no permanent relationship to the subscriber.

When operational, the subscriber is identified by the transmission of a global International Mobile Subscriber Identity (IMSI) number, which is stored in the SIM card. However, the call follow up on a device is done through a Temporary Mobile Subscriber Identity (TMSI) number, which is a randomly allocated number given to the mobile handset the moment it is switched "on". The number is local to a location area so it is updated each time the mobile terminal moves to a new geographical area. Mobiles detect local area codes. When a mobile finds that the location area code is different from its last update, it sends another to the network. This update includes a location update request together with its previous location, and its TMSI. This mechanism allows a continuous tracking of the mobile device that can be paged with broadcast messages. All updates regarding location are reflected in the VLR database. The network can also change the TMSI of the mobile at any time. It normally does this to avoid the subscriber being identified and tracked by eavesdroppers.

PCS, the architecture and specifications for the second generation of mobile systems, was born in 1994. Along with it came the FCC spectrum allocation of the 1900 MHz radio band. That band was assigned for new digital mobile phones in the United States and Canada to augment the existent cellular 850 MHz spectrums which had became too crowded. Sprint set up the first PCS network in the USA, a GSM 1900 network in the Washington-Baltimore area. Sprint adopted CDMA technology and sold it to Omnipoint.

This, in turn, was ultimately acquired by T-Mobile. Currently DAMPS, CDMAone (IS-95), and GSM 1900 are all working in the PCS spectrum. Details of the cellular and PCS spectrums will be given in the next section.

4.4 Cellular Mobile Radio Spectrum

In the USA, the Federal Communications Commission (FCC) is responsible for spectrum allocation. Coverage for the initial cellular systems was divided into metropolitan and rural service areas, each having two cellular operators working in two frequency blocks, the A-band and B-band. The A-band was allocated to non-wire-line systems, and the B-band to wire-line systems. The initial cellular spectrum allocation in the United States is given in Figure 4.4.

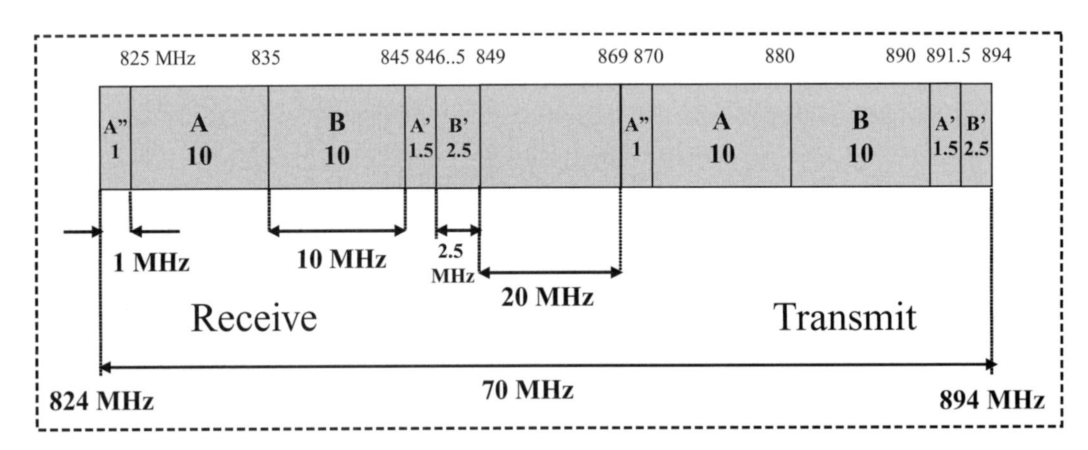

Figure 4.4 Cellular Spectrum Allocations in the US

The cellular spectrum has allocated a band of 70 MHz which is spread between 824 MHz and 894 MHz and which includes two licensable bands of 20 MHz each (transmit and receive). There are 306 Metropolitan Service Areas and 428 rural service areas assigned. There is also an unlicensed block of 20 MHz between 849 MHz and 869 MHz.

In the second generation of mobile systems, known as Personal Communications Service, (PCS), the move from analog to digital systems, and the addition of new services, has required a new spectrum designation as presented in Figure 4.5.

In the United States, the PCS communication spectrum covers 140 MHz spread between 1850 and 1990 MHz. There are three licensable bands of 10 MHz (D, E, and F) and three licensable bands of 30 MHz (A, B, and C). These transmit and receive bands were auctioned in 1995. There is an unlicensed block of 20 MHz dedicated to wireless LANs, PBXs, and WPAN technologies.

PCS licenses are defined as Major Trading Areas (MTA) and Basic Trading Areas (BTA). In each MTA there are several BTAs. In the USA, 93 MTAs and 487 BTAs were defined with two operators licensable for each band. Therefore, a total of 186 MTA licenses were awarded for the realization of PCS, each with a spectrum of 30 MHz to

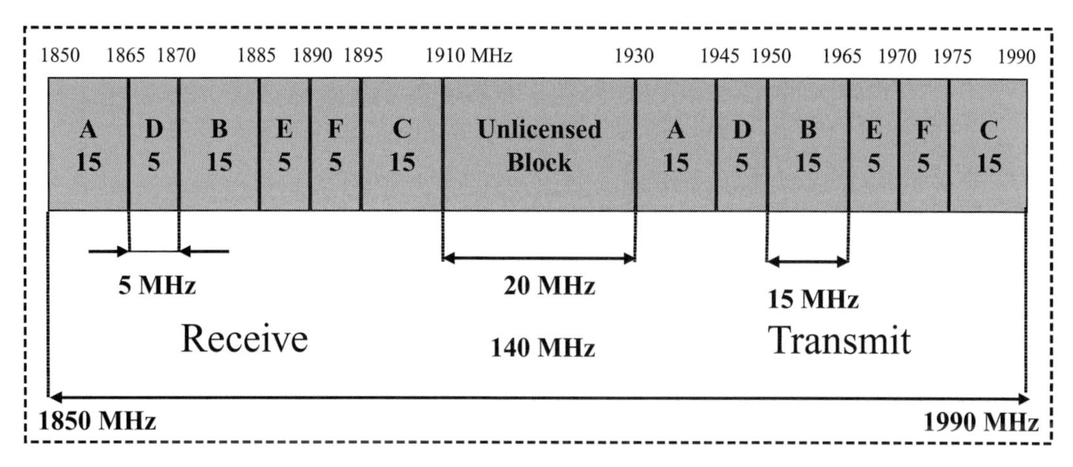

Figure 4.5 PCS Spectrum Allocations in the US

use. 1948 BTA licenses were granted to use with the C band having 30 MHz available. The D, E, and F blocks have only 10 MHz available [9].

International Telecommunication Union (ITU) and various standards organizations around the world have agreed on a global approach to the spectrum allocation of frequencies for mobile communications, both terrestrial and satellite. This is connected to the third generation of mobile communication known as International Mobile Telecommunications 2000 (IMT-2000). ITU-T spectrum allocation, compared with the spectrum used in some major regions or countries, is summarized in Figure 4.6.

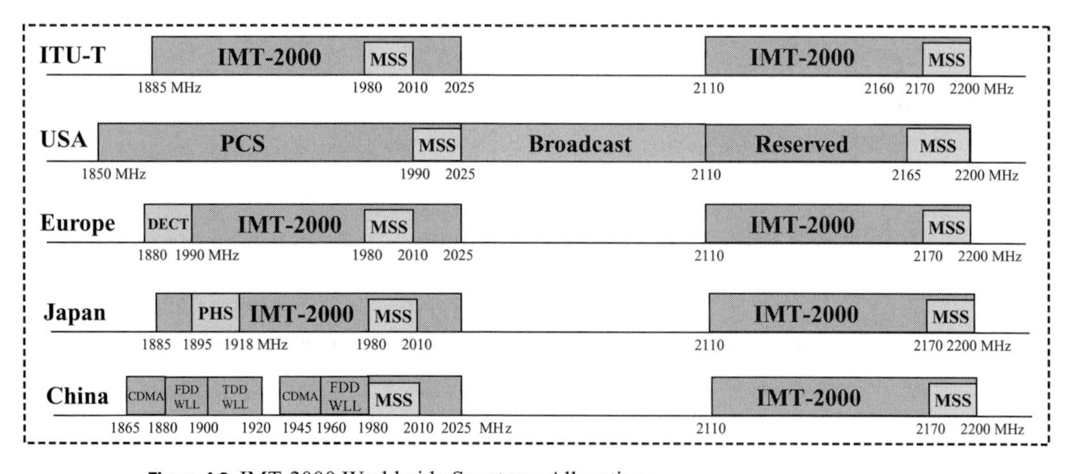

Figure 4.6 IMT-2000 Worldwide Spectrum Allocations

In the ITU-T Recommendations two major bands are designated to be used globally, the 1885–2025 MHz band and the 2110–2200 MHz band. Within those two bands, a portion is allocated to Mobile Satellite Services (MSS) spectrums in two parts, uplink and downlink, that correspond to C band satellite spectrums. In the USA the first IMT-2000 band includes, roughly, the second generation PCS spectrum while the second band was opened for auction. In Europe, the lower IMT-2000 band contains the Digital

European Cordless Telecommunications (DECT) band, while in Japan it contains the Personal Handyphone Service (PHS) band. In China, the lower IMT-2000 spectrum contains a mixture of CDMA, Frequency Division Duplex (FDD), and Time Division Duplex (TDD) Wireless Local Loop (WLL) bands.

4.5 Handoff/Handover and Roaming in Mobile Networks

The fundamental idea of mobile communications is that continuous communication is maintained while the mobile unit is moving through the coverage area. The mobile unit can be a customer talking while walking or talking while driving (not recommended). The handoff or handover (interchangeable terms) is the process of transferring ongoing calls from one cell site to another cell site as the mobile unit is moving through the service area. This is depicted in Figure 4.7. The transfer can also be caused by temporary congestion within a particular cell when call processing must be handed off to a less busy cell. Handovers may also take place when the caller is switched from one type of technology to another (for example, GSM to CDMA) or from one mobile operator to another as part of roaming.

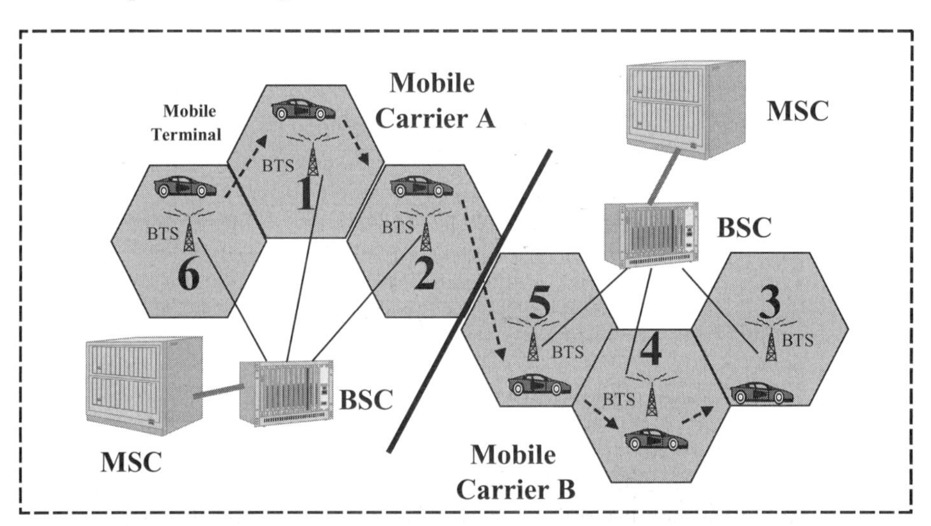

Figure 4.7 Handoffs/Handovers between Mobile Cells and Carriers

There are two types of handoffs. The **hard handoff** takes place when the connection to the original cell site is dropped before the connection to a new site is established. This is a normal occurrence in a FDMA-based mobile technology (like AMPS) and a TDMA technology (GSM, D-AMPS) because the adjacent cells have different frequency allocations (transmit and receive). Since this process is quite short, less than 250 ms, it is hardly noticeable in voice conversations. However, it is a different story when data is transmitted over the same systems. The **soft handoff** occurs when the connection to a new cell site takes place before the connection to the original cell site is broken. That is possible with CDMA technologies that do not require different frequency allocations in adjacent cells. In CDMA-based mobile communications, it is possible for the mobile unit to be connected simultaneously to two or more cell sites or cell sectors since independent transmitters within the same cell will have antennas pointed in different directions.

Both types of handoff just described are considered **horizontal handoffs** because they take place in the same type of network, the mobile cellular network. There is also a **vertical handoff** where communication is transferred from cellular radio network to a fixed wireless network such as wireless local area network. This will be discussed in detail as part of fixed-mobile convergence in subsequent chapters.

In most cases, the mobile communication component that is responsible with the handoff is the BSC. A BSC handles tens of BTS. Each BTS is identified by a unique "location area" that is advertised periodically to the adjacent BTS. The BSC handles allocation of radio channels, receives signal strength measurements from the mobile phones, and controls handovers from base station to base station. The handoff takes place as the mobile terminal travels out of a cell site and the signal strength of radio link between terminal and BTS serving the cell is below standard level. The cell site will request a handoff that is coordinated by the leading BSC. A special case of handoff takes place when the mobile terminal is moving out of one carrier serving area into other carrier serving area. This transfer is coordinated through mobile switching centers.

The general term **roaming** is applied when a mobile terminal is used outside the home location where the subscriber is registered. Roaming is assured by contiguous coverage in the service area, or between different service providers as long as service agreements exist for roaming between those carriers. The ultimate example of roaming is provision of mobile service in another country when the technologies used are compatible or the handset is multi-band. Initially, roaming was accompanied by additional charges and tariffs for roaming were part of agreements between regional carriers. Now, as the major service providers have almost nation-wide coverage, roaming fees have been gradually dropped from service plans. International roaming is facilitated by the commonality of a particular technology such as GSM which is supported by 80% of worldwide operators. If you have a tri-band, quad-band or a five-band handset you may establish a mobile connection virtually anywhere in the world. The key to GSM roaming is the International Mobile Subscriber Identity (IMSI) number.

Roaming with horizontal handoff may also take place within a Wireless Mesh Network (WMN) as mobile terminals are transparently serviced by different access points as they move in the WLAN coverage area. This situation will be analyzed in subsequent chapters dealing with convergence aspects between fixed wireless networks and mobile cellular networks.

4.6 Cellular Mobile Networks Classification

In this chapter and the first we introduced wireless networks and mobile networks along with some of the basic architectural developments used to implement them. We also addressed radio access methods, giving an indication of the large variety of cellular wireless networks that now exist. Mobile cellular networks can be classified based on six major criteria: technical development, wireless radio link access methods, data communication support, geographical coverage, level of mobility, and spectrum allocation. A diagram depicting the variety of cellular mobile communications systems is presented in Figure 4.8.

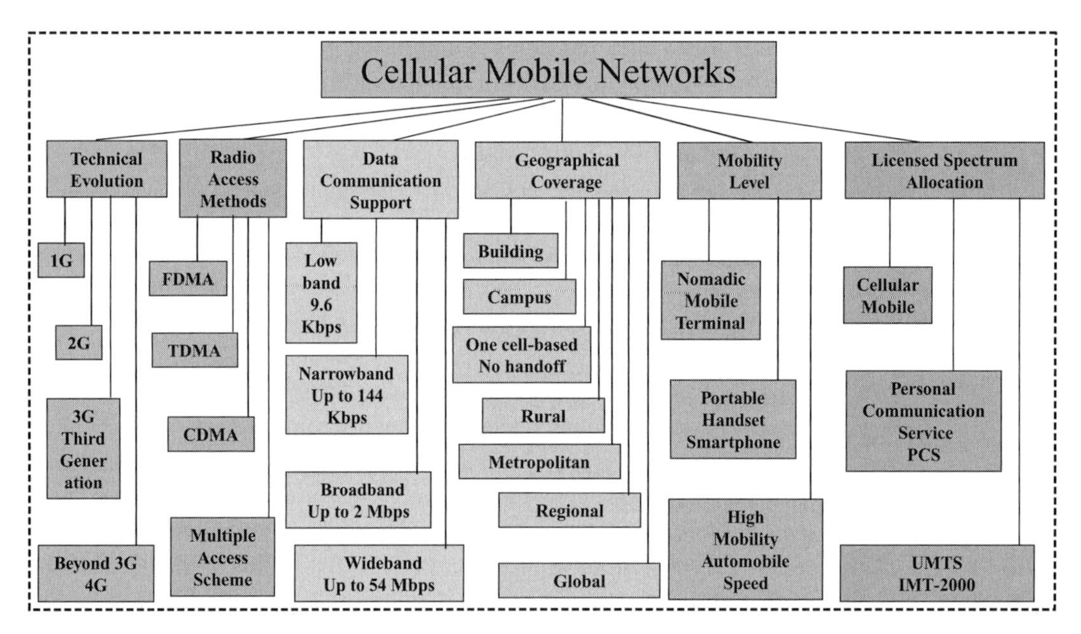

Figure 4.8 Cellular Mobile Networks Classification

In the span of two decades, cellular mobile networks have evolved from simple mobile networks to wide band mobile networks From generation 0 (0G), offering services such as MTS and IMTS, to First Generation (1G), analog cellular networks such as AMPS, to second generation (2G), digital cellular networks such as GSM. Then, the Third Generation (3G), broadband cellular mobile networks such as Wideband CDMA (W-CDMA), and now Beyond Third Generation (B3G) with services such as High Speed Downlink/Uplink Packet Access-HSDPA/HSUPA UMTS/IMT-2000 extensions.

Cellular mobile networking technologies, as defined by the wireless radio link access methods, have evolved from FDMA to TDMA to CDMA. Each of the new technologies provided capabilities to have higher densities of users in the available spectrum, reduced power, and higher quality of service. Some cellular mobile networks allow multiple access methods. Within each method there are variations based on synchronization such as Time Division Synchronous CDMA (TD-SCDMA) or Asynchronous CDMA (DS-CDMA), and spectral band (1.25 MHz as in CDMAone or 5MHz as in W-CDMA).

Initially, the focus was on voice communications. Data communication support was not yet top service objective. The first generation of systems provided a meager 9.6 Kbps data rate for services such as Cellular Data Packet Data (CDPD) or High Speed Circuit Switched Data (HSCSD). Starting with the second generation, the focus on data support became an imperative, so successive enhancements such as GSM Packet Radio Service (GPRS), Enhanced Data Rates for GSM Evolution (EDGE), and CDMA provided narrowband data services up to 144 Kbps. The third generation was linked to the challenge of providing broadband services up to data rates of 2 Mbps. This trend continues with B3G and 4G generations targeting data rates up to 54 Mbps.

One way to classify cellular networks is on geographical coverage. For example, coverage can be limited to a single building or perhaps a campus, or it can be defined in terms of the radius of the cell. Rural areas with sparse population will have large cells, whereas densely populated metropolitan areas will have coverage with physically smaller cells. There is good regional coverage for densely populated areas such as the east and west coasts of USA, Japan, Taiwan, or South Korea and global coverage such as European Community countries and land mass. A unique type of coverage is that provided by one cell-based mobile communications system such as Japan's Personal Handyphone Service (PHS) designed for pedestrian users in densely metropolitan areas, where there are no handoffs from cell to cell.

Another key aspect differentiating cellular networks is terminal mobility, the very essence of mobile networks. Level of mobility can be as basic as the nomadic feature of mobile terminals associated with computers to the portability of handsets and smartphones working in multiple spectral bands across multiple carriers, to the mobility of terminals that move in the coverage area with speeds up to 400 km per hour.

Last on the list of classification criteria is the licensed spectrum allocation. In the previous sections, we indicated that cellular mobile technologies working in the initial cellular spectrum from 850 MHz to 910 MHz were typical for the first generation. Other classes of mobile networks are those in the second and third generations working in the PCS licensed spectrum. These represent technologies such as GSM 1900, CDMA one, and CDMA 2000. A newly licensable spectrum was opened to networks and services aligned with UMTS IMT-2000 architecture and services.

There are several other criteria that can be used to classify cellular mobile networks. These include the kind of data modulation that is used (QPSK, BPSK, 16 QAM), channel coding schemes (1/2, 1/3, 1/4 convolutional coding), carrier spacing (25 KHz, 100 KHz, 200 KHz), the type of speech coding (ADPCM), antenna diversity, power control, level of mobile IP support, duplex scheme (FDD or TDD), frame length, synchronization (between BTSs), and mobile terminal detection method.

4.7 GSM Packet Radio Service Network Architecture and Components

The idea of providing mobile data services along with voice services preoccupied the designers of the first generation (1G) mobile networks. Examples of the services provided in these networks include: AT&T's and Verizon's Cellular Data Packet Data (CDPD), former Bell South's Mobitex, and ARDIS's RD-LAP. The most recognizable is CDPD, a 1G service used with AMPS mobile technology. In this approach, packetized data is sent during the periods of silence during conversations. With a nominal data rate of 19.2 Kbps about 2.4 Kbps can be achieved at peak usage of channels. To support higher data networking capabilities, GSM has evolved from a voice-oriented circuit switched network to a mixed voice/data network by adding technologies such as High Speed

Circuit Switched Data (HSCSD), or network configurations known as **General Packet Radio Service** (GPRS). **Enhanced Data for GSM Evolution (EDGE)** was added later. GPRS extends GSM voice-based services to allow mobile terminal users to transmit packet-based data and even to reserve bandwidth through guaranteed classes of service. The GPRS architecture and components are shown in Figure 4.9.

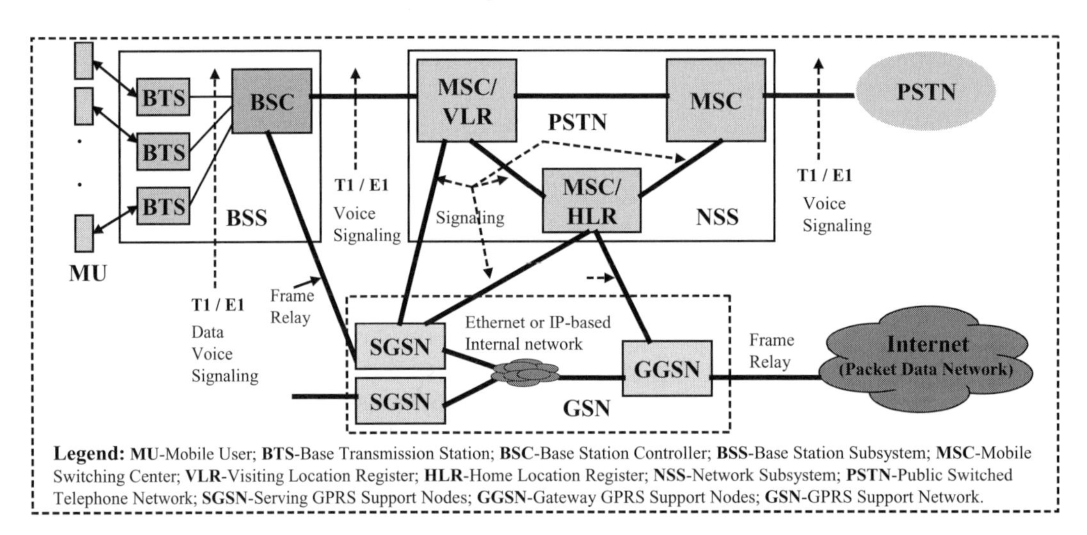

Figure 4.9 Core GSM/GPRS Network Architecture and Components

GPRS is a data service achieved by directing data communications requests to a separate packet switched data network that consists of two components: Serving GSM Support Nodes (SGSN) and Gateway GSM Support Nodes (GGSN). Multiple SGSNs are used to handle connectivity to BSCs. There are fewer GGSNs than SGSNs to connect the GPRS Support Network (GSN) to the outside world, primarily the Internet. GPRS was initially developed by ETSI in tandem with GSM 2G and is specified in Releases 97 and higher. Later work was taken over by 3GPP. A similar data construct was used to augment the functionality of D-AMPS, also known as IS-136 technology.

A critical component of GPRS is the BSC. BSCs handle dozens of BTSs, providing important functionality such as the allocation of radio channels, analysis of measurements from mobile terminals, and proper handoffs. BSCs also act as switches, directing voice calls to the MSC along with the proper signaling.

GPRS uses a combination of Frequency Division Duplex (FDD) and FDMA. When data service is requested, the user is assigned two channels; one for transmit (Tx or uplink) and one for reception (Rx or downlink). These channels operate on separate frequencies. Several users share the channel in a statistically multiplexed mode. Packet length is limited to a GSM time slot. Initially, the GSN network was designed to support several connectionless/connection-oriented network protocols such as X. 25 PLP, OSI CLNP, and IP. In practice, IP V4 was the network protocol that was used. There are numerous alternatives used at Data Link and Physical layers: Ethernet, Frame Relay, and ATM.

GPRS Phase 1 offered data services in the range from 9.6 Kbps to 57.6 Kbps. It is used for SMS, MMS, and Wireless Application Protocol (WAP)-based services such as email and web access. GPRS Phase 2 implementation, also known as EDGE, augmented data rate capabilities up to 144 Kbps, matching ISDN basic rate capabilities. More details about GPRS will be provided when we analyze quality of service of mobile networks.

4.8 Cellular Mobile Standards and Standards Organizations

The aim of wireless cellular mobile networking is to provide global services for voice, data, video, and multimedia communications. This goal cannot be achieved without international standardization that includes radio access links, the use of certain spectrums, signal strength, power control mechanisms, antenna design, and connectivity to PSTN and the Internet. Both voice and data must be handled in mixed wireless and wired environments. Standardization work is performed with the goal of creating a consensus among the standards organizations that include the vendors and users of the equipment and services. A diagram depicting the range and hierarchy of standards organizations is presented in Figure 4.10.

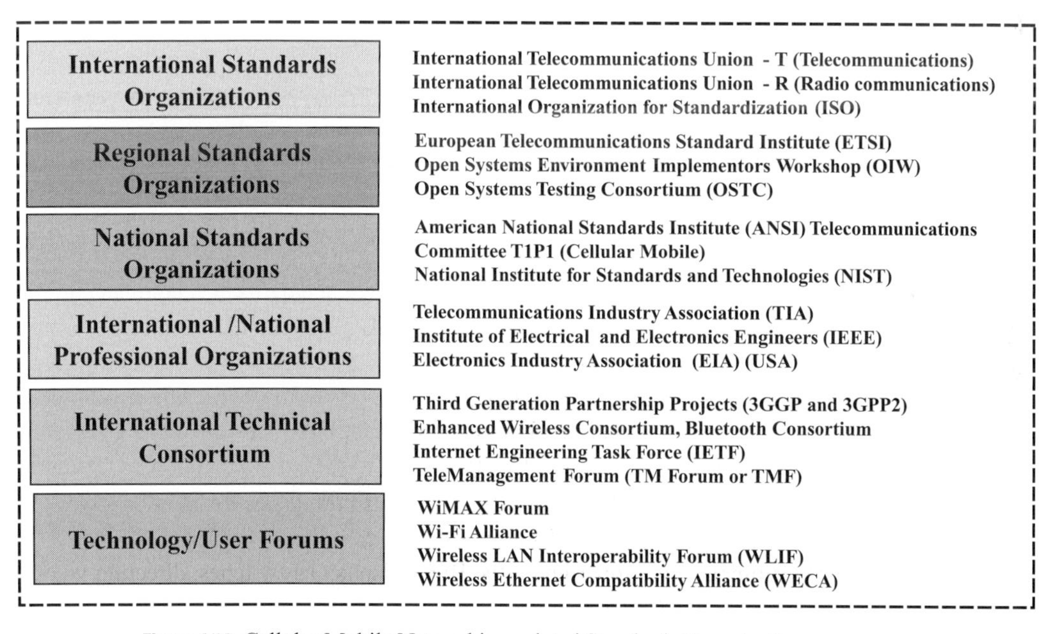

Figure 4.10 Cellular Mobile Networking-related Standards Organizations

At the very top of the list are the international standards organizations. They, of course, design globally recognized standards. Very powerful standards are promoted by regional standards organizations such as ETSI, or national standards organizations such as ANSI. Professional organizations such as the IEEE develop important standards such as LAN and WLAN. In many instances, standards specifications are worked out in international technical consortiums such as IETF, 3GPP/3GPP2, and the TM Forum. Speedy feedback from the users in the field comes from technology/user forums that submit pointed core specifications to technical consortia, and even to international standards bodies.

Standards specifications are the product of a phased and sometimes lengthy process. Simple proposals and draft specifications go through a complex process to reach the final stage of becoming an internationally recognized document. This process is indicated in Figure 4.11.

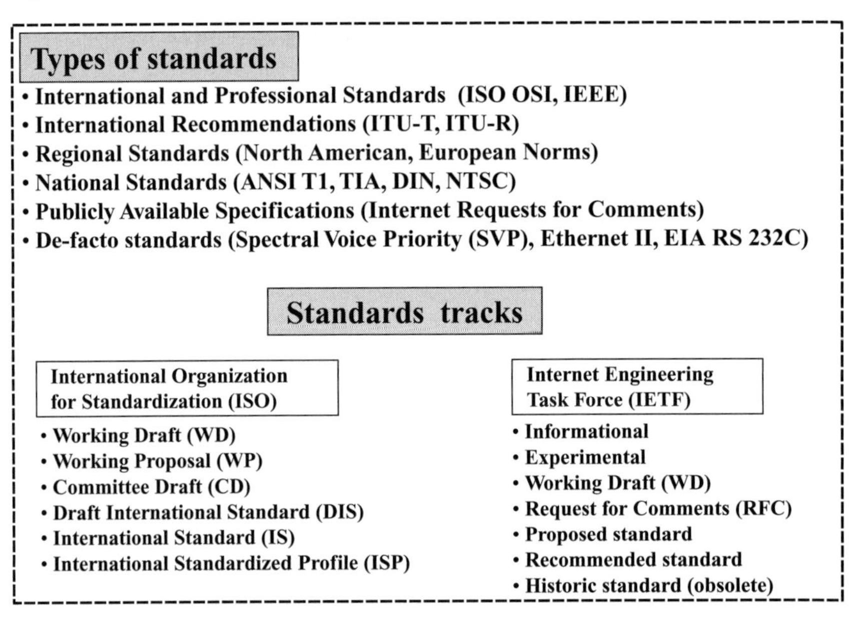

Figure 4.11 Types of Standards and Standards Processes

Standards specifications range from the international standards or recommendations issued by internationally recognized standards organizations to de-facto standards pushed by powerful vendors/mobile carriers, or manufacturers. Strong global recognition is given to regional or national standards such as GSM standards issued by ETSI or those promoted through TIA. Publicly available specifications such those issued by the Internet Society's IETF working groups often become recognized international standards. One example of a de-facto standard is Spectral Voice Priority (SVP) promoted by SpectraLink (now Polycom). Lacking IEEE QOS standard for VoWLAN, SpectraLink created the SVP specifications that are used to configure WLAN Access Points (APs). APs recognize voice packets and prioritize them.

Cellular mobile networking is a huge field so specific organizations and specific standards are created to cover the uniqueness of cellular radio mobile networks. The organizations and standards that provide 2G cellular networking specifications in the PCS spectrum are presented in Figure 4.12. The most recognizable standards are IS-54 and its successor IS-136 for cellular TDMA wireless technologies and IS-95 and its successor IS-95C for cellular CDMA technologies. These standards were developed by EIA/TIA in one of the TR-45 Committee working groups. The PCS specifications were developed in a T1P1 Committee, an accredited ANSI standardization body. Since the FCC did not specify a standard for PCS/PCN radio interface, several solutions were adopted by necessity by these standards groups. These include D-AMPS IS 54/IS 136, CDMA IS 95/IS 136, and GSM 1900.

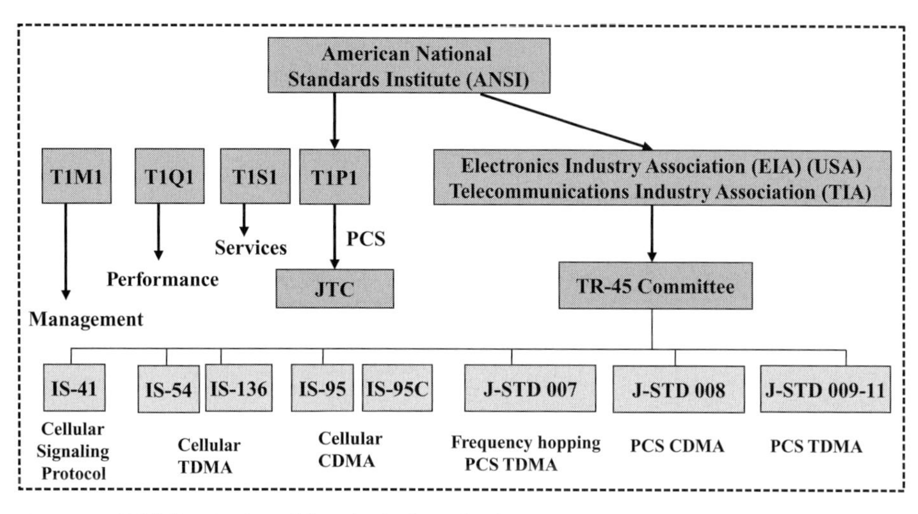

Figure 4.12 PCS Standards and Standards Organizations

Standardization of GSM, the dominant cellular mobile networking technology, started in the early 1980s with the work of the Groupe Spécial Mobile (GSM), which is a part of the European Posts and Telegraph association. The first GSM network was installed in 1988 in Finland. In 1989, GSM standardization was transferred to the ETSI. Phase 1 of the GSM specifications was published in 1990.

Currently, there are eight main flavors of GSM specs: GSM 450, GSM-480, GSM-850, GSM 900 (P-GSM), GSM 900 (E-GSM), GSM 900 (R-GSM), DCS-1800 or GSM-1800, and GSM 1900 (PCS). See Table 4.1 [10].

Table 4.1 GSM Frequency Bands and Channel Numbering

GSM System	Band	Uplink	Downlink	Channel Numbers
GSM 400	450 MHz	450.4–457.6	460.4–467.6	259–293
GSM 400	480 MHz	478.8–486	488.8–496	306–340
GSM 850	850 MHz	824–849	869–894	128–251
GSM 900 (P-GSM)	900 MHz	890–915	935–960	1–124
GSM900 (E-GSM)	900 MHz	880–915	925–960	975–1023
GSM 900 (R-GSM)	900 MHz	876–915	921–960	955–973
GSM 1800 (DCS)	1800 MHz	1710–1785	1805–1880	512–885
GSM 1900 (PCS)	1900 MHz	1850–1910	1930–1990	512–810

Mobile technologies advance at a rapid pace. This requires more than just consensus on standards, and that is speedy delivery of specifications, to build the products and to provide service applications. In this respect, two major organizations have helped ensure the successful future of third generation mobile network developments: the Third Generation Partnership Project (3GPP) and its counterpart known as 3GPP2.

3GPP was initiated in 1998 by ETSI. ETSI's objectives were to speed up the development of UMTS IMT-2000-based 3G mobile network specifications and to accelerate 3G

GSM worldwide implementations. It was joined in the same year by four other standards organizations: T1 Committee (USA), TTC and ARIB (Japan), and TTA (South Korea). Participation in 3GPP is open to partner members that represent carriers, manufacturers, marketing organizations, and individual members. A project coordination group supervises technical work that is done in four specific areas: Core Network (CN), Radio Access Network (RAN), Terminals (T), and Service and Systems Aspects (SA). That is all the defining components of UMTS architecture and interfaces.

3GPP2 was initiated in 1999 by ANSI to advance non-GSM cellular mobile networking. Emphasis was to be on CDMA technologies. It was joined by TIA (USA), TTC and ARIB (Japan), and TTA (South Korea). A steering committee acts as work coordinator for six groups: Core Network (CN) evolved from ANSI IS-41 specifications, CDMA2000, Wireless Packet Data Network (WPDN), Radio Access Network (RAN), A-Interface (between RAN and CN), and Service and Systems Aspects (SA). These groups are the same defining components of UMTS/IMT-2000 architecture and interfaces, but using different names and codes.

Recognizing that there would be different approaches and work emphasis in two partnership projects, an Operators' Harmonization Group (OHG) was formed. OHG membership is composed of representatives from major operators and manufacturers in North America, Europe, and Asia. A major work objective was harmonization of CDMA proposals to build 3G UMTS/IMT-2000 infrastructures as envisioned by the ITU-T. Three major CDMA variations were considered: Direct Spreading (DS), Multi-carrier (MC), and Time Division Duplex (TDD). The scope was transparency, interoperability, and seamless handoff between these three modes of operations. Another objective was compatibility between GSM Mobile Application Part (GSM/MAP) and ANSI-41. More details will be provided in subsequent sections.

4.9 Cellular Mobile Applications and Services

Historically, mobile communications applications were built around voice communications services. However, in the past decade, most innovations and value added services have been focused on data communications-related applications and services. At the start, modest modem-like 9.6 Kbps data rates, today broadband data rates in the range of multiple Mbps. At the beginning, advancements in digital signal processing, among them digitization, data compression, and voice recognition. Then, adoption of color Liquid Crystal Display (LCD) followed by computer miniaturization and new antenna design. Last, the adoption of soft-handover technologies that allowed smaller handsets with highly sophisticated service features. For example, handsets are capable of displaying captured still images, video communications broadcasts, and even high definition television.

The handset is a consumer electronics masterpiece that can be used as a digital camera with multiple mega pixel resolution, an iPod and MP3 music player, a miniature monitor to display TV contents, a GPS, and a personal digital assistant device. Also, it employs all the cherished features such as miniature keyboard, address list, scheduler, clock, and alarm clock. All of these features operate in a high mobility environment at car race speeds. A short list of terrestrial mobile applications and services follows.

- **Voice communication services in large coverage areas using multiband handsets**
  - Campus, metropolitan, rural voice services;
  - Regional, nation-wide, global voice services;
  - Voice mail, call forwarding, call waiting;
  - Automatic Number Identification (ANI);
  - Voice message services (recording, unified messaging, ring tones);
  - Instant messaging;
  - Interactive Voice Response (IVR), voice recognition;
  - Voice over IP.
- **Data communication services up to broadband data rates**
  - Internet access;
  - Short Message Services (SMS);
  - Multimedia Message Services (MMS);
  - Web browsing;
  - Small files transfer;
  - Access to small information contents (directories, inventories, schedule);
  - Access to large information contents (news, sport events, entertainment);
  - Music streaming, music download;
  - Financial transactions (credit card authorization);
  - Mobile electronic commerce (M-Commerce);
  - Mobile Virtual Private Network (VPN);
  - Instrumentation;
  - Gaming.
- **Video communication services up to video broadcast reception**
  - Videophone;
  - Still pictures, imaging;
  - Web cameras;
  - Video clips;
  - Video streaming, video download;
  - Videoconference;
  - Video content display.
- **Convergent communication services using multi-mode handsets covering voice, data and video services**
  - Convergence with WLAN services (802.11x);
  - Convergence with WPAN services (Bluetooth);
  - Convergence with WiMAX;
  - Convergence with near field sensor services (UWB, RFID).
- **Personal communications services**
  - Location service;
  - Presence service;
  - GPS service;

  - Navigation service;
  - Peer-to-Peer communication services.
- **Ad-hoc communication services**
  - Military, police actions;
  - Exhibitions;
  - Conferences;
  - Sports events.
- **Cell-based broadcast services**
  - Announcements;
  - Advertising.
- **Emergency call services**
  - E911.

In addition to all the positive features of mobile communication services and applications, there are some negative aspects. For example:

- **Ability to track a person's movement** (privacy and personal freedom issues);
- **Ability to capture mobile communications** (unwarranted collection of secure, private, and critical information, plus eavesdropping on private conversations and information exchange);
- **Phishing** (collecting personal information);
- **Unwanted advertisements** (commercials and other interruptions that come along with digital content and subscription to more sophisticated services);
- **Nuisance** and **Annoyance** (ubiquitous use of cell phones in public sphere, meetings, shows, conferences, parties).

4.10 Cellular Mobile Networks Evolution

Cellular networks have evolved in their 20 years of existence through several generations (1G, 2G, 2.5G, 2.75G, 3G) via adoption of more efficient access methods and migration paths to support higher data rate communications capabilities. This is depicted in Figure 4.13. The first generation belongs to analog, voice-oriented technologies. The second generation is based on digital technologies and low data rate capabilities. The intermediary generations, sometimes called 2.5G or 2.75G, had augmented data rate capabilities, used a new spectrum band, and were based on a set of architectural and radio interface specifications collectively known as PCS. Third generation capabilities are focused on broadband data capabilities and sophisticated data and video services.

1G mobile analog system **AMPS**-Advanced Mobile Phone Service, and **NAMPS**-Narrowband AMPS, are early FDMA-based mobile systems used in the USA. TACS-Total Access Communications Systems and Japanese TACS (JTACS) are former AMPS implementations in the UK and Japan. They operated in the 900 MHz band. Ultimately, AMPS was replaced by its digital counterpart D-AMPS, also known as NADC-North American Digital Cellular, while TACS and JTACS were replaced by GSM.

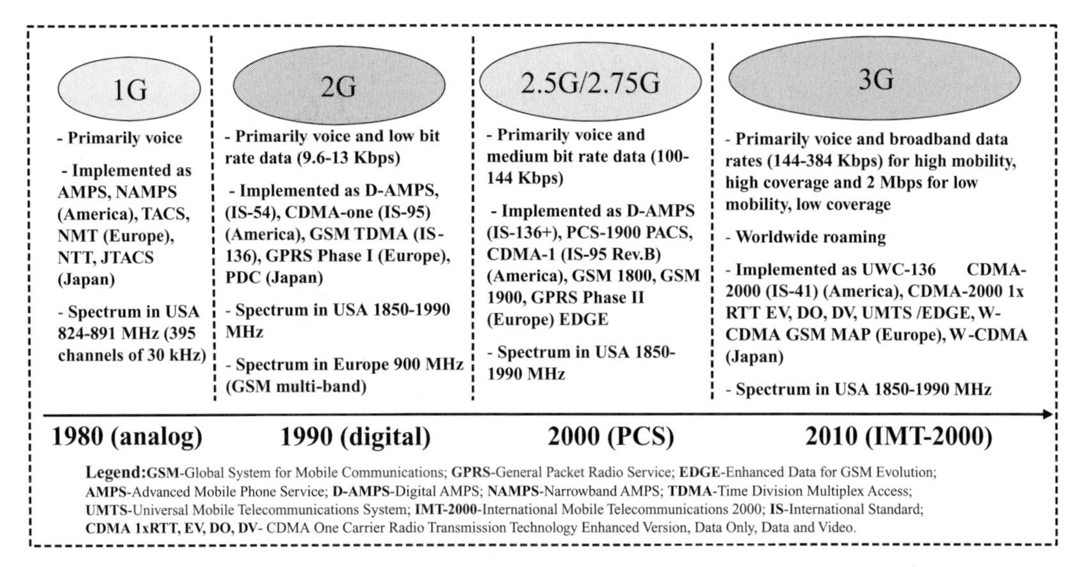

Figure 4.13 Cellular Mobile Networks Evolution

In the second generation, dominating roughly the 1990s, data services were introduced with rates as high as 9.6 Kbps. Notable implementations in this generation were: GSM TDMA, also known as IS-136, CDMAone, also known as IS-95, D-AMPS, also known as IS-54, and PDC-Personal Digital Cellular, Japan. These networks operated in the 800 MHz and 1.5 GHz bands. The GSM was augmented later with GPRS-GSM Packet Radio Service, a dedicated data communication network with connectivity to the Internet.

The transition towards 3G included the 2.5G and 2.75G intermediary steps. During these stages data communication rates were raised up to 144 Kbps. There were also enhancements such as IS-136+ for D-AMPS and CDMA-1, also known as IS-95B. GSM opened operations in the 1800 MHz band in Germany and the UK as well as in the USA 1900 MHz PCS band (PCS-1900 or DCS-1900). During phase 2 of the GPRS implementation, also known as EDGE, data rates were increased to 144 Kbps. This matched for the first time ISDN basic rate capabilities. The 1900 PACS-Personal Access Communication System was Bell Laboratory's combination of cordless phone and PCS cellular systems working in the 1900 MHz band.

During the third generation, data rate capabilities were raised into the broadband range, up to 2 Mbps for low mobility and low coverage mobile terminals. This increase in data rate was provided by the UMTS/IMT-2000 architecture and specifications. There were several intermediate steps along the way, before the higher data rates were finally realized. Some of the most notable of these steps were the W-CDMA GSM/MAP in Europe and the Universal Wireless Communications-IS136 and CDMA 2000 in the USA. Service variations were known as CDMA 1xRTT, EV-DO, DV, CDMA One Carrier Radio Transmission Technology Enhanced Version, Data Only, Data and Voice, and W-CDMA in Japan.

A concentrated view of the migration path of current 2G, 2.5G, and 2.75 G wireless mobile systems towards 3G systems is shown in Figure 4.14.

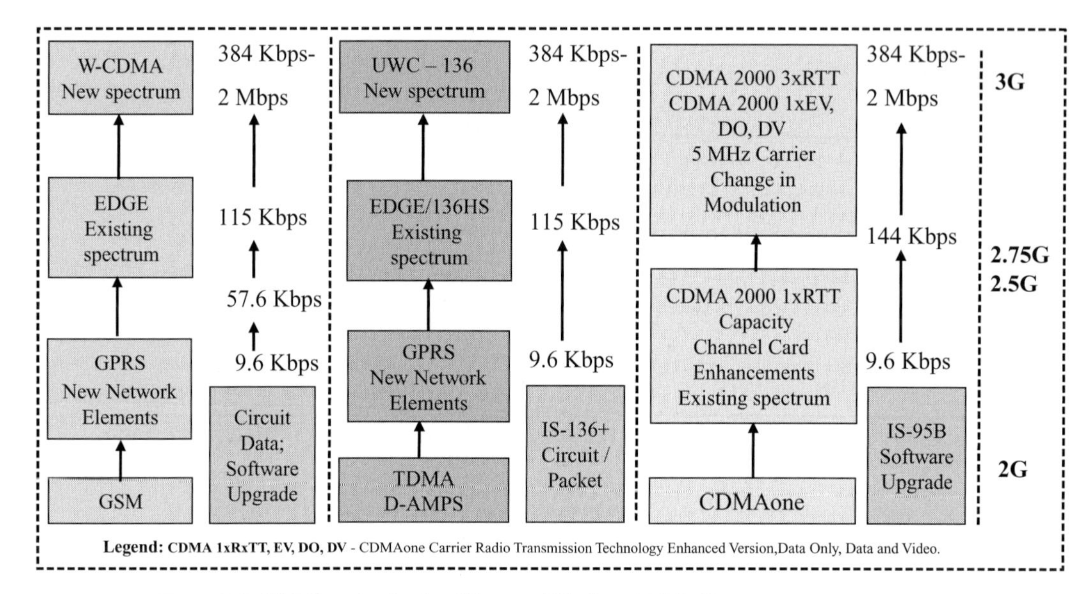

Figure 4.14 3G Migration Path of Current Wireless Mobile Systems

Given the mixture of different types of existing wireless communications systems (GSM, CDMA, and TDMA), there are three possible migration paths to 3G networks/services. In this migration, the final winners are Wideband CDMA (W-CDMA) and CDMA 2000 1x Enhanced Version (EV) along with Data Only (DO) and Data and Voice (DV) variations.

4.11 GSM and CDMA Cellular Networks Comparison

From the previous sections, it is clear that there are two dominant cellular mobile build-ups, GSM and CDMA. Each of these comes with a suite of advanced technologies, specific radio link interfaces, spectrum bands used for communications, and a school of design that has developed sophisticated services beyond traditional voice communications. Both with a common goal, to amass in a finite spectrum band multiple users and multiple services. Before analyzing the strengths and weaknesses of these two approaches, it would useful to have a comparative perspective of mobile networks as they evolved from analog to digital to PCS and UMTS.

There are five major wireless telephony systems in use or evolving: mobile telephony, analog cellular telephony, digital cellular systems, digital PCS/PCN systems, and 3G UMTS/IMT-2000 systems, mostly achieved as evolution paths towards 3G of previous digital cellular systems. The five systems are compared with wireless Personal and Local Area Networks (PAN/LAN) characteristics to establish a reference point in the fixed wireless world, as represented by Wi-Fi and Bluetooth technologies. This comparative analysis is shown in Table 4.2.

To understand the basis up which we compare the mobile technologies that make up the PCS/PCN and 3G UMTS/IMT-2000 networks and services, it is useful to present

Table 4.2 Comparisons of Cellular Mobile PCS with other Wireless Technologies

Service	Radio link type	Architecture	Capacity (# of simultaneous calls)	Access method
Mobile Telephony	Analog AM	AM radio frequency reuse	12 channels	FDMA
Cellular Telephony (analog)	Analog FM	Cell-based frequency reuse	Tens of thousands (depending on the number of cells)	FDMA
Cellular Telephony (digital)	Digital PM	Cell-based frequency reuse	TDMA, 3–5 times analog cellular CDMA, 1–20 times	FDMA/TDMA FDMA/TDMA/CDMA
PCS/PCN	Digital PM	Microcell-based frequency reuse	Hundred of thousands (depending on the number of microcells)	FDMA/TDMA FDMA/TDMA/CDMA
PAN/LAN	Digital Baseband	Piconet-based (cubicle, home)	Hundred of billions	Bluetooth radio 802.11 MAC

some projections regarding GSM and CDMA-based technologies. In mid 2005 there were 1,734 billion mobile users worldwide. Of that, the GSM portion represented 1,312 billion users. In 2006 the number of worldwide mobile users surpassed the two billion landmark. The countries with the most users were: China with over 400 million, USA with over 200 million, Japan over 95 million, Russia over 80 million, Brazil over 75 million, Germany with more than 70 million. Projections for 2010 are that GSM and W-CDMA networks will have 3.2 billion worldwide users, with 80–85% of them on GSM networks. In the USA and Canada, the revenue from data services represent roughly 10–15% of the total, while in Asia that portion is up to 45–50%. On the road to 3G, in 2010, it is expected that 60% of the users will be using W-CDMA, 30% EDGE, and 10% classic TDM-GSM. Europe is fully GSM-based (92% with SIM card capability that allow users to move from one operator to other) while in the USA there are five different standards in use and most handsets are "locked" to block the use of the same handsets in alternate mobile domains and with alternate mobile operators. The main technical features of GSM are shown in Table 4.3.

In GSM, regardless of the frequency band used, voice and data are carried over 200 KHz channels. Using TDMA access methods, 8 time slots are used in each channel, and 8 simultaneous voice users are supported per channel. Gaussian Minimum Shift Keying (GMSK) modulation is used with a frame size of 1250 bits, so each user gets 156.25 bits per slot. Excluding the control bits, 144 bits of useful information remain available, leading to a total of 31.2 Kbps transmission rate.

A high-level listing of the strengths and weaknesses of second generation of GSM is presented in Table 4.4.

The initial data rate supported in GSM was 14.4 Kbps. This was augmented later to 53.6 Kbps in the first GSM extension, GPRS. The second extension, known as EDGE,

Table 4.3 GSM Basic Technical Features

GSM Systems	Main Features
Main bands used	850, 900, 1800, 1900 MHz
Multiple access techniques	TDMA/FDMA
Modulation technique	GMSK/8-PSK (EDGE)
Duplex techniques	FDD (Uplink/Downlink)
Channel bandwidth	200 KHz
Users per channel	850 MHz (124), 900P (124), 900E (174), 900R (194), 1800 (374), 1900 MHz (299)
GSM peak data rate	14.4 Kbps
GPRS peak data rate	53.6 Kbps uplink, 115 Kbps downlink
EDGE peak data rate	384 Kbps
W-CDMA peak data rate	2 Mbps

Table 4.4 GSM Strengths and Weaknesses

GSM	Evaluation/Description
Strengths	– Only one TDMA user can use the allotted spectrum channel at any given time. – GSM is a mature technology with worldwide acceptance and implementations. – Universal market acceptance by both manufacturers and users. – Evolution path to 3G via GPRS, EDGE and W-CDMA mobile technologies. – Easy switch between carriers, grace to Subscriber Identity Module (SIM) card. – Easy worldwide roaming because of global coverage. – Limited deterioration of signal inside buildings, ability to use repeaters.
Weaknesses	– GSM use is limited by a fixed maximum cell site of 35 km. – Possible interference with electronic devices because of TDM pulse-based nature of radio access. – Intellectual property is concentrated among a few industry participants, limiting new entries and competition among mobile equipment manufacturers. – GSM requires multi-band handsets because of multiple frequency bands used. – Limited number of users per frequency band.

has raised date rates to a theoretical 384 Kbps. The third generation GSM, adopting W-CDMA technology, provides rates up to 2 Mbps.

The other dominant cellular mobile technology is based on CDMA radio access links. The CDMA family of technologies evolved from the first generation, as standardized by TIA under the name of IS-95. Work on this standard began in 1992. It used a larger bandwidth and took advantage of spread spectrum technologies, producing a set of specifications soon after. In 1994, the first set of standards was replaced by IS-95 Release A, thus becoming the first CDMA standard solution for the first generation of mobile

Table 4.5 CDMA Main Features

CDMA Systems	Main Features
Main bands used	900 MHz (UL 824–849, DL 869–894) 1900 MHz (UL 1850–1910, DL 1930–1990)
Multiple access techniques	CDMA/FDMA
Modulation technique	QPSK/OQPSK
Duplex techniques	FDD (Uplink/Downlink)
Channel bandwidth	1.25 MHz
Users per channel	Depends on service
CDMAone peak data rate	14.4 Kbps (IS-95 Release A)
CDMAone peak data rate	115.2 Kbps (IS-95 Release B)
CDMA2000 peak data rate	307.7 Kbps (CDMA 2000 1x)
CDMA2000 peak data rate	2.4 Mbps (CDMA 2000 3x)

networks. There are other CDMA systems based on proprietary solutions such as those proposed by DoCoMo (Japan) and Lucent Technologies (USA). There were also other proposals that were based on combinations of the TDMA and CDMA technologies. The main standard features of the second generation of CDMA, known as CDMAone, are presented in Table 4.5, above.

CDMA is the most widely deployed technology in the USA. Sprint and Verizon Wireless operate major CDMA networks. They were initially based on CDMAone technologies and later migrated to CDMA 2000 1xRTT. Following that, there were two more variations: CDMA 2000 1xEV-DV and the CDMA 2000 1xEV-DO which were mentioned earlier.

In CDMA, regardless of the frequency band used, voice and data are carried over a wide bandwidth channel of 1.25 MHz channel for uplink and a 1.25 MHz channel for downlink. Digital Walsh coding is used that spreads the initial allocation of 19.2 Kbps for a basic voice communication to cover the entire 1.25 MHz bandwidth, by a factor of 64. Both transmitter and receiver are uniquely identified by using individual Walsh codes on each channel of the spread codes.

An important aspect of CDMA is the need to continuously control the signal power. Signals should arrive at the BTS with roughly equal power to avoid overwhelming other channels using the same carrier. To solve this problem, BTS adjusts the mobile station power signal 80 times per second. More advanced CDMA systems, such as CDMA 2000 1xRTT, adjust the power level 800 times per second. By reducing interference between channels, one can have more users or a higher data bandwidth available. The strengths and weaknesses of second generation of CDMA mobile networking are presented in Table 4.6, next page.

Promotion of the CDMA 2000 family of technologies is done by CDMA Development Group (CDG). They reported that by mid 2006 CDMA 2000 services were used by 200 operators in 75 countries serving 275 million subscribers. These figures are by far higher than W-CDMA implementations, the 3G evolution of GSM networks. However, given

Table 4.6 CDMA Strengths and Weaknesses

CDMA	Evaluation/Description
Strengths	– Multiple CDMA users are spread over a very wide shared spectrum. – Best use of frequency spectrum; number of users per MHz of bandwidth. – There are no limits on the number of concurrent users. – There are no limits on the distance a BTS tower can cover. – It consumes less power to transmit and receive and covers larger cell areas. – Soft-handoff built in CDMA technology, so better QOS is guaranteed. – Improved QOS and robust operation in multipath and fading environments. – Evolution path to 3G via CDMA2000 and W-CDMA.
Weaknesses	– Requires continuous power control to equalize channels on the same carrier. – Most CDMA networking technologies are patented and must be licensed from one vendor Qualcomm raising fears of fair treatment. – CDMA does not work well in hilly areas because of reduced BTS tower heights to avoid interference between towers. – Reduced worldwide presence, smaller pool of handset vendors, and more expensive handsets.

the use of GSM by more than 2 billion worldwide subscribers, this situation might change quite soon once the industry starts rolling new implementations. In April 2005 there were only 70 operators in 30 countries using 3G out of the 130 operators who have applied for licenses as 3G mobile operators. An optimistic view of 3G advancement expects to count 1 billion users as of 2010.

In the works are B3G solutions such as High Speed Downlink/Uplink Packet Access (HSDPA/HSUPA) which will provide data rates from 2 Mbps up to 14.4 Mbps. They are characterized by reduced latency and increased capacity (in downtown areas the density of microcells and users will increase five fold). HSPDA is already a reality in the industry (lab and manufacturers) and it went commercial in 2006. It will be facilitated by mobile cards on PCs and four-band handsets (850/900/1800/1900 MHz).

4.12 UMTS/IMT-2000 Architecture and Components

Universal Mobile Telecommunications Systems (UMTS) is the generic name for the third generation of GSM cellular radio mobile systems. More precisely, UMTS is the European vision of International Mobile Telecommunications 2000 (IMT-2000). It represents the ITU-T initiative to conceive, design and implement 3G mobile systems for the year 2000 and beyond, promising data rates or bandwidth up to 2 Mbps. The very fact that in 2007 we are still using UMTS/IMT-2000 as reference architecture is an indication of the soundness of the original concept. The UMTS architecture is presented in Figure 4.15.

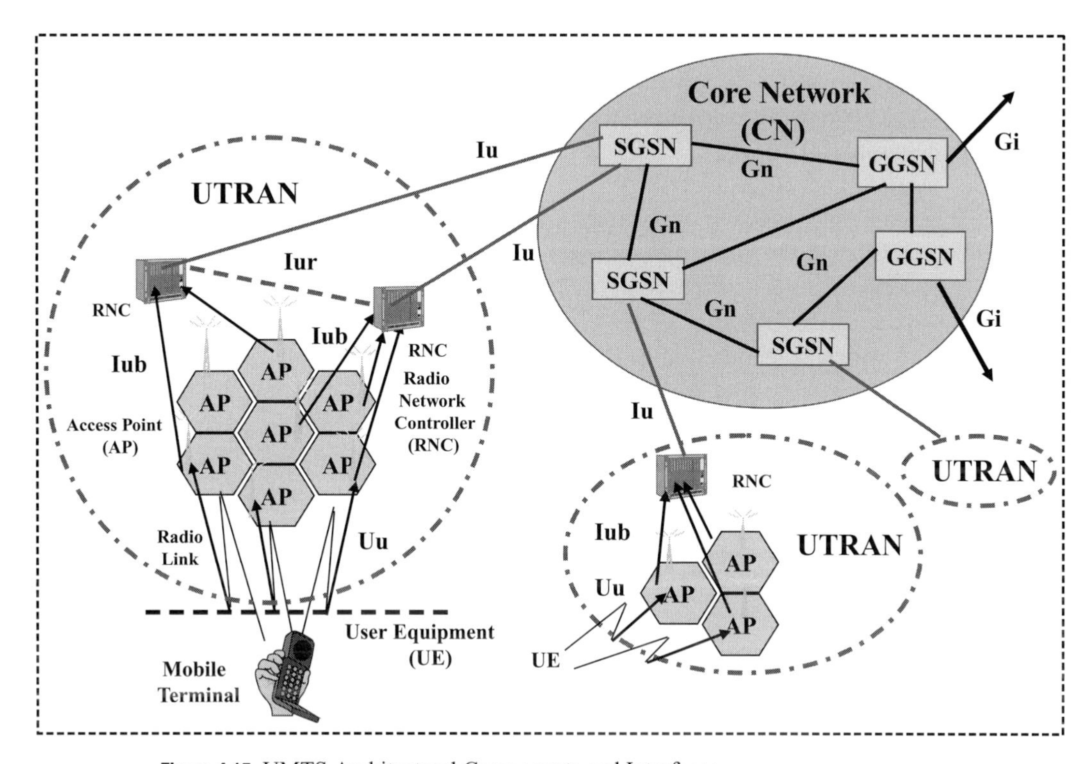

Figure 4.15 UMTS Architectural Components and Interfaces

UMTS consists of two major types of networks: Universal Terrestrial Radio Access Network (UTRAN) and Core Networks (CN). UTRAN provides connectivity with the mobile users. CN provides connectivity to multiple UTRANs and to the outside world represented by the PSTN or the Internet [11].

A Mobile Terminal (MT), represented by User Equipment (UE), communicates with one or more Access Points (APs) over the wireless radio link interfaces, coded **Uu** (User-UTRAN). An AP communicates with multiple users. User packets are segmented into transport blocks. Multiple transport blocks from multiple users are assembled into a larger frame and passed to a Radio Network Controller (RNC) over the **Iub** interface. Since a MT can communicate with multiple APs, which are in turn controlled by different RNCs, one of them will act as a serving RNC. This RNC is responsible for selecting the best frame (error free) passing it to the CN through an **Iu** interface. The entire collection of APs and RNCs represents a Radio Network Subsystem (RNS) or in the parlance of UMTS, a UTRAN. Multiple UTRANs communicate with the CN. Communication between RNCs is done through the **Iur** interface. In practice, the link between an AP (BTS) and a RNC (BSC) is realized with fixed wired point-to-point T1 lines or Frame Relay links.

The core networks can be divided into two network domains: a circuit switched network and a packet switched network. The circuit switched domain provides extensions and connections of CN to the voice oriented PSTN. The packet switched network provides extensions and connections of CN to the data oriented Internet. Therefore, Iu interfaces have two flavors: **Iu-CS** to connect an RNC to a circuit switched network, and **Iu-PS** to connect an RNC to a packet switched network. In Figure 4.15 only the packet switched CN domain is represented. It consists of two components introduced earlier as part of the description of the GSM/GPRS network: Serving GSM Support Nodes (SGSN) and Gateway GSM Support Nodes (GGSN). Multiple SGSNs handle connectivity between the CN and multiple UTRANs. Only a few GGSNs provide further connectivity to the outside world playing the role of routing nodes. In practice, CN can be achieved as a simple LAN, or a more complex network such as a TCP/IP-based WAN Internet or an ATM network.

A simplified diagram of the UMTS architecture and interfaces is given in Figure 4.16. In this diagram [12], the RNC itself is split into two domains. One of these is the controlling domain (CRNC) and the other is the serving domain (SRNC). The SRNC passes the information to the Core Network (CN). It is envisioned that in the future CN will be an IP-based network with connectivity to the Internet. This vision is going to be put in practice by implementing the IP-based Multimedia Subsystem (IMS). A description of IMS will be detailed in Chapter 17.

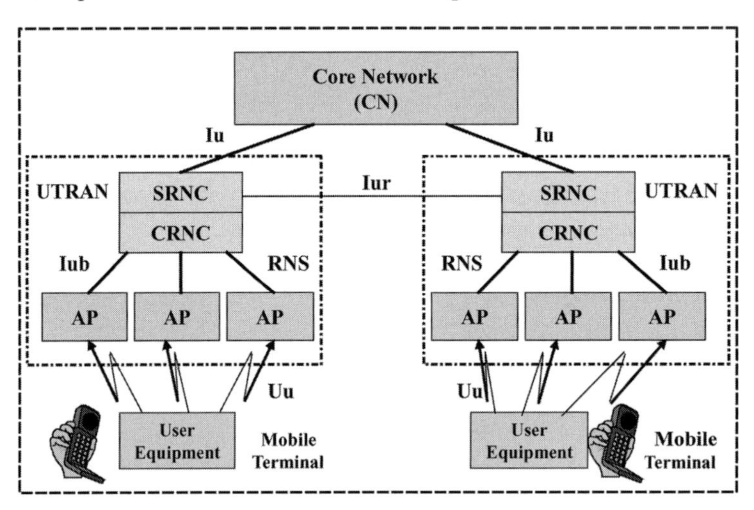

Figure 4.16 A Simplified View of the UMTS Architecture and Interfaces

From the title of this section, and from previous introductions of UMTS and IMT-2000 architectures, we mentioned the similarities between these two concepts. Both UMTS and IMT-2000 have features that could support their adoption as a 3G mobile network standard. However, UMTS was promoted by ETSI, as a leading standards organization under the 3GPP umbrella project. The underlying technology is GSM even though the radio access interface is based on W-CDMA.

By contrast, IMT-2000 was promoted by ANSI, under the 3GPP2 umbrella project. Its underlying technology is CDMA and the radio access interface of choice is CDMA 2000. The implementations of UMTS and IMT-2000 concepts are visible in the USA mobile world, where AT&T/Cingular is promoting the 3GGP standard, while Verizon and Sprint are promoting 3GPP2 specifications.

Architectural similarities between 3GGP and 3GPP2 are evident despite the use of different terminology for system components and interfaces. However, digging deeper one can note technical differences that derive from the use of different radio access technologies. These differences are in the synchronization channel, framing, and use of a common assignment channel. A translation of terms used by 3GPP and 3GPP2 to describe 3G architectural components and interfaces is given in Table 4.7.

Table 4.7 Cross-reference between 3GPP and 3GPP2 Terminology

Component Description	3GPP	3GPP2
Radio Interface between UTRAN and User Equipment	Uu	Um
Interface between AP (BTS) and RNC (BSC)	Iub	Abis
Interface between two RNCs (BSCs)	Iur	A3 and A7
Interface between RNC (BSC) and CN	Iu	A
Radio Network Controller (controls multiple BTSs)	RNC	BSC
Access Point (communicates with multiple users)	AP	Node B
Subsequent Releases of mobile specifications	Release 99, 001, 002	Release A, B, C
Future Releases	Release n	Release x

Any transfer of information across UMTS components and interfaces requires support for common, standardized, layered communications stacks. One of the scenarios used is based on the Internet UDP Transport Layer and IP Network Layer protocols. The User Equipment mobile terminal and the Access Point part of the Radio Network Subsystem communicate across a W-CDMA radio link using specific Medium Access Control (MAC) and Radio Link Control (RLC) fields which represent Data Link (layer 2) subcomponents.

GPRS Tunneling Protocols (GTP) are used to tunnel the whole protocol stacks across RNSs and core networks. Two distinct sessions are used, one across Iu interfaces between RNS and SGSN, the other across Gn interfaces within the core network. The use of GTP is depicted in Figure 4.17. APs and RNCs are not represented in this diagram.

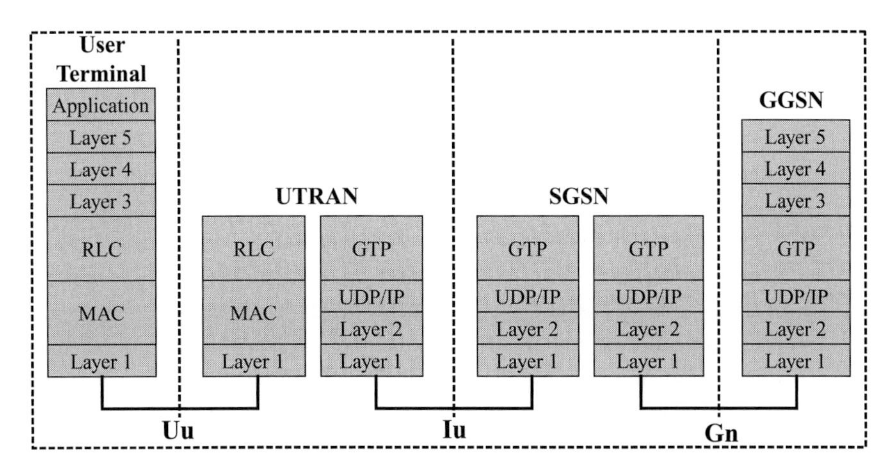

Figure 4.17 UMTS Layered Communication Stacks for Data Communications

Other scenarios are based on IP and Multi-Protocol Label Switching (MPLS). User packets are encapsulated in UTRAN into GTP frames and forwarded to the CN over a transport network that can be an all IP network or an ATM network.

4.13 Mobile Internet Protocol

The Mobile Internet Protocol (Mobile IP or m-IP) is a standardized variation of IP Network Layer protocol that allows mobile terminals to move across networks while still maintaining the same IP address. Mobile IP was standardized as part of the RFC 3344 documents that obsolete RFC 3220 and RFC 2002.

To better understand the mechanism and the specific framing used in m-IP, a short review of classic IPV4, Internet Protocol (IP) version 4, is necessary. This is shown in Figure 4.18.

<table>
<tr><td colspan="8">1 2 3 4 5 6 7 8 1 2 3 4 5 6 7 8 1 2 3 4 5 6 7 8 1 2 3 4 5 6 7 8</td></tr>
<tr><td>Version
4 bits</td><td>IHL
4 bits</td><td colspan="2">Type of service (TOS)
8 bits</td><td colspan="4">Total length
16 bits</td></tr>
<tr><td colspan="4">Identification (unequivocally)
16 bits</td><td>flag
3 bits</td><td colspan="3">Fragment offset
13 bits</td></tr>
<tr><td colspan="2">Time to live
8 bits (# of hops)</td><td colspan="2">Transport Protocol
8 bits</td><td colspan="4">Header Checksum
16 bits</td></tr>
<tr><td colspan="8">Source address
32 bits</td></tr>
<tr><td colspan="8">Destination address
32 bits</td></tr>
<tr><td colspan="3">Options
variable (less than 32 bits)</td><td colspan="5">Padding
variable (together with Options 32 bits)</td></tr>
<tr><td colspan="8">User Data
variable (depends on the size of 'Maximum Transmit Unit' accepted for a medium)
data field and header maximum 65,535 octets</td></tr>
</table>

Figure 4.18 Internet Protocol (IP) Data Unit Structure

An explanation of the significance and functionality of the IP protocol fields follows:

- **Version of the protocol**: IPv4 or IPv6;
- **Internet Header Length (IHL)**: given in 32 bit words (the minimum is 5 i.e. a header of 20 octets);
- **Type of service**: indicates Quality of Service (QOS) (reliability, throughput, precedence, and delay);
- **Total length**: of the data unit including the header (in octets);
- **Identifier**: assigned by the originator, which allows reassembling of messages along with source and destination addresses, SA and DA;
- **Flags**: contains control bits that indicate if the message was fragmented;
- **Fragment offset**: indicates the position of the fragment in the datagram;

- **Time-to-live**: indicates the maximum number of hops that can be involved in routing the packet;
- **Transport Protocol**: field identifies the type of transport protocol TCP, UDP, ICMP which uses the IP;
- **Header checksum**: is ones's complement arithmetic sum computed on the header;
- **Source Address (SA)**: is the IP address of the originator, 4 bytes;
- **Destination Address (DA)**: is the IP address of the target destination;
- **Options field**: represent a collection of options requested by the sender;
- **Padding**: is a series of bits addition to assure that the option field is filled up to 32 bits;
- **User Data**: contains a multiple of 8 bits with a length determined by the Maximum Transmission Unit that is 65,535 bytes.

One of the Internet design assumptions is that devices remain fixed and attached to one subnet that is part of a larger network. Data packets are routed through the network based on the IP addresses that are allocated to each device. Therefore, the source and destination of those packets are uniquely identified. When a device is moved from one network to another network, a new IP address is acquired for that device. A common mechanism to allocate a new address, which may only be a temporary one, is the use of Dynamic Host Control Protocol (DHCP). IP address allocation depends on the organization hosting that device or on the Internet Service Provider (ISP) that provides network service access. The IP address of the device doesn't change as long as it remains connected to that network.

With the movement of a mobile terminal on TCP/IP-based networks there is the need to constantly change IP addresses which in turn implies TCP connections will be broken and new ones established. To avoid such a situation, a mechanism is needed that allows retention of the initial IP addresses while the terminal is roaming across multiple networks. This mechanism is called Mobile IP. Mobile IP datagram flow is presented in Figure 4.19.

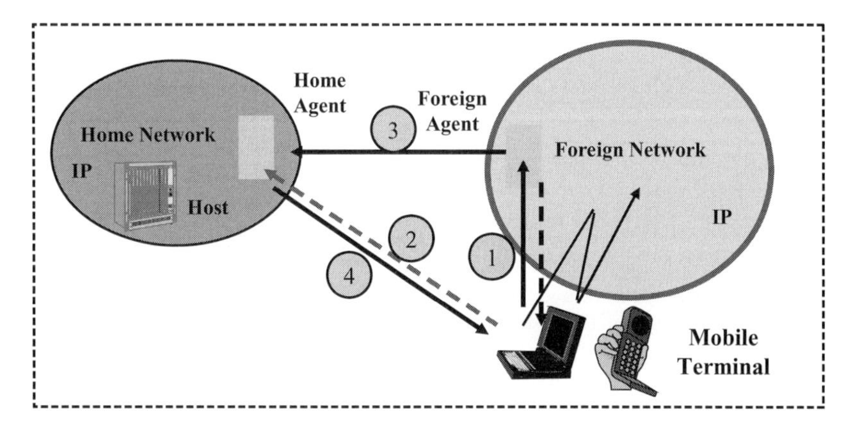

Figure 4.19 Mobile IP Datagram Flow

In the normal situation, when a mobile terminal communicates with the host from the home network, it will send and receive packets using a "permanently" allocated IP address. When the mobile terminal moves away from the home network into a foreign one, it will register with the foreign agent (1) and will acquire a temporary Care-of-Address. It will then let the home agent know this new address (2). Packets will be sent to the home host through a foreign agent (3). When packets are received in the home network, intended for the mobile terminal, the home agent will send them to the mobile terminal via the foreign agent (4). If the mobile terminal moves rapidly, it may have several Care-of Addresses at the same time.

When packets are received at the home address, the home agent forwards them to the foreign agent using the care-of-address and the foreign agent passes this information to the mobile terminal. To hide the home address, packets sent between the home and foreign agents containing the home address will be encapsulated by adding a new IP header, while preserving the home, permanent IP address, and will be tunneled through. At the end of the tunnel, the packets are decapsulated and delivered to the nomadic terminal. Mobile agents (home or foreign) constantly advertise their services. Another way to identify foreign mobile agents is through discovery processes. The registration of a mobile terminal with the home agent is done prior to roaming. An example of the IP tunneled PDU Header for registration request is given in Figure 4.20 [13].

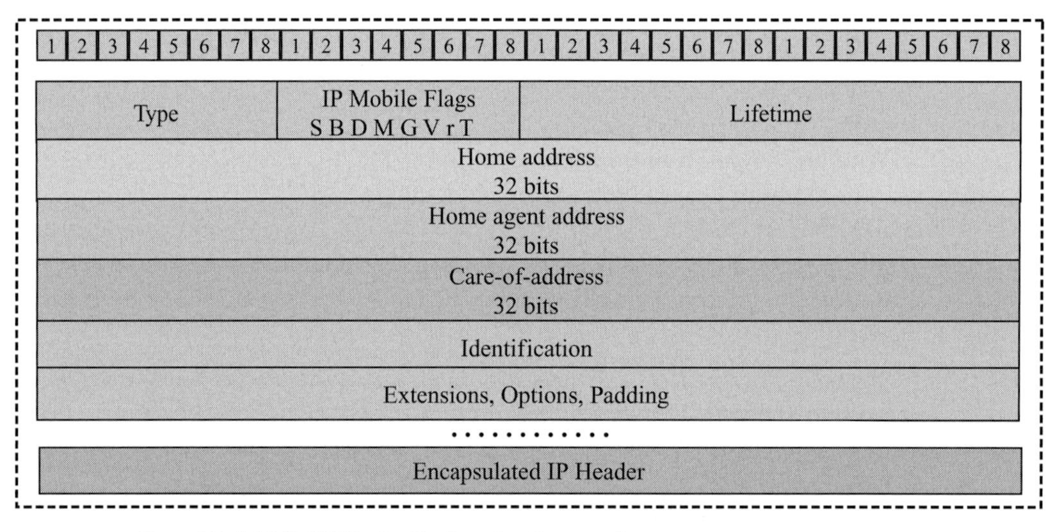

Figure 4.20 Mobile IP Header Registration Request Format

The mobile terminal initiates the registration process after it has acquired a care-of-address from the foreign agent through a discovery process or, has obtained it from the DHCP. The UDP Transport Layer is used (port 434.) An explanation of the significance and functionality of the Mobile IP Header format and the individual fields used for a registration request follows:

- **Type**: of IP mobile message (registration request, reply, reverse tunneling, etc);
- **Flags**: contain control bits related to the specifics of Mobile IP messages;

  - **R**: -Registration is required with a foreign agent;
  - **B**: -Busy. The foreign agent will not accept registrations from additional nodes;
  - **D**: -Collocation. It indicates that the mobile node is collocated with its care-of-address agent for the purpose of multicast and broadcast delivery;
  - **M**: -Minimal encapsulation. This agent implements receiving tunneled datagrams that use minimal encapsulation (RFC 2004);
  - **G**: -GRE encapsulation. This agent implements receiving tunneled datagrams that use GRE encapsulation (RFC 1701);
  - **V**: -Van Jacobson header compression is desired;
  - **R**: -Reserved. Sent as zero; ignored at reception. SHOULD NOT be allocated for any other uses;
  - **T**: -Foreign agent supports reverse tunneling.
- **Lifetime**: of registration measured in seconds. It indicates the longest time the agent is willing to accept any registration request;
- **Home address**: indicates the IP address assigned for an extended period of time to a mobile node. It remains unchanged regardless of Internet attachment;
- **Home agent address**: indicates the IP address of the home agent, part of the home network that facilitates the delivery of datagrams to the mobile terminal;
- **Care-of-address**: indicates the termination point of a tunnel toward a mobile node. It can be the foreign agent IP address with which the mobile node is registered or a collocated care-of-address which has been externally obtained (DHCP);
- **Extensions, Options, Padding**: represents a collection of fields containing M-IP header extensions, options requested by the sender, and a series of bits added to assure that the datagram is filled up to an even 32 bits;
- **Encapsulated IP Header**: represents the original IP header followed by the original packet.

Mobile IP is suitable for use across both homogeneous (inside cellular networks or WLANs) and heterogeneous convergent networks. This means there is transparent mobility between cellular networks and other fixed wireless networks. In all the schemes to communicate across networks from a mobile terminal, there is an authentication process needed to verify the identity of the originator of messages. These processes were not detailed in this limited introduction of mobile IP protocol specifications.

4.14 Signaling in Cellular Mobile Communications

Global telecommunications across multiple carriers, countries, and continents would not be possible without a standardized Common Channel Signaling (CCS) system. CCS and its companion protocols, Signaling System Number 7 (SS7 or SS#7), represent a conceptual architecture and a set of components and interfaces that constitute a real packet data network connecting all types of wired networks. It is the network that enables call set up and termination and delivery of many advanced features and services. It is a dedicated, separate network that does not carry the actual voice calls or data traffic.

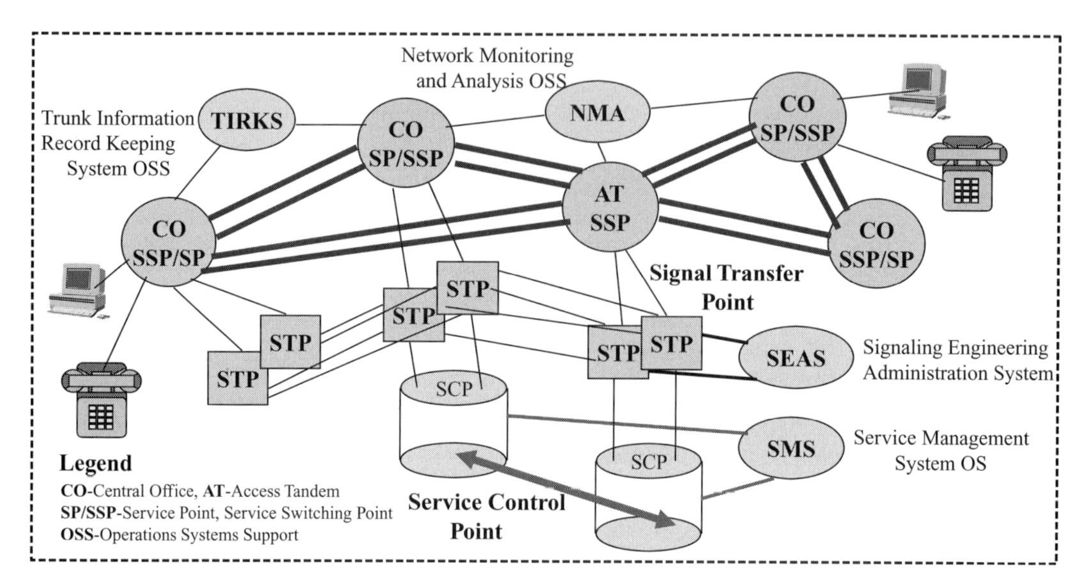

Figure 4.21 Common Channel Signaling SS7 Network Architecture

The architecture of the CCS SS7 network as applied to the wired telecommunications networks is presented in Figure 4.21.

The CCS SS7 network consists of three major components: Signaling Point/Signaling Switching Point (SP/SSP), Signal Transfer Point (STP), and Service Control Point (SCP). SSPs are central office digital switches that originate and terminate the calls once a request for a call is received from the subscriber via the local loop access network. Most SSPs are actual Points of Presence (POP) or service points to the subscribers. Some of them act only as Access Tandem (AT) offices interconnecting central offices without providing services to the subscribers. STPs are hardware/software packet switches that act as relays/routers, passing signaling messages/requests along signaling links to set up and terminate calls between originating and destination SSPs. SCPs are centralized databases that contain customer-specific information that is needed to route calls. This information is also used to validate advanced services such as 800/900 services, local number portability, and Custom Local Area Signaling Services (CLASS) such as caller ID, call forwarding, and automatic call back. Both STP and SCP are connected in redundant configurations, physically separated, given their vital role in handling the routing of calls.

CCS SS7 is a network by itself, and it is managed using specialized Operations Systems Support (OSS) applications such as Service Management System (SMS) and Signaling Engineering Administration System (SEAS). Additional management systems such as Trunk Information Record Keeping System (TIRKS) and Network Monitoring and Analysis (NMA) OSS, are used to monitor and control the central office equipment and the inter-office trunking facilities.

In the first chapter, we emphasized the tight relationship that exists between wired and wireless networks. The wired PSTN represents a giant conduit that connects various cellular mobile entities and connects wired subscribers to wireless subscribers and vice versa. Therefore, to provide global wireless-wired communications, cellular mobiles should use signaling protocols that transparently interconnect with the wired signaling network. This collection of protocols, and the protocol stack that is used, is known as the Signaling System # 7.

The SS7 protocol stack consists of four layers. The first three, 1 (Physical), 2 (Data Link), and 3 (Network), are collectively known as Message Transfer Parts (MTP1, 2, 3). They have layer functionality similar to the first three layers of the OSI protocol stack. SS#7 Layer 4 corresponds to layers 4 through 7 of the OSI stack. It consists of the Signaling Connection and Control Part (SCCP) and various user parts such as the Telephone User Part (TUP), ISDN User Part (ISUP), Transaction Capabilities User Part (TCAP), Intelligent Network Application Part (INAP), and Mobile Application Part (MAP). MAP was developed to provide signaling into cellular radio networks. The SS#7 protocol stack is presented in Figure 4.22.

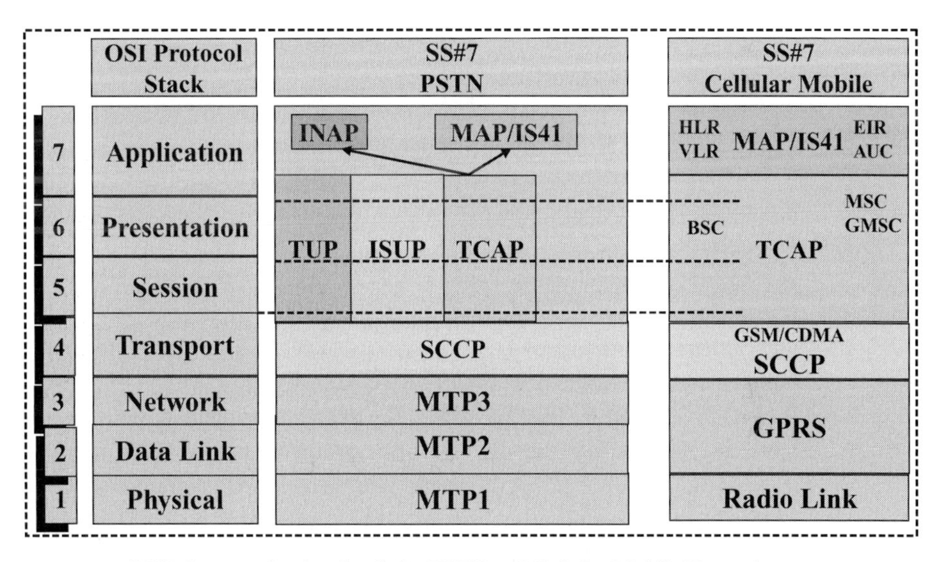

Figure 4.22 SS#7 Communication Stack for PSTN and Cellular Mobile Networks

The functions of the Message Transfer Parts (MTP1/2/3) include those in the Physical radio link, Data Link Layer error detection and correction, and Network Layer message handling and routing. These functions are typical for GPRS subnetworks that provide data service connectivity to the Internet. SCCP is a sub-part of SS#7 Layer 4. It provides end-to-end addressing and resolves the differences in approach found in networks based on GSM and CDMA. Both of these networks use TCAP to create database queries and to invoke advanced network functionality. They also provide links to intelligent networks (INAP) and mobile services (MAP/IS41). TCAP is used for SS7 signaling by Base Station Controllers (BSC), Mobile Switching Centers (MSC), and Gateway Mobile Switching Centers (GMSC). These components of mobile networks send queries to the Home Location Register (HLR), Visiting Location Register (VLR), and Equipment Identity Register (EIR) databases. Messaging is also sent to the Authentication Center (AUC).

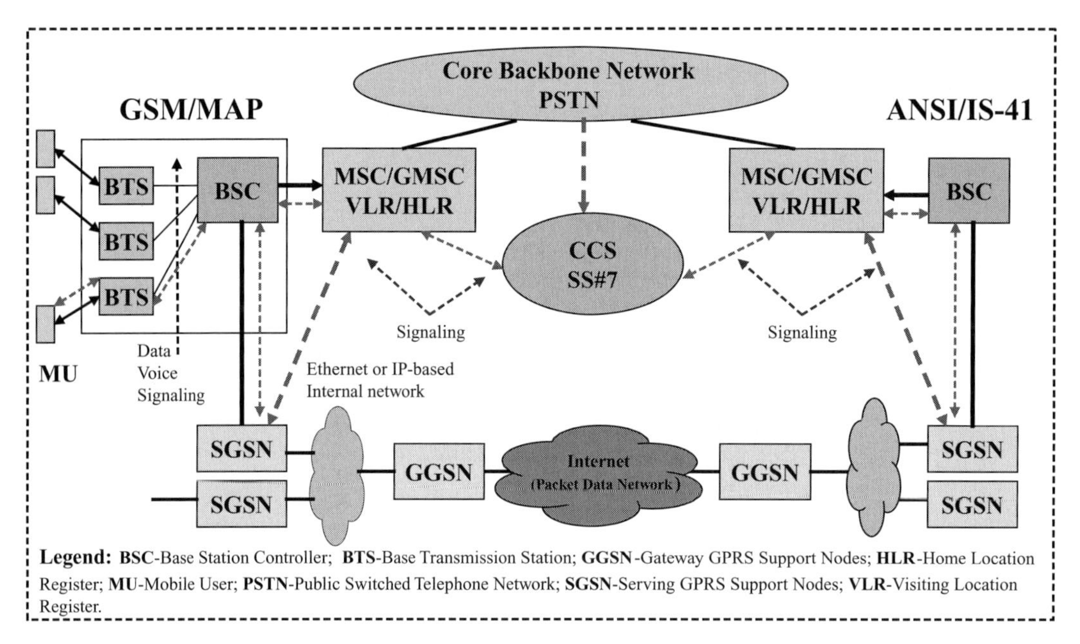

Figure 4.23 Cellular Mobile and PSTN Signaling Network

Signaling carried by the SS#7 protocols in and across wired and wireless networks is shown in Figure 4.23.

The **Mobile Application Part** (**MAP**) is a GSM protocol based on the SS#7 stack. It provides queries and messaging to handle calls, to manage users' mobility, to check user authentication, to provide user services such as SMS, and for general operation and maintenance. Messages are carried by TCAP. MAP is based on ETSI/3GGP standards (GSM UMTS Release 99 TS29.002). MAP is implemented by GSM and 3G WCDMA mobile systems. Use of standard signaling protocols such as Session Initiation Protocol (SIP) in 3G and B3G mobile networks will be analyzed in subsequent chapters.

IS-41 is a SS#7 application layer protocol standardized by ANSI (ANSI-41). It is used in conjunction with TCAP and provides similar functions to MAP. IS-41 is the standard used by AMPS (analog), IS-136 (TDMA) and CDMA networks.

4.15 Leading Cellular Smartphone Technical Specifications

Up to this point, discussion about handsets or cell phones has been very limited. They have only been presented as small boxes capable of wireless communications and we called them, in generic terms, mobile terminals. The first devices considered as mobile units would be unrecognizable today as such because of the drastic changes that have taken place in their physical appearance and functional capabilities. Mobile terminals are no longer just dumb handsets, they are in every sense intelligent devices. The discussion that follows is oriented to cellular mobile phones as opposed to cordless phones, satellite phones, or two-way walkie-talkies.

The real transformation started with the second generation of mobile communications. At that point, mobile phones became slim hand-held devices, weighing less than 100 g. They used smaller but longer life batteries, more efficient electronics, and require less power because of higher cell density. Today, 3G hand sets provide Internet connectivity, text messaging capabilities, they can download and play music, can take and send digital pictures, serve as personal organizers, and even downstream live radio and video broadcasts. There is another generation of convergent handsets coming that have incorporated Wireless USB, Bluetooth, Wi-Fi, and even WiMAX capabilities. They will be discussed in subsequent chapters.

Some mobile operators provide prepaid services where one can purchase services in advance and therefore avoid monthly bills. Other mobile services are based on monthly payments that reflect the initial contractual agreement and actual usage in the number of calls made and text messages sent. Some vendors lock the cell phones so they cannot be used with other mobile carriers. Some operators subsidize the cost of handsets to retain subscribers under contract for a certain amount of time (usually two years).

Criteria used for evaluation of mobile handsets can be divided into the following groups:

- Multiband connectivity: Dual band, tri band, four bands, or even five working frequency bands;
- Basic computing platform: Operating system, processor, keyboard, memory, multitasking capabilities;
- Power: Maximum transmission power;
- Battery: Removable lithium, rechargable, hours of operation, lifetime;
- Appearance: Standard front display, data card type, clam shell double display PDA type;
- Convenience: Speakerphone, hands free speakers and voice recorder;
- Multimode connectivity: Bluetooth, Wi-Fi, WiMAX, UWB;
- Internet connectivity: Internet access, maximum downlink data rate, maximum uplink data rate;
- Application access: Web browsing, email, Instant Messenger, Skype, scheduler;
- Multimedia capabilities: Web camera, music download and storage, music and video streaming, Web cast;
- Special features: iPod combination, GPS, navigation system;
- Source of interference to other systems: Home/car stero systems, television;
- Locking systems: Vendor controlled, SIM card.

Table 4.8 provides a synopsis of technical specifications of four leading handset: Nokia 9300 in two models, Samsung i730, and Treo 650 used by Cingular, Verizon, and Sprint subscribers.

When it comes to the usage of mobile phones, the highest penetration is 170% of the total population in Luxembourg followed by Hong Kong with a 120% rate. In 2005 the total number of mobile users exceeded 2.14 billion and there are expected to be 3.2 billion in 2010. As of 2006, 80% of world's population had cellular mobile coverage with 90% expected in 2010.

Table 4.8 Leading Cellular Smart Handset Technical Specifications

Features	Nokia 9300 (Cingular)	BlackBerry (Cingular)	Samsung i730 (Verizon)	Treo 650 (Sprint)
Cellular connectivity	GSM 0.85, 1.8, 1.9 GHz EDGE/GPRS class 10	1xRTT; 850 MHz, 1.9 GHz GSM 0.85, 0.9, 1.8, 1.9 GHz GPRS class 8	1xRTT, EV-DO: 850 MHz, 1.9 GHz	1xRTT; 0.8, 1.9 GHz GSM 0.85, 0.9, 1.8, 1.9 GHz EDGE/GPRS class 10
Wi-Fi capability	No	No	Yes	No
Bluetooth capability	Yes	Yes	Yes	Yes
Infrared capability	Yes	No	Yes	Yes
Operating system	Symbian 7.0, Nokia 80	RIM proprietary	Windows Mobile 2.0	Palm OS 5.4 Garnet
Keyboard	Clamshell design interior keyboard	Small keys on the front of device	Slider design exposes keyboard small keys	Clamshell design interior keyboard
Processor	ARM 9 150 MHz	Not disclosed	Intel PXA 272, 520 MHz	Intel PXA 272, 520 MHz
Multitasking	Yes	Yes	Yes	No
Memory	80 MB user data/application	32 MB flash memory	86 MB user data/application	23 MB user data
Volatile memory	No	No	Yes	No
Battery	Removable Li-ion	Removable Li-ion	Removable Li-ion	Removable Li-ion
Front display	128x128	240x160	240x320	320x320
Camera	No	No	No	Yes
Speakerphone	Yes	No	Yes	Yes
E-mail client	Internet TCP/IP, Blackberry client	Internet TCP/IP, Exchange Lotus Notes, Domino	Internet TCP/IP, ActiveSync	Internet TCP/IP, ActiveSync

Major mobile phone manufacturers include Apple, Fujitsu, Kyocera, 3G, LG, Motorola, NEC, Nokia, Panasonic, Philips, Samsung, Sanyo, Sharp, Siemens, Sony Ericsson, T&A Alcatel, and Toshiba. Currently, Nokia is the world largest manufacturer of mobile phones with a global market share of 36% (as of Q4 of 2006). Each company has a variety of models in shapes, forms, and functionality covering the 2G, 2.5G, and 3G UMTS generational stages. The sets work in one or multiple frequency bands with various embedded features.

A sample of this variety is offered by Nokia announcements that include models, as of 4Q 2006, such as Nokia 6300, a thin phone with appealing design; the Nokia 6290, a 3G model with a number of practical new features and Quick Cover access keys; and the Nokia 6086 model, a quad-band GSM and UMA-enabled camera phone. A joint Nokia's announcement with T-Mobile refers to the sleek and powerful Nokia 6133 handset with a 1.3 Megapixel camera and an integrated music player. Other recent announcements include "Active" phones with the Nokia 5500 Sport Music Edition and Nokia 2626, a colorful mobile phone designed for style-conscious consumers. There were also the announcements of two CDMA entry models, the Nokia 1325 and Nokia 1265, which feature a number of desirable features such as hands free speakers and voice recorders. Regarding special deals, a recent announcement from Nokia and Yahoo on the extension of their partnership, offers Yahoo branded services, including Yahoo! Mail and Messenger on Nokia's wide range of mobile phones operating on the Series 40 platform. Table 4.9 displays the results of some high-level performance testing of a few handsets.

Table 4.9 Leading Cellular Smart Handset Performance Test Results

Smart Handsets Metrics	Nokia 9300	RIM 7290	Samsung i730	Samsung i730	Samsung i730	Treo 650
Network used	Cingular	T-Mobile	Verizon	Verizon	Hotspot	Sprint
Data service used	**EDGE**	**GPRS**	**CDMA 2000 1xEV-DO**	**CDMA 2000 1xRTT**	**Wi-Fi**	**CDMA2000 1xRTT**
Peak user throughput (kbps)	187	38	360	112	764	116
Ability to operate as mobile modem	Yes	No	No	No	No	Yes
Browser ranking based on speed	2nd	4th	1st	Not tested	Not tested	3rd
Minutes to type a test sentence	1.8	1.7	2.5	N/A	N/A	2.0

Legend: GPRS-General Packet Radio Service; **CDMA 2000 1xRTT**-Code Division Multiple Access One Carrier Radio Transmission Technology; **EDGE**-Enhanced Data for GSM Evolution; **CDMA 1xEV-DO**-CDMA 1x Enhanced Version, Data Only. ***Note***: A test sentence is used to qualify the ease of typing.

Among the issues facing subscribers and operators regarding the use of handsets:

- Transmission power in the handset is limited to a maximum of 2 watts in GSM 850/900 and 1 watt in GSM1800/1900;
- Overweight handsets with poor battery life;
- Problems with handover from UMTS to GSM and vice versa;
- Initially, poor coverage due to the time it takes to build a network;
- Narrow spectrum available that limits the number of simultaneous users.

5 Cellular Mobile Radio Networks Management and Services

5.1 Cellular Mobile Network Management Services

In Chapter 2 we discussed network management concepts in general terms. We introduced a network and systems management model, looked at the design of management platforms, and presented key milestones in management systems evolution. Two management interfaces with associated protocols and services were analyzed: Internet SNMP-based management and OSI CMIP/CMISE-based management. The concept of layered management and systems management functional decomposition, as promoted through the TMN architecture, was also introduced.

In many ways, wireless networks resemble wired networks, since both contain grand management domains such as the networks themselves, systems, services, policies, and protocols. However, there are differences given the nature of cellular mobile communications. One major difference comes from the use of wireless links that connect mobile terminals to the network. These links are vulnerable to environmental conditions and to eavesdropping. Another difference is the mobility of the users. Mobility requires continuous tracking of users and reallocation of network and systems resources, i.e., network management. Clearly, this is not a requirement of fixed, wired networks. Mobile management services can be grouped following the individual components subject of management, as follows:

- **Wireless Networks Resource Management**
  - MS, BTS, BSC, MSC, GMSC, SGSN, GGSN;
  - WiMAX SS, WiMAX BS, AP, OT, RFID Reader;
  - HES, PES, Satellite Tracking Systems.
- **Wireless Systems Resource Management**
  - Applications;
  - Storage systems.
- **Mobility Management**
  - Registration, Authentication;
  - Hand-off;
  - Roaming.
- **Mobile Call Management**
  - Call Origination, Call Destination;

  - Call Delivery, Call Blocking;
  - Signaling System.
- **Customer Profile Management**
  - HLR Management;
  - VLR Management.
- **Account Management**
  - Billing;
  - Tariffs;
  - Compensation.
- **Security Management**
  - Encryption;
  - Secure Hash Algorithms;
  - Wi-Fi Protected Access (WPA1 and WPA2);
  - Wired Equivalency Privacy (WEP).

5.2 Mobile Networks Management Requirements

A set of high-level network and systems management requirements, applicable to wireless cellular mobile communications networks, is summarized as follows:

- Provide remote access to and configuration of mobile terminals;
- Monitor and control mobile networks and systems components;
- Report management-related information, backup and restore lost information;
- Secure data transfer, user access and management operations;
- Provide real-time management and management automation.

It is highly desirable to design and use open management systems characterized by the following management requirements:

- Open, standard management interfaces and protocols;
- Dedicated mobile networks Management Information Base (MIB);
- Modular, management platform-based, architectural components;
- Scalable management for large mobile networks;
- Support for common graphical user interfaces;
- Interoperability with other management systems;
- Open, standard Applications Programming Interfaces;
- Management software applications development tools.

5.3 Cellular Mobile Networks Service Providers

There is a continuous shifting in market share, acquisitions, and consolidation in the cellular telephony industry, both among the manufacturers and the major wireless telephone service providers. As the number of mobile subscribers heads towards the 3 billion landmark, markets in Europe, USA, Japan, South Korea, and some other Asian pockets, will tend to have a slower growth. This will be in contrast with still fast growing markets such as some African countries, China, and South America. A snapshot of the USA market, as of late 2007, indicated that the total number of subscribers surpassed the 250 million mark. Of this number, the top four carriers had a combined 211 million users. The list of major US mobile operators along with some features of their operations is presented in Table 5.1.

Table 5.1 Cellular Mobile Service Providers in the USA

Company	US Subscribers (voice) (millions)	Churns (%)	US Coverage (%)	2G Access	2.5G (late 2002)	3G (2004/2006)
AT&T Mobility (www.cingular.com)	65.7 million (September 2007)	2.6%	98%	GSM D-AMPS	GPRS/ EDGE	EDGE UMTS
Verizon Wireless (www.verizonwireless.com)	63.7 million (October 2007)	2.2%	95%	CDMA (IS-95a)	CDMA 1xRTT	**CDMA 1xEV-DO (2005)**
Sprint Nextel (www.sprintpcs.com) (www.nextel.com)	55 million (August 2007)	2.7%	90%	CDMA (IS-95a) iDEN	CDMA 1xRTT	**CDMA 1xEV-DV (2006)**
T-Mobile (www.tmobile.com)	27 million (August 2007)	3.1%	85%	GSM	GPRS EDGE	UMTS HSDPA

Major USA Carriers: Note: All these companies are the results of many more acquisitions of regional carriers and consolidations.

- **AT&T Mobility** (Joint venture SBC Communications (formerly Cellular One) and Bell South Mobile; later acquired AT&T Wireless; solely owned by AT&T Mobility after SBC changed its name to AT&T);
- **Verizon Wireless** (Nynex Mobile, GTE Mobilnet) with 45% owned by Vodaphone Group;
- **Sprint Nextel Corporation**, after they merged;
- **T-Mobile**, previously known as VoiceStream Wireless (acquired by Deutsche Telekom), has a worldwide presence.

Major Overseas Carriers:

- China Mobile (spun off from China Telecom, over 350 million subscribers);
- Vodaphone (206 million as of March 2007);
- China Unicom (over 156 million subscribers);
- Telefonica/Movistar/O2 (155 million subscribers as of September 2007)
- America Movile (Mexico, Latin America, 144 million subscribers);
- Germany T-Mobile global (113 million subscribers);
- Orange (98 million subscribers as of December 2006).

5.4 Cellular Mobile Networks Management Products

Because of the significant differences with traditional wired networks, mobile cellular networks brought new challenges to systems management. The reason is the constant mobility of handsets. However, from the list of management platforms that we presented in Chapter 2, one product stands out in its ability to satisfy many of the mobile system management requirements. This is the HP TeMIP management platform. TeMIP was originally developed by Digital Equipment Corporation (DEC), evolving from a product called EMA, Enterprise Management Architecture, and from DEC mcc. DEC was acquired by Compaq, which in turn, was acquired by HP. This explains the name HP OpenView TeMIP, since it was branded together with HP OpenView, a very popular HP mid range management platform.

Currently, TeMIP version 5.3 or above, is the telecom industry's leading management platform for very large enterprises and carrier networks. Delivered by a single vendor, it provides management across infrastructures such as wireless, wireline, IP, and IT environments. To do this, the platform uses interfaces such as Transaction Language 1 (TL1), Common Object Request Broker Architecture (CORBA), OSI CMIP/CMISE, ASCII, and SNMP [14]. It is a highly scalable, open management platform that connects through access modules to more than 200 standard telecom and computing interfaces and to numerous third-party applications. It was adopted by over 160 telecom operators. Eight out of the10 largest carriers use TeMIP for their network management. To understand the complexity of this management platform, a high-level architecture of TeMIP is provided in Figure 5.1.

What makes TeMIP a management platform is the development environment it supports. It consists of Framework, OSI management, ASCII, and TL1 toolkits. The toolkits allow development of management applications based on standard interfaces or on TeMIP platform service APIs. The Graphical User Interface (GUI) is based on OSF/Motif and X Window Systems specifications and gives users the ability to get a common look at all managed resources. The **Presentation Modules** (PM) include an Icon Map and a Command Line Interface (CLI). Several **Functional Modules** (FM) provide Event Logging, Alarm Handling, and Reports Generation (trouble ticketing system) through Distributed Notification modules. The Performance Analyzer module collects statistics from TCP/IP Internet and DECnet nodes using RMON probes. Statistics collected include throughput, counts of various counts parameters, average values, and utilization data. The information is collected through a variety of **Access Modules** (AM) based on standard interfaces such as OSI CMIP/CMISE and SNMP, legacy interfaces such as TL1, and proprietary interfaces. Communication services with OSI agents, SNMP agents, and various Network Elements are provided through various communications stacks, both standard and proprietary.

At the center of the platform, there is a **Core Executive**, an Object Request Broker (ORB) middleware aligned with OSF CORBA specifications, called Information Manager. An object-oriented database serves as the Management Information Repository. Framework Security Services provide access control, filtering, and users/operators

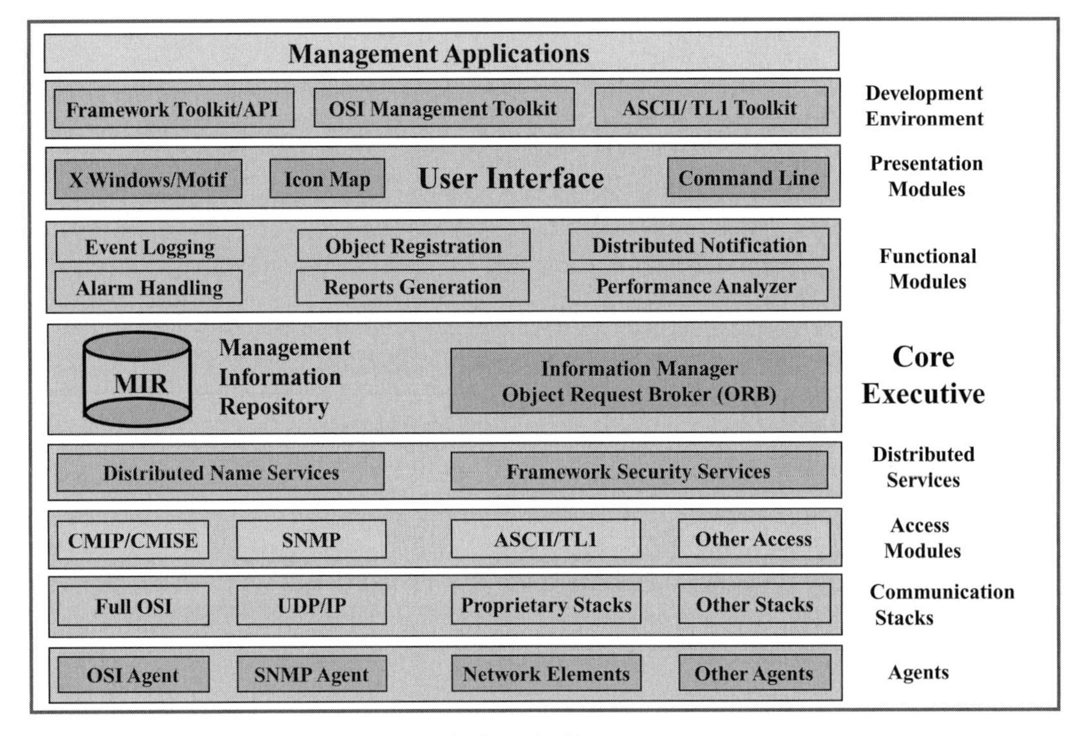

Figure 5.1 TeMIP Management Platform Architecture

profiling along with security APIs. The TeMIP platform can be used in a centralized architecture, either as a single manager, manager of managers, or in a distributed architecture, that has multiple cooperating managers, distributed applications, and Distributed Name Services.

During the development of TeMIP, which spreads over a decade, a **TeMIP Expert** module was added. It acts as an event correlation system to establish the root-cause of critical information and can aid in reducing the number of alarms handled by operators by 85%. It is a rule-based software application that can prioritize and automate management operations, pointing to recommended solutions, and even proactively identify and prevent network problems. TeMIP Expert augments the powerful **TeMIP Fault Management** modules that provide event logging, filtering, and alarm handling in tandem with trouble ticketing, service activation, and inventory applications. The **TeMIP Fault Statistics** module allows creation of statistical reports and data analysis to identify fault patterns. Of the TeMIP management platforms that have been developed, we describe two of them.

The first was implemented by Vodaphone/Mannesmann mobile communication. It consists of a **Network Management Center** (NMC) and eight regional **Operations Management Centers** (OMC), each equipped with full blown TeMIP management platforms. This implementation is shown in Figure 5.2.

Communications between the NMC and OMC is based on the TMN Q3 and OSI CMIP/CMISE standard interfaces. Only critical events such as alarms are elevated to the NMC level along with initial configuration information. Each regional OMC acts

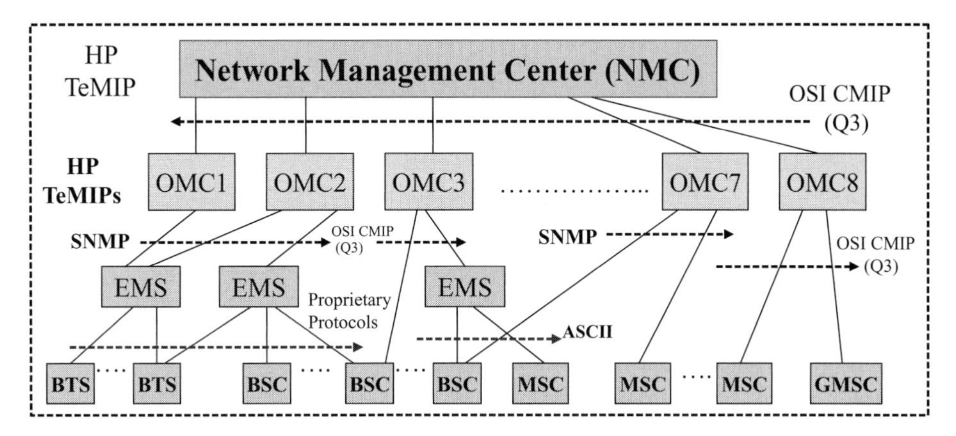

Figure 5.2 Vodaphone/Mannesmann Mobile Network TeMIP-based TMN

as centralized management system responsible for managing a portion of the mobile cellular network.

Management is performed by multiple Siemens and Ericsson Element Management Systems (EMSs). In 2003, these systems managed 73 MSCs, 111 BSCs, and 8324 BTSs, supporting over 20 million subscribers. A mixture of TMN Q3 OSI CMIP-based, SNMP, TL1/ASCII-based, and proprietary interfaces are supported on the communication links between OMCs and EMSs, and between EMSs and Network Elements. The architecture allows each OMC to take over the management of some adjacent EMSs when major failures occur. Given the size of the network, multiple operators, at each layer of manger of managers topology, are required.

The second platform, the **Siemens Mobile Integrator**, is also a TMN OSI-based TeMIP management platform that consists of a Mobile Integrator acting as a Network Management Layer centralized manager. A family of Element Management Systems (EMSs) and a mixture of TeMIP and other proprietary Siemens and Ericsson EMSs are connected to TeMIP integrator through dedicated access modules (TMN Q3 OSI CMIP, SNMP, TL1, or ASCII based). This is shown in Figure 5.3.

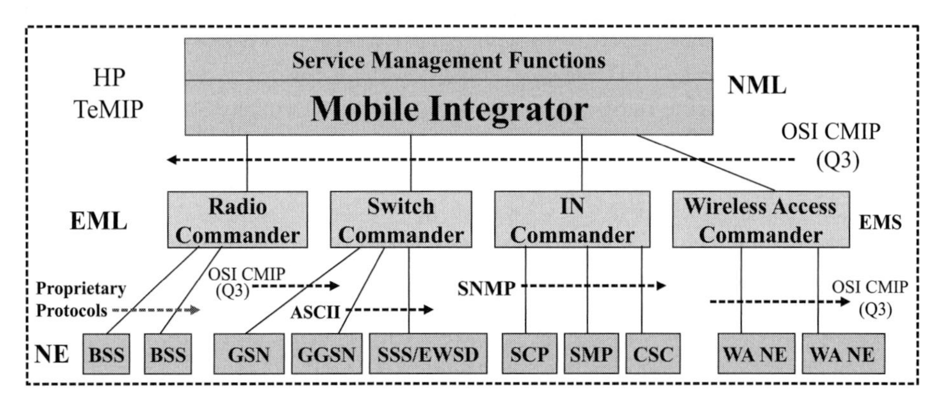

Figure 5.3 Siemens Mobile Integrator TeMIP-based Management Solution

Base Station Subsystems (BSS), General Packet Radio Service Support Nodes and Gateways (GSN and GGSN), Digital Electronic Switching Subsystem (SSS/EWSD), SCP (Service Control Point), SMP (Service Management Point), Customer Service Control (CSC), and Wireless Access (WA) components represent some of the variety of NEs and the associated interfaces supported.

5.5 Service Level Management in Mobile Networks

In Chapter 3 we presented the service management concept, in general terms. We introduced a service management model and we looked at the system management components: class of service, quality of service, and service level agreements. Also, we presented key metrics used in assessing quality of service and the general tools used to measure, and collect QOS metrics. Ultimately, we introduced the service level management concept Telecommunications Operations Map (TOM) that was promoted by the Telemanagement Forum along with a list of service level management products.

Cellular mobile communications, given the nature of mobile communications, poses a complicated matter when it comes to service management. First, because of the high mobility of the user and the continuous need to keep track of the changing transmission parameters, there will be rapid variations in the observed QOS. Second, given the considerable number of cellular mobile operators in the world (close to a thousand) there are serious technical challenges in the delivery of a satisfactory QOS across heterogeneous networks and a multitude of intermediary service providers.

To discuss service management in cellular mobile wireless environments we select, as a reference, the UMTS IMT2000 architecture that was presented in Chapter 4. A simplified view of UMTS includes the Mobile Terminal (MT) user equipment, the Radio Access Network (UTRAN), and the Core Network (CN). The CN is a packet data communication network that consists of a multitude of CN Nodes connected to the external network through CN Gateways. This is shown in Figure 5.4.

The goal of service management is to provide acceptable end-to-end QOS between mobile terminals. Implementations of those UMTS specifications that were initiated back in Release 99 of UMTS, have the capability to provide good QOS on networks that span from the MT to CN gateway. The external network, Internet in most likelihood, is outside of the UMTS realm. Within UMTS QOS we have two distinct sections, the RAN QOS which is dependent on the radio link interface between MT and UTRAN, and the CN QOS that is dependent on UTRAN and CN networks. A further dissociation can be done in the CN Service QOS that includes the interface between UTRAN and the CN and the actual CN.

As indicated in Figure 5.5, a simplified service management model looks only at two distinct points, the Service Access Point (SAP) and the Network Service Access Points (NSAP). The model is applicable to any generation of mobile networks, including multi-standard mobile network environments such as the traditional GSM, CDMA, and UMTS, or even WiMAX. SAP and NAP are terms introduced in the ISO OSI Reference Model standards. Service Access Point is the interface between two adjacent communication

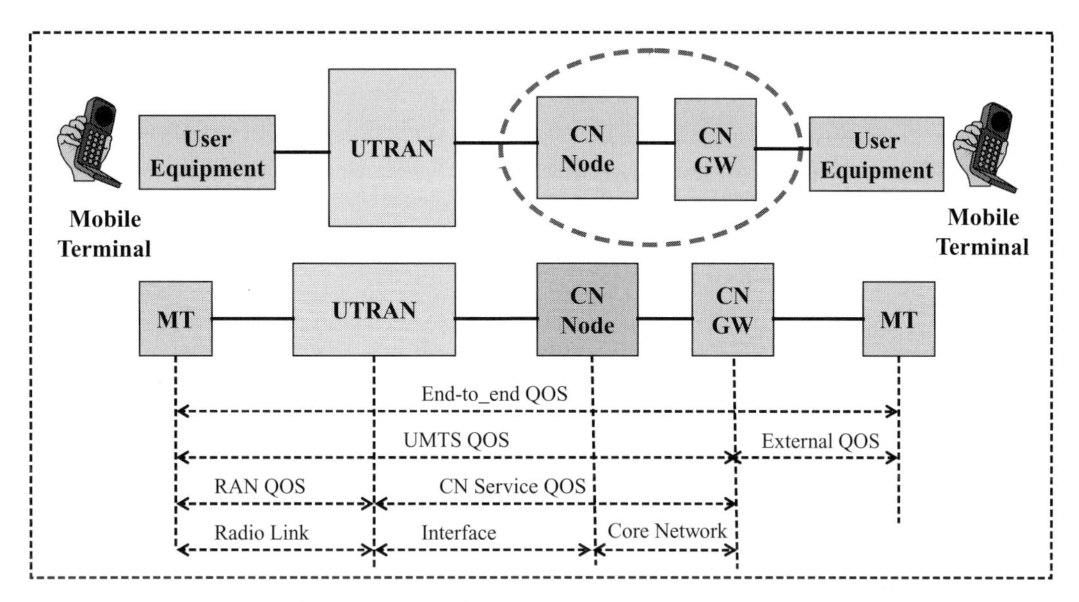

Figure 5.4 QOS in UMTS Networks

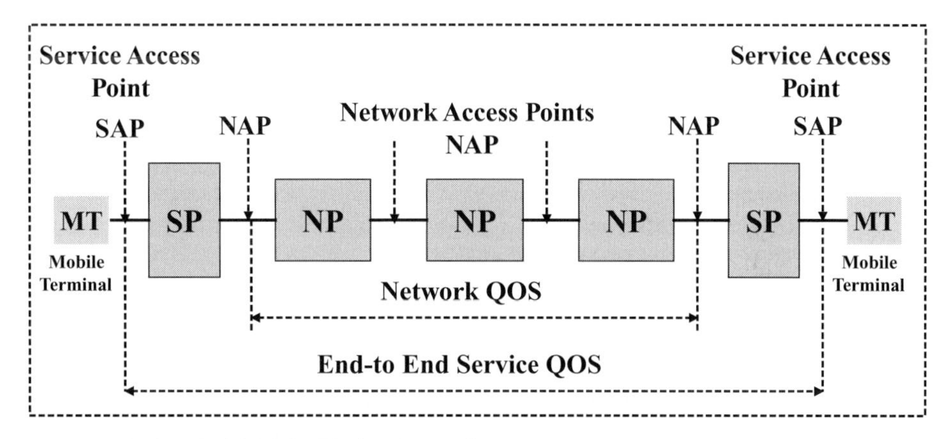

Figure 5.5 End-to-End QOS in Wireless Networks

Layer Service Entities. A system A in order to communicate with system B, at a particular layer level, should know the SAP of the remote system for that particular layer. NSAP is the service provided by the network layer to the layer above, i.e. Transport Layer.

UMTS Release 99 specifications differentiate traffic flows in mobile networks into four traffic classes according to their real-time processing needs: **conversational, streaming, interactive**, and **background or best effort** [15]. Some characteristics of these types of traffic flow are indicated in Table 5.2.

Given the characteristics of applications supported, there are some differences in the attributes used by individual classes of traffic. A list of transmission attributes and their relevance to a particular traffic class are indicated in Table 5.3.

Table 5.2 UMTS QOS Traffic Classes

Traffic Class	Basic Features	Additional Features	Example of Applications
Conversational	Preserves time relation (variation) of conversation between end-systems	Conversation pattern requires stringent low delays and jitters	Voice conversations
Streaming	Preserves time relation (variation) of streams between end systems	Moderate delays and packet loss are admitted	Video/Audio streaming
Interactive	Preserves the payload conveyed between end-systems	Requests are followed by response patterns	Web browsing
Background Best Effort	Preserves the payload conveyed between end-systems	The destination is not expecting the data conveyed	E-mail

Table 5.3 UMTS Traffic Class Attributes

Traffic Class Attributes	Conversational Class	Streaming Class	Interactive Class	Best Effort Class
Maximum peak bit rate	x	x	x	x
Guaranteed mean bit rate	x	x		
Residual bit error rate	x	x	x	x
Maximum SDU size	x	x	x	x
SDU format information	x	x		
Delivery of corrupted SDUs	x	x	x	x
SDU error ratio	x	x	x	x
In-sequence SDU delivery	x	x	x	x
Transfer delay	x	x		
Traffic handling priority		x		
Allocation/Retention priority	x	x	x	x

The meaning of most of the attributes used as headers of this table were presented earlier. Additional explanation follows: SDU (Service Data Unit) format information provides the list of available SDU sizes to be selected by a particular application; SDU error ratio indicates the percentage of SDUs lost or corrupted; Transfer delay refers to the maximum delay of 95% of the delivered SDUs; Traffic handling priority indicates the use of specific SDUs to indicate prioritization; and Allocation/Retention priority indicates the importance of allocating resources to handle a certain traffic flow.

For practical purposes, individual QOS parameters for wireless cellular mobile networks can also be organized, in accordance with transmission parameters (bit, packet, physical) or applications parameters as follows:

- **Bit Level** parameters
  - Data Rate
  - Bit Error Rate (BER)
  - Signal to Noise Ratio (SNR)
- **Packet Level** parameters
  - Throughput
  - Delay
  - Jitter
  - Packet Loss
- **Radio Cell level** parameters
  - Call blocking probability
  - Call dropping probability
- **Application Level** parameters
  - End-to-End Response Time
  - End-to-End Throughput

In this section, QOS analysis was focused exclusively on the need to provide good end-to-end data delivery QOS across UTRAN and CN networks. Equally important are the factors that contribute to the overall quality of voice services. When we refer to voice services we include here the requirement to have satisfactory voice quality, call quality, service quality, and quality of value added services. **Voice quality** is a subjective perception that is based on personal experience and judgment. It usually is described as good, satisfactory, poor or unacceptable. **Call quality** includes the quality of dial tone and the delay in setting up the call. **Service quality** includes the traffic congestion, calls interrupted, and calls dropped. Quality of value added services includes voice mail prompts, size of mail boxes, and ease of use of interactive features.

The quality of a voice call is determined by some primary factors: loudness, distortion, noise, fading, and crosstalk. **Loudness** is determined by the strength of the original signal and its amplification. **Distortion** indicates perceptible changes of the original signal beyond normal digitization. **Noise** refers to the background noise and its effect on the overall intelligibility of the speech. **Fading** indicates changes in the signal level during the call. **Crosstalk** indicates audible interference from other channels that might allow listening to separate conversations in some cases.

Voice quality can also be evaluated by using other factors such as end-to-end delay, jitter, echo, and silence suppression. **End-to-end delay** between terminals includes propagation delay and processing delay. In PSTNs, within the USA, a delay of 30–50 ms is the norm. In VoIP applications, delay can reach 150 ms. In most satellite communications, delay can reach a quarter of second. Most cumbersome for digitized voice quality is the **jitter** caused by variations in delay. Values higher than 30 ms are perceived as poor voice quality. **Echo**, in the sense of hearing your own voice repeated, indicates poor echo cancellation performance. **Silence suppression** is caused by bad timing of active voice detection mechanisms that allow premature clipping of the beginning and ending of words.

As subjective as it is, the **Mean Opinion Score** (MOS), as defined in the ITU-T P800 standard, remains the currently accepted way of "measuring" voice quality. It requires a minimum of 30 people to listen to sections of speech for 8–10 seconds under controlled conditions. Scores from 1 to 5 are attributed to these speech sequences. Toll quality voice calls typically have a MOS of 4.4–4.5. Calls below a MOS of 3.5 are considered poor to unacceptable. For ratings below 2.6, callers usually drop the calls and look for other alternatives of communications. A typical MOS for voice conversations across mobile networks is around 3.8 [16].

There are two ways to test voice quality. **Passive testing** is based on analyzing just portions of calls received. No additional traffic is generated. **Active testing** compares received calls of speeches with a copy of the original speech. Additional traffic is generated in this case that can affect overall performance [17].

5.6 GSM/GPRS Data Networking

GSM is one of the standard digital cellular networks developed in accordance with ETSI specifications. It is used by over 160 countries in Europe, Asia, Australia, Africa, and by some US mobile operators. **GSM Packet Radio Service** (GPRS) extends GSM voice-based services to allow users to transmit packet-based data and to reserve bandwidth through guaranteed classes of services. **Enhanced Data for GSM Evolution** (EDGE) goes further with a new design that provides even higher data rates. The basic components of a GPRS/EDGE network are shown in Figure 5.6.

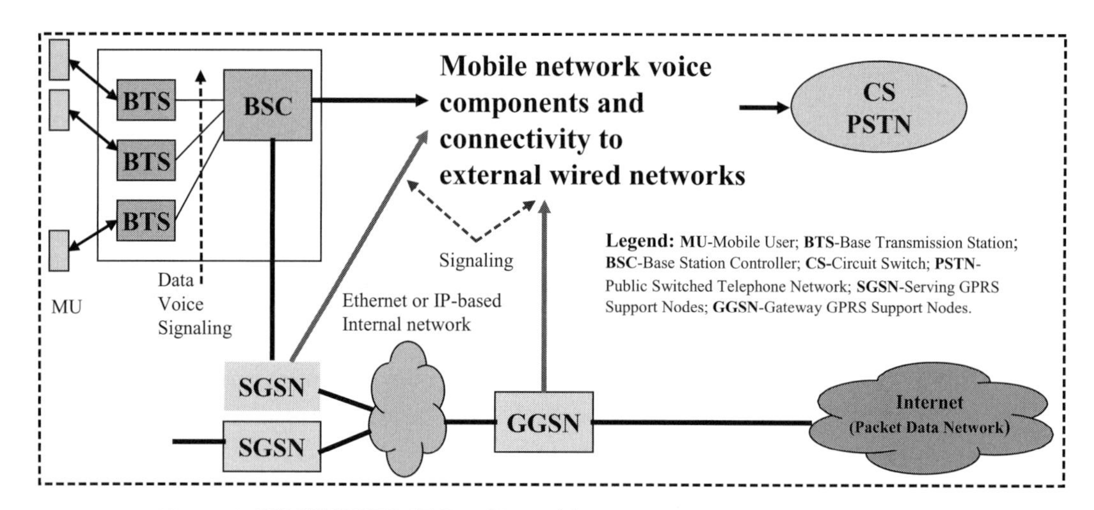

Figure 5.6 GSM/GPRS/EDGE Data Networking

Traditional GSM provided basic data access at 9600 bps over its radio scheme time slots. **High Speed Circuit Switched Data** (HSCSD) has squeezed higher data rates out of the same design by using available time slots from multiple TDMA channels. **General Packet Radio Service** (GPRS) extends the traditional GSM infrastructure by adding a special packet network. To do this, it combines up to eight GSM time slots in both the uplink and downlink directions to support a maximum data rate of 170 Kbps. GPRS is a 2.5 G technology, a major step in the evolution of GSM towards 3G. EDGE goes even further by using special techniques that can provide data bandwidths up to 384 Kbps. EDGE is considered a 2.75G technology.

GSM only provides circuit-mode connectivity while GPRS provides end-to-end packet-mode connectivity using a completely separate Internet IP-based transmission system. This means that multiple users can share the same available bandwidth. Users that send packets intermittently (email for example) may use the available bandwidth if there is no contention from other users. The transmitted packets have constant lengths corresponding to the duration of a GSM time slot. Downlink packets to the mobile units are delivered on first-come first-served basis. Uplink packets use the slotted Aloha technique which provides a reservation prior to transmission to avoid contentions.

As noted earlier, GPRS networks consist of two main components, **Serving GPRS Support Nodes** (SGSN) and **Gateway GPRS Support Nodes** (GGSN). Multiple SGSNs provide user security access functions, packet switching, and routing. GGSNs provide connectivity and communication with the Internet. Links between SGSNs and GGSNs are provided by an internal IP-based data network or Ethernet LANs. GPRS connections are set up as wireless modems with Internet connectivity.

Given the variety of USA and worldwide cellular telephony service providers, in context of the overlapping of three generations of technologies, there is a great dispersion in the overall data service theoretical performance and the expected user throughput. This dispersion starts with GPRS, goes through EDGE, and, currently, ends with UMTS-High Speed Downlink Packet Access (HSDPA). This is shown in Table 5.4.

Table 5.4 USA/Worldwide Cellular Telephony Data Service Performance

Data Technology	Generation	Peak network downlink speed	Average user throughput for file downloads	Examples of Service Providers
GPRS	2.5G	115 Kbps	30–40 Kbps	AT&T Mobility T-Mobile, Vodaphone
CDMA-1xRTT	2.5G	153 Kbps	50–70 Kbps	Sprint PCS Verizon Wireless
EDGE	2.5G-3G	473 Kbps	100–130 Kbps	AT&T Mobility T-Mobile, Vodaphone
UMTS-WCDMA	3G	384 Kbps–2 Mbps	220–320 Kbps	AT&T Mobility Orange, SK Telecom
CDMA-1xEV-DO	3G	2.4 Mbps	400–700 Kbps	Verizon Wireless
UMTS-HSPDA	3G-4G	14 Mbps	550–1100 Kbps	European Operators AT&T Mobility

Four **Coding Schemes** (CS), which are variations of **convolutional coding** with **error correction** capabilities, are used for the radio interfaces. GPRS defines several QOS profiles that are used to allow the network to prioritize traffic and to give users the desired service level.

5.7 GPRS Classes of Services

When GPRS was standardized by ETSI, four classes of **Coding Schemes**, CS-1, CS-2, CS-3, and CS-4, were defined for the system. Each of these coding schemes is based on various rates of convolutional coding plus error correction. CS-4 is the least redundant and is used when the user is in the proximity of BTS and the Signal-to-Noise Ratio (SNR) is high. Using CS-4, the maximum data rate achievable per time slot is 21.4 Kbps and the theoretical maximum data rate is 171.2 Kbps. However, the geographical area covered by a cell is only 25% of the normal coverage. The most robust coding scheme is CS-1. In this case, the data rate per time slot is just 9.05 Kbps and maximum data rate is 72.4 Kbps. In contrast to CS-4, the area of coverage is 98% of normal coverage. CS-1 is used when the user is the furthest from the BTS. The characteristics of GPRS coding classes are shown in Table 5.5.

Table 5.5 GSM GPRS Classes of Services

Coding Scheme	Code Rate	Radio Block Payload Size	Radio Block Size	Data Rate per Time Slot	Maximum Data Rate
CS-1	–	181 bits	456 bits	9.05 Kbps	72.4 Kbps
CS-2	2/3	268 bits	456 bits	13.4 Kbps	107.2 Kbps
CS-3	3/4	312 bits	456 bits	15.6 Kbps	124.8 Kbps
CS-4	1	428 bits	456 bits	21.4 Kbps	171.2 Kbps

GPRS is provided to the subscriber's MU in three **capability classes**, A, B, and C:

- **Class A** allows GPRS (data, SMS) and GSM (voice) services simultaneously. Only a few handsets are Class A;
- **Class B** allows GPRS and GSM services to work alternately, one service at a time. When a voice call is set up or an SMS message is sent, the GPRS connection is not available. However, the GPRS connection can be active while the handset receives a voice call. Most handsets are class B;
- **Class C** requires manual switching between GSM and GPRS services.

To simultaneously provide GSM and GPRS services, Class A devices would need two pairs of frequencies (uplink and downlink). To avoid this situation, a new generation of class A handsets has been developed to support the dual transfer mode while using only one pair of frequencies [18].

GPRS data rate is also determined by the number of assigned slots for downlink and uplink. The asymmetric distribution of slots for the uplink and downlink reflects the asymmetric nature of interactions with most applications, where the uplink data stream consists of queries, and downlinks consists of bulk downloads of large informational contents. The GPRS Multislot Classes are shown in Table 5.6.

Most handsets are class 8 or class 10. Examples of class 8 handsets (4 downlink slots, 1 uplink slot, 8–12 Kbps Send, 32–40 Kbps Receive) are Ericsson (T39, R520), Motorola (v60i, v66i), Samsung (Q200, S100), Siemens (S45, ME45, M50). Classes 10 and 12

Table 5.6 GPRS Multislot Classes

Multislot Class	Downlink Slots	Uplink Slots	Active Slots
1	1	1	2
2	2	1	3
3	2	2	3
4	3	1	4
5	2	2	4
6	3	2	4
7	3	3	4
8	4	1	5
9	3	2	5
10	4	2	5
11	4	3	5
12	4	4	5

are implemented in high-end handsets, like Nokia N93, and PC cards. Most operators price GPRS based on the megabytes of data transferred. Some operators offer flat rate GPRS access.

5.8 GPRS QOS Profiles

Several Quality of Service (QOS) profiles are defined for GPRS that are used to prioritize data traffic according to the user desired service level. Mobile stations request particular QOS profiles during the Packet Data Protocol (PDP) context activation phase, and SGSN replies with the corresponding QOS profile available on that network.

GPRS QOS uses well-established end-to-end QOS metrics such as throughput, packet loss, latency, and jitter that are as defined in RFC 1242. Most data applications use the TCP Transport Layer protocol and its metrics: roundtrip time, retransmission, throughput, and window size. A summary of the GPRS QOS profiles components is presented below.

- Precedence Class: Three classes are defined, i.e., 1, 2, and 3; where class 1 is the highest;
- Delay Class: Four guaranteed mean and 95th percentile delay classes are defined, i.e., 1, 2, 3, and 4; where class 4 is only "best effort";
- Reliability Class: Three reliability classes are defined in handling corrupted and duplicated packets, i.e., 1, 2, and 3; where class 3 represents the lowest reliability allowed;
- Throughput Class: Up to 9 peak throughput classes and up to 31 mean throughput classes are defined, where the larger the number the higher the throughput provided.

There are four QOS **delay classes** supported in GPRS specifications. Each of these classes guarantee mean and 95th-percentile delays. These delays include the access and scheduling delays incurred for use of mobile resources. Delay is critical in real-time

Table 5.7 GPRS Delay Classes

		SDU Size: 128 bytes		SDU Size: 1024 bytes	
		Mean Transfer Delay (sec)	95 Percentile Delay (sec)	Mean Transfer Delay (sec)	Mean Transfer Delay (sec)
⟶	1	<0.5	<1.5	<2	<7
Delay Classes	2	<5	<25	<15	<75
	3	<50	<250	<75	<375
⟶	4	Unspecified – Best Effort			

applications such as encoded digital voice and video over mobile networks. The GPRS delay classes and the allowed bounds for the classes are shown in Table 5.7.

As a general note, the overall latency in mobile networks is quite high. During testing, it is common to see round-trip ping times in the range from 600ms to one second. One of the reasons for these high delays is a lack of prioritization, with voice traffic having precedence over data transmission.

There are three **reliability classes** of service defined for maximum admissible packet loss. Reliability class depends on the ability of applications to handle corrupted or duplicated packets. The highest reliability class is class 1, which assumes just one packet lost or corrupted in a billion packets sent. The lowest reliability class allows one lost or corrupted packet out of 100 packets sent. The GPRS reliability classes are shown in Table 5.8.

Table 5.8 GPRS Reliability Classes

		Probability of			
		Lost Packet	Duplicated Packet	Out of sequence Packet	Corrupted Packet
⟶	1	10^{-9}	10^{-9}	10^{-9}	10^{-9}
Reliability Classes	2	10^{-4}	10^{-5}	10^{-5}	10^{-6}
⟶	3	10^{-2}	10^{-5}	10^{-5}	10^{-2}

There are several **peak** and **mean throughput classes** specified for GPRS networks. Peak throughput applies to short time intervals where the transfer rate is at maximum. Mean throughput applies to data rate transfers averaged over extended periods of time, which may include idle periods. The peak and mean throughput classes are indicated in Table 5.9.

Factors Affecting GPRS QOS

Given the nature of mobile communications, with mobile users continuously on the move, GPRS QOS and overall network performance can fluctuate wildly. Packets can take different paths thus causing changes in individual packet latency and high levels of

Table 5.9 GPRS Throughput Classes of Services

	Peak Throughput Class	Peak Throughput (bytes/sec)	Mean Throughput Class	Mean Throughput (bytes/sec)
⟶	1	1000	1	100
Throughput Classes	2	2000	2	200
	3	4000	3	500
	4	8000	4	1000
⟶	....	....	....	....
	7	64000	17	20 000 000
	8	128000	18	50 000 000
	9	256000	31	Best Effort

jitter. Therefore, it is difficult to guarantee classes of service in accordance with defined GPRS QOS profiles.

- Data applications such as file transfer, web surfing, and even email use the TCP Transport Layer Protocol where network performance is determined by specific metrics: roundtrip time, throughput, retransmission, and TCP window size. Because of its capability to provide a highly reliable end-to-end service, TCP is inherently a much slower transport protocol than User Datagram Protocol (UDP), which only provides a best-effort service.
- Provision of a high classes of service based on the CS-1 coding scheme requires more built-in redundancy which may result in slowing data transfer rates and, therefore, overall throughput. Although retransmission allows application level data packet recovery, and thereby improves reliability, it can worsen the overall throughput, latency, and, of course, jitter. This may affect encoded voice packets, although a few missing frames do not render users' conversations unintelligible.
- Quantifying QOS performance metrics by using intrusive testing methods such as transmitting specialized QOS test packets may affect overall performance. When traffic on the network is a significant percentage of its capacity, the impact of test data on performance will become more significant.

5.9 Service Management Products in Mobile Networks

Network system parameters for Wireless SLAs should cover both the service provider network (backbone, connectivity, access) and end-to-end user applications. SLA monitoring requires primary tools and probes to collect QOS performance statistics according to QOS metrics for mobile services. The primary measurement tools are based on SNMP polling and the use of RMON probes. The most important service level metrics include network and access availability, end-to-end delay, and packet loss. They are defined and calculated as indicated below.

- **Network Availability** (measured over a month);

$$\frac{(\text{Total hours in a month}) \times (\text{Number of sites}) - (\text{Network Outage Time})}{(\text{Total hours in a month}) \times (\text{Number of sites})}$$

- **Access Availability** (percentage of time (%) a given service is operational);

$$\frac{(\text{Elapsed Time}) - (\text{Outage Time})}{(\text{Elapsed Time})}$$

- **End-to-End Delay** (the amount of latency for packets transiting mobile networks);

$$\frac{(\text{Total packets end-to-end}) \times (\text{End-to-End delay for the packets})}{(\text{Number of packets sent})}$$

- **Packet Loss Ratio (PLR)** (the percentage of the total number of end-to-end lost packets); packet loss may be caused by congestion, buffer overflows, or code violations.

$$\frac{(\text{Ingress Packet Count}) - (\text{Egress Packet Count})}{(\text{Ingress Packet Count})}$$

In Chapter 3, we provided a list of service management products. From that list we present one of the more successful service management suites, the Micromuse Inc. Netcool, now part of IBM.

Micromuse Netcool provides a comprehensive umbrella suite of processes and applications targeting real-time Fault Management and Service Level Management (SLM). More than 120 off-the-shelf probes and monitors are available to collect event data from network and systems management platforms, various databases, QOS measurement tools, application suites, and Internet network probes. A list of the most important platforms, databases, and monitors/probes working with Netcool is shown in Figure 5.7. Micromuse Netcool was acquired by the IBM Software Group in 2006 and it is part of IBM-Tivoli brand Netcool management portfolio [19].

Original data manipulation is done by using **Netcool FilterBuilder** and **Netcool ViewBuilder** modules. The enablers are myriads of probes and monitors which collect events and fault messages from the UNIX and the NT operating systems environments, the Internet, and other wide area networks.

Micromuse **Netcool/OMNIbus** is the core application of the Netcool suite. It collects, filters, and processes real-time data from numerous protocols and management environments. It gives operators in a Network Operations Center (NOC) the ability to create "views" of services, situations as they should normally function, and from there, through comparison, to derive faulty or degraded functionality.

At the core of OMNIbus is the **Netcool ObjectServer**, a high speed memory-resident database, where managed entities (servers, mainframes, NT systems, UNIX applications, routers, switches, SNMP agents) are treated as generic objects, thereby eliminating dependency on any particular interface or protocol. Network and service management related events can be augmented with business and operational perspectives. It is based on an open client-server architecture running under Unix, NT, and Java. The Micromuse Netcool/Omnibus architecture is shown in Figure 5.8.

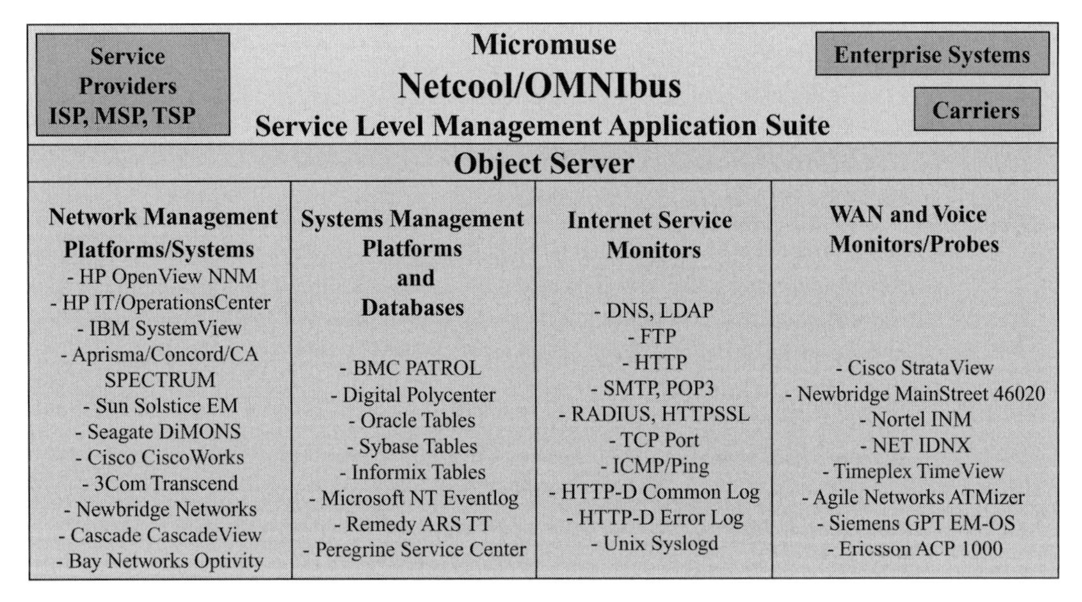

Figure 5.7 Micromuse NetCool SLM Application Suite

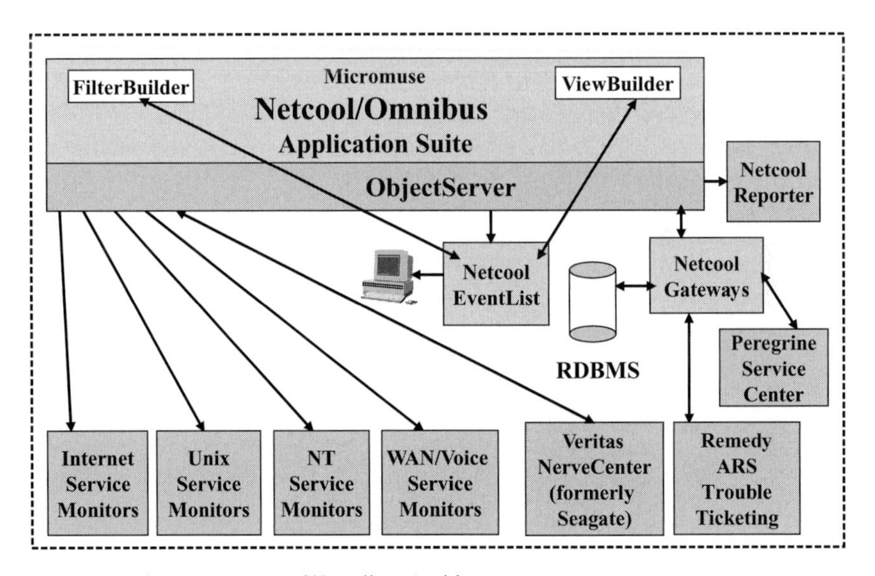

Figure 5.8 Micromuse Netcool/Omnibus Architecture

The relational database is optimized for high volume inputs; potentially hundreds of thousands of raw events in large networks. Typical service management metrics collected are:

- **Physical port metrics** (port failure, recovery, count of occurrences, performance events);
- **Server metrics** (node up/down, processes running, count of ping fails);

- **LAN** and **WAN segment metrics** (circuit failures, recoveries, group of events, end-to-end availability, reliability of FR/ATM switches);
- **E-mail metrics** (state of processes, load average, messages delayed, message size, SMTP queue);
- **FTP metrics** (download/upload files, throughput, volume);
- **Database transfer** (time/transfer rates).

Several Netcool ObjectServers can be chained in a peer-to-peer or hierarchical configuration for unlimited scalability. The Netcool Gateway functions are intended to connect Netcool ObjectServers to the Remedy Action Request System (ARS) trouble ticketing system, to the Peregrine Service Center, and, through Microsoft Open Data Base Connectivity (ODBC) to Informix, Oracle, and Sybase RDBMSs. The function of the gateway is to share ObjectServer data with other systems.

A typical implementation in a Network Operation Center (NOC) requires major Netcool components such as: **Netcool Object Server, Netcool Reporter, OpenView NNM** probes, **ASCII** probes, **Netcool Service Monitors** (SNMP trap receiver, DNS, FTP), UNIX SysLog Daemon probes, and **Netcool Gateways** (pass event messages to a historical data store, DBMS such as Oracle, Sybase). The HP OpenView Network Node Manager (NNM) is used as a polling engine along with CISCO's Works. Micromuse has over 1800 customers. Major NOC implementations include AOL America, British Telecom, Deutsche Telekom, Cable and Wireless, Charles Schwab, Digex, EarthLink, GE Appliances, ITC, DeltaCom, J.P. Morgan Chase, Swiss Com Mobile, US West, AirTouch, AT&T, T-Mobile Austria, and Verizon.

Part III

Fixed Wireless Technologies: Networking and Management

6 Wireless Local Area Networking

6.1 Wireless LAN Architecture

Traditional Local Area Networks (LANs) use wired cables (copper-based coaxial or twisted pairs and fiber optic) as the communications media. Extending on the success and resilience of LAN technologies, **Wireless LANs** (WLANs) use **short-range Radio Frequency (RF) communications** to connect the WLAN components. Although communication within a WLAN can be done using wireless infrared links, our focus will be on WLANs that are based on radio links. WLAN technologies bring the users one step closer to the technical ideal of communications "anywhere, anytime, any technology".

A typical WLAN configuration consists of **Wireless Terminals** (WT), standard laptops, desktops, PCs, and PDA clients equipped with RF transceivers (PC wireless cards). These units communicate with a WLAN **Access Point** (AP), a device that communicates with WTs across **radio links**. The WLAN infrastructures of single or multiple APs connect WLANs to a wired network that may consist of wired LANs, switches, and routers. The WLAN architecture is depicted in Figure 6.1.

A wireless **Fire Wall** security gateway may separate the WLAN from the rest of the network. A dedicated **Remote Access Dial-In User Service** (RADIUS) server provides **Authentication, Authorization, Access** (AAA) security services.

Working in the low microwave frequency range, below 10 GHz, the coverage area is limited by the allowable power radiation, interference from other radio frequency sources, physical obstacles such as metal walls, and multi-path propagation effects due to reflections. Implementation of WLANs has started in retail, manufacturing, education, and health care and moved rapidly into common areas such as airports, small/home offices, Internet cafes, and in residential homes.

WLAN advantages include network mobility (computing terminals can move outside of their wired working environment), flexibility (easy ad-hoc implementation with a single AP deployed), scalability (easy addition of new users), cost savings (compared with the cost of running physical cables), and higher productivity. WLAN disadvantages include limited range (because of restrictions in maximum power admitted and other factors), compromised security (wireless packets can be easily intercepted, hence the need for encryption), lowered reliability (radio signals exposed to propagation and interference effects), and limited data rate (compared with the wired LAN counterparts currently working in the Gbps range).

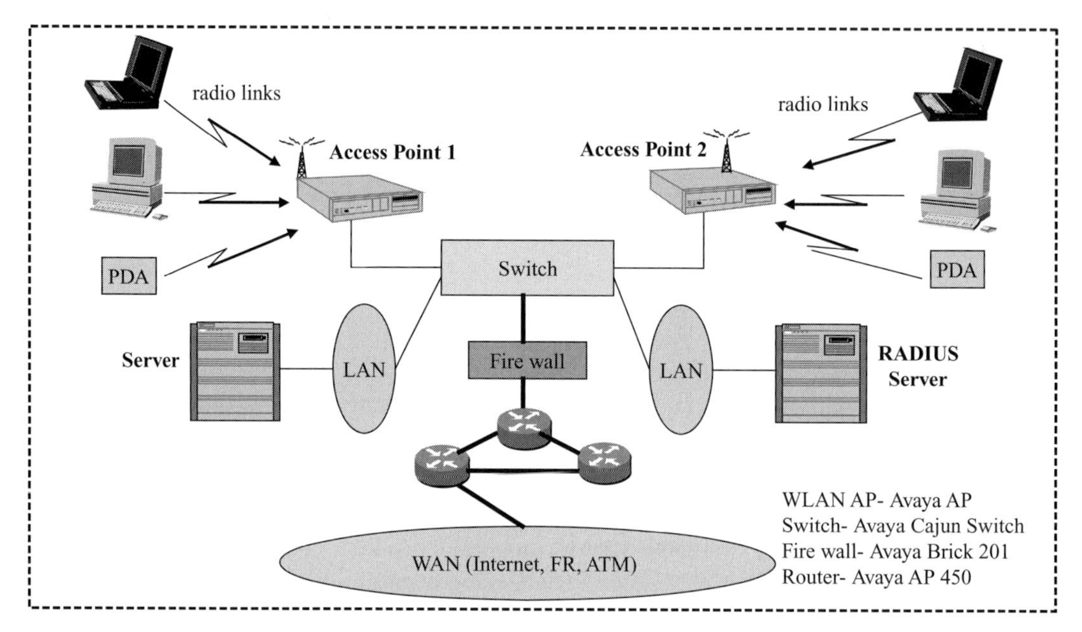

Figure 6.1 Wireless LAN Architecture

WLAN standardization in the USA is done by Institute of Electrical and Electronics Engineers (IEEE) working groups while in Europe it is done by the European Telecommunications Standard Institute (ETSI). The goal is to make WLAN as ubiquitous as the wired LAN in speed, mobility, and overall flexibility regarding data rates, frequencies used, modulation techniques, and access methods.

There are three main WLAN specifications, each with a different degree of acceptance and implementation: **IEEE 802.11a, IEEE 802.11b**, and **IEEE 802.11g**. The most popular standard is 802.11b, also known as **Wireless Fidelity (Wi-Fi)** or Wireless Ethernet. Currently, the Wi-Fi term is attributed to the entire IEEE 802.11 a/b/g family of specifications. Therefore, we can discuss peer-to-peer WLANs, residential Wi-Fi, hotel Wi-Fi, enterprise Wi-Fi, metropolitan city-wide Wi-Fi, and convergent cellular mobile Wi-Fi. WLANs operate in one of the unlicensed spectrum bands at either 2.4 GHz, 5 GHz, 1.9 GHz, or 900 MHz (902–928 MHz). Coverage area is in the ranges of 3m to100m from the wireless access points.

6.2 WLAN Networking Solutions Comparison

Although WLAN technology has been known for two decades, its implementation was slowed for some time by high component costs. The first generation of WLANs (2000–2002) was based on the IEEE 802.11a and ETSI HyperLAN1 standards. They required addition of WLAN capabilities to existent computer terminals in the form of wireless PC cards. The second generation (2003–2004) was based on IEEE 802.11b and 802.11g, and ETSI HyperLAN2 standards, computing devices having built-in WLAN capabilities at the level of radio link interfaces and microprocessors such as Intel's Centrino. The third generation (2005 and beyond) is focused on providing Voice over WLAN (VoWLAN) capabilities, WLAN QOS convergent mobile WLAN solutions, and mesh WLAN metropolitan area networking.

Characteristics of the main WLAN technologies are presented in Table 6.1. The table is followed by an overview presentation of additional details and specifications. All WLANs presented are based on the well-known Ethernet Carrier Sense Multiple Access/Collision Avoidance (CSMA/CA) technique.

Table 6.1 Comparisons of the Main WLAN Networking Solutions

Standard	Physical capacity	Max throughput	Maximum range	Modulation type	Channel BW	Frequency band	No. of channels			QOS
							US	Asia	EU	
802.11b	11 Mpbs	6 Mbps	100 m	DSSS	25 MHz	2.4 GHz	3	3	4	Y
802.11a	54 Mpbs	31 Mbps	80 m	OFDM	25 MHz	5 GHz	12	4	0	Y
802.11g	54 Mpbs	24 Mbps	150 m	OFDM/DSSS	25 MHz	2.4 GHz	3	3	4	Y
HiperLAN2	54 Mbps	31 Mbps	80 m	OFDM	25 MHz	5 GHz	12	4	15	Y
5-UP	108 Mpbs	72 Mbps	80 m	OFDM	50 MHz	2.4 GHz	6	2	7	Y

Legend: DSSS-Direct Sequence Spread Spectrum; 5-UP-5 Unified Protocol; OFDM-Orthogonal Frequency Division Multiplexing.

The **802.11b WLAN** was designed to operate in the 2.4 GHz (more precisely 2.4–2.485 GHz) **Industrial Scientific Medical** (ISM) band. Initially, it started with data rates of 1 Mbps and 2 Mbps, and later, 1999, the rate was extended to 5.5 Mbps, and, finally, it has reached the 11 Mbps mark. Promotion of 802.11b standards-based products is handled through interoperability testing under the auspices of the **Wireless Ethernet Compatibility Alliance** (WECA) or **Wi-Fi Alliance** [20] by branding the products passing the tests as "Wi-Fi" certified (over 3500 products certified since March 2000).

802.11b WLAN uses either the **Direct Sequence Spread Spectrum** (DSSS) or the **Frequency Hopping Spread Spectrum** (FHSS) access technique. DSSS splits the 2.4 GHz available bandwidth into 14 sub-channels of 22 MHz each. By using a redundant bit pattern, it can provide one mechanism for error detection and correction. In FHSS, the 2.4 GHz available bandwidth is spread into 75 sub-channels of 1 MHz each, shared by multiple transmitters and receivers. By using a hopping pattern agreed upon at the beginning of the session by the transmitting parties, it provides good protection against noise. If the signal strength and quality becomes a problem, the 802.11b wireless cards can operate at fall-back data rates of 5, 2, and 1 Mbps. There are interference issues since the 2.4 GHz band is shared with wireless phones, Bluetooth, and microwave ovens.

The **802.11g WLAN** was designed to operate in the same crowded 2.4 GHz **Industrial Scientific Medical** band. The IEEE 802.11g standard was adopted in 3Q 2003. 802.11g WLANs use either the DSSS or the FHSS access technique. Therefore, it is fully backward compatible with 802.11b WLAN standards. This assures an easy migration from 802.11b to 802.11g solutions. 802.11g WLANs operate at a maximum speed of 54 Mbps and at a maximum data rate of 24.7 Mbps. This allows operation with more simultaneous calls (15–20 calls) per AP compared with eight calls on 802.11b. Promotion of 802.11g standards-based products is also handled by Wi-Fi Alliance [20].

IEEE 802.11a, the first IEEE WLAN standard, adopted with amendments in 1999, was designed to operate in the **5 GHz Unlicensed National Information Infrastructure** (UNII) spectrum in two different bands: a 200 MHz (5.15–5.35 GHz) and a 100 MHz (5.725–5.825 GHz) band. 802.11a WLANs use an encoding scheme called coded **Orthogonal Frequency Division Multiplexing** (OFDM) which breaks the high speed data carrier into lower sub-carriers that are transmitted in parallel. This approach avoids interference problems due to the difference in delay of signals arriving at the receiver. The lower 802.11a 200 MHz band is divided into 8 independent clear channels of 20 MHz each. The 20 MHz channels are divided into 52 sub-channels (carriers) each about 300 KHz wide. It has 12 co-channels, 8 assigned for indoor WLANs, and 4 channels for point-to-point installations.

There are three "power domains". The first 100 MHz of the lower band is restricted to a maximum power output of 50 mW. The second 100 MHz band allows a maximum of 250 mW power output. The third 100 MHz tier, assigned for open outdoor applications (e.g., parking lots, warehouses), allows 1W power output. In contrast, 802.11b allows 1 W in USA but most devices limit the output to 30 mW to minimize heat dissipation and conserve battery power. It is not interoperable with 802.11b and 802.11g.

Initially, the 802.11a standard was not accepted in Europe. Instead, another solution was pursued, **HiperLAN2** standard. HiperLAN2, as an alternative to 802.11a specifications, was promoted by the ETSI Broadband Networks Access Group (BRAN). The HiperLAN2 WLAN was designed to operate in the 5 GHz unlicensed spectrum shared in Europe by military and civilians for satellite communications and ground tracking stations. Consequently, to avoid interferences between these applications, two other specifications needed to be supported: Dynamic Frequency Selection (DFS) and Transmit Power Control (TPC). With the dominance of IEEE 802.11 b/g, and the acceptance of 802.11 in Europe, the attraction to HiperLAN2 is diminishing. The promotion of HiperLAN2 standards specifications is handled through the HiperLAN2 Global Forum.

To achieve high-speed data rates, HiperLAN2 uses the same encoding scheme as the 802.11a WLANs, i.e., **Orthogonal Frequency Division Multiplexing** (OFDM). However, in contrast to 802.11a WLAN, HiperLAN2 uses a different MAC address and a QOS mechanism that is closer to the ATM approach rather than Ethernet. To address this HiperLAN2 and 802.11a divergence, Atheros Communications has proposed a **Unified Protocol 5** (5-UP) that allows interoperations at various data rates at the level of each individual 20 MHz channel of the 5GHz spectrum band.

Many current implementations are using dual band (2.4 GHz and 5 GHz), tri-mode (a, b, g) Access Points. In 2005, as part of the Wi-Fi Alliance, work has started to provide specifications for **Control and Provisioning of Wireless Access Points** (CAPWAP).

6.3 WLAN IEEE Standards

In addition to the three main WLAN standards, 802.11a, 802.11b, and 802.11g they have developed, the IEEE is working on many other specifications aimed at solving recognized weaknesses in the existing standards or in enhancing features. These specifications are in various stages of development, drafting, or balloting. A short description of the purpose of each of these specifications follows. All, with the exception of 802.11F and 802.11T, are amendments to the original IEEE 802.11 WLAN standard.

802.11d: Intended to make 802.11b suitable to other frequencies, where the 2.4 GHz band is not available for the purpose of international (country to country) roaming (in this regard, there is an ITU-T Recommendation).

802.11e: Will provide QOS capabilities to WLAN networks. The MAC sub-layer will be replaced with a coordinated TDMA scheme and an error correction mechanism will be added.

802.11F: Will address the roaming capabilities of users between two switched LAN segments or wireless access points in a manner similar to cellular handover (inter-Access Point protocol).

802.11h: Addresses power control and radio channel selection for 802.11a WLAN products to make these standards acceptable in Europe.

802.11i: Addresses WLAN security by adopting a new standard, rather than WEP, called Advanced Encryption Standard (AES), the US government encryption algorithm.

802.11j: Intended to solve the issue of 802.11 and HiperLAN2 coexistence in the same 5 GHz spectrums and made them acceptable in Japan.

802.11n: Will provide high speed WLANs that will support enterprise class bandwidth-intensive applications. More details will be provided in the next section.

802.11r: These specifications will address fast roaming capabilities of users between WLAN APs, allowing VoWLAN users to roam inside buildings.

802.11s: These specifications are intended to address WLAN mesh networking. More details will be provided in the next section.

802.11T: These specifications will address wireless performance predictions by recommending test methods and specific performance metrics.

802.11v: These specifications will provide amendments to the IEEE 802.11 PHY/MAC layers to enable management of attached stations (centralized or distributed), including SNMP-based management.

More details will be provided in subsequent sections of this chapter that deals with WLAN Management (802.11v), WLAN Security (802.11i), and WLAN QOS (802.11e).

6.4 IEEE 802.11n and 802.11s WLAN Standards

A new, promising development in WLAN implementations is the addition of another standard, IEEE 802.11n [21], targeting higher data rates and extended range of operations as compared with the 802.11a/b/g standards. IEEE 802.11n has a target data rate of 540 Mbps and a real data rate of 200 Mbps. Anyhow, speed in excess of 100 Mbps and range of operation within a radius of 70 meters indoors and 250 meters outdoors are the big attractions. 802.11n WLANs will work in the unlicensed 2.4 and 5 GHz bands.

IEEE Task Group n (TGn) was formed in January 2004 to develop 802.11n. The work schedule called for a draft standard by mid 2006 and a full standardization by April 2008. Although the current specifications are still in the Draft 2.0 format, the pre-standard has generated a great interest from chip makers, laptop vendors with built-in wireless adapters, manufacturers of access points, and integrators of enterprise-wide WLAN solutions. Given the high level of consensus around the current draft, Wi-Fi Alliance, in tandem with the IEEE 802.11n Working Group, gave a formal approval of current specifications to allow building pre-802.11n products. As of August 2007, over 70 products had been announced for "compliance" with the 802.11n Draft 2.0 standards.

Three technologies made IEEE 802.11n advancements possible. The first was Multiple Input Multiple Output (MIMO) antennas. These antennas allow transmission and reception of multiple streams of data, in configurations known as 2x2 and 3x3 transmitters and receivers. Frame aggregation allows the wireless terminal supporting these specs to aggregate multiple (2, 3, 4) frames at the MAC layer, thus, using more efficiently the existing channel.

The second development is **channel bonding** that allows transmission of two nonoverlapping channels at the same time. The third is **payload optimization** which aggregates packets to maximize throughput. The advantages gained by 802.11n are clear because higher throughput provides more bandwidth for voice communications over WLANs. It also opens the door to quality video streaming over WLANs and extension of these features into dual-mode Wi-Fi cellular mobile handsets. The Enhanced Wireless Consortium (EWC) was formed (currently with over 80 members) to help accelerate the IEEE 802.11n development and to promote technology specifications for interoperability across WLAN products as promoted by Wi-Fi Alliance [20].

Among the early promoters of this technology are chipmakers Intel (Centrino/Santa Rosa), Atheros, and Broadcom using proprietary channel bonding and frame bursting techniques. Also interested are Linksys, Netgear, Belkin, and D-Link, manufacturers of wireless access points and routers. Other promoters, laptop maker Dell with draft 802.11n card Inspiron, Apple with iMac desktop computers and MacBook Pro laptops with 802.11n wireless cards as well as systems integrators such as Meru Networks.

Another development in WLAN implementation is the addition of the **IEEE 802.11s** standard. This standard targets extension of WLAN technology into **Wireless Mesh Networking**. That means, extending the initial single-hop communication to a multi-hop communication. The 802.11s standard is expected to be ratified in 2008. The committee that evolved the standard has been in existence under a different name since 2003, demonstrating the fact that technical advancement is often faster and more demanding than the process of standardization. Just witness the spreading of municipal and metropolitan mesh networks implementations.

As with other standards, the high level of consensus around the current draft has led manufacturers and system integrators such as Motorola and Metro-Fi to start building products and providing implementation based on pre-standard specifications. Motorola stated that they will align MeshConnex WLAN Mesh System with the current draft and later with the final standard. In fairness, the first WLAN mesh implementations have preceeded the formation of IEEE Task Group 802.11s. In these cases, the implementations were the drivers of the standards. Regardless, interoperability problems are expected between products supporting WLAN Mesh systems and between implementations.

Four major aspects are addressed in the IEEE 802.11s draft standards: topology, routing, channel allocation, and backhauling. **Topology** deals with mesh node discovery and layering to establish the hierarchical relationship between nodes. Some nodes act as access nodes, other are involved in concentrating/distributing the traffic, and a third category is to provide gateway functions to the outside world. **Routing** involves frame formatting, addressing, and the use of algorithms that facilitate path selection and forwarding of information on a hop-basis while still providing adequate performance. The Hybrid Wireless Mesh Protocol (HWMP) was embraced as the mandatory algorithm inspired by IETF's Ad-hoc On-Demand Distance Vector Routing (RFC 3561). Some alternate algorithms are also considered. **Channel allocation** deals with the use of single or dual radio links using the unlicensend 2.4 GHz and 5.8 GHz frequency bands. **Backhauling** addresses solutions to connect WLAN Meshes to the outside world, using gateway nodes or portals connected through wired or wireless links.

Other areas addressed by the 802.11s draft standards are: powering, traffic monitoring, security, and network management. **Powering** deals with the power provided to outdoor WLAN nodes, a prevalent solution being the Power over Ethernet (PoE). **Traffic engineering** describes routing and congestion resolution mechanisms, using hop-by-hop congestion control mechanisms that were specified in the IEEE 802.11e standard. WLAN Mesh **security** will follow the already in place IEEE 802.11i standard by adding multi-hop end-to-end security and a mutual authentication scheme for adjacent mesh nodes. Last but not least, a centralized **network management** system is required to manage a highly distributed network for faults, congestion, and to assure adequate QOS across the whole network.

6.5 Wi-Fi Multimedia Specifications

Given the slowness in ratification of some of the IEEE 802.11 series of standards, the Wi-Fi Alliance has developed a subset of 802.11 specifications, commonly known as **Wi-Fi Multimedia** (WMM). The most important features of WMM are:

- **Multilevel priority support**: Differentiates the high priority voice traffic from low priority data traffic;
- **Admission control**: Rejects new calls when access point capacities are reached;
- **Automatic power saving**: Schedules the handsets to sleep and to wake up every 10 ms instead of being powered continuously;
- **Scheduled access**: An advanced feature of IEEE 802.11e that dictates to associated clients when to communicate. This allows better control of QOS if there is enough WLAN capacity and physical coverage;
- **Wi-Fi Protected Access** (WPA): A stopgap set of security specifications to use until 802.11i standards are adopted. WPA, now, provides support for both Wi-Fi Protected Access (WPA) and WPA2 certification as part of overall security measures.

6.6 FCC Released Unlicensed Spectrum Specifications

The demand for wireless communications has induced FCC to open up parts of the radio spectrum for use by unlicensed operators, organizations, and individual users. These spectrum bands can be shared by WLANs, WiMAX, and Bluetooth users as well as by business and consumer applications such as cordless phones, microwaves, and garage openers. The official spectrum allocation for IEEE 802.11 standard specifications in the USA is based on the unlicensed Industrial, Scientific, Medical (ISM) spectrum, opened for use in 1985. It has distinct spectrums in the 2.4 GHz, 5GHz, and 900 MHz bands. WLAN implementations are in the 2.4 GHz and 5 GHz bands.

- IEEE 802.11a uses the unlicensed spectrum between 5.15 GHz and 5.35 GHz and the spectrum between 5.725 GHz and 5.825 GHz.
- IEEE 802.11b uses the unlicensed spectrum between 2.4 GHz and 2.4835 GHz.
- IEEE 802.11g uses the unlicensed spectrum between 2.4 GHz and 2.4835 GHz.
- IEEE 802.11n will use the unlicensed spectrum between 2.4 GHz and 2.4835 GHz, the spectrum between 5.15 GHz and 5.35 GHz, and the spectrum between 5.725 GHz and 5.825 GHz.

IEEE 802.11 spectrum allocations for all IEEE WLANs, along with other important features, are presented in Table 6.2.

All three IEEE WLANs 802.11 a/b/g use the same multiple access CSMA/CA, the same uplink/downlink Time Division Duplex (TDD), and the same channel bandwidth, 20 MHz. However, various modulation techniques are associated with each of these WLANs: Binary Phase Shift Keying (BPSK), Complementary Code Keying (CCK), Quadrature Phase Shift Keying (QPSK), Differential Quadrature Phase Shift Keying (DQPSK), 16 Quadrature Amplitude Modulation (16QAM), 64 Quadrature Amplitude Modulation (64QAM), Orthogonal Frequency Division Multiplex (OFDM), and Packet Binary Convolutional Coding (PBCC).

These modulation techniques combined with a particular convolutional coding scheme determine overall differences in peak data rate provided by each of these WLANs. **Convolutional Coding** (CC) is an error protection-coding scheme that encodes data bits into a continuous stream. It is an error correction code in which every m-bit information symbol to be encoded is transformed into an n-bit information symbol ($n > m$) where the transformation is a function of the last k symbols (error detection scheme used in radio/satellite communications). Another major difference between WLANs is the available number of non-overlapping channels. 802.11a has 12 channels while 802.11b and 802.11g have only three available channels.

6.7 WLAN Security Aspects

All computing and communication systems face some degree of security risks. Wireless LANs, by the nature of propagation of RF waves, inside and outside of "confined" locations, present to the designers and the users major security challenges. It is impossible to restrict unauthorized access to the radio signals since any wireless terminal supporting

Table 6.2 IEEE WLAN Spectrum Allocation and other Features

Features	802.11a	802.11b	802.11g	802.11n
Frequency Range	5.15–5.35 GHz USA 5.47–5725 GHz Europe 5.725–5.825 USA/China	2.40–2.485 GHz North America Europe 2.471–2.497 Japan 2.4465–2.4835 France 2.445–2.475 Spain	2.40–2.485 GHz North America Europe 2.471–2.497 Japan 2.4465–2.4835 France 2.445–2.475 Spain	2.4 GHz or 5 GHz
Modulation	BPSK, QPSK, 16QAM, 64QAM, OFDM	BPSK, DQPSK-header BPSK, QPSK-payload CCK, PBCC	BPSK, DQPSK, QPSK, 16QAM, 64QAM, OFDM, CCK, PBCC	OFDM
Multiple Access	CSMA/CA	CSMA/CA	CSMA/CA	-
Duplex Uplink/Downlink	TDD	TDD	TDD	TDD
Channel Bandwidth	20 MHz	20 MHz	20 MHz	10, 20, or 40 MHz
Number of Channels	12 Non-overlapping	11 Overlapping 3 Non-overlapping	11 Overlapping 3 Non-overlapping	11, 14, 24
Peak Data Rate	54 Mbps	11 Mbps	54 Mbps	Up to 540 Mbps

802.11 specifications can listen to all traffic within its range of reception. In particular, there are four major security aspects at stake: **Authentication, Authorization, Access** (AAA), and **Privacy**. These aspects should be part of any comprehensive security policy that encompasses all mobile devices and access points.

Authentication: A process used to determine user identities, i.e., to prove to the system that the user is exactly whom he/she claims to be. This process ensures that only legitimate users can access the systems and services. The "user" is a generic term for an organization, person, computing device, or a software process that should be authenticated. Simple authentication identifiers are names accompanied by attributes such as passwords. An open system authentication allows any mobile station to freely join an access point with no restrictions.

Authorization: The process of verification that an entity is authorized to use services because it is either a known subscriber or has pre-paid for those services. Authorization is done by checking, most likely, a remote database. Authorization assumes that an entity in the system has already gained access to some of the resources.

Access Control: A process to allow or restrict access to the services the user is authorized to use, once authenticated. Most security systems use access control lists that contain the services available and the users or computers that are allowed to use these services. Access control lists based on MAC addresses are vulnerable because they can be spoofed and in larger networks represent an administrative headache.

Privacy: A process that protects information against eavesdropping and unauthorized access. Privacy is achieved by encrypting all voice and data information. Different techniques and algorithms are used for encryption. It is assumed that both the sender and the receiver know the encryption key to cipher and decipher the encrypted information. Encryption works if we have the assurance that information has not been altered during its transitions across radio links.

In the first generation of WLANs the following security features were used:

- **Open System Authentication**: No restrictions to access and use of an AP;
- **Shared Key Authentication**: Both sender and receiver use the same key to encrypt and authenticate the mobile station;
- **Wired Equivalency Privacy** (WEP): An algorithm for frame encryption.
- **32-bit Integrity Check Value (ICV)**: Appended to the information transmitted and encrypted using WEP;
- **Initialization Vector (IV)**: A 24 bit field appended to the key before encryption.

WEP is a Data Link Layer algorithm, more exactly a Medium Access Control (MAC) sub-layer algorithm that uses the Rivest Cipher 4 (RC4), an encryption/decryption mechanism used also in Cellular Digital Packet Data (CDPD). WEP is used to encrypt data only between wireless terminals and APs. WEP specifies a 40-bit key for encryption, implemented in the wireless NIC adapter. To gain Wireless Fidelity (Wi-Fi) certification from the Wireless Ethernet Compatibility Alliance (WECA), Wi-Fi certified products must support at a minimum the 40-bit WEP encryption.

Nevertheless, it was only a matter of time before the WEP keys were cracked exposing the vulnerability of the selected mechanism and of WLANs as a whole. The reason is three fold. First is the shortness of the 40-bit key, which is too small and easy to be cracked with software tools found on the market. Second is the use of only four static keys between clients and access points. The last reason is the use of CRC-32 (Cyclic Redundancy Check) which is an error detection and correction mechanism rather than a security feature. Addition of the 24-bit Initialization Vector (IV) proved a temporary aid against compromising of the WEP keys. Soon, it was discovered that freely available AirSnort software could break the Initialization Vector, given its shortness. A similar fate faced North American GSM services that were hacked and cracked by researchers from the Israel Technion Institute in a scientific demonstration.

After WLAN vulnerabilities were exposed, WEP was slightly improved by having the encryption key changed before the IV is repeated or by extending it from 40 bits to 104 bits. The RC4 cipher algorithm requires that a key's length be a multiple of 64 bits, where 24 bits are used for IV and either 40 bits or 104 bits are left for the encryption key.

Major changes in security came through in 2003 when **Wi-Fi Protected Access (WPA)**, an interim replacement of WEP, promoted by the Wi-Fi Alliance, became mandatory for all WLAN products. Among the improvements were an extension of the encryption key from 40 bits to 128 bits, extension of IV from 24 bits to 48 bits, and use of an authentication key of 64 bits based on Temporal Key Integrity Protocol (TKIP). Other security improvements were in the key distribution area by adoption of the key hierarchy specified in the IEEE 802.1x standard, and use of the Extensible Authentication Protocol (EAP) for key distribution. EAP supports multiple authentication methods such as Lightweight Extensible Authentication Protocol (LEAP), a proprietary extension from Cisco Systems, a major player in WLAN networking. The evolution of WLAN security standards together with their main features, are presented in Table 6.3.

Table 6.3 WLAN Security Standards Evolution

Features	WEP	WPA	802.11i (WPA2)
Cipher Algorithm	RC4	2RC4 (TKIP)	AES-CCMP (Rijndael)
Encryption Key	40-bit	128-bit (TKIP)	128-bit (CCMP)
Initialization Vector (IV)	24-bit	48-bit (TKIP)	48-bit (CCMP)
Authentication Key	None	64bit (TKIP)	128-bit (CCMP)
Integrity Check Value (ICV)	CRC-32	TKIP	CCM
Key Distribution	Manual	802.1X (EAP)	802.1x (EAP)
Key Unique to:	Network	Packet, Session, User	Packet, Session, User
Key Hierarchy	No	Derived from 802.1x	Derived from 802.1x
Cipher Negotiation	No	No	Yes
Ad-hoc (P2P) Security	No	No	Yes (IBSS)
Pre-Authentication	No	No	Using 802.1x (EAPOL)

Following the lessons learned from the weaknesses of WEP, the new set of **IEEE 802.11i** standards (amendments), also known as **Wi-Fi Protected Access 2** (WPA2), went even further. Security features included in this standard are:

- A User Level Authentication that is based on IEEE 802.1x security standard carried through a Remote Access Dial-In Service (RADIUS) server.
- Use of powerful encryption protocols.
  - Counter Mode with the Cipher Block Chaining Message Authentication Code (CBC MAC) Protocol (CCMP) for authentication and integrity combined with the Advanced Encryption Standard (AES), the US government official encryption standard; or the
  - Temporal Key Integrity Protocol (TKIP) in combination with the RC4 cipher.
- Use of dynamic keys based on the 802.1x revised key distribution mechanism and carried by the Extensible Authentication Protocol (EAP).
- Pre-authentication based on the 802.1x standard and the Extensible Authentication Protocol over LAN (EAPOL)
- Support for pre-shared keys or keys hierarchy, also derived from 802.1x standard.
- Support for fast roaming.

Key Hierarchy means sharing several known keys among participants. Keys are specific to a particular user or to a particular transmission such as broadcast. At the top of the hierarchy are the public keys. Cipher negotiation between clients and APs is required given the mixture of security mechanisms, for example, WEP-based, TKIP-based, or AES-based, employed by different generations of devices. The WPA2 802.11i based standard includes a Peer-to-Peer ad-hoc authentication, where no AP or RADIUS server is involved, called Independent Base Service Set (IBSS). Also, IEEE 802.11i includes a pre-authentication using EAP over LAN (EAPOL) that allows the terminal to pre-authenticate with APs to facilitate roaming and fast roaming.

The security process of the second generations of WLANs, based on WAP and WAP2, includes the following steps:

- 802.11 Association, when a secure connection between mobile station and the AP takes place;
- 802.1x Authentication, when a series of messages are exchanged between the Supplicant Agent on the mobile station and the Authenticator Agent on the AP;
- A 4-way Pairwise Key Handshake (PKH); distribution and maintenance of these keys among numerous clients being a serious issue.

The advanced WAP2 features were incorporated, along with the IEEE 802.1x standard, in WLAN devices starting with 2005. The features were also included in WLAN radio chips that included some proprietary extensions, such as those from Cisco Systems. These developments are not fully compatible with the former Wi-Fi devices, so the acceptance in legacy systems remains a major issue. The assumption, made at the beginning of this section, about network and mobile device availability should not be taken for granted. Denial-of-Service attacks, radio jamming (even though it is illegal), and software viruses are real possibilities when it comes to providing security to any mobile system, including WLANs. One security solution is through a management plan that combines wireless intrusion detection systems with mobility management. All these aspects become critical in convergent systems where security vulnerabilities in the fixed wireless systems might open the gates into the core mobile and wired networks.

6.8 Wireless LAN Adapters for Computing Devices

WLANs have become ubiquitous in many enterprises, organizations, small offices and homes. The explanation is simple: convenience, adequate performance, and consumer electronics-level prices for WLAN infrastructure. Critical components for this success are client wireless card adapters for computing devices, PCs, laptops, desktops, workstations, and even servers. WLAN card adapters can be attached or built-in, and are available in CardBus, Personal Computer Memory Card International Association (PCMCIA) or Peripheral Component Interconnect (PCI) formats. WLANs are expected to provide data transmission rates and services comparable to wired LANs. Will the wireless adapters be limited to one of the standards, 802.11 a, b, g, or will they support all IEEE 802.11 types of WLANs and how reliable and how secure are WLANs compared with the traditional LANs? The answers to these questions depend very much on the radio links and the quality of wireless adapters on one end and the access points on the other end.

There are numerous manufacturers and vendors of wireless LAN adapters. In many cases the vendor is also the provider of WLAN infrastructure, i.e., a family of adapters and corresponding APs that reflect generational changes in WLAN technologies. Examples of the lines of WLAN client adapters supported by several vendors are shown in Table 6.4.

Table 6.4 WLAN Adapters for PCs and Laptops

Vendor	Family of Products	802.11a/b/g Models	802.11b/g Models	802.11a Models	Special Features
Cisco Systems	Aironet LAN Client Adapters	802.11a/b/g CardBus 802.11a/b/g PCI	Aironet 340 series	802.11a CardBus	CiscoWorks Wireless Solutions
Dell	Wireless Card Utility	Latitude, Inspiron Notebooks	Optiplex, Dimension Desktop AXIM Pocket PC	-	Wireless Ready Notebooks, Desktop Pocket PC
Intel	PROS Wireless Network Connection	3945ABG 2915 ABG	2200 BG	-	LANDesk Network Management
Proxim	ORINOCO Client Adapters	11 a/b/g Combo card 11 a/b/g PCI card	11b/g PC Card 802.11 b/g USB	-	ProximVision Network Management
3Com	AirConnect OfficeConnect	Wireless 11a/b/g PC 11a/b/g XJACK Antenna	Wireless 108 Mbps 11 g XJACK PC	-	AirConnect Wireless AP

In addition, there are several dozen other vendors/manufacturers that offer a variety of WLAN adapters. Some of these are shown here.

- **AirPort**: Extreme Card 802.11g;
- **Avaya/Lucent**: Silver 802.11b WiFi PC card;
- **Avaya Wireless**: PC Card WiFi for Laptop;
- **Dell**: True Mobile 1150 mini PCI Wireless PCMCIA NIC card for notebooks;
- **Enterasys/Cabletron**: Roamabout 802.11;
- **HP/Compaq**: iPAQ 11 Mbps Wireless PC Card (HNW-100);
- **HP/Compaq**: WL100, WL110 Wireless PC Card;
- **Linksys**: Dual-band A+G Notebook Adapter;
- **Linksys**: Dual-band A+G PCI Adapter;
- **Linksys**: 802.11n PCMCIA Wireless Adaptor with MIMO;
- **Lucent**: WaveLAN Gold, Silver and Bronze;
- **Microsoft**: Wireless Notebook Adapter MN-520 (510, 130, 120, 111);
- **Netgear**: 54 Mbps PCI Adapter and 54 Mbps CardBus;
- **Nortel Networks**: WLAN Adapters 2201 (a/b), 2202 (a/b/g);
- **Nokia**: C110, C111 Wireless LAN Cards;
- **Planet Technology**: G54 Mb PC CardBus Adapter;
- **Planet Technology**: G54 Mb 32 bit PCI Adapter;
- **Proxim Wireless**: ORINOCO 11 PCI card, PC card, WUSB;
- **Proxim Farallon**: Skyline 802.11b PC Card for notebooks;
- **Samsung**: MagicLAN SWL-2000 N and SWL-2100 N;
- **Siemens**: SpeedStream Wireless PCMCIA Card (SS1021).

These WLAN adapters, by adopting open standards, have many features in common:

- The common modulation technique is CCK for 802.11b, CCK/OFDM for 802.11g, and OFDM for 802.11a adapters;
- The maximum data rate is 11 Mbps for 802.11b/g adapters and 54 Mbps for 802.11a adapters;
- Most adapters employ dual diversity antennas and operate up to 100m from the AP, in office areas, and 300m, in open areas;
- Most adapters provide connectivity to 10/100 Base-T Access Points;
- All Wi-Fi certified adapters support standardized security features such as WEP, WPA, and WPA2;
- Many adapters have embraced 802.11v amendments that include SNMP-based management capability;
- Most adapters operate under Microsoft Windows operating system with the exception of those Network Managers that support Linux;
- Most adapters are based on the Atheros Communications chipset.

Beside these similarities, there are some relatively minor differences in the security features supported in special in the area of authentication. For example, there are several authentication protocols such as EAP, LEAP, EAP-TLS, and PEAP. Another area of differentiation is in the security package supported given that the 802.11i standard has partial implementations of security features.

6.9 WLAN Systems Controllers

The explosive growth of WLAN technology in the past five years, both in number of WLAN networks and density of those networks, has made traffic management and maintenance of a satisfactory level of performance more difficult. This has been compounded by the addition of Voice over WLAN, with its special demands regarding use of the bandwidth and Quality of Services. The solution has been the introduction of a centralized WLAN radio traffic controller. This new generation of WLANs is called **WLAN System** and the new network element is the **WLAN Controller** [22]. The first implementation of a WLAN system controller was by Symbol Technologies. As of 2007, most of the high end WLAN Access Point manufacturers and WLAN integrators have their own controller. The Wireless Systems Controller Architecture is shown in Figure 6.2

The controller is a dedicated hardware and software appliance designed to manage access points. It can be implemented as a standalone device or it can be embedded into switches or routers. Its most common functions are: RF power management, channel selection, mobility control, load balancing, AP firmware loading, and mutual authentication of AP and controller. These functions are pertinent to APs. Additional functions such as encryption and data forwarding can also be performed in the controller. For these functions, a standard protocol for the communications between controller and AP is required. In 2005, NTT DoCoMo and Cisco submitted a proposal to the Control and Provisioning of Wireless Access Points (CAPWAP) consortium to standardize this link,

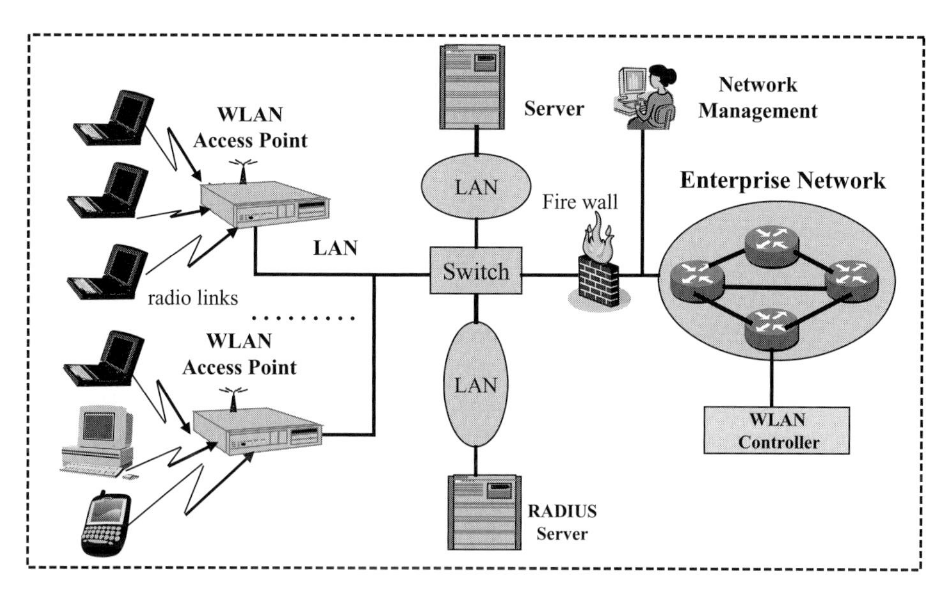

Figure 6.2 Wireless Systems Controller Architecture

called Lightweight Access Point Protocol (LWPP). As of September 2007, the standard has not been ratified [22].

A short list of major WLAN Systems follows:

- **Airspace** (now Cisco): 3500 WLAN Controller, (www.airspace.com);
- **Aruba Networks**: Aruba MMC-6000 Modular Mobility Controller, MMC-3000, MC-2400, Aruba Mobility Management System, (www.arubanetworks.com);
- **Cisco Systems**: Cisco 4400/2100 Controllers, Cisco WCS, (www.cisco.com);
- **Colubris Networks**: Colubris Intelligent MultiService System (CIMS), (www.colubris.com);
- **HP: ProCurve**: Switch and Access Controller Module, (www.procurve.com);
- **Meru Networks**: Meru Wireless Controller 3000, (www.merunetworks.com);
- **Motorola**: Canopy Wi-Fi, (www.motorola.com);
- **NEC**: Univerge WL3000series Controllers, (www.necunifiedsolutions.com);
- **Proxim Wireless**: ORINOCO Smart Wireless Controller, (www.proxim.com);
- **ReefEdge Networks**: WLAN System Controller, (www.reefedge.com);
- **Siemens**: MobileConnect Controller MC 50, 250, 1500, (www.siemens.com)
- **Symbol Technology** (now Motorola): AirBeam Controller, (www.sysmbol.com);
- **Trapeze Networks**: Mobility Exchanges Controllers MX-200, MX-400, Trapeze Mobility System Software, (www.trapezenetworks.com).

6.10 Advantages and Disadvantages of WLAN Technologies

The WLAN concept and design largely follows the steps taken by one of the most successful technologies developed in the past quarter of century, the ubiquitous Local

Area Network (LAN). Therefore, WLAN performing as a LAN without wires, has assured a very large market, from home networking to large enterprises. However, the novelty of wireless links is just one part of the story because the terminals are not just un-tethered but they are allowed a certain degree of mobility. The advantages and benefits of using WLAN technologies and networking include:

- Use of the techniques and experience accumulated by LAN technology;
- Many computer terminals and some handsets come with built-in Wi-Fi;
- Allows nomadic terminal moves and a good degree of mobility;
- Operation in unlicensed spectrum, avoiding the burden of huge expenses;
- Based on international standards, augmented by industry alliances;
- Supports data and voice communication at acceptable QOS levels;
- Extensive vendor acceptance and very large markets;
- Wi-Fi Alliance certification program in place for 802.11x-based products;
- Mature technology with a decade of accumulated experience;
- Easy and quick device-based implementation; and
- A cost effective solution in comparison with its wired counterpart.

But, there are some limitations or disadvantages in using WLANs:

- Short range of communications, generally less than 100 meters;
- High power consumption of terminals and access points;
- Overall performance trails wired LAN performance;
- No worldwide consistency in spectrum allocation;
- The relatively low data transmission rates make WLANs unsuitable for video and multimedia applications;
- There are major security concerns since early WEP-based encryption has been cracked;
- Omni-directional communication makes WLAN vulnerable to eavesdropping;
- There is an unintentional and intentional use of access points by outside parties;
- The use of a shared unlicensed spectrum makes WLANs prone to interference;
- Performance is low in high density areas such as libraries, hotspots, apartment buildings; and
- WLANs require careful engineering to avoid multipath interferences and blind zones.

6.11 WLAN Access Points and WLAN Service Platforms

The critical node of any WLAN is the Access Point. The AP concentrates communications from multiple wireless terminals, tracks their mobility, and connects the WLAN to a wired LAN, typically represented by a switching device. With exploding growth in the WLAN market, a large variety of APs have been designed to support business needs. WLAN AP products can be divided in three classes: SOHO, enterprise, and specialized Access Points. Further, several vendors provide integrated WLAN platforms with management capabilities, scalable to support large organizations and city-wide mesh networks. In some instances, management functions are performed by the AP itself. In other instances, the WLAN platform includes a powerful switching device that performs management functions.

Examples of Enterprise Class Access Points include:

- **Aruba Networks**: Aruba AP-70, AP-80M, AP-124 802.11n (a/b/g);
- **Cisco Systems**: Aironet 1100 (b), Aironet 1200 (a/b/g) series Access Points;
- **Colubris Networks**: Secure Wireless AP (b);
- **Enterasys**: RoamAbout RBT-4102, RBT-1002, RBT-602 (b/g);
- **Extreme Networks**: Altitude 350–2 AP (b/g), Summit Wireless Mobile Switch;
- **Foundry Networks**: Iron Point 200 AP (a/b/g), Iron Point-FES AP;
- **Intel Corporation**: PRO/Wireless LAN AP;
- **Lucent Technologies** (now Alcatel): AP-500, AP-1000;
- **Meru Networks**: Meru AP150;
- **Nokia**: A32 AP 2 (only WLAN AP);
- **Nortel**: WLAN Access Port 2230;
- **Proxim Wireless**: ORINOCO AP-4000 Enterprise Access Point (a/b/g);
- **Proxim Wireless**: ORINOCO AP-700 Enterprise Access Point (a/b/g);
- **Samsung**: MagicLAN SWL-3000AP;
- **Symbol Technologies**: AP-5131, AP-5181, Wireless Switch WS 5100, 2000;
- **Value Point Networks**: Super AP510 (b/g), Super AP600;
- **Value Point Networks**: Super AP600 802.11a Rugged Outdoor AP/Bridge;
- **Trapeze Networks**: Mobility Points AP MP-371/372, MP-422, MP-620.

A group of selected APs that are part of a larger enterprise WLAN design and implementation is presented in Table 6.5.

SOHO Class Access Points

- **Apple**: AirPort Graphite;
- **Cisco Systems**: Aironet 350 (b) series APs;
- **D-Link** (now Cisco): DWL-1000AP;
- **Intermec Technologies**: Mobile LAN Access WA 21, WA 22;
- **Lynksys**: Dual-band WAP55AG AP (a/g), WAP54G (g) AP;
- **NetGear**: ProSafe Dual Band Wireless AP, Prosafe 802.11g AP;
- **Nokia**: MW 1642;
- **SMC Networks**: SMC Elite Connect wireless AP.

Specialized Class Access Points

- **BreezeCOM**: WBC-DS.11;
- **Cirond**: AirPatrol Mobile;
- **Cirond**: WINC Manager 2.0;
- **Cisco Systems**: Aironet BR 340 Series;
- **SMC Networks**: SMC Wireless Hotspot Gateway;
- **Value Point Networks**: Wireless Controller WC 3000.

In Table 6.5 we presented several high-level evaluation criteria. These criteria were focused on compliance with IEEE WLAN standards, security standards, support for

Table 6.5 WLAN Access Points Evaluation

Vendor	Platforms and Products	Data Link Protocols	Networking Standards	Maximum Data rate	Special Features
Aruba	Aruba MMC-6000 WLAN		802.1x, 802.11a/b/g		Aruba Mobility Mg. System
	Mobility Controller		802.3af, 802.3u		CLI, Web GUI, SNMP V3
	Aruba AP-70, AP-80M	802.11a/b/g	802.11e QOS	54 Mbps	Power over Ethernet
Cisco	Cisco 4400/2100 Controllers		802.1x, 802.11a/b/g		Cisco Wireless Controller Sys.
	Aironet 1240AG Series AP	802.11a/b/g	802.3af, WPA,WPA2,	54 Mbps	CiscoWorks Mg. System,
	Aironet 1250 Series APs	802.11n	WEP, TKIP, LEAP	300 Mbps	PoE, Cisco SCCP, Wi-Fi WMM
Enterasys	RoamAbout Switches	802.11a/b/g	802.1x, 802.11a/b/g		RoamAbout Switch Manager,
	RBT-8400, 8210, 8110		802.3af, 802.11i	54 Mbps	NetSight Mg. System.
	AP RBT 4102/1002, 1602	802.11 b/g	802.11e, WMM		Power over Ethernet
Proxim	ORINOCO Smart Wireless		802.1x, 802.11a/b/g		ProximVision
	Controllers	802.11a/b/g	802.3af	54 Mbps	Network Management
	AP AP-400, AP-700		802.11e		Power over Ethernet
Symbol Motorola	Wireless RF Switch 7000	802.11n	802.1x, 802.11a/b/g	300 Mbps	Motorola RF Mg. Suite.
	Wireless Switch WS 5100		802.3af		Power over Ethernet.
	AP AP-5131, AP-5181	802.11a/b/g	802.11i, WPA2	54 Mbps	Wi-Fi WMM
Trapeze	Mobility System Software		802.1x, 802.11a/b/g		RingMaster Mg. System
	Mobility Exchange MX-400,	802.11a/b/g	802.3af, 802.11.e	54 Mbps	Power over Ethernet
	Mobility Points MP-371/372		WPA, WPA2, TKIP		Wi-Fi WMM, Spectralink SVP
Nortel Networks	WLAN 2300 Solutions		802.1x, 802.11a/b/g		WLAN Management Software
	AP-120, AP-2330, AP-2332	802.11a/b/g	802.3af, EAP	54 Mbps	2300, PoE

PoE, and management capabilities. In addition, there are other features that should also be considered when evaluating WLAN Access Points:

- Antenna type: Dual band, omnidirectional, diversity;
- Interference mitigation: Scanning channels for signal noise and load, automatic channel assignment, and power regulation;
- Dynamic RF capabilities: Load balancing, self-healing;
- Roaming capability: Support for 802.11r, fast roaming, fast handoff;
- Access control preserved across subnets;
- Integrated RADIUS server;
- Rogue AP intrusion detection, location, and containment;
- Inclusion of a user and group-based access control list;
- Protocol Analysis capability;
- Support for WLAN QOS: 802.11e standard or SpectraLink VoIP support.

6.12 WLAN Layered Communications

The 802.11 standards provide detailed specifications for the Medium Access Control (MAC) Layer and the Physical Layer. The general MAC Protocol Data Unit (PDU) or MAC Frame consists of a sequence of fields in a specific order, each field having a well-defined function down to the level of individual bits. This is the format passed down to the Physical Layer to be transmitted in the wireless medium.

The MAC frame contains a MAC Header, a Frame Body, and a Frame Check Sequence (FCS). The MAC Header consists of the following fields: Frame Control, Duration/ID, and Address. The Frame Body is a variable-length field that contains information specific to the frame type. The FCS is a 32-bit IEEE Cyclic Redundancy Code (CRC) used to detect and correct transmission errors. The MAC frame format is presented in Figure 6.3.

The frame control, a 2 byte field, consists of several sub-fields to indicate protocol version, frame type and subtype, direction of the frame (to or from destination), management, and security (WEP). Duration/ID is a 2 byte field that indicates the Association Identity (AID) value (1–2007 range) of the Power Save Poll (PS-Poll) control frame types. The duration value is determined by the frame type. For frames transmitted during the Contention-Free Period (CFP), the duration value is 32768.

There are four address fields: Source Address (SA), Destination Address (DA), Transmission Station Address (TA), and Receiving Station Address (RA). Depending on the frame type, two or four address fields are used (address 1, 2, 3, and 4).

There are three major types of MAC frames; **Control Frames** (e.g., Request to Send-RTS, Clear to Send-CTS, Acknowledgment-ACK, Power-Save Poll (PS-Poll), **Data Frames**, and **Management Frames** (e.g., beacon, association, reassociation).

Given the variety of LAN and WLAN environments, IEEE decided to divide the Data Link Layer into two sub-layers. Thus, the common upper sub-layer division, the **Logical Link Control** (LLC), was added, as specified in the **IEEE 802.2** standards. It is common to all LAN and WLAN technologies. LLC is an encapsulation

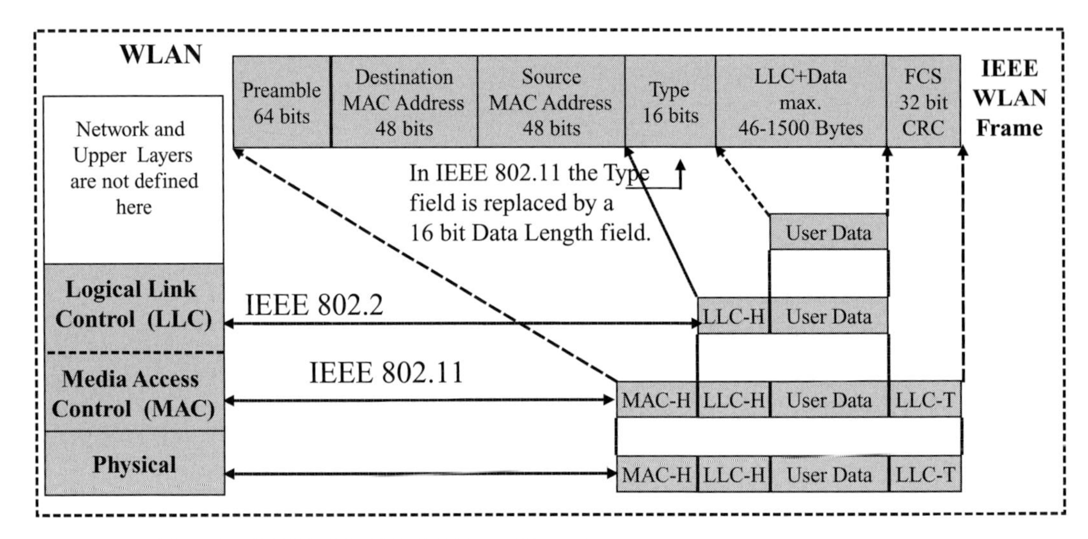

Figure 6.3 WLAN 802.11 MAC Frame Format

protocol that operates between the Data Link Control header/trailer (in WLANs the MAC sub-layer header) and the Network Layer header (Internet IP, or other standard or proprietary Network Layer protocols). It consists of three fields: Destination Service Access Point (DSAP), Source Service Access Point (SSAP), and Control field, as indicated in Figure 6.4.

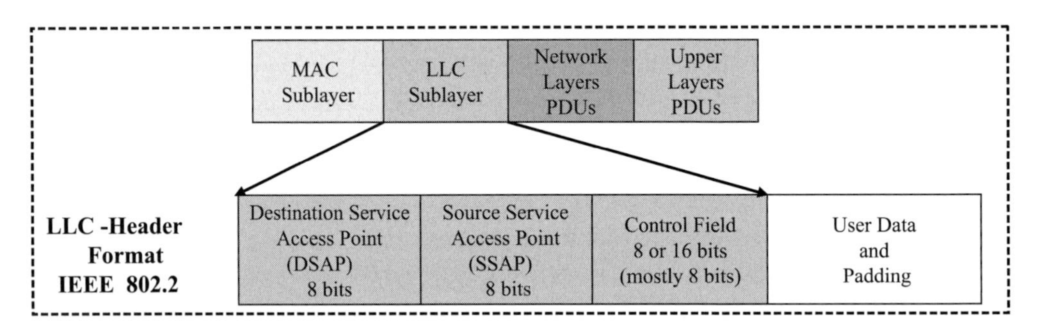

Figure 6.4 Logical Link Control Frame Format

The transfer of information across radio link interfaces between WLAN terminals and Access Points requires a layered communications stack, such as those defined by IEEE 802.11 standards. See Figure 6.5. One possible scenario is that based on the Internet IP Network Layer and the TCP/UDP Transport Layer protocols.

6.13 WLAN Management Requirements

It is generally accepted that once a wireless network exceeds 10 Access Points there is a need for some sort of management application or a management platform. Managed

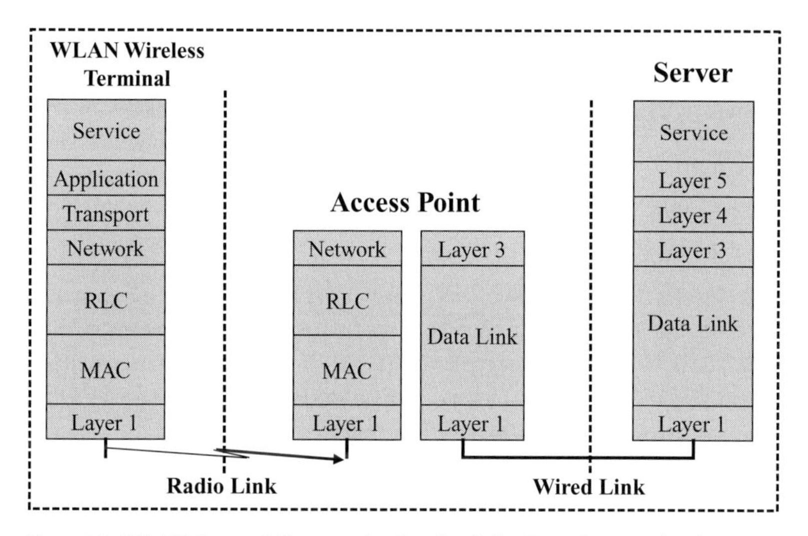

Figure 6.5 WLAN Layered Communication Stack for Data Communications

components include, beside APs, computing devices with built-in or attached WLAN adapters, and wireless switches. WLAN management requirements include:

- Management of multi-band APs: Many APs support 802.11a/b/g standards;
- Centralized monitoring and control: Configuration and profiles management, packet capture and analysis;
- Radio management services: Physical radio links, interference, MIMO antennas;
- Advanced mobility services: Given the nomadic nature of WLAN users;
- Security management: Authentication, encryption in accordance with IEEE 802.11i, Wireless Protected Access 2 (WPA2) specifications, and multi-level administrative access;
- Quality of Service management: 802.11p and 802.11q as well as new developments providing support for Voice over WLANs.

The **IEEE 802.11v** document provides amendments to the IEEE 802.11 Physical and Medium Access Control (PHY/MAC) layers that enable management of attached stations in centralized or distributed fashions (e.g. monitoring, configuring, and updating). The new IEEE 802.11k Task Group is responsible for defining messages that will be used to retrieve information from mobile stations. However, the ability to configure a station is not in its scope. The same task group will also create specifications for an Application Management Information Base (AP MIB).

The need for 802.11v specifications is driven by the current IEEE 802.11 standards that imply that stations may be managed via SNMP. However, the use of SNMP specifications introduces the following problems:

1. Very few stations in the market include SNMP capabilities;
2. Use of the secure SNMP protocol (e.g. SNMPv3) requires significant pre-configuration of the mobile station and, currently, it has had a limited implementation;

3. Management of stations may be required prior to the establishment of an IP connection. There are cases where a device must be managed because it cannot get IP connectivity. Therefore a standardized approach to manage stations is needed;
4. The 802.11 AP design has significantly increased complexity and the number of value-added features, but these additions cannot be supported by the current MIB. The task group needs to either expand the existing MIB or create a new MIB to support these new devices.

The **Internet Engineering Task Force** (IETF) is working to give standards-based support in the area of APs, namely "**Control And Provisioning of Wireless Access Points**" (CAPWAP). The original CAPWAP Work Group charter included drafting a problem statement and a taxonomy of architectures. The new charter for the CAPWAP WG proposed building upon the original charter to develop a protocol that provides interoperability among WLAN back-end architectures. The intent of the CAPWAP protocol is to facilitate control, management and provisioning of WLAN Access Points. There is a need for specifying the services, functions and resources related to 802.11 APs to allow for interoperable implementations.

The CAPWAP Architecture Taxonomy draft references two classes of centralized WLAN architectural solutions, namely the **Local MAC** and the **Split MAC**. The protocol defines the CAPWAP control plane, including primitives to control data access. An effective centralized CAPWAP architecture will impact how WLAN data traffic is managed over the back-end network. This is achieved by controlling how data is forwarded and by specifying data payload formats that will ensure interoperability between CAPWAP vendors.

6.14 WLAN Management Products

The IEEE is close to finalizing 802.11v specifications that define management requirements and technical specifications for products providing WLAN management. This has spurred the development of several products in the form of management platforms or applications, as shown below.

- **AdventNet**: ManageEngine Wi-Fi Manager, (www.adventnet.com);
- **Agilent**: Wireless Service Manager, (www.agilent.com);
- **AirDefense**: AirDefense Guard 4.0, (www.airdefense.net);
- **AirMagnet**: AirMagnet Distributed Management 4.0, (www.airmagnet.com);
- **AirSpace** (now Cisco): Wireless Enterprise Platform, (www.airspace.com);
- **AirTight**: SpectraGuard, (www.airtight.com);
- **AirWave**: AirWave Management Platform (AMP), (www.airwave.com);
- **Aruba**: RF Director, Mobility Management System, (www.arubanetworks.com);
- **Computer Associates**: Unicenter TNG, (www.ca.com);
- **Cisco Systems**: Cisco Works Wireless LAN Solutions, (www.cisco.com);
- **Cisco Systems**: Cisco Structured Wireless-Aware Network (SWAN);
- **Foundry Networks**: IronPoint Wireless Products Management Solutions, (www.foundrynetworks.com);

- **Intel Corporation**: PRO/Wireless LAN AP, (www.intel.com);
- **HP**: OpenView Management Platform, (www.hp.com);
- **Proxim Wireless**: ProximVision Management System, (www.proxim.com);
- **Qualcomm**: Qconnect Managed Wireless Service, (www.qualcomm.com);
- **Newbury Networks**: WiFI Watchdog 3.0, (www.newburynetworks.com);
- **Network Chemistry**: Wireless Intrusion Protection System 2.0, (www.networkchemistry.com);
- **Network Instruments**: Expert Observer 9.0, Advanced Wireless Probe, (www.networkinstruments.com);
- **Trapeze**: Mobility System Software and RingMaster 3.0, (www.trapeze.com);
- **Symbol Technologies**: SpectrumSoft™ Wireless Network Management System (WNMS), (www.symbol.com);
- **WaveLink**: (an application that works with HP NNM), (www.wavelink.com);
- **Wild Packets**: AiroPeek NX 2.0, RF Grabber 1.1, (www.wildpackets.com).

6.15 Voice over WLAN Architecture

A new development on the road to convergence of data and voice communications is the addition of voice to WLANs. **Voice over Wireless LAN** (VoWLAN) technology was designed to allow voice communications over existing WLANs. The IEEE has developed standard specifications, and products were certified by the Wi-Fi alliance.

A typical VoWLAN consists of **Wireless Terminals** (WT), standard laptops, desktops, PCs, PDA clients, and specialized dual mode handsets. All communicate with a WLAN **Access Point** (AP) across **radio links**. The VoWLAN architecture is depicted in Figure 6.6. The architecture may include a security firewall and RADIUS servers.

This architecture resembles the WLAN architecture presented in Figure 6.1, with one exception, addition of support for voice as represented by the set of handsets. Ultimately, these handsets, when convergence is achieved between WLAN and cellular mobile networks, will be dual mode handsets. In the same vein, it is highly desirable that voice communication be based on Voice over IP (VoIP) implementations. Given the intolerance of voice transmission to delay and jitter, there is a strong need for a QOS mechanism that gives voice priority over data. This should be implemented in a queuing mechanism in both access points and switches.

Roaming is another issue because the user is moving within the coverage area between access points. Currently, reauthorization and authentication is required while roaming between APs and no interruption of communications or dropping of packets is acceptable. Other concerns when providing VoWLAN are security for both voice and data against intrusions, interoperability among multiple vendors, and adequate available WLAN bandwidth.

Therefore, an ideal VoWLAN implementation should include support and control of VoIP, roaming, handoff management within and outside the WLAN (cellular mobile), and security features including VPN terminations.

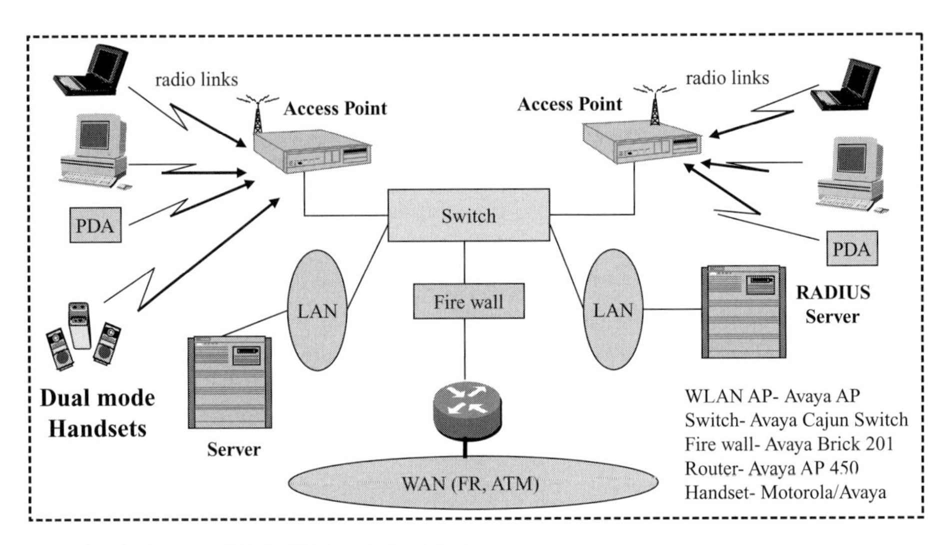

Figure 6.6 Voice over WLAN Network Architecture

6.16 Challenges of Transmitting Voice and Video over WLAN

Given the speed with which WLAN implementations are spreading at all levels – enterprise, small offices, and homes, and the general support for VoIP throughout the industry, interest and support for VoWLAN is a natural continuation. However, there are challenges ahead, as described below.

- **An extended WLAN infrastructure design**: Given the habits of mobile talkers and the lack of tolerance for delays/interruptions or low quality of voice calls, there is a need to provide complete RF coverage throughout the whole facility, including elevators, staircases, storage closets, restrooms, etc.;
- **Roaming across VoWLAN domains**: Standards need to be developed to assure transparent handoffs between access points (similar to those in cell networks) such as 802.11k (radio resource management), 802.11r (fast roaming), and 802.11v (SNMP-based network management);
- **New VoWLAN standards for power usage**: Given the limits on current handsets output power, (which ranges from 20–50 mW as compared to 50–100 mW power of PC WLAN), it will be necessary to raise AP output powers to assure full coverage;
- **Limited capacity of APs handling VoWLAN**: The number of simultaneous active calls possible per AP (6 to 20) is limited and depends on the particular type of 802.11 implementation;
- **Quality of Service in VoWLAN**: Final standards for QOS in WLANs, IEEE 802.11e, are needed to prioritize and differentiate active voice calls from data calls. Packet loss, latency, and jitter are accentuated by the limited bandwidth in a shared access medium [23];
- **Support for location discovery**: Most systems can report which AP a handset is using and the vendors are expected to include E911-like services as the technology matures;

- **Load balancing**: There is a need to avoid congestion by automatically redirecting calls from overwhelmed access points to less congested access points;
- **Matching cellular handsets**: The development of VoWLAN handsets should be driven by integration of new features in the more advanced (in form and functionality) cellular handsets to create truly dual mode or multi-mode convergent handsets;
- **VoWLAN** is also referred as wVoIP, VoFi, VoWi-Fi or Wi-Fi VoIP. The concept is several years old, but, at the beginning, the handsets were too pricey and the standards for roaming, QOS, and power usage, were not fully implemented.

In 2004 about 150,000 units of pure VoWLAN were sold. The estimate for 2009 is 17 million units, while the estimate for dual-mode handsets ranges from 30 to 100 million units. This number is still small compared with the 700 million mobile phones sold in 2004 and the projected 1 billion in 2009. Therefore, the penetration of the VoWLAN market in 2009 is estimated to be around 3–5%. It is projected that the VoWLAN market will take off in the 2007–2008 time frames.

Vertical industries that combine the constant needs of accessibility and mobility are strong candidates in implementing VoWLAN. Other strong candidates are universities having rich WLAN environments but who are waiting for a more mature QOS technology and for reduced handset and services costs. Full WLAN coverage area across smaller townships or even metropolitan zones is another candidate. VoWLAN offers solutions of mobile connectivity within building structures where it is known that cell service is poor if not non-existent, provided that a WLAN infrastructure is in place.

Standards for roaming across VoWLAN domains call for interruptions caused by handoffs to be less than 50 ms (see SONET ring recovery). Note that in WLANs, handsets are responsible for initiating and controlling the roaming, as opposed to the BSC/BST in cellular networks. Currently, the AP handover may require anywhere from 25 ms to 3 seconds. The 802.11k standard for radio resource management will specify that VoWLAN handsets have a table of access points that includes a service ID, channel assignment, signal strength, and usage load. The 802.11k standard will supplement two other draft standards, 802.11r (fast roaming) and 802.11v (SNMP-based management).

To reduce capacity problems, it is desirable to separate the use of WLANs (802.11a for voice) and 802.11b for data (for example Motorola CN620 dual-mode handset). Secondly, there is a need to reduce WLAN cell size by increasing the number of APs. Third, using a call admission control mechanism such as SpectraLink, widely adopted by most of the VoWLAN vendors, a busy signal will be launched, and the algorithms specified in 802.11k and 802.11v will help spread the load among the APs.

All the major players in WLAN that were listed earlier have a vested interest in promoting VoWLAN. Those vendors/manufacturers who integrated voice support and especially VoIP support into their initial WLAN design from the beginning will have a better start and a larger market share. Introduction of the Session Initiation Protocol (SIP) in conjunction with VoIP is another important step on the road to IP-based Multimedia Subsystem (IMS) integration, the global solution for WLAN and cellular mobile convergence. A list of some VoWLAN vendors and their products with some important characteristics is presented in Table 6.6.

Table 6.6 Voice over Wireless LAN Products Evaluation

Vendor	WLAN products	Voice QOS standard	Security Protocols/ Standards	Interoperability with other products	Management Fast Roaming Mechanism
Aruba	AP 60/65/70 2400/6000 Wireless Switches/Controllers	SVP Wi-Fi WMM	WEP, WPA, IPSec, 802.11i	All Wi-Fi Certified WLAN Products, Avaya	Mobility Management System 802.11i Preauthorization only; 802.11r when ratified
Avaya	Access Points AP-7/8 3631, 3641, 3645 VoWLAN Handsets	SVP, WMM plus 802.11e, AVPP	WPA & WPA2 with PSK 802.11i	Cisco, Meru Networks, Aruba Networks, Vocera	Multi Service Network Manager 802.11i Preauthorization only; 802.11r when ratified
Cisco	Aironet 1000 Series Cisco IP Phone 7920 Cisco 4400 Controller	SVP, Wi-Fi WMM	Cisco EAP, Cisco LEAP; Cisco EAP-FAST 802.11i, WPA2	Almost all Wi-Fi Certified WLAN Products	CiscoWorks Cisco Fast Secure Roaming 802.11r when ratified
Proxim	ORINOCO AP-4000 Controller	SVP, 802.11e Wi-Fi WMM	WEP, 802.11i	Alcatel, NEC, Nortel Internal testing only	ProximVision NMS 802.11i Preauthorization; 802.11r when ratified
Symbol	AP5131, AP5181 WS 2000/5100 switches	SVP Wi-Fi WMM	WPA, WPA2, 802.11i	3Com, Airspace, Aruba Avaya, Cisco, LinkSys Proxim	Motorola RF Management Suite 802.11i Preauthorization; 802.11r when ratified
Trapeze	MP AP MXR-2 MX-200, MX-400 Controllers	SVP Wi-Fi WMM	WPA, WPA2, WEP 802.11i	Enterasys, Nortel, 3Com D-Link	RingMaster NMS 802.11i Preauthorization; 802.11r when ratified

Legend: SVP-SpectraLink Voice Priority; WMM-Wireless MultiMedia; WEP-Wired Equivalent Privacy; WPA-Wi-Fi Protected Access; PSK-Private Security Key; AVPP-Avaya Voice Priority Processor.

A critical component of any VoWLAN product is support for QOS. Lacking a QOS standard for VoWLAN, SpectraLink created the Spectral Voice Priority (SVP) specifications. These specs allow configuration of Access Points (APs) to recognize voice packets and prioritize them by setting the AP back-off value at zero for voice packets and moving voice packets to the head of the transmission queue. SVP has been adopted by many vendors and has become a de-facto VoWLAN standard. However, after ratification of 802.11i standards, support for SVP may diminish. More details will be given in the next section.

Table 6.6 also shows the importance of support for standard security features: Wired Equivalent Privacy (WEP), WPA, and WPA2. Currently, most products support all security standards. Another important feature is a roaming capability based on early 802.11r specifications or on proprietary mechanisms. Last, but not least, a feature to prove interoperability with other WLAN products, verified by test as part of the Wi-Fi certification process. The Syracuse University Real-World Laboratory is one of the independent organizations performing WLAN testing.

6.17 WLAN QOS and VoWLAN QOS Metrics

IEEE Task Group "e" (TGe) has developed the 802.11e MAC specs that will support QOS in WLAN environments [23]. Two operating modes from the original 802.11 standards will still be supported: Enhanced Distributed Channel Access (EDCA) and Hybrid Coordination Function (HCF) Controlled Channel Access (HCCA). HCF replaces the old 802.11 Point Coordination Function (PCF) that has never been implemented. It includes a simple "listen-before-talk" mechanism that minimizes the chance of packet collisions in accordance with the Ethernet-like Carrier Sense Multiple Access/Collision Avoidance (CSMA/CA) environment. Also, it allows polling client stations during contention-free periods.

The IEEE 802.11e standard was ratified in late 2005. In the meantime, since then, the **Wi-Fi Alliance** has developed a subset of 802.11e called **Wi-Fi Multimedia** (WMM). Four priority classes are recognized for **voice, video, best effort, and background**. These WLAN priority classes map into Ethernet 802.1d prioritization tags as specified for wired LANs. This assures a consistent QOS mechanism across wired and wireless networks. Wi-Fi Alliance has already certified QOS product compliance to a subset of the draft specifications as implemented on Access Points, laptops, and LAN phones.

The limit values for latency, jitter, and packet loss that qualify QOS for voice services as either Good, OK, or Bad, are shown in Table 6.7.

In Section 5.5 we introduced the **Mean Opinion Score** (MOS), as defined in the ITU-T P800 standard, a subjective way of "measuring" voice quality. An "objective" tool for voice quality measurements is the ITU-T P. 862 standard **Perceptual Evaluation of Speech Quality** (PESQ) [24]. PESQ provides a quality score, known as PESQ score, which conforms to ITU-T P. 862. PESQ scores use a scale from -0.5 to 4.5, though in most cases they are between 1 and 4.5. These scores correlate with subjective quality scores used by MOS.

Table 6.7 Voice over Wireless LAN Quality of Service

Parameter	Description	Good	OK	Bad	Adverse effects
Latency	Packet delay in reaching the destination caused by traffic contention, queueing, and propagation time.	<150 ms	150–300 ms	>300 ms	High latency causes "talking over" type of conversations unless each party waits for the other to finish before they speak.
Jitter	Variability in arrival delay of successive packets caused by traffic contention on certain routes and bursts of data.	<20 ms	20–50 ms	>50 ms	High jitter causes unacceptable quality of conversations because of choppy voice and glitches.
Packet Loss	Caused by uncorrectable errors in transmission. Data packets are recovered by requesting a retransmission. Lost voice packets are discarded.	<1%	1–2.5%	>2.5%	High percentage of packet loss causes audible "voice clipping" unexpected pauses, and audible white noise.

Wi-Fi Multimedia (WMM) specifications provide prioritization categories by allocating different fixed and random waiting periods for various wireless applications. The fixed waiting interval is called **Arbitration Inter-Frame Space Number** (AIFSN). The random interval is called the **Contention Window** (CW). This is indicated in Table 6.8.

Table 6.8 Wi-Fi Multimedia QOS Specifications

Category	AIFSN	Contention Window	Total initial waiting time
Voice priority	2 slots	0–3 slots	2–5 slots
Video priority	2 slots	0–7 slots	2–9 slots
Best Effort priority	3 slots	0–15 slots	3–18 slots
Background priority	7 slots	0–15 slots	7–22 slots

Both intervals are expressed in time slots and each helps to avoid collisions of packets belonging to the same category while giving a chance to each multimedia application to exchange information with minimum delay. The time slot for 802.11b is 20 μs and for 802.11a and 802.11g it is 9 μs.

There are many WLAN local service providers but only a few have or plan to extend their coverage to be nationwide. This coverage would entail tens of thousands of hot-spots in the USA and hundreds of thousands worldwide. Examples of Wi-Fi hot spot service providers are: AT&T, Boingo Wireless, BT, Fiberlink, iPass, T-Mobile, Vanco, Verizon Business, and Wayport, Inc.

7 Wireless Personal Area Networking

7.1 Wireless PAN Architecture

A typical residential network infrastructure, shown in Figure 7.1, consists of three types of network: a local wired loop that connects the residence to the serving telephone company (voice/data or xDSL), a coaxial cable that connects the residence to a cable company (video, data, voice with an alternative for video broadcast using a satellite network), and the AC power lines that connect the residence to the power company.

A **home network**, also known as a **Personal Area Network** (PAN), is intended to integrate and standardize the use and interaction of home end-devices and appliances. PAN facilitates and supports the interconnection of multiple computing data devices and peripherals (printers, scanners, digital cameras, video cameras), voice and video communications, music distribution (MP3 and CD players), and use of surveillance devices to command and control appliances (meter reading, temperature and light regulation). Wireless Personal Area Network (WPAN) is the new generation of PANs that uses wireless communication to connect the PAN components. A typical WPAN infrastructure consists of a **concentration point** in the form of a **Residential Gateway** (RG) with a **connectivity component** to a broadband network in the form of a **Network Gateway** (NG). This is shown in Figure 7.2.

Both wired and wireless communications capabilities can be used for connectivity between the RG and the NG. Connection types include unshielded twisted pairs, coaxial cable, fiber optic, or power lines. Similarly, connectivity between the RG and various devices could be wired or wireless with a dominance of wireless links in the WPAN. This includes wireless NIC cards, set-top boxes, wireless faxes, and a home management system as a concentration point for home appliances and environmental control devices.

To create a WPAN infrastructure, a standardized mechanism is required to connect devices to the residential gateway using the wireless medium. This mechanism should be supported or enabled by all the devices, including the residential gateway. A comparison of the current WPAN architecture, a wired PAN, and the future WPAN architecture to support all the data, voice, video, and telemetry digital applications is given in Table 7.1.

From this table, we see that a PAN can be built using a wireless environment, as opposed to a wired environment. In the future, it is expected that fixed wireless networks data rates will match data rates achieved in wired environments, with the advantages of providing high mobility and greater flexibility, at moderate costs.

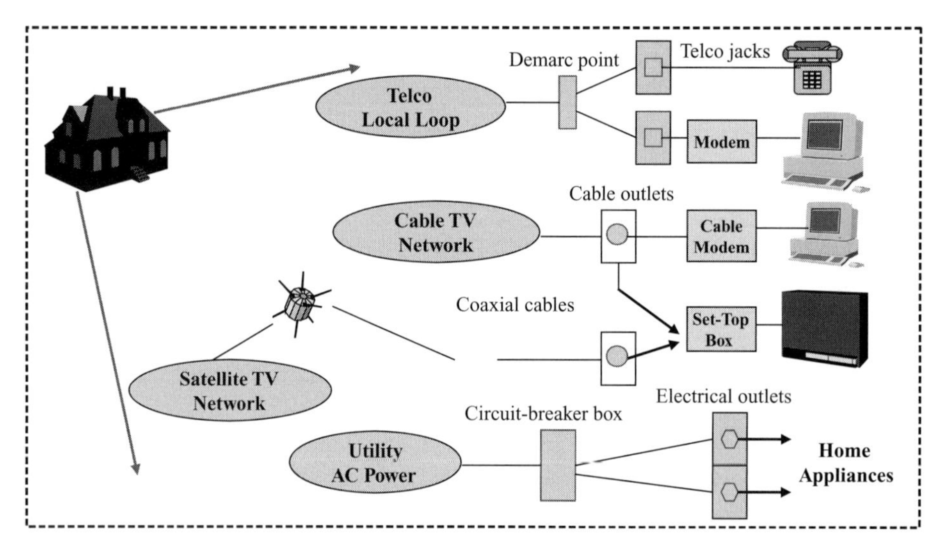

Figure 7.1 A Residential Network Infrastructure

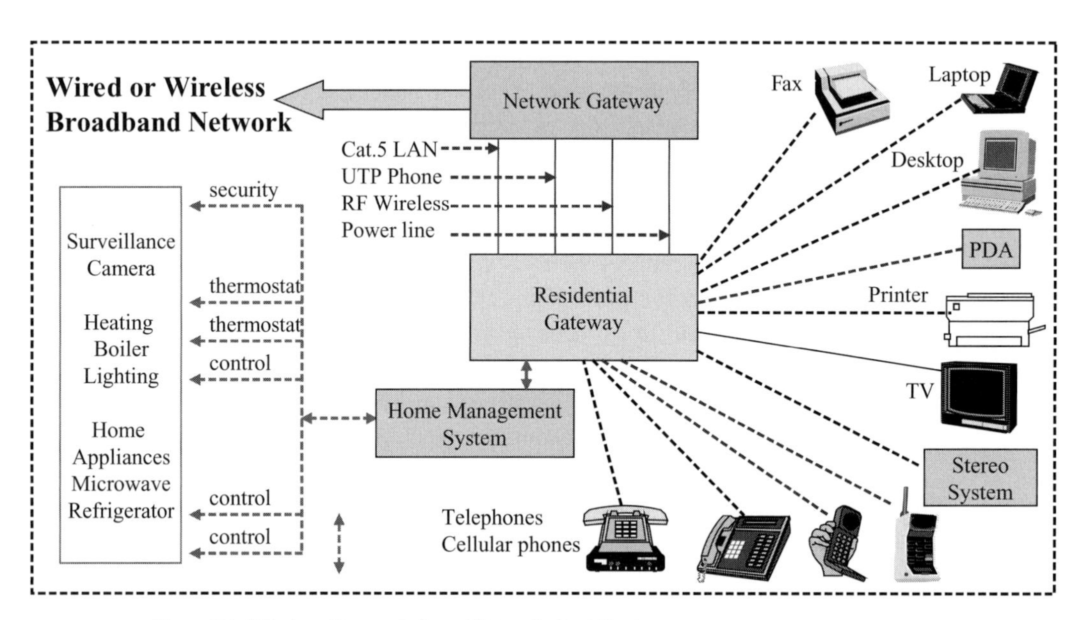

Figure 7.2 Wireless Personal Area Network Architecture

7.2 WPAN Networking Solutions

There are several competing solutions that can be used to build and manage a WPAN infrastructure. These are shown in Table 7.2. One group of candidates are those used in WLANs, i.e., IEEE 802.11 a/b/g/ and their European counterparts, the Broadband Radio Access Networks (BRAN) and its successors Hiper LAN1, and HiperLAN2. Two solutions, Bluetooth and ZigBee, designed specifically for WPAN, dominate the

Table 7.1 PAN and WPAN Comparative Infrastructures

Characteristics	Existing Wiring	Wireless	New Structure
PAN concept	Interconnection of stationary end-systems and appliances.	Interconnection of stationary and mobile devices.	New construction or home remodeling.
Useful lifetime	Short	Long	Very long
Insertion points	In each room multiple electrical outlets; not all the rooms with phone jacks; one or two cable outlets in each home.	Anywhere within home; compensates the lack of phone and cable outlets in each room.	Anywhere needed
Security	Secure	Less secure	Very secure
Current data rates	10 Mbps	6–11 Mbps	100 Mbps
Future data rates	10–100 Mbps	31–100 Mbps	1000 Mbps or more
Standards	ITA/EIA-570A, IEEE 1394, EIA-600 VESA/IEEE 1394b	IEEE 802.11a/b/g WLAN IEEE 802.15.1 WPAN/Bluetooth IEEE 802.15.4 WPAN/ZigBee	Well defined global standards
Cost	Low	Moderate	Very high (installation)

WPAN infrastructure development. They have been standardized by IEEE as 802.15.1 [25], and 802.15.4 [26], respectively. More details will be provided about these two technologies in subsequent sections of this chapter.

Two other technologies with limited applicability and implementations, can also be considered as part of WPAN solutions, the **Digital Enhanced Cordless Telecommunications** (DECT) and **Home RF2**.

Several types of modulation are used in WPAN solutions, ranging from Direct Sequence Spread Spectrum (DSSS), to Dual Multitone (DMT), to Orthogonal Frequency Division Multiplexing (OFDM), and to Gaussian Phase-Shift Keying (GPSK). The maximum data rate provided by these technologies varies from 128 Kbps up to 54 Mbps. The range of action for most of these technologies is less than 100 meters. Common implementations for Bluetooth work in a range up to about 10 meters.

7.3 Bluetooth WPAN Architecture

Originally **developed by Ericsson** as a cable-replacement technology, the Bluetooth wireless technology allows devices to be connected over short-range wireless links. The applicability of Bluetooth, within its range of distance, is almost limitless and is captured by standardized **Bluetooth profiles**.

Bluetooth wireless technology is based on specifications for a low-cost, low-powered radio and associated protocol stacks that can be used to provide **short-range wireless links** between notebook/laptop computers, mobile phones, PDAs, and other personal portable electronic devices. The specifications, currently at **version 2.0**, are developed by

Table 7.2 WPAN Technologies Evaluation

Standard	Application	Main Characteristics	Frequency band	Modulation type	Maximum data rate	Standard body	Certification body
802.11b/g	Wireless LAN	25 MHz BW 100 m range	2.4 GHz	Direct Sequence Spread Spectrum	11 Mbps 54 Mbps	Institute of Electrical and Electronics Engineers (IEEE)	Wireless LAN Interoperabilty Forum (WLIF)
802.11a	Wireless LAN	25 MHz BW 100 m range broadband	5 GHz	DMT/OFDM	54 Mbps		
BRAN Hiper-LAN1 Hiper-LAN2	High-speed Multimedia LAN	25 MHz BW 80 m range	5 GHz	GPSK/OFDM	24 Mbps/31 Mbps	European Telecommunication Standards Institute (ETSI)	
DECT	Voice & data for SOHO	Integrated voice & data	1.9 GHz	GPSK	1.15 Mbps	Electronics Industry Association (EIA)	
Bluetooth	WPAN All devices	Low-cost 3–100 m	2.4 GHz	Frequency Hopping Spread Spectrum	1.1 Mbps	Bluetooth Consortium (1.1/2.0) and IEEE 802.15.1	
ZigBee	Industrial Automation	Low-power 10–75 m	2.4 GHz	Frequency Hopping Spread Spectrum	128Kbps	ZigBee Alliance (1.0/1.1) IEEE 802.15.4	
HomeRF2	Short-range 50 m	Combines DECT and 802.11b	5 GHz	Frequency Hopping Spread Spectrum	10 Mbps	Home RF Working Group	

Legend: BRAN-Broadband Radio Access Networks; DECT-Digital Enhanced Cordless Telecommunications; FH-Frequency Hopping; DMT-Discrete Multitone; OFDM-Orthogonal Frequency Division Multiplexing; GPSK-Gaussian Phase-Shift Keying.

the **Bluetooth Special Interest Group** [27]. The original IEEE **802.15.1** [25] standard is based on Bluetooth **version 1.1**. Another IEEE standard **802.15.1a** corresponds to Bluetooth **version 1.2** [28]. It contains QOS enhancements and is backward-compatible with version 1.1.

Bluetooth wireless technology operates in the **2.4 GHz Industrial, Scientific and Medical** (ISM) band (2.4 to 2.4835 GHz). This spectrum is divided into 79 subchannels, 1MHz each, with hopping from channel to channel 1,600 times per second. Transmitting and receiving devices must synchronize to the same hop sequence to communicate. The maximum theoretical data rate of early versions was 1.1 Mbps. Maximum throughput is 400–700 Kbps, depending on the channel configuration. Bluetooth-enabled devices need to be fairly close together, typically no more than **10 meters** (30 feet) [28]. Unlike Infrared Data Association (IrDA) devices, a Bluetooth link does not require the devices be lined up precisely within line-of-sight of each other. For example, it is possible for a PDA and a portable computer to link to each other even if the PDA is inside a briefcase or a person's pocket.

A typical Bluetooth-based WPAN architecture is presented in Figure 7.3. Bluetooth-enabled devices are organized into groups of 2–8 devices and linked together in an ad hoc WPAN called a **piconet**. A piconet consists of a single master device and one or more slave devices. Up to 255 different idle or parked devices can be accessed and activated by the master. Two or more piconets can be interconnected to form a **scatternet** through bridge devices that play a double master-slave role. Within the same piconet, the master device performs service discovery, authentication, and establishes and terminates connections.

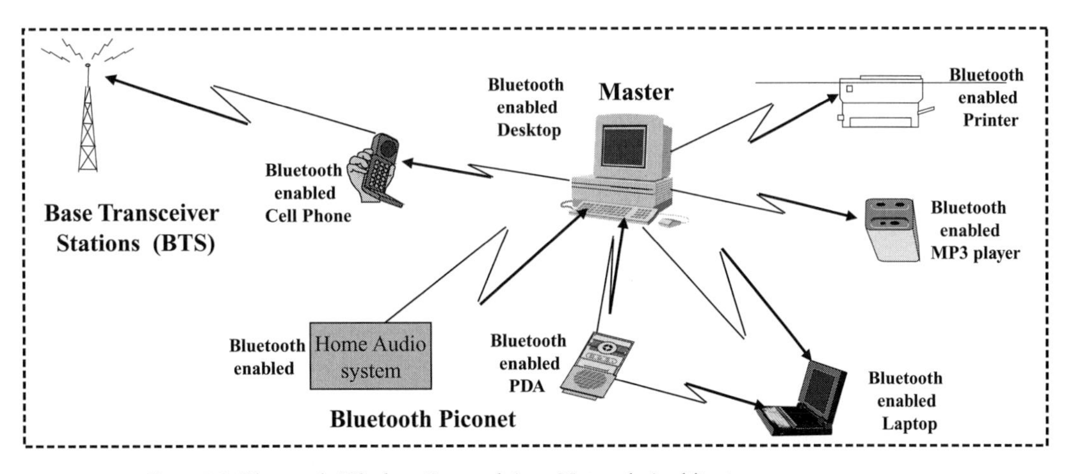

Figure 7.3 Bluetooth Wireless Personal Area Network Architecture

A connection is established between Bluetooth-enabled devices after a set-up phase during which the devices provide critical information to each other such as the device name and class, and the list of services supported. Specific information about devices such as technical features, manufacturer, and clock offset might also be provided. In the same piconet, the master laptop or desktop computer establishes a connection with a cell phone that acts as a modem, dialing the mobile service provider to access the Internet.

The radio and baseband controller are the heart of Bluetooth hardware solutions. Because of the controller's small size, low cost, and low power requirements, it can be incorporated into electronic devices or appliances such as cell phones and PDAs with their small form factors and low power requirements. It can also be implemented on Universal Serial Bus (USB) devices such as specialized PC cards or "dongles".

7.4 Bluetooth Protocol Architecture

Each Bluetooth device employs a **protocol stack** realized by hardware and software components interconnected through USB or PC Card physical buses. The Bluetooth Protocol Architecture is presented in Figure 7.4.

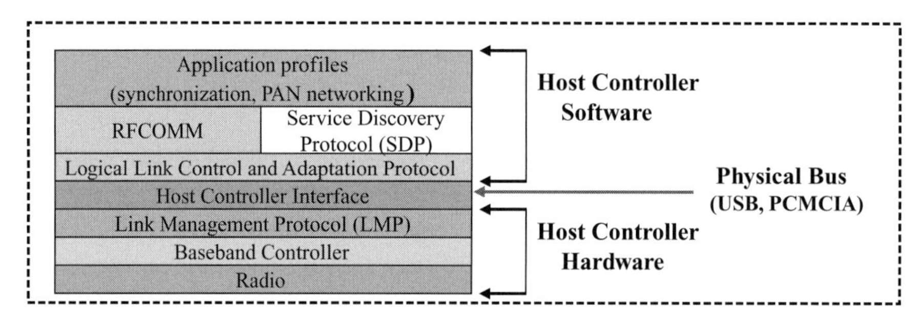

Figure 7.4 Bluetooth Protocol Stack

The hardware portion consists of the **radio baseband controller** and **Link Manager Protocol** (LMP), which is used to set up and control the link. LMP also implements the Bluetooth Data Link-level security. The upper Data Link sublayers of the stack consist of the **Logical Link Control** and **Adaptation Protocol** (LLCAP). This protocol is used to segment and reassemble data into packets for transmission. LLCAP interfaces with **client protocols** such as the **Bluetooth Service Discovery Protocol** (SDP), which enables applications to discover which services are available on a Bluetooth device. LLCAP also interfaces with **RFCOMM**, which enables a Bluetooth device to emulate a serial port. Finally, **application "profiles"** define how user scenarios (such as dial-up networking and synchronization of data between two devices) are accomplished. A profile specifies mandatory options and parameters for each protocol.

To **establish a connection**, a user initiates a Bluetooth connection between two devices (such as a notebook computer and PDA) that are close to each other. The link is used to perform a specific task such as transferring a file between the two devices. To initiate a connection, the user opens Bluetooth client software on one of the devices, usually the portable computer, and "discovers" nearby devices. In this case, it is the PDA that is configured to be "discoverable". For security purposes, a device can be configured so that it cannot be discovered by other Bluetooth devices. The following **discovery and pairing** processes can be initiated:

- **Name discovery**: The name of the Bluetooth device is detected. Default device names are designed to easily identify the device;
- **Service discovery**: Specific services available from the device are discovered.

These services are usually based on profiles defined in the Bluetooth specifications. In addition to low-level services associated with establishing connections, there are profiles for services such as serial port connectivity, dial-up networking, faxing, file transfers, and synchronization. Both devices must support a common Bluetooth profile such as synchronization. "Pairing", also referred to as "bonding", takes place when two devices form a private connection. Pairing is optional and is not required in all scenarios. For security purposes, a PIN (referred to here as a "passkey") is usually required before the connection is allowed. For example, to control access to a Bluetooth printer, the assigned PIN must be entered when pairing with that printer). Once the devices' services are discovered, the user can initiate a specific service such as a serial port connection.

7.5 Bluetooth Profiles

Bluetooth devices can communicate if they support the same Bluetooth SIG adopted profile. Bluetooth standards also allow developers to create other applications to work with devices that conform to Bluetooth specifications. The list of Bluetooth profiles follows. For some profiles a short explanation is attached. Most profile titles are self-explanatory [28].

0. Advanced Audio Distribution Profile (AADP): Used for mobile phone to headset high quality audio streaming;
1. Audio Video Remote Control Profile (AVRCP);
2. Basic Imaging Profile (BIP);
3. Basic Printing Profile (BPP);
4. Common ISDN Access Profile (CIP): Used to access ISDN data and signaling services;
5. Cordless Telephony Profile (CTP): Used for cordless telephones;
6. Device ID Profiles (DIP);
7. Dial-up Networking Profile (DUN): Standard access to Internet through dial-up modems;
8. Fax Profile (FAX);
9. File Transfer Profile (FTP);
10. General Audio/video Distribution Profile (GAVDP);
11. Generic Access profile (GAP);
12. Generic Object Exchange Profile (GOEP);
13. Hard Cable Replacement Profile (HCRP);
14. Hands Free Profile (HFP): Used to provide hands-free communication between a mobile phone and a car set device;
15. Human Interface Device (HID): Used to support communication for wireless mice, keyboards, joysticks;

16. Headset Profile (HSP): Allows use of a headset to extend/complement a mobile phone with all its features;
17. Intercom Profile (ICP);
18. Object Push Profile (OPP): Used to upload pictures, schedules;
19. Personal Area networking (PAN): Allows buildup of a piconet in a typical PAN environment;
20. Phase Book Access Profile (PBAP);
21. Serial Port Profile (SPP): Used to replace the RS 232 functionality in dial-up and fax communications;
22. Service Discovery Application Profile (SDAP);
23. SIM Access Profile: Used for SIM card recognition in external devices for mobile communication;
24. Synchronization Profile (SYNC): Allows synchronization of devices prior to information exchange;
25. Video Distribution Profile (VDP);
26. Wireless Application Protocol Bearer (WAPB): Used to carry WAP over Point-to-Point Protocol (PPP) over Bluetooth.

Bluetooth supports a variety of operating systems such as Linux, MAC OS X V10.4, Palm OS, Windows Mobile (WM) and Windows Vista.

Some profiles have multiple functions built into the specifications. To perform a particular task, each profile uses assigned options and parameters at each layer of the Bluetooth stack. For example, the Basic Imaging Profile (BIP) includes the following distinct functions:

- Image Push: To send images from the device;
- Image Pull: Browsing and retrieving images from a remote device;
- Image Printing: Send photo images to the printer in a known format such as those adopted by Kodak, Canon, or Fujitsu;
- Automatic Archive: Automatic backup of all images from a target device such as digital camera to a laptop;
- Remote Camera: Remote activation of a camera;
- Remote Display: Send images to a video projector.

Because the Bluetooth wireless technology shares its unlicensed spectrum with other 2.4 GHz technologies such as cordless phones, microwave ovens, and 2.4 GHz Wi-Fi networks (802.11b and 802.11g), there may be problems caused by interference when these networks operate in close proximity with each other. A common scenario is a portable computer which connects to the Internet network via a Wi-Fi connection and is synchronized with a handheld device via Bluetooth. The impact varies, depending on factors such as the distance from the 802.11b access point, the duration and timing of the Bluetooth and Wi-Fi transmissions, and the power of the devices.

These potential interference problems are mitigated by the "bursty" nature of Bluetooth and Wi-Fi connections. Bluetooth wireless technology is mainly used for short periods of time to perform operations such as synchronizing mobile devices to a PC several times a day. Thus, the opportunity for interference with an 802.11b connection is limited to these short periods of time. However, in a business environment, with many active Bluetooth devices, the interference could be noticeable.

7.6 Bluetooth Standards and Applications

Bluetooth development is an ongoing process, marked by a succession of several new versions. Some versions became reference points in the Bluetooth evolution, Version 1.2 and 2.0, specifications used by most of the Bluetooth products for example.

Version 1.0 was the first Bluetooth version adopted, followed by version 1.0B. It had limited features and interoperability problems. It quickly became obsolete.

Version 1.1 remedied the major issues of earlier versions; introduced encryption while keeping support for non-encrypted channels; and adopted the Received Signal Strength Indicator (RSSI).

Version 1.2 adopted Frequency Hopping Spread Spectrum (FHSS) to avoid interference; provided higher bandwidth (up to 720 Kbps), supported faster connections, incorporated a better retransmission mechanism for error recovery; included well-defined Bluetooth profiles and a discovery mechanism; and it is backward compatible with Version 1.1.

Version 2.0+EDR (Enhanced Data Rate) is considered the second generation of Bluetooth. It is three times faster than Version 1.2 (2.1 Mbps); has lower power requirements, has improved BER performance, provides extended range of support with directional antennas; and is backward compatible with Version 1.2.

Version 2.1+EDR, adopted in 2007, brings improvements in security by periodically changing encryption keys; offers more sophisticated binding to avoid unwanted connections. It is expected to have reduced power consumption by sniffing idle periods; and implements QOS improvements to enable superior audio and video communication compared with traditional "best effort" mechanism. The QOS improvements will be limited to piconets.

Version 3.0 draft (code Seattle) is a far looking version which will incorporate Ultra-Wide Band (UWB) technology. Higher data rates are expected to support quality audio and video communication using the Multi-Band Orthogonal Frequency Division Multiplexing (MB-OFDM) modulation version of UWB adopted by the Bluetooth SIG in 2006.

The Bluetooth SIG has also adopted specifications used in Infrared communications. One is the Object Exchange (OBEX) which is used to facilitate data and file transfer between devices. It uses a binary Type-Length-Value (TLV) format as headers, similar to those used in abstract syntax notation. These specs were created and are currently maintained by the Infrared Data Association. OBEX functionality is built into the application layer and runs on top of a connection-oriented Transport Layer protocol.

Common **Bluetooth devices** are mobile phones, notebooks/laptops, printers, PDAs, headsets, digital cameras, USB memory sticks, remote control devices, smart keys, keyboards, mice, joysticks, video game controllers, PC cards, modems, wireless Access

Points, and GPS navigation systems. The most popular **Bluetooth applications** are hands-free hand sets, hands-free car sets, wireless communications between PCs and mice, keyboards, joysticks, printers, file transfer and data transfer between computers, business cards exchange, PDA downloads, digital photo transfer, remote control, music files and MP3 audio streaming, synchronization of calendars, dial-up networking, faxing, and serial port connectivity.

A wide range of Bluetooth protocol stacks and associated software applications are provided by third-party software companies. For example, Dell recently launched Bluetooth WPAN technology in its latest Dell™ Latitude™ and select Inspiron™ portable computers and the Axim™ X5 Pocket PC. Market research firms predict that Bluetooth wireless technology will experience a high growth rate over the next few years. Sales of the Bluetooth chip sets embedded in Bluetooth devices increased from 10.4 million units in 2001 to 510 million units in 2006 and this trend will continue in 2007.

7.7 Bluetooth Security

It is well known that all types of communication face some degree of security concern and threats that range from issues of privacy, to identity theft, to denial of service attacks. This is also true for Bluetooth technology and networking. Therefore, these networks and applications require security awareness and security measures.

Vulnerability comes from the use of wireless communication where eavesdropping can be easily performed, packets can be captured, and, if no security measures are taken, Bluetooth phones can be hacked. Security solutions are derived from measures that range from pairing devices, use of authentication mechanisms with longer passwords or pin numbers (8 to 12 characters), and encryption of passwords and user data.

The most common Bluetooth configurations are point-to-point links between well-known pair devices (e.g., phone and headset or car set, Bluetooth-enabled phones, or between Bluetooth master and slave stations). This pairing means easy recognition of the partner in a Bluetooth exchange of information based on common profiles and perhaps even the same pin number. In addition to pairing, encryption algorithms can be applied along with security key management. The short range of Bluetooth links also helps minimize security threats.

Two major security keys are used in Bluetooth technology. One is the link security key, made up of a 128-bit random number. The primary role of this key is to determine if the two communicating devices had a previous relationship and, if no such relationship existed, a key would be generated. In addition, there is an encryption key derived from the current link key used whenever the encryption procedure is requested. The second major key, the combination key, is generated as a combination of two connecting Bluetooth devices. Thus, for each new combination of devices a new key will be created.

The unit key is generated when a Bluetooth device is installed, remaining the same for the lifetime of the Bluetooth device (master station or slave stations). Thus, the unit key is not considered a safe key for encryption. Instead, the combination key should be used, recognizing that memory requirements will increase as more devices are linked.

Another major threat to Bluetooth security and availability is interference from overlapping wireless networks and accidental RF noise. This may happen because Bluetooth operates in a frequency band shared by other technologies. Also, industrial processes such as arc and spot welding, motor drives, switching gear, relays, and power converters can generate a wealth of fundamental and harmonic emissions that extend into the GHz range. These security concerns must be addressed through careful engineering in the design and installation phases, followed by security threats monitoring.

7.8 Advantages and Disadvantages of Bluetooth Technology

Bluetooth is one of the most successful wireless technologies developed in the past decade. It has taken the wireless world by storm. Developed initially as a cable replacement technology for WPANs, it quickly found niche markets in virtually every field of consumer electronics, computers, and even in business and industrial applications. The number of organizations affiliated with the Bluetooth SIG is approaching 3000, and the number of units shipped in 2007 exceeded one billion. The advantages and benefits of using Bluetooth technology and networking include:

- It is based on wireless technology, thus implying a level of mobility;
- Use of unlicensed spectrum, thus avoiding some large expenses;
- It includes automatic discovery mechanism of other Bluetooth devices;
- It was designed as a low consumption technology;
- It is based on frequency hopping technology, therefore less prone to interference;
- It is based on international standards, helped by an industry consortium;
- The new standard developments are compatible with earlier versions;
- It supports data and voice communications at reasonable data rates;
- It defines standardized profiles that help application developers and facilitate interoperability;
- It is strongly driven by cellular mobile networks developments;
- Extensive vendor acceptance and large markets;
- Maturity reached in a short span of time with much accumulated experience;
- Easy and quick device-based implementation with no line-of-sight required; and
- Cost effective solutions, helped by minimal processing and memory needs.

Like any technology, regardless of its success, there are some limitations and disadvantages in using Bluetooth:

- Short range of communications, generally less than 100 meters;
- Low data transmission rates; unfit for video and multimedia applications;
- Omni directional communication, making it vulnerable to eavesdropping;
- Use of common unlicensed spectrum, making it prone to interference; and
- Lack of standardized algorithm and protocols for Bluetooth networking.

7.9 Bluetooth Products

A very small sample of some **Bluetooth SIG-based** products (mainly dual mode mobile phones, notebooks, PDAs, and a variety of consumer electronics such as headsets, computer mice, Bluetooth-enabled printers) follows: Most current implementations are based on SIG Bluetooth Version 1.2, 2.0+EDR (Enhanced Data Rate), and Version 2.1+EDR.

- **3Com**: USB Bluetooth Adapter, Wireless Bluetooth PC card, (www.3com.com);
- **Anycom**: EDR-AP Bluetooth Access Point, Bluetooth PC Card 2000, CF-2001 Compact Flash Card, BIWAS-20 stereo speakers, (www.anycom.com);
- **Apple Computers**: iPhone, Powerbook (version 2.0) (www.apple.com);
- **Axim**: X5 Pocket PC, (www.axim.com);
- **D-Link**: DBT-120 Personal Air Bluetooth USB Adapter, (www.dlink.com);
- **Dell Computers**: Latitude and select Inspiron laptops (www.dell.com);
- **HP/Compaq Computers**: Evo N600C Notebook, iPAQ H3870 Pocket PC, DeskJet 995 C Bluetooth printer, Pavilion ZT1000 Notebook, (www.hp.com);
- **IOGear**: GB V311, GB 221 Bluetooth USB, GBE 201W7 Headset, GME 228BW6 Bluetooth laser mouse, GBIPODKIT headphone, (www.iogear.com);
- **LG Electronics**: LG VX 8350, VX 8700, VX 9400 mobile phones, LAC-M8600, LAC M9600 audio amplifier, (www.lge.com);
- **Microsoft**: Bluetooth keyboard and mouse, (www.microsoft.com);
- **Motorola**: MOTORAZR V3m, MOTORAZR V9m, MOTOSLVRL 7c Mobile phones, V710 handset, (www.motorola.com);
- **Nokia Corporation**: Nokia 6315i, 2366i, 6102i, 6355 mobile phones, (www.nokia.com);
- **Palm**: Palm Treo 680, Palm Treo 750, Palm 700wx, (www.palm.com);
- **Research in Motion**: BlackBerry Pearl, Curve 8300, 8310, 8820, (www.rim.com);
- **Samsung Electronics**: SGHu540, SGHu410, SGHa 990 mobile phones, (www.samsung.com);
- **Sony Ericsson**: W580i, W960i, K850i mobile phones, Car Speaker Phone HCB-100E, Car hands-free set HCB 400, Headset HBH-610a, Bluetooth Watch MBW-150, (www.sonyericsson.com);
- **TDK Systems**: Go Blue PC Card, (www.tdksys.com);
- **Toshiba**: Tecra 9000 Notebook, Stereo Headset, (www.tais.com);
- **Xircom (now Intel)**: Credit Card Bluetooth Adapter, (www. xircom.com).

7.10 ZigBee Network Architecture

Another WPAN implementation with multi-tenant and industrial applications is based on the ZigBee specifications. **ZigBee** relies on the **IEEE 802.15.4 standards** [26] for wireless networking, control, and monitoring of devices found in a limited proximity. Low data rate radio communications, ultra low power, and few firmware resources are used. The IEEE WPAN 802.15.4 standard specifies multiple channel availability in three bands: 16 channels at 250 Kbps each in the 2.4 GHz ISM band, 10 channels at 40 Kbps in the 915 MHz ISM band, and one 20 Kbps channel in the European 868 MHz band. The ZigBee version 1.0 specifications were ratified in December 2004 and are available to the members of **ZigBee Alliance** [29], a consortium of over one hundred member organizations. A competing alternative to ZigBee is Nokia's WiBree. In 2007 WiBree Forum merged with Bluetooth SIG.

ZigBee targets **devices/sensors** and **applications** that have small amounts of data to transfer, and are turned off or idle most of the time but ready to communicate or respond when needed. These include HVAC sensors, lighting switches, emergency lights, heating controllers, smoke detectors, automatic meter readers, and home security systems. Any Microprocessor Control Units (MCU) micro-sensors and actuators that are used in home, building, and industrial automation that do not have high memory, data storage, or special security needs are also possible applications. A generic ZigBee networking architecture is provided in Figure 7.5.

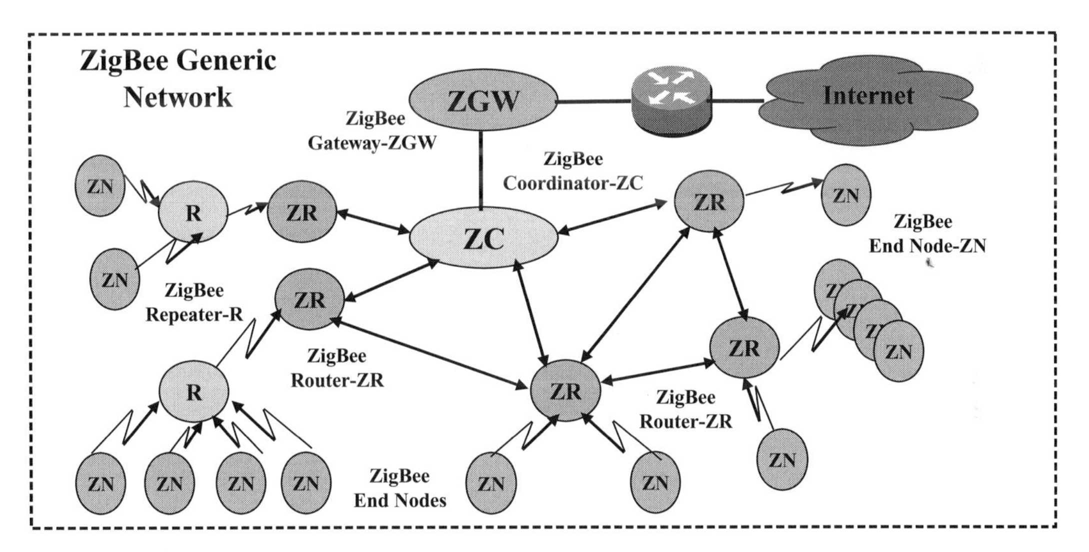

Figure 7.5 ZigBee Network Architecture

A generic ZigBee network architecture consists of five types of interconnected devices:

- **ZigBee End Nodes** (ZN): Provide physical interfaces to collect information from home, commercial, or industrial sensors and to transmit this information via a RF link to a hierarchical device or to a peer end-node device. They have minimal memory and are usually in a sleeping or idle state;
- **ZigBee Routers** (ZR): Relay information to a hierarchical component or to a peer router;
- **ZigBee Coordinator** (ZC): Provides a centralized control of the ZigBee network as a whole with the ability to change information with multiple ZigBee routers or end nodes. It is responsible for forming the network. Following that, it acts as a router. It

has a larger storage capability and provides a link or a bridge to a data network infrastructure. Security services are also provided by the ZC;

- **ZigBee Gateway** (ZGW): Provides connectivity through a serial, RS 232 or USB interface, to a data communication network or to a computing platform for programming and management. It can be either a fixed or mobile standalone device or incorporated into a traditional switch, hub or IP router;
- **ZigBee Repeaters** (R): Relay the information from far-reaching end nodes planted in "dead zones" beyond the range of normal communications.

A ZigBee network is essentially a cluster of ZigBee end nodes and routers with one ZigBee coordinator in a mesh configuration. The mesh topology includes point-to-point and point-to-multipoint connections. There are two types of links: **tree connectivity links** are used to link ZNs to a router or repeater, and **mesh routing links** to route information upward to the ZC. Several clusters can be controlled by a single ZC. Mesh networking specifications are based in the IEEE 802.15.4 standard.

ZigBee networks do not just replace wires; they provide wireless distributed monitoring and control of ad-hoc networks with self configuration and even self-healing capabilities based on specially designed algorithms. To become successful, ZigBee, like all other technologies, should be reliable, scalable, and secure in its deployment.

7.11 ZigBee Protocol Architecture

IEEE 802.15.4 standards provide specifications of the RF link, Physical Layer and Data Link Layer MAC sublayer frame formats. The RF link includes RF channelization within the allocated frequency spectrum, modulation type, spread spectrum coding, and symbol/bit rate specifications. In addition, to help chip and reference design developers, the ZigBee Alliance has defined a **ZigBee Compliant Platform** that covers network, security (AES-128 algorithm), and application layer functionalities. There is also a ZigBee **Certification Program** that can be used to assess conformance to the standards, mainly through interoperability testing between products and reference implementations. ZigBee frequency and channel allocations are given in Table 7.3.

Table 7.3 ZigBee Frequency Allocation

Frequency bands	868.3 MHz	902–928 MHz	2.4–2.483.5 GHz
Coding	DSSS	DSSS	DSSS
Number of channels	1	10	16
Modulation	BPSK	BPSK	QPSK
Bandwidth	600 KHz	2000 KHz	5000 KHz
Data rate	20 Kbps	40 Kbps	250 Kbps
Symbol rate	20 Kbps	40 Kbps	125 Kbps
Unlicensed spectrum	Europe	North America	Worldwide

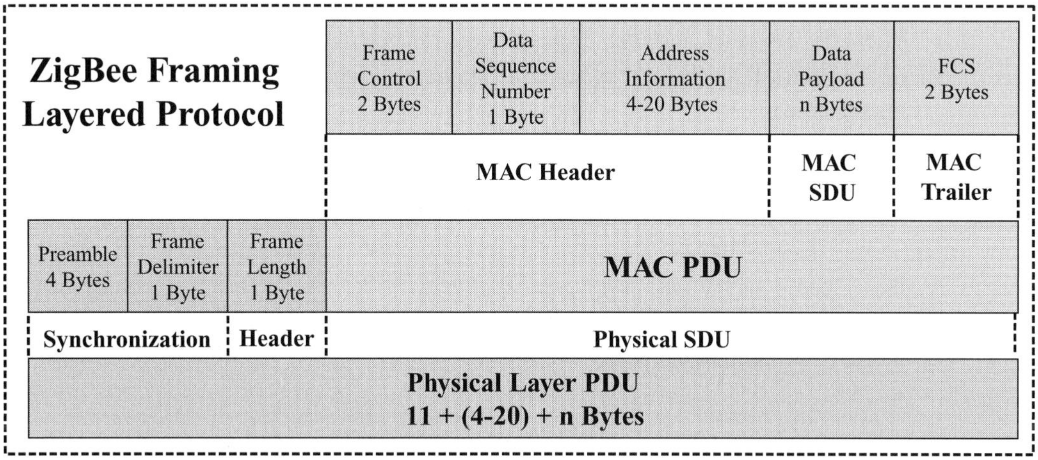

ZigBee Framing Layered Protocol

Frame Control 2 Bytes | Data Sequence Number 1 Byte | Address Information 4-20 Bytes | Data Payload n Bytes | FCS 2 Bytes

MAC Header | MAC SDU | MAC Trailer

Preamble 4 Bytes | Frame Delimiter 1 Byte | Frame Length 1 Byte | MAC PDU

Synchronization | Header | Physical SDU

Physical Layer PDU
11 + (4-20) + n Bytes

Figure 7.6 ZigBee Framing Protocol Architecture

Currently, ZigBee is a tri-band technology using unlicensed spectrums; differing from region to region. The Physical and MAC sublayer PDU formats are given in Figure 7.6.

Four types of ZigBee frames are defined: Data, Acknowledgment, Beacon, and MAC Commands. The **Data** frames are used to carry the data payload containing sensor monitoring or controlling information. **Acknowledgment** frames are used to confirm to the sender that the information was received correctly. **Beacon** frames are used by the ZigBee coordinator, and, occasionally by ZigBee routers with power saving mode, to establish the network and make known their presence. **MAC Command** frames are used to send low-level commands between ZigBee end nodes.

IEEE 802.15.4 standard defines two types of devices: full function devices and reduced function devices. Full function devices support most services offered while reduced function devices support only a limited number of services. Evidently, the first category will use the power supply of the component where the sensor is attached while in the second case the power will be supplied by a long-life battery; these devices being mostly off when idle. The same standards stipulate an address field that can be up to 20 bytes long. In practice, a two to eight byte field is entirely sufficient.

The 802.15.4 specs provide support for both beacon-enabled and non-beacon enabled networks. In a non-beacon network, an unslotted CSMA/CA technique is used that assures quasi-equal access to the ZC. It is assumed that the ZRs continuously listen to the beacon so they should have a robust battery or a continuous power supply. In this case, sensors will be attached to native power devices such as light switches or electrical engines.

In beacon-enabled networks, ZRs periodically transmit beacons to make their presence known to other routers or to the clustered nodes. ZNs sleep between beacons to conserve battery life and to minimize the duty cycle. Beacon interval can range from 15.36 ms to 250 seconds for ZigBee networks operating in the 250 Kbps bandwidth. Beacon intervals are even larger, from 48ms to 786 seconds, for networks operating at 20 Kbps. Beacons are set on a fixed timing schedule and do not use the CSMA/CD mechanism. The same is true for acknowledgement frames. In case of low-latency near real-time sensors or micro electrical mechanical systems, a Guaranteed Time Slot (GTS) mechanism can be implemented.

7.12 Advantages and Disadvantages of ZigBee Technology

Developed over the past five years, ZigBee as a WPAN wireless technology fills in the niche markets of home, building, and industrial automation that require very low power consumption. Some advantages and benefits of using this technology include the following:

- It is based on wireless technology;
- It uses batteries and power consumption is very low;
- It provides two-way communication with no line-of-sight limitations;
- The basic network design is a highly versatile self-organizing mesh configuration;
- It is highly scalable, allowing one coordinator to handle up to 65,000 devices;
- The cost of end devices is low, helped by minimal code required (20–70 KB);
- Use of unlicensed spectrum bands of frequencies;
- It includes an automatic discovery and configuration/reconfiguration;
- It is based on international standards, with input from industry consortia; and
- A new ZigBee Professional standard, developed in 2007, has been added.

Despite its acceptance in the field of building and industrial automation there are some notable limitations and disadvantages in using ZigBee:

- There is no full backward compatibility between ZigBee 2004, ZigBee 2006 and ZigBee Pro standard alternatives;
- Use of the common unlicensed spectrum makes ZigBee vulnerable to interference;
- The cost per unit is still relatively high, impeding mass implementations;
- Low data rates limit the amount of information that can be transmitted and stored;
- The design and implementation requires careful engineering of radio links; and
- Communication is limited to short ranges, generally less than 70 meters.

7.13 ZigBee Products

The large number of available ZigBee products, ranging from ZigBee chipsets, systems-on-chip, ZigBee protocol analyzers to ZigBee integrators, are one of the indications of the increasing role of this wireless technology in WPAN networking and industrial automation. One group of products qualify as ZigBee Compliant Platforms – a testing program for chip suppliers or platforms that are intended to be used as building blocks for end products. Another group of products that are "ZigBee Certified" are end products built on the ZigBee Compliant Platform and tested for certification in accordance with the

ZigBee Alliance testing program. All testing programs ensure that ZigBee devices delivered to the market will successfully coexist in ZigBee network environments designed by the end user. Products currently available include [29]:

- **AeroComm**: ZB2430 ZigBee Your Way transceiver, (www.aerocomm.com);
- **Agilent Technologies**: N4010–102 for ZigBee Wireless Network Analyzer, (www.agilent.com);
- **AirBee Wireless**: ZNS 2004–2006 ZigBee protocol stack, (www.airbee.com);
- **Atmel Corporation**: AT86RF230 ZigBee transceiver, AVR microcontroller, (www.atmel.com);
- **Chipcon**: CC2420 SoC (now Texas Instruments), (www.chipcon.com);
- **Ember Corporation**: EM250 ZigBee system-on-chip, EM260 ZigBee network co-processor, EmberZNET 1.0/2.0, InSight Development Kit, (www.ember.com);
- **Exegin Technologies**: Q51 ZigBee Protocol Analyzer, (www.exegin.com);
- **Freescale Semiconductors**: MC13192 SoC, MC1322x ZigBee platform, ZigBee Protocol Analyzer, (www.freescale.com);
- **Frontline Test System**: FTS4ZB ZigBee Protocol Analyzer, (www.fte.com);
- **Helicomm Inc.**: Helicomm EZDK ZigBee development kit, (www.helicomm.com);
- **Jennic**: ZigBee JN5121, JN 5139 Single Chip Microcontroller, (www.jennic.com);
- **LG Electronics**: ZigBee-based HomNet networking system, (www.lge.com);
- **MeshNetics Corporation**: ZigBit platform, ZigBee-Ethernet Gateway, (www.meshnetics.com);
- **Microchip Technologies**: AN965 Microchip Stack for ZigBee protocol, Zena Wireless Network Analyzer, (www.microchip.com);
- **NEC Engineering**: NEC Engineering Platform with Freescale SoC for ZigBee-ZB24FM embedded modules, (www.nec-eng.com);
- **OKI Electric Industry**: OKI AME51, OKIzNED 2006 platforms, (www.oki.com);
- **Renesas Technology**: Renesas ZigBee Solution, ZigBee Product Development Kit (ZPDK) RZB-ZPDK, (www.renesas.com);
- **Rohde and Schwartz**: Wireless Network Analyzer, (www.rohde-schwartz.com);
- **Silicon Laboratories**: Silabs ZigBee development board, (www.silabs.com);
- **Texas Instruments**, Chipcon CC2431 SoC for ZigBee, (www.ti.com).

Given their importance in the industrial and facility environments, ZigBee devices must be **reliable**, **secure** in their communication (encryption) in the range from 1 to 100m, and operate with very **low power consumption**. Common technical requirements for ZigBee compliance include: peak data rate of 128 Kbps, 128-bit symmetric key encryption, 30 ms network join time, 15 ms active slave channel access time, 32 KB stack size, 8 Kb Static Random Access Memory (SRAM) for additional program storage, and 2–3 year battery life time.

Most ZigBee-related products and implementations, chipsets, reference design, and OEM design for manufacturers, are based on the ZigBee platform promoted by ZigBee Alliance. They are based on Version 1.0 specifications.

- **Atmel Corporation**: ZigBee Hardware Platform (ZigBit AT 86RF230 ZigBee radio transmitter and Atmega 12810 MCU with transmitter, (www.atmel.com);
- **Ember Corporation**: ZigBee chip set, ZigBee devices, ZigBee implementations, (www.ember.com);
- **Fanna Technologies**: MikroMesh System (Mikro Coordinator, Mikro Gateway, Mikro Router, Mikro Mesh Units (MMU) ZigBee end nodes for PLC and tracking devices), (www.fanntech.com);
- **Renesas Technology** (joint venture Hitachi and Mitsubishi): Microcontrollers, flash MCU for ZigBee nodes, routers, coordinators), (www.renesas.com);
- **AirBee**: Reference Design (AirBee Protocol Stack, AirBee ZigBee Agent, AirBee ZigBee network management system), (www.airbee.com);
- **STMicroelectronics**: Microcontrollers for ZigBee end nodes, (www.microelectronics.com);
- **Max Stream**: ZigBee OEM Module, (www.maxstream.com);
- **MeshNetics Corporation**: ZigBit-based OEM Module (ultra-compact 2.4GHz ZigBee/802.15.4 module for ZigBee Mesh Networking Applications, ZigBeeNet full stack and eZeeNet partial protocol stack), (www.meshnetics.com);
- **Nokia Corporation**: **Wibree** (ZigBee competing solution), (www.nokia.com).

7.14 Comparison of Bluetooth and ZigBee

Although both Bluetooth and ZigBee were developed as wireless solutions for environments that, in general terms, qualify as personal area networks, there are substantial differences between these two technologies. One difference is the fact that ZigBee complements Bluetooth by taking the concept further from wireless home and consumer electronics into building, and industrial automation. Second, the technologies were developed 4–5 years apart and that is well reflected in the strong, even ubiquitous, Bluetooth implementations compared with the few ZigBee implementations. A high level comparison between Bluetooth and ZigBee is given in Table 7.4.

In addition to IEEE standards, the advancements of Bluetooth technology are promoted by the Bluetooth Consortium and Bluetooth Special Interest Group. In case of ZigBee, standard advancements and the certification process is promoted by the ZigBee Alliance.

7.15 Power Line Communications Architecture

Power Line Communications (PLC), also known as Power Line Networking (PLN), is a networking technology that uses power distribution lines and wires to provide data, voice, and video services. This approach to networking is presented here along with the wireless technologies for two reasons. One, of course, is that PLC can be a viable communication solution for home and multi-tenant units, part of a Personal Area Network (PAN). And, second, some of the devices used for PAN communications are combined wired and

Table 7.4 Bluetooth and ZigBee Comparison

Main Features	Bluetooth	ZigBee
Technology timing	Started in 1994, v1.0 1998	Started in 1999, v1.0 2005
Standards	802.15.1	802.15.4a and 802.15.4b
Frequency bands	2.4 GHz	863 MHz, 915 MHz, 2.4 GHz
Modulation techniques	FHSS (MB-OFDM)	DSSS (BPSK, QPSK)
Data rate	720 Kbps – 2.1 Mbps	20 Kbps, 40 Kbps, 250 Kbps
Memory	128 Kbytes to Giga Bytes	128 KBytes, 256 KBytes
Normal range of action	1m (1mW), 10m (2.5mW), 100m (100mW)	3–75 m
Power consumption	Moderate	Very low
Battery Life	Moderate duration (months)	Long duration (years)
Scalability	Limited	Unlimited
Topology	Point-to-Point, to-Multipoint	Tree, Star, Mesh
Operational license	Unlicensed band	Unlicensed bands
OEM availability	Vendor offer	ZigBee Alliance full OEM offer
Management capability	SNMP/Web-based	Web-based
Installation time 12 nodes	2 hours	2 hours
Cost	Moderate	Low

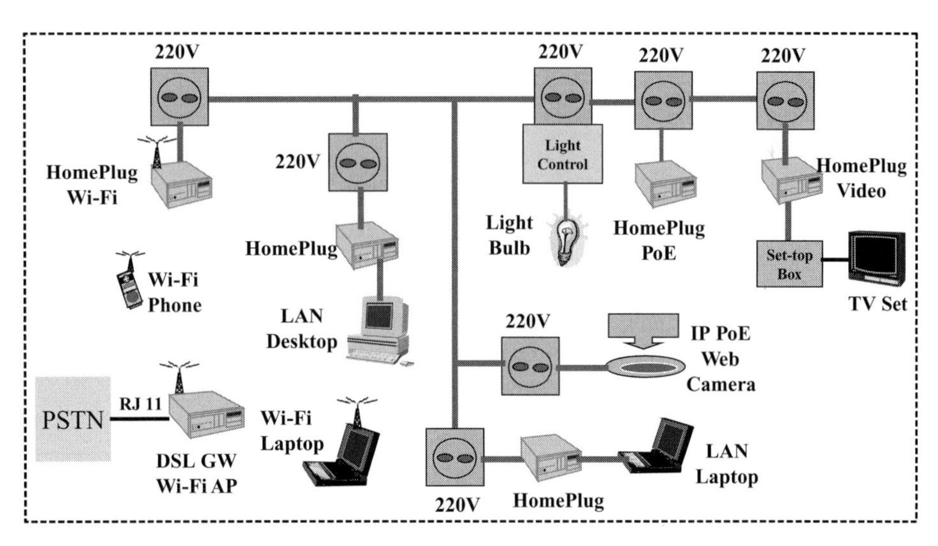

Figure 7.7 Power Line Communications-based PAN

wireless devices, the latter the subject of this chapter. An example of PLC-based home networking is shown in Figure 7.7.

DSL-based broadband connectivity provides a link to the PSTN and Internet (RJ 11 interface). The CPE DSL modem is combined with a wireless AP. The PLC-based network starts with a PLC Homeplug device, inserted into the power line network (represented by the thick red line). The Homeplug also supports IEEE WLAN standards (Wi-Fi). The network itself is a LAN wired with power lines, in this case 220V AC lines. Other wireless devices such as a VoIP-based Wi-Fi phone and a Wi-Fi laptop communicate directly with the AP as in a regular WLAN. Devices such as LAN-based desktops and laptops connect to the LAN by taking their power supply from one of the Homeplugs inserted into the 220V outlets. Among special PLC devices are Homeplugs combined with a Light Control device that allows switching light bulbs on and off. Another special device is the HomePlug PoE (Power over Ethernet) that provides PLC-based connectivity to an IP-based PoE Web camera used to watch children or to provide minimal security. Last, video transmission, carried over DSL from a video server can provide PLC-based communications through the Homeplug Video and a set-top box.

PLC technology consists of superimposing a modulated and encoded high frequency analog signal over alternating current power lines working at 50 Hz or 60 Hz. Two systems are proposed:

- Broadband Power Line (BPL): Data rate higher than 1 Mbps.
- Narrowband Power Line (NPL): Data rates in a lower range, typically, 128 Kbps.

Several standard organizations are involved in developing technical specifications for PLCs. These include IEEE, ETSI, the Homeplug Powerline Alliance (HPA), and the Open PLC European Research Alliance (OPERA). HPA was founded in 2000 and includes over 70 members; among them Comcast, Earthlink, GE, Intel, Linksys, Motorola, Radio Shack, and Samsung. The most important specifications guiding PLC implementation are:

- **HomePlug V1.0**: Data transmission support;
- **HomePlug AV**: HDTV and VoIP support;
- **HomePlug CC**: Command and Control (monitoring and controlling devices);
- **IEEE P1675**: Standards for Broadband Power Line Hardware [30];
- **IEEE P1775**: Standards for Powerline Communications Equipment [31];
- **IEEE P1901**: Draft: Broadband over Power line Networking: MAC and Physical Layer specifications.

There are many manufacturers of the Homeplug devices used in PLC-based home networking and which provide support for video streaming. A short list follows:

- **Cisco-LinkSys**: (www.cisco.com), (www.linksys.com);
- **Devolo**: (www.devolo.com);
- **D-Link**: (www.d-link.com);
- **Gigafast**: (www.gigafast.com);
- **Intel Corporation**: (www.intel.com);
- **LG Electronics**: (www.lgelectronics.com);
- **Topcom**: PowerLAN 400 Turbo Ethernet, (www.topcom.be);
- **NetGear**: (www.netgear.com);
- **LEA SAS**: NetPlug AV 200, NetPlug Turbo WiFI, Elektra GW, (www.leacom.fr);

- **Samsung Electronics**: (www.samsungnetwork.com);
- **Spidcom Technologies**: SPC AV Chip, SPiDMonitor, (www.spidcom.com);
- **Zyxel**: HomePlug AV PLA-401, Turbo PLA-150 WiFi, (www.zyxel.com).

7.16 WPAN Management and WPAN QOS

There are four different classes of solution for WPANs: First, the IEEE 802.11 WLANs solutions as applied to WPANs; second, the 802.15.1 Bluetooth, families of applications; third, the IEEE 802.15.4, ZigBee, solutions; and fourth, the PLC families of solutions. Each class includes specifications for parameters and system features such as range of spectrums, range of action, modulation type, data rates, topologies, and scale of components and applications. But the most important aspect of management is the limitation in power consumption, memory space and microprocessor ability to handle the sophisticated interactions that characterize monitoring and controlling capabilities.

The IEEE 802.11 a/b/g solutions appear the most homogeneous class and its management requirements and solutions were extensively analyzed in the previous chapter. The Bluetooth class, although based on a single standard, has several versions of implementations and there are plans to move the technology into the ultra wideband domain. That will require drastic changes in RF link characteristics. The ZigBee class is split into IEEE 802.15 4a and IEEE 802.15 4b versions, and it also is on the path to UWB. The PLC-based solutions differ from IEEE recommendations and follow specifications derived from two major alliances formed around European and North-American designers and manufacturers.

Most current Bluetooth implementations are built around point-to-point applications involving pair devices. In this case, the management is an intrinsic part of the RF link. Since very few implementations of Bluetooth-enabled devices are clustered in genuine piconets or scatternets, there is no need to have scale-based management. Therefore, in Bluetooth technologies testing and measurement of the RF link in the development phase of chipsets and reference designs represents an important aspect of management. Testing includes transmitting and receiving antennas. Most of the time, it is a sufficient way to assess compliance to the standards aimed at assuring the performance required and minimization of interference with the ambient environment. Vendors providing Bluetooth testing include:

- **Agilent Technologies**: RF Testing including Bluetooth digital modulation and FHSS, (www.agilent.com);
- **Anritsu**: Bluetooth Tester (only RF link), (www.anritsu.com);
- **CATC/Lecroy Corporation**: Bluetooth Tracer, Bluetooth Analyzer, (www.lecroy.com);
- **CETECOM (USA) Test House**: BITE (Bluetooth Qualification Tester), Blue Unit Tester (Protocol/Profile Tester, RF Engineering Tester), (www.lecry.com);
- **MobiWare**: BPA-D10 Bluetooth Protocol Analyzer, (www.mobiware.com);

Table 7.5 WPAN Applications and Performance Metrics

Digital Application	Actual Services	Data Rate	QOS Performance
Data Applications	– Internet access (email, web browsing, chat)	6 Mbps or less	
	– File, printer, scanner, and fax sharing	10 Mbps	Best effort
	– Broadband data sharing		
Voice Applications	– Conventional voice services	4.8–64 Kbps	– Availability 99.99%
	– Additional voice features (call waiting, speed dialing, call forwarding, three way call, etc.)		– Call blocking 1% – Reduced jitter and delay
	– Videoconferencing, video telephony	128–384 Kbps	
Video Applications	– Conventional video broadcast	6 MHz analog	– NTSC standards
	– DVD players	6 Mbps	
	– Video server (compressed HDTV)	20 Mbps compressed	– HDTV standards
	– Broadband video	6 Mbps compressed	– MPEG 3/4
Audio Applications	– Digital audio broadcast	128 Mbps	– Audio standards
	– CD player, MP3 player	1 Mbps	
	– Media server, loudspeakers		– MPEG 2
Telemetry and Appliance Control	– Home security (video cameras)	Low data rates	– Best effort
	– Lighting, electricity, gas control	0.1–4 Kbps	– Availability 99%
	– Appliance monitoring and control		

- **Rohde & Schwartz**: PTW60 Tester, Bluetooth R&D Qualification and Maintenance, (www.rohde-schwartz.com);
- **SGS Consumer Testing Services**: Bluetooth Qualification Testing (Radio, Protocol and Profile Conformance and Interoperability Testing), (www.sgs.com);
- **Tektronix**: BP 100 Bluetooth Protocol Analyzer, (www.tektronix.com).

The situation is somewhat different, and more complex, when developing management solutions for ZigBee and PLC-based WPANs. More details about management of ZigBee and PLC-based WPAN networks will be provided in Chapter 12 where case studies of WPAN and mobile networks convergence will be analyzed.

It is anticipated that, in the long run, WPANs will support data, voice, video, and automation through several digital applications. Each of these applications will have specific performance requirements and Quality of Services (QOS) parameters. This is indicated in Table 7.5. These parameters are applicable to those WLAN environments based on Wi-Fi specifications. This is also true for PLC-based networks which are essentially just Ethernet LANs. Specific QOS aspects in LAN and WLAN were presented in Chapter 6. QOS is critical for real-time applications such as voice and video communications.

For networking technologies such as Bluetooth and ZigBee, given their usual point-to-point topologies the need for QOS is strongly determined by the connected external systems. Therefore, quality of service is typically embedded into the RF link and in the software and hardware handling the ingress and egress of native data, voice, video, or multimedia information. The character of service in WPAN is less evident, so there is no reason to sign a SLA between service providers and users.

8 Wireless Metropolitan Area Networking

8.1 Wireless MAN Technologies

Wireless Metropolitan Area Network (WMAN), like WLAN and WPAN, is a generic term for networking confined to a geographical area and a set of specific networking technologies that provide wireless communications in metropolitan areas. WMAN is a new technology that will be a supplement to well-known wired technologies such as Resilient Packet Ring (RPR), Synchronous Optical Network/Synchronous Digital Hierarchy (SONET/SDH), SONET over IP, Gigabit Ethernet, and Wavelength Division Multiplexing (WDM). The area of coverage of WMAN falls between WLAN/WPAN, which are customer premises networks, and Wireless Wide Area Networks (WWAN), which are associated with cellular radio mobile networks. Methods of access to WMANs have some resemblance to those of broadband wired access technologies such as Digital Subscriber Line (DSL) and Data over Cable Service Interface Specifications (DOCSIS).

Conceptually, WMAN networks provide services to metropolitan or regional areas, either urban or rural, within a radius of 50 km. They can be used to connect WLANs/WPANs and provide access to data, voice, video, and multimedia services. Although WMANs provide city-wide coverage, the area may be as small as a university campus or even a group of buildings. WMANs belong to a network operator or a service provider, in many cases a wireless extension of services provided by wired or wireless carriers. WMANs can be implemented using a variety of wireless technologies: **Local Multipoint Distributed Service** (LMDS), **Multi-Channel Multipoint Distributed Service** (MMDS), **Free Space Optics** (FSO), **Wireless Local Loop** (WLL), and **Wireless Interoperable Metropolitan Area Exchange** (WiMAX).

Although this chapter will focus mainly on WiMAX architecture, standards, solutions, products, and implementations, as the ultimate solution in WMAN, we would like to give a clear perspective of the road that led to this technology. It is a lesson in technology evolution, where a standard that is finally accepted results from several attempts and missteps, all of which contributing in some way to the perfection of previous proposals. It is also a good lesson in technological advancement facing opposition from legacy technologies, alternative technologies, and wrong time-to-market.

WMAN technologies must satisfy a wide range of service requirements such as broadband access capability, reliability, scalability, security, quality of services, manageability, and cost effectiveness, all at a time when multimedia traffic is exploding at exponential rates. Special applications extend beyond traditional Internet access, and they include VoIP, inexpensive video conferencing, audio and video streaming, and gaming. Individual subscribers as well as small and medium size businesses operate networks of this kind. All these aspects are both challenges and opportunities to designers, developers, and service providers.

The steadily increasing traffic loads in MANs has produced a new breed of service provider, one that is focused on providing larger available bandwidth, enhanced network control, cost-effective technologies, and the convergence of data, voice and video into a massive, combined data stream. In addition, the developments of new technologies that address the operational and transmission requirements of WMANs are providing new opportunities for incumbent network operators throughout the world.

WMAN architecture is relatively simple. It consists of CPE Subscriber Stations (SS) connected to radio Base Stations (BSs) through a Wireless Local Loop (WLL) interface operating in the radio/microwave range of frequencies. Architectural details and technical specifications of WMAN technologies are provided in subsequent sections.

8.2 Local Multipoint Distributed Services

Local Multipoint Distributed Service (LMDS) is a high-speed wireless broadband access link that operates using microwave signals to connect subscribers and business customers. It **requires line-of-sight** paths between transmitters and receivers as well as directional antennas. It is sensitive to heavy rainfall, and operates in a range up to 5 km.

LMDS frequencies occupy a very large band of spectrum located in sections of the 26–29 GHz (worldwide) and 31 to 31.3 GHz band (in the USA). LMDS can operate in a bandwidth as large as 1.3 GHz. This is in stark contrast with cellular mobile, which consists of 25 MHz bandwidth, or with PCS, which utilizes a 30 MHz bandwidth. A typical LMDS network architecture is shown in Figure 8.1.

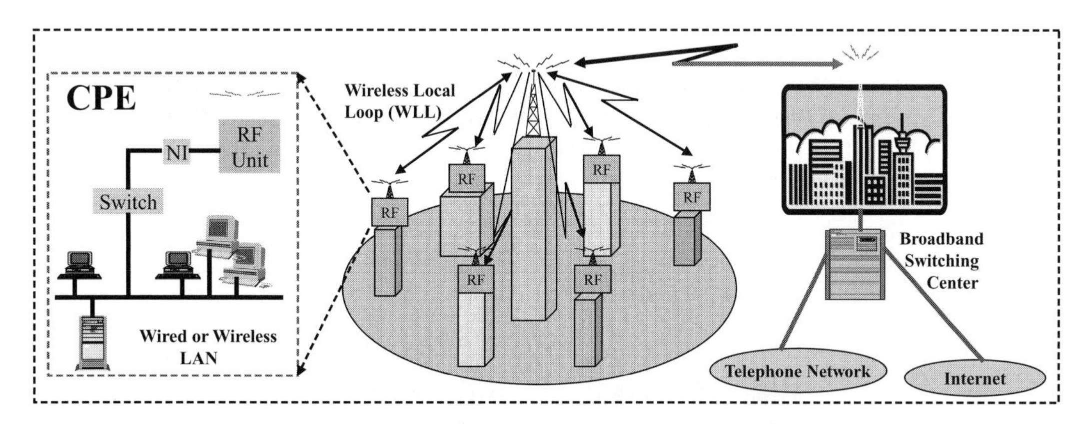

Figure 8.1 LMDS Broadband Access Network Architecture

Main architectural components of a LMDS are the CPE Network Interface (NI), CPE Radio Frequency (RF) equipment, Wireless Local Loop (WLL), RF hub (multi-antenna receivers), connections to other RF hubs, a broadband switching center, and fiber link connections to the Internet and PSTN.

LMDS network implementations may consist of multiple hubs that deliver point-to-multipoint services to subscribers. Each hub serves a small area, about 4 km in radius. This small coverage area requires a large number of LMDS antennas or multi-antenna receivers. Signal strength is adjustable according to atmospheric conditions. Once deployed and tuned, LMDS achieves 99.99% network availability in ideal conditions. A short list of **LMDS equipment manufacturers**, with the last five vendors acting also as systems integrators, follows (there were no standards in place till late 2001):

- **Alcatel**: 7390 LMDS, (www.alcatel.com);
- **Triton Network Systems**: Invisible Fiber Internet, (www.triton-network.com);
- **Ericsson**: Mini-Link BS, (www.ericsson.com);
- **Lucent**: WaveLAN, (www.lucent.com);
- **Nortel Networks**: Reunion System, (nortelnetworks.com);
- **Hughes Network Systems**: AB9400, (www.hns.com);
- **Siemens**: WALKair 3000, (www.siemens.com).

The initial high cost of CPE, hub centers, and spectrum, helps explain the large number of bankruptcies that have occurred among smaller service providers and limited number of implementations that exist today. Early **LMDS service providers who are now out of business**: XO Communications, Adelphia Communications, WinStar, Telligent, and Advanced Radio Telecom (ART).

8.3 Multi-Channel Multipoint Distributed Services

Multi-Channel Multipoint Distributed Service (MMDS) is another high-speed wireless access technology that, in contrast with LMDS, does not require a stringent line-of-sight alignment between transmitters and receivers. A typical MMDS architecture is shown in Figure 8.2.

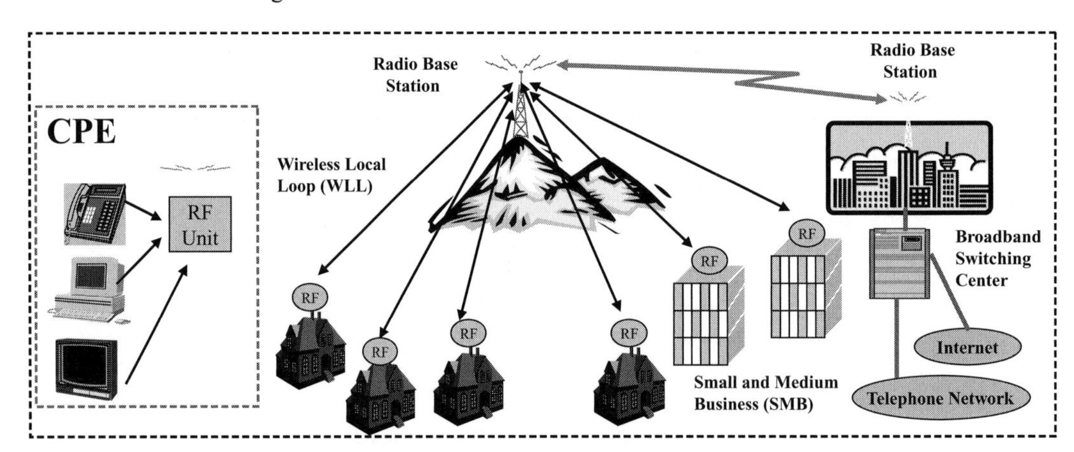

Figure 8.2 MMDS Broadband Access Network Architecture

The main components of a MMDS are a CPE Network Interface, the CPE Radio Frequency (RF) equipment, the Wireless Local Loop (WLL), a RF hub, wireless or terrestrial connections to a broadband switching center, and the fiber link connections to the Internet and PSTN.

MMDS started as a broadband access solution in 1999 when the FCC opened an additional spectrum for licensing allowing two-way communications. It was embraced by all the interexchange carriers as a way to bypass Regional Bell Operating Companies (RBOCs). However, it had only limited success due to high cost and poor performance. MMDS technology targets residential and small/medium size business customers. It can operate up to 100 km at data rates of 10Mbps.

MMDS frequencies, in the 2.1–2.7 GHz band, were used to provide one-way, analog wireless broadcast services. In 2006, frequencies in the 2.11–2.155 GHz were removed from use by MMDS and became part of the Advanced Wireless Services (AWS) auction. The practical result was a gradual dismantling of MMDS services in the USA. Broadband Wireless Internet Forum (www.bwif.org) and Wireless Communications Association International (www.wcai.org) pushed the implementations of MMDS.

Advantages of using MMDS include:

- Existence of under-utilized spectrum that will, once completely digital, become increasingly valuable and flexible;
- Easy and quick system implementation-installation of a transmitter on a high tower and a small receiving antenna on the customer's site;
- Maturity – since MMDS services have been around for 20 years, there is a wealth of experience, at least in respect to the one-way distribution technology.

A short list of **MMDS equipment manufacturers** follows. Although more successful than LMDS, it was only partially embraced by the market.

- **ADC Telecommunications**: Cell Span MMDS and Axity Broadband Wireless Access System, (www.adc.com);
- **Advantech**: Rackmount Integrated Broadband Transmitting Systems, (www.advantech.com);
- **Andrew Corporation**: High Power Amplifiers and MMDS System Components, (www.andrew.com);
- **Cisco Systems**: WT-2700 Fixed Wireless Access Products, (www.cisco.com);
- **Hybrid Networks**: Base Station & Broadband Routers, (www.hybrid.com);
- **Iospan Wireless**: AirBurst MIMO-OFDM Non-LOS, (www.iospanwireless.com);
- **NextNet Wireless**: Expedience, (www.nextnetwireless.com).

Early MMDS service providers with limited services, some in experimental phase:

- **Alaska Wireless Cable**; Internet and TV, (www.awcable.com);
- **American Rural Telcos**: Minerva Network Systems IP Television Platforms, (www.minervanetworks.com);
- **BellSouth**: Alongside its landline cable service, (www.bellsouth.com);
- **Nucentrix**: Heartland Cable Television, (www.nucentrix.com);

- **US Sprint**: Broadband Direct Service, (www.sprintbroadband.com);
- **Worldcom**: The major holder of MMDS licenses (www.wcom.com);
- **NTL Ireland**: TV Broadcast (www.ntl.ie).

8.4 Free Space Optics Metropolitan Access

Free Space Optics (FSO) is a wireless communication technology using line-of-sight infrared light beams to provide high-speed data, voice and video transmission between commercial buildings in metropolitan areas. Also, FSO can complement (backup) telecom network extensions. Because of high availability, bandwidth scalability, and deployment simplicity, FSO is a viable and cost-effective alternative for wireless broadband access networks. FSO components are: signal converters, Optical Transceivers (OT), optical links, and connections to broadband switching centers. The FSO network architecture is provided in Figure 8.3.

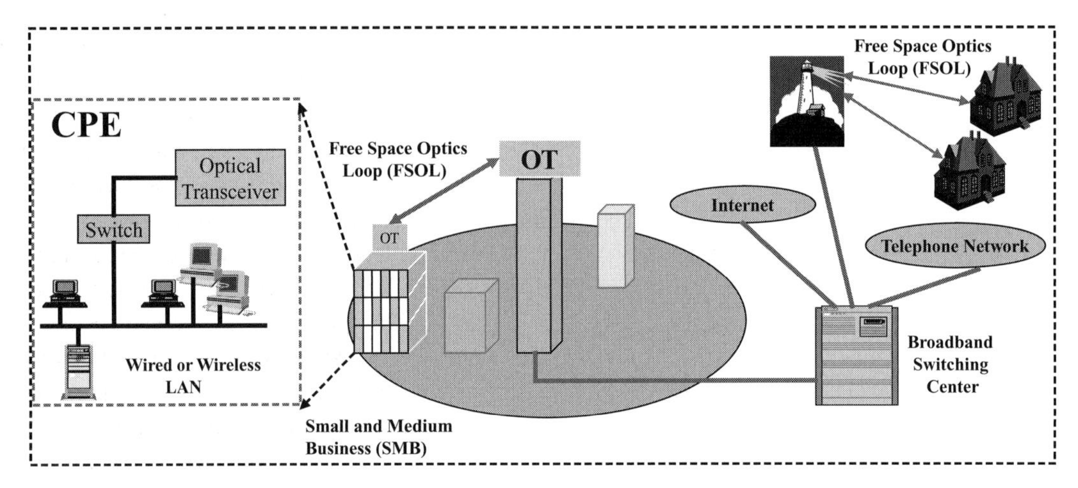

Figure 8.3 Free Space Optics Wireless Access Network Architecture

FSO became a successful access alternative when auto-tracking systems were added to keep the optical transceivers aligned during periods when there was motion caused by disturbances in the environment. The main FSO applications are: metro network extensions, enterprise LAN-to-LAN connectivity, wireless backhaul, and as a supplement to LMDS for high-speed data, voice, and video.

There are several **advantages** in using FSO. These include: very high data rate, quick and simple installation, no licenses are needed, cost effectiveness (lasers or LEDs as light sources), less subject to interference and eavesdropping, and data rates that are not affected by range. FSO can be combined with licensed and unlicensed wireless radio access networks. FSO accepts signals from fiber optic physical layers (SONET, DWDM) and data rates can go up to 10 Tbps.

There are some **FSO disadvantages**: limited link margin (20–40 dB), prone to strong attenuation (1dB/km clear air); severely affected by fog, clouds, and rain; signal strength is limited (if laser-based), beam wandering; beam spreading, and safety aspects (laser). To judge **FSO performance**, the following metrics are used: range, bandwidth, availability, scalability, and flexibility. A short list of FSO systems manufacturers follows:

- **Plaintree**: WaveBridge 10 Mbps and 100 Mbps, (www.plaintree.com);
- **IR LAN (Unity Wireless)**: (www.irlan.co.il), (www.unitywireless.com);
- **Canon**: Canonbeam DT-100 Series (110, 120, 130), (www.canon.com);
- **fSona**: Sonabeam M Series (52, 155, 622, 1250), S Series (155, 622, 1250), and E Series (52, 155), (www.fsona.com);
- **LightPointe**: FlightStrata 155/G/HD, FlightLite 100/100E Outdoor Optical Transceivers, (www.lightpointe.com);
- **Attochron**: DTech Ultrafast Optics, (www.attochron.com);
- **TeraBeam Corp. (Proxim Wireless)**: FSO Broadband, (www.terabeam.com);
- **Utfors AB Sweden**: FSO Broadband Access, (www.utfors.com);
- **Electric Power Research Institute** (EPRI), (www.epri.com).

The **FSO Alliance** (www.wcai.com/fsoalliance) consists of 50+ members including all the major manufactures: AT&T, Teligent (www.teligent.com), Texas Instruments (www.ti.com), WinStar (www.winstar.com), XO Communications (www.xo.com). Some of these companies are no longer in business. Since 2002, the FSO Alliance has been affiliated with the Wireless Communication Association International (WCAI) (www.wcai.org).

8.5 WiMAX-Wireless Metropolitan Area Network Architecture

Worldwide Interoperability for Worldwide Access (WiMAX) is the code name given by the WiMAX Forum [32] to the wireless metropolitan technology and network design that is based on the IEEE 802.16d-2004 version of Fixed Wireless MAN standard specifications. WiMAX also refers to the IEEE 802.16e-2005 version of the Mobile Wireless MAN standard specifications. From now, we will use the WiMAX term interchangeably for both standards and we indicate any specific differences as we advance through this chapter. A typical WiMAX architecture that supports both fixed and mobile wireless versions is shown in Figure 8.4.

The fixed WiMAX network architecture consists of WiMAX CPE Subscriber Stations (SS) connected to radio Base Stations (BSs) through a Wireless Local Loop (WLL) air interface. The architecture is based on the WiMAX 802.16-2004 standard. SSs can be installed as indoor or outdoor units. A wireless point-to-point backhaul connection to another BS can also use the fixed WiMAX air interface. The BSs are connected to broadband switching centers, and from there, through a cable or fiber link, to the public Internet and telephone network.

The mobile WiMAX network architecture consists of WiMAX CPE Subscriber Stations (SS), in this case a laptop or a WiMAX phone, connected to a Base Station (BS) through a Wireless Local Loop (WLL) air interface. The architecture is based on the WiMAX 802.16e standard. Generally, a BS radio sector is shared by multiple WiMAX mobile users. However, the resultant reduced bandwidth allocation is hardly noticeable because the backhaul connection has a very large bandwidth, 100 Mbps. Multiple Input

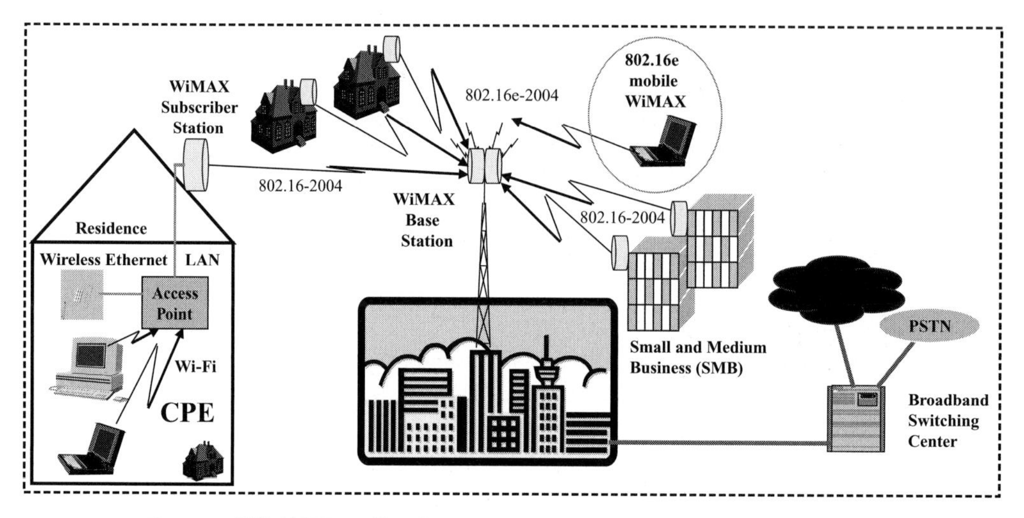

Figure 8.4 WiMAX Broadband Access Network Architecture

Multiple Output (MIMO) antennas are supported to increase the overall bandwidth and mobility.

WiMAX standards stipulate the spectrum to be used. However, there is no globally licensed spectrum for WiMAX. Therefore, one can find applications that work at 2.3–2.5 GHz, 3.5 GHz, 5 GHz, or 5.8 GHz. It is expected that WiMAX services will be provided in the 1.7–2.1 GHz band which was acquired by carriers and service providers (Sprint/Nextel, Clearwire) as part of the AWS auction that took place in the USA in 2006. Because there is such a diversity of spectrums in which WiMAX can operate interoperability problems might arise.

The primary **Fixed WiMAX application** is broadband wireless access, as an alternative to the DSL and cable modem technologies. Access types include metropolitan wireless networking, enterprise building/location to building/location connectivity, wireless backhaul to both wired and wireless networks, and a supplement to high-speed data, voice, and video technologies. The primary **mobile WiMAX application** is broadband wireless access to nomadic, mobile devices such as computing devices (laptops, PDAs, notebooks, mobile PCs) and smart WiMAX phones.

WiMAX can operate in the 700 MHz, 1.9 GHz, 2.1 GHz, and 3.5 GHz licensed bands as well in the 900 MHz, 2.4 GHz, 5.4 GHz, and 5.8 GHz unlicensed bands.

8.6 WMAN Networking Solutions Technical Comparison

A comparison of the WMAN network architectures, introduced earlier in this chapter, is shown in Table 8.1. They are presented along with satellite-based technologies, and for contrast, with ISDN narrowband and broadband technologies. All are wireless access technology alternatives competing for residential and business broadband local loop services.

This table has some historic connotations since LMDS and MMDS wireless MAN technologies are no longer being implemented. By comparison, satellite and FSO-based

Table 8.1 Wireless Metropolitan Access Technologies Comparison

Wireless Local Loop Access Alternatives	Maximum downstream rate	Maximum upstream rate	Operational spectrum	Shared bandwidth	Maximum distance	Availability Comments
LMDS	1.5–622 Mbps	1.5–622 Mbps	28–40 GHz (USA)	Yes	3 miles Line-of-sight	Business applications; Career Class; Typical-Point to Point; Experimental technology.
MMDS	384–512 Kbps	256–384 Kbps Burst 10 Mbps	2.1–2.7 GHz (USA)	Yes	35 miles Non-line-of-sight	Residential and small business; Typical-Point to Multipoint; Carrier class; Matured technology.
WiMAX (fixed wireless)	64 Kbps–54 Mbps	64 Kbps–54 Mbps	2–11 GHz (USA) 2.5, 3.5, 5.8 GHz	Yes	30–50 miles Line-of-sight	Residential and small business; Typical-Point to Multipoint; Backhaul Carrier class; New technology.
Satellite (PES-HES)	256 Kbps–60 Mbps	9.6–512 Kbps	2–6 GHz C 11–14 Ku	Yes	Line-of-sight	Rollouts since 1998; Direct PC one-way broadband.
Free Space Optics (FSO)	10 Mbps–2.5 Gbps	10 Mbps–2.5 Gbps	THz	Yes	1 mile Line-of-sight	Rollouts since late 1990; Inexpensive; No license is needed.
Narrow ISDN BRI-basic service	128 Kbps	128 Kbps	No	No	4 miles	CO switch-based; Widely available; Insufficient bandwidth.
Wide ISDN PRI-service	1.544–2.048 Mbps	1.544–2.048 Mbps	No	No	4 miles	CO switch-based; Widely available; Carrier class bandwidth.

broadband access continue as niche alternative solutions. FSO remains the most powerful alternative, given its ability to support high data rates, but it is limited to strict line-of-sight transmissions with a range of just one mile. Therefore, WiMAX remains the only viable and growing alternative for mobile broadband access.

8.7 WiMAX Standardization

Fixed WiMAX, 802.16d, employs the Orthogonal Frequency Division Multiplexing Access (OFDMA) scheme with 256 sub-carriers. In turn, mobile WiMAX, 802.16e, employs a different modulation scheme, Scalable OFDMA (S-OFDMA). This was done to improve near line-of sight coverage by adopting adaptive MIMO antenna systems, which help to mitigate multi-path interference. Since the two modulation schemes are incompatible, upgrade from fixed to mobile WiMAX will require changing the RF components.

Initially, in 2001, **802.16** standardization focused on the **10–66 GHz** range (**LMDS**), thus requiring Line-of-Sight (LOS) communications. Single-Carrier (SC) modulation was selected, hence the "Wireless MAN-SC". The BS transmitted a TDM signal with time slots allocated to each SS. An uplink that uses TDMA technology has two alternatives:

- Time Division Duplexing (TDD), when uplinks and downlinks share the same channels but do not transmit simultaneously; and
- Frequency Division Duplexing (FDD), when uplinks and downlinks operate on separate channels, either simultaneously or alternatively (half duplex FDD).

The **802.16a** amendment project (2003) dealt with Non-Line-of-Sight (NLOS), licensed bands in the **2–11 GHz** range (roughly the **MMDS** and European **HiperMAN** systems). It allowed several interface specifications: Single-carrier-based format (WirelessMAN-SC2), Wireless OFDM with 256-point transform, TDMA access (mandatory for license exempt bands), and Wireless MAN-OFDMA with 2048-point transform. Finally, the **802.16d** version, now known as fixed WiMAX, was issued in 2004. It contained the specifications needed to build WiMAX products. The initial WiMAX implementations and services are based on this standard. 802.16d-2004 obsoletes all previous versions of the 802.16d standards. Its official name is "Air Interface for Fixed Broadband Wireless Access System", IEEE Standard 802.16-2004 [33].

The **802.16e** amendment project was brought to fruition in December 2005. It adds a very important feature, the mobility of subscriber stations, hence the name of this technology mobile WiMAX. It introduces changes in the protocol architecture that make the two versions, fixed and mobile WiMAX, incompatible. The official name of the standard is "Air Interface for Fixed and Mobile Broadband Wireless Access System", IEEE Standard 802.16e, 2005 [34].

8.8 Mobile WiMAX 802.16e Main Features

Mobile WiMAX, based on the IEEE 802.16e standard, is an enhancement of fixed WiMAX. It creates the best basis on which to reach convergence of WMAN and cellular mobile networks. A summary of the main features of mobile WiMAX follows. Convergence aspects will be discussed in Part IV of this book.

- **Mobility**: Both soft and hard handover capabilities are included with latencies less than 50 ms. This allows support for time-sensitive applications such as VoIP and IPTV;
- **Broadband Data Rates**: Adoption of scalable channelization (from 1.25 MHz to 20 MHz) and advanced modulation and coding schemes; In a 10 MHz channel,

it allows peak downlink data rates up to 63 Mbps and uplink data rates up to 28 Mbps;

- **Quality of Service**: WiMAX is one of the few standards that incorporated in the initial design service flows that map to Differentiated Services Code Points to enable end-to-end IP-based QOS;
- **Security**: Authentication, encryption, use of various user names and passwords, and flexible security key management;
- **Scalability**: Use of adaptive MIMO antennas and channelization to provide scalability regarding deployment in dense urban areas and rural areas.

8.9 WiMAX Protocol Architecture

Key to many WiMAX features is the adoption of Scalable-Orthogonal Frequency Division Multiplexing (S-OFDM). In this technique, the bandwidth is divided into multiple sub-carriers. The active carriers are grouped in subsets of sub-carriers called sub-channels. This sub-channelization is applied to both the downlink and uplink. Channelization eliminates intra-cell interference. The high rate input data stream is divided into several sub-streams of lower rates. Each sub-stream is modulated and transmitted through separate orthogonal sub-carriers. Sub-channels can be aggregated to provide a larger bandwidth for individual users. Channel bandwidth can vary between 1.25 MHz and 20 MHz. However, only two values, 5 MHz and 10 MHz, are currently considered for WiMAX profiles that will be tested for certification [35].

WiMAX supports both Time Division Duplex (TDD) and Full and Half-Duplex Frequency Division Duplex (FDD). The current profile points only to TDD in cases where uplink is followed by downlink transmission. The advantage is that only one channel is needed. Several modulation techniques can be used but only three are mandatory: QPSK, 16 QAM, 64QAM. Convolutional coding with various coding rates can be used as needed (1/2, 2/3, 3/4, 5/6) [36].

In addition to the Physical Layer characteristics, a schematic layout of WiMAX Medium Access Control sublayer framing is presented in Figure 8.5.

The most important fields in the Data Link Layer PDU are:

- **Preamble**: Used for synchronization between transmitter and receiver;
- **Frame Control Header**: Provides information about the MAP configuration, length, coding scheme, and sub-channeling availability;
- **DL-MAP**: Provides sub-channel and control information about DL frames;
- **UL-MAP**: Provides sub-channel and control information about UL frames;
- **DL Burst**: The downlink subframes burst portion;
- **UL Ranging**: A special frame used by mobile stations to provide power closed-loop time, frequency, and power adjustment along with requests for bandwidth;
- **UL Fast Feedback**: Used by mobile stations to provide feedback channel information;
- **UL Burst**: The uplink subframes burst portion.

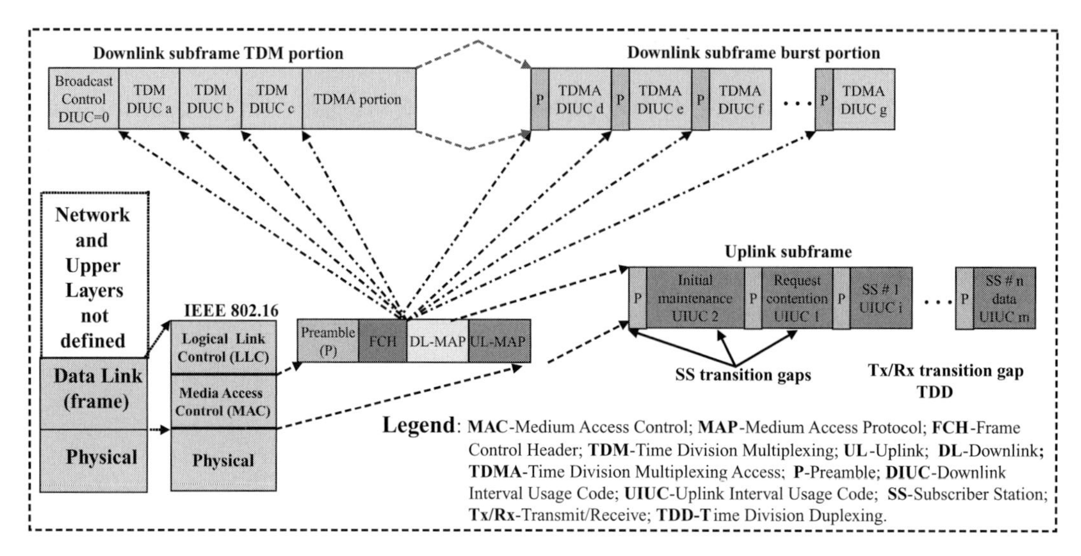

Figure 8.5 WiMAX Protocol Architecture

8.10 WiMAX Security

Given the inherent vulnerability that comes with wireless communications, the success of WiMAX technology will strongly depend on secure communications and the topology of WiMAX networks. Three topological architectures are supported in the WiMAX 802.16e standard: Point-to-Point (PTP), Point to Multipoint (PMP), and the Mobile Multi-hop Relay (MMR) mesh configuration. In the first two topologies, communication takes place between a BS and SSs that are found within the transmission range: either in line-of-sight or near line-of-sight. In mesh configurations, each node is capable of taking the dual BS/SS role by relaying information to the next hop. The downlink channel (from BS to SS) is separated either in frequency or time from the uplink channel (from SS to BS) [37]. Both Time Division Duplex (TDD) and Frequency Division Duplex (FDD) operations are supported.

To provide a secure WiMAX network environment, the following security features should be supported by WiMAX applications:

- **Authentication**: Identification of valid users of the system in both the downlink and uplink directions, i.e., BS initiated authentication and SS initiated authentication;
- **Authorization**: The permission given to authenticated subscribers to use services according to their individual profile;
- **Availability**: Maintenance of secure communication despite attempts to intrude upon the systems, including protection against Denial of Service (DoS) attacks;
- **Confidentiality**: Protection of sensitive information through encryption;
- **Integrity**: Protection of information integrity from unexpected alterations caused by transmission impairments or malicious actions.

Adequate QOS should be guaranteed in terms of availability, bandwidth, delay, jitter, and packet loss. These features are even more critical in the MMR environment where there are multiple relaying and gateway nodes used to backhaul the WiMAX network to public wired or wireless network environments.

8.11 Advantages and Disadvantages of WiMAX Technology

WiMAX is based on the accumulated experience of the past decade in wireless metropolitan networking. That experience includes failures of metropolitan solutions such as LMDS and MMDS. But the success and the promise of this technology is explained by the capability of WiMAX to serve both fixed and mobile clients. In this way, WiMAX has moved from the realm of technologies that were classified as "fixed wireless" into the mobile world. Further, by providing long-range communications, WiMAX challenges traditional cellular mobile networks in metropolitan areas. Thus, WiMAX has become a contender in the race to provide 4G mobile services [38]. The advantages and benefits of using the WiMAX technology and networking follows:

- It provides wireless broadband access, competing with DSL, cable modem, and 3G and Beyond 3G technologies;
- It supports a high level of mobility;
- There are no line-of-sight communications requirements so it can be implemented easily;
- It uses licensed and unlicensed bands of the spectrum;
- It can be used as a backhaul technology for cellular mobile networks and remotely located individual and group users;
- It is based on IEEE internationally recognized standards, for both fixed and mobile solutions;
- It supports data, voice, and multimedia communications;
- It is strongly driven by the expansion of nomadic laptops and PDAs;
- It is largely supported by all major computer vendors and application providers;
- It provides long-range communications up to 30 miles;
- It has a built-in QOS mechanism that can be used for both voice and multimedia transmissions;
- It is scalable regarding the number of users and coverage area; and
- It exhibits a very good spectral efficiency (ratio of bits transmitted to the channel bandwidth).

However, there are some limitations and disadvantages in using WiMAX:

- Overall performance is affected by the degree of mobility, line-of-sight paths, and environmental conditions;
- There is a limited level of support from major cellular mobile operators (they feel their business and revenue are threatened);
- There is no uniform global license spectrum for WiMAX;

- The use of common unlicensed spectrums makes it prone to interference;
- It has relatively high power consumption that requires separate power feeders;
- The new standard development is not backward compatible with earlier versions.

8.12 WiMAX and 3G Cellular Mobile Comparison

Since its standardization as a fixed WMAN technology, WiMAX has raised questions regarding the relative position of WiMAX to WWAN cellular mobile networks. This issue has been discussed even more since the adoption of mobility as part of the WiMAX technology. Therefore, a technical comparison between WiMAX and 3G cellular network technologies is necessary [39]. This analysis will be continued when the convergence between WiMAX and cellular mobile networks is analyzed in subsequent chapters.

All of the cellular mobile technologies have improved dramatically in the past five years to provide broadband data services at the highest possible data rate and throughput. Two technologies will be analyzed to illustrate somewhat different trends in 3G implementations. One is the 3GPP adoption of WCDMA (GSM camp) with two advancements, the High Speed Downlink Packet Access (HSDPA) and the High Speed Uplink Packet Access (HSUPA). The other technology is the CDMA2000 family (CDMA camp) which is 3GPP2 output with advancements such as 1xEV-DO Rev. A (uplink enhancement) and 3xEV-DO Rev. B (multi-carrier enhancement). 3xEV-DO Rev. B-based commercial products are expected sometime in 2008.

A high level comparison of 802.16e mobile WiMAX, W-CDMA HSDPA/HSUPA, and the CDMA2000 1xEV-DO Rev. A cellular mobile technologies comparison is provided in Table 8.2. Modulation types, downlink and uplink parameters, coding, and handoff characteristics are analyzed.

8.13 WiMAX Applications, Products and Service Providers

Wireless broadband metropolitan access via fixed and mobile WiMAX communications provides support for services to small, medium, and large multi-tenant units, and small and medium-size businesses. WiMAX includes the following applications:

- High speed broadband Internet access, Email, Web browsing, File transfer;
- LAN-to-LAN interconnections, campus and buildings interconnection;
- Voice and video streaming, IPTV;
- VOIP and video conferencing;
- Multiplayer interactive gaming;
- Backhaul channel for wired Internet Service Providers;
- Backhaul connectivity for cellular mobile traffic from remote, rural areas;
- Disaster Recovery using WiMAX as a backhaul channel;
- Wireless metro networking; and
- Broadband access alternative in areas where DSL and cable modem are not available.

Table 8.2 Mobile WiMAX and 3G Cellular Mobile Technologies Comparison

Features	1xEV-DO Rev. A	HSDPA/HSUPA	Mobile WiMAX
Base standards	CDMA 200/IS-95	WCDMA	IEEE 802.16e
Duplex method	FDD	FDD	TDD
Downlink access	TDM	CD-TDN	S-OFDMA
Uplink access	CDMA	CDMA	OFDMA
Channel bandwidth	1,25 MHz	5 MHz	Scalable 1.25–20 Mbps
Downlink frame size	1.67 ms	2 ms	5 ms TDD
Uplink frame size	6.67 ms	2.1 ms	5 ms TDD
Downlink modulation	QPSK/8PSK/64QAM	QPSK/16QAM	QPSK/16QAM/64QAM
Uplink modulation	QPSK/8PSK/BPSK	BPSK/QPSK	QPSK/16QAM
Downlink peak air data rate	3.1 Mbps	14 Mbps	46 Mbps DL, 32 Mbps UL (10 MHz)
Uplink peak air data rate	1.8 Mbps	5.8 Mbps	7 Mbps DL/UL = 3; 4 Mbps DL/UL = 1
Coding	Turbo	Convolutional Coding (CC), Turbo	Convolutional Coding (CC), Turbo
Hybrid ARQ	Fast 4 channel	Fast 6 Channel Asynchronous CC	Multichannel Asynchronous CC
Scheduling	Fast scheduling DL	Fast scheduling DL	Fast scheduling DL & UL
Handoff	Virtual soft-handoff	Network hard handoff	Optimized hard handoff
Tx Diversity and MIMO	Simple open loop diversity	Simple open/closed diversity	Space Time Block Code (STBC)
Beamforming	No	Yes (dedicated pilots)	Yes

Following is a list of **WiMAX equipment manufacturers** and **carriers** that use the technology or act as systems integrators:

- **Alcatel/Lucent**: Using Alvarion OEM-based Breeze Max components integration in Canada, (www.alcatel.com);
- **Airspan Networks**: MacroMAX and MicroMAX BSs and EasyST SS: Deployed in Germany and USA, (www.airspan.com);
- **Alvarion**: Breeze Net BS series, Broadband Data CPE, Broadband Voice Gateway, Broadband Network Gateway, BreezeVIEW SNMP-based NMS, (www.alvarion.com);
- **Aperto Networks**: 2.5, 3.5 and 5.8 GHz PacketWave family of BS and Subscribers Stations, (www.apertonet.com);
- **Atmel Corporation**: AT86RF535B SS, (www.atmel.com);
- **Cambridge Broadband**: WiMAX VectaMAX, VectaStar CPE using Sequans' WiMAX, System-on-a-Chip, VectarStar EMS (standalone and HP OpenView), (www.cambridgebroadband.com);
- **Edge-Core**: 802.16e WA81111 SS, WiMAX WA8121 PCMCIA card, (www.edge-core.com);

- **Fujitsu**: WiMAX chip set, (www.fujitsu.com);
- **Intel Corporation**: Rosedale WiMAX chip set, (www.intel.com);
- **Motorola Corporation**: Canopy migration to WiMAX and wi4 WiMAX BS and CPEi600 SS with commercial grade PC cards, (www.motorola.com);
- **Nortel Networks**: Austar (Australia) Regional WiMAX Network, Chungwa Telekom (Taiwan), (www.nortelnetworks.com);
- **Proxim Wireless**: Tsunami MP.11 5012, 2454 (2.4 GHz, 5.3 GHz, 5.5 GHz, 5.8 GHz), Tsunami MP 16 3500 (3.5 GHz), MeshMAX 3500 tri-radio, WiMAX and Wi-Fi mesh access point series, (www.proxim.com);
- **PMC Sierra** and **picoChip**: 802.16e chipsets, (www.pmc-sierra.com);
- **Redline Communications**: RedMAX SU-I, SU-O) indoor/outdoor SS, AN-100U BS, AN-100 UX; Deployed in Pakistan, (www.redlinecom.com);
- **Samsung Electronics**: SPH-P9000 WiMAX terminal, (www.samsung.com);
- **Sequans Communications**: SQN 1010-RD (FDD) SS, SQN 2010 BS, (www.sequans.com);
- **Siemens**: SkyMax radio base stations and radio modems, (www.siemens-mobile.com);
- **Orthogon Systems**: OS-Spectra 300 Ethernet bridge, a platform operating in the 5.8 GHz (coverage 125 miles at 300 Mbps), (www.orthogonsystems.com);
- **SR Telecom**: Broadband Fixed Wireless Access Symmetry Platform, SSU4000, (www.srtelecom.com);
- **WiNetworks**: Win-MAX E series WiN5000 SS, WiN7000BS, (www.winnetworks.com);
- **ZyXel Communications**: MAX-100 PCMCIA cards, MAX-200 Series indoor SS, MAX-300 series outdoor SS, (www.zyxel.com).

Early WiMAX Service Providers:

- **Telabria**: First WiMax Service in South East England, (www.telabria.com);
- **Iberbanda**: Aperto PacketWave components, Spain, (www.iberbanda.com);
- **Chibardun Telephone Cooperative**: WiMAX services in Wisconsin (12 MHz channel in the 700 MHz band), SOMA Networks, (www.chibardun.com);
- **ClearWire**: Partnership with Sprint/Nextel for wide WiMAX deployment in USA; Deployments in Denmark, Belgium, Ireland, (www.clearwire.com);
- **Irish Broadband**: Fixed WiMAX services using Alvarion and Navini Networks gear, (www.irishbroadband.ie);
- **Motorola Corporation**: MOTOwi4 WiMAX access, (www.motorola.com);
- **Sprint/Nextel**: US cities and market (based on LG and Samsung, ZTE/Intel components), (www.sprint.com);
- **Towerstream**: Fixed WiMAX in major USA metropolitan areas and markets, (www.towerstream.com);
- **CETECOM Labs**: WiMAX Testing and certification, (www.cetecom.es);
- **Wibro-based**: South Korea Telecom (SKT), WiMAX broadband services;
- **Finland** (SuomiCom), **Georgia** (QTEL Global Networks), **India** (Dishnet), **Pakistan** (Vateem Telecom/Motorola), **Puerto Rico** (Volare PR), **Russia** (Nextnet Wireless),

Singapore (SingTel), **UK** (Urban WiMAX Plc), **USA** (ClearWire, NextWave Wireless) WiMAX trials and services.

South Korea's Electronics and Telecommunications Industry led by Samsung Electronics and ETRI has developed a version of WiMAX, called WiBro. Intel and LG Electronics have an agreement to make WiBro fully compatible with the WiMAX 802.16e standards.

8.14 WiMAX Management Requirements and Management Products

Following principles developed so far in this chapter, network and systems management requirements, applicable to any wired and wireless telecommunications network, can be summarized as follows:

- Ability to monitor, and control WiMAX base stations and subscriber stations, either fixed, portable, or mobile;
- Management of WiMAX connectivity to external wired/wireless network;
- Mobility management (handoff and power management);
- Ability to report management-related information;
- Security management (operations, authentication, encryption); and
- QOS management (data services, QOS parameters).

The WiMAX Forum [32], a non-profit trade association, founded in 2003, plays an important role in the implementation of IEEE WiMAX standards. Currently, it is comprised of over 350 member companies representing service providers, equipment vendors, manufacturers, and chip designers. The WiMAX Forum collaborates closely with ETSI's Broadband Radio Access Network (BRAN), thus assuring full harmonization with the European version of wireless MAN architecture, called HiperMAN.

The WiMAX Forum's Technical Working Groups are also responsible for the certification process that involves developing definitions of systems performance and publishing of conformance profiles. These profiles are a subset of those standards that contain clearly defined mandatory and optional features along with conformance and interoperability testing suites. The first WiMAX certification testing laboratory was established at CETECOM Labs (Centro de Tecnologia de la Comunicationes), Malaga, Spain. CETECOM is an experienced testing center involved with GSM/GPRS, EDGE, W-CDMA, Wi-Fi, Bluetooth, RFID, and PSTN. It provides services related to conformance testing, test development, interoperability testing, and training (www.cetecom.es).

Other certification labs will be established to test those mobile WiMAX, 802.16.e products that are based on Wave 1 and Wave 2 chipset profiles. It is expected in the 2007–2008 timeframe that mobile WiMAX specifications will be embedded in chipsets as part of nomadic, portable, and mobile devices such as laptops, PDAs, and mobile handsets. Given the technical superiority of mobile WiMAX, a migration of fixed WiMAX products based on 802.16d standards will be necessary since mobile WiMAX is not backward compatible with the fixed WiMAX standards.

Several management products have been developed to support WiMAX-based networking and services:

- **Alcatel/Lucent**: Network Operation Center, Navis NMS, (www.alcatel.com);
- **Alvarion**: AlvariSTAR NMS, (ww.alvarion.com);
- **AirSpan Networks**: Netspan SNMP-based NE Manager, (www.airspan.com);
- **Iskratel**: SI3000 WiMAX NMS and WiMAX Provisioning, (www.iskratel.com);
- **Motorola**: Prizm EMS (Canopy migrating to WiMAX), (www.motorola.com);
- **Proxim Wireless**: ProximVision 25, 100, 200 NMSs, ORINOCO Smart Wireless Suite Management System for converged WiMAX and Wi-Fi networks and products, (www.proxim.com);
- **Redline Communications**: RedMAX NMS, (www.redlinecommunications.com);
- **WaveIP**: GigaAccess NMS 802.16e QOS-based, (www.waveip.com);
- **WiNetworks**: WiN5000 SS, WiN7000BS Local Management SNMP2-based, (www.winnetworks.com).

WiMAX NMSs provide Graphical User Interfaces (GUI) to manage entire WiMAX Application Service Networks (ASN) with tens and hundreds of SSs and BSs. They allow monitoring and control of channelization, transmission power, and encryption methods. They are SNMP-based managers with agents deployed in both BSs and SSs based on standard MIBs and enterprise specific MIBs. More sophisticated NMSs provide more than monitoring and diagnostics, i.e., provisioning, configuration, trouble reporting, and even policy-based management. Very few are providing integrated QOS management as a basis for SLA management.

8.15 WiMAX Quality of Services and QOS Metrics

QOS management is a feature built into the WiMAX standards by defining QOS classes and QOS parameters. QOS classes are related to time sensitive applications such as those involving voice and video communications. WiMAX Forum and IEEE 802.116e QOS classes [34] are presented in Table 8.3.

The first group of QOS classes is related to voice and video support. It includes unsolicited grant services that provides VoIP QOS to extended real-time polling service that provides support for interactive voice. The second group of QOS classes provides support for various forms of data communications. Each class has several qualifiers associated with it. It was left to the next phase of profiling to attach quantitative values to data rates and tolerance to jitter as presented in Table 8.4.

Key to the achievement of WiMAX broadband QOS are the scheduling services built into the MAC sub-layer. The main characteristics of WiMAX scheduling services are:

- **Fast Data Scheduler**: This is the capability of the BS to allocate resources to bursty traffic and to dynamically change channel conditions. Data packets are associated with service flows (data, voice, or video) and QOS parameters. The SS, in its uplink information, provides fast-feedback information through a dedicated frame;

Table 8.3 WiMAX QOS Classes and Applications

QOS Classes	Applications	QOS Specifications
Unsolicited Grant Service (UGS)	VoIP	• Maximum Sustained Rate • Maximum Latency Tolerance • Jitter Tolerance
Real-Time Polling Service (rtPS)	Streaming Audio & Video	• Minimum Reserved Rate • Maximum Sustained Rate • Minimum Latency Tolerance • Traffic Priority
Extended Real-Time Polling Service (rtPS)	Voice with Activity Detection (VoIP)	• Minimum Reserved Rate • Maximum Sustained Rate • Maximum Latency Tolerance • Jitter Tolerance • Traffic Priority
Non-Real-Time Polling Service (nrtPS)	File Transfer Protocol (FTP)	• Minimum Reserved Rate • Maximum Sustained Rate • Traffic Priority
Best Effort Service (BE)	Data Transfer, Web Browsing	• Maximum Sustained Rate • Traffic Priority

Table 8.4 WiMAX Forum Applications Classes and QOS Parameters

Class	Applications	Bandwidth	Latency	Jitter
1	Multiplayer Interactive Gaming	Low 50 Kbps	Low < 25 ms	N/A
2	VoIP and Video Conferencing	Low 32 to 64 Kbps	Low < 160 ms	Low < 50 ms
3	Streaming Audio, Music & Video	Low to High 5 Kbps to 2 Mbps	N/A	Low < 100 ms
4	Web Browsing Instant Messaging	Moderate 10 Kbps to 2 Mbps	N/A	N/A
5	Media Content Downloads	High Over 2 Mbps	N/A	N/A

- **Frame Scheduling**: WiMAX provides traffic selection in both directions, DL and UL. Critical for this scheduling is the uplink feedback through ranging field containing bandwidth request;
- **Dynamic Resource Allocation**: WiMAX provides both time- and frequency-based allocation for both uplink and downlink directions on a frame basis. The allocation

information is delivered in MAP header messages. The time-based resource allocation can be done from a slot-based to an entire frame-based allocation. The allocation is facilitated by well-defined QOS classes, as presented in the previous table;

- **Frequency-based Scheduling**: WiMAX provides selective frequency-based scheduling by allowing operations on different sub-channels. The frequency–diversity scheduling includes pseudo-random permutations of users across the bandwidth and upfront allocation of resources to a sub-channel;
- **QOS-oriented Scheduling**: Schedulers choose the corresponding data service and QOS parameters on a connection-by-connection basis for both uplink and downlink. That helps resource allocation and overall performance.

9 Wireless Near-Field Sensor Networking

9.1 Near-Field Sensor Technologies

Near-Field Sensors (NFS)-**based Networking** (NFSN) is a generic term for wireless transmission technologies and networks confined to a short proximity between transmitters and receivers, generally a few centimeters, exceptionally 1–3 meters. Conceptually, NFC/NFSN covers digital data communications and multimedia information exchange and can be implemented using a variety of wireless technologies such as **Radio Frequency Identification (RFID), Near-Field Communications (NFC)**, and **Ultra Wide Band** (UWB).

NFSN solutions, regardless of the technology that is adopted, must satisfy service requirements such as broadband transmission capability, reliability, scalability, simplicity, security, quality of services, manageability, and cost effectiveness. Special applications extend NFSN into movements of physical goods and persons, and even commercial and industrial operations. The array of applications ranges from individual customers/subscribers automatic identification to tracking commercial and industrial operations and chain of supplies of medium and large size businesses, all aimed at enhancing overall productivity. This presents challenges to designers of sensors and receivers, NFSN developers and integrators.

Although NFSN and its representative technologies, RFID, NFC, and UWB, are far from being wide-spread mass applications, the signs of their presence are visible in applications involving major retail organizations in their tracking of major items using RFID technologies, in physical persons that accept RFID tag implants to follow-up on health status information, and in spectrum allocation for UWB technologies across major markets. Specialists expect future growth in manufacturing of the components and systems needed to read a multitude and a variety of tags as well as an explosion of data traffic generated in conjunction with tracking in time and space of these tags. Another aspect, which will be discussed in subsequent chapters, is the convergence of NFSN with other wireless networks, in general, and with cellular mobile networks, in particular. This means addition to the ubiquitous handset or cell phone, the centerpiece of mobile communications, of NFC, RFID, and UWB capabilities in pursuit of multi-mode devices that can work transparently across all types of wireless networks.

NFSN architecture is relatively simple and consists of two-way wireless communications between transmitters and receivers found in short proximity, sometimes in close touch. Communication takes place across a wireless interface operating in the radio/microwave range of frequencies. Architectural details and technical specifications of NFC/NFSN technologies are provided in subsequent sections.

9.2 RFID Networking Architecture, Components, Frequencies

Radio Frequency Identification (RFID) is a short-range RF wireless, tag-based identification technology. RFID networking implies a multi-tier architecture that consists of a **pure RF segment**, a **RFID middleware infrastructure**, and **external applications and databases**. The RF segment consists of **RFID tags** and the **RFID reader**. Since communications across the RF segment is two-way communications, where the RFID tag has a dual role as transmitter and receiver, the RFID tag is also referred to as a RFID transponder. A high-level RFID architecture is shown in Figure 9.1.

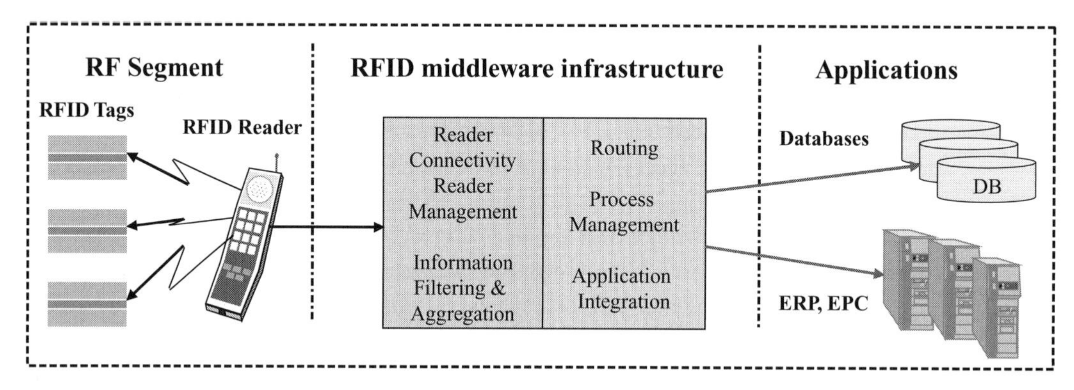

Figure 9.1 A High Level Passive RFID Networking Architecture

The portion of middleware infrastructure connected to the RFID reader provides core functions such as reader connectivity, device management, and information filtering and aggregation. The portion of middleware infrastructure connected to the enterprise systems provides higher-level functions such as routing, application integration, and process management. The external applications consist of process/operation management, such as management of Electronic Product Code (EPC), Enterprise Resource Planning (ERP), trading partner management, and communications with internal or external databases. RFID technology is considered an enhancement and in many instances a replacement of line-of-sight laser-based bar code systems.

Use of RFID and spectrum allocation is regulated from country to country. The most common RFID frequencies are listed in Table 9.1.

Table 9.1 RFID Range of Frequencies

RFID Transponders								
LF		HF		VHF	UHF			
125 KHz	134 KHz	8.2 MHz	13.56 MHz	433 MHz	868 MHz	915 MHz	2.45 GHz	5.8 GHz

Depending on the applications, RFID transponders operate in the Low Frequency (LF), High Frequency (HF), Very High Frequency (VHF), and Ultra High Frequency (UHF) bands. The most common RFID implementations are in the HF domain, at 13. 56 MHz, and in the UHF domain, at 868 MHz and 915 MHz. Data transmission rates are in the range of 10 to 40 Kbps.

9.3 Classification of RFID Tags

RFID is a tag-based technology where the tag or contactless card can be attached to a whole world of things, from individual items, to collective packages, to sensors of various kinds, to documents, and even to living beings such as humans and animals. Virtually any item imaginable, having a certain size, can be tagged. The RFID tag consists of a microchip and an antenna system, all in the simplest form, especially when the cost of the tag is a major issue. Innovations in the design of tags of various sizes and shapes, and the design of antennas, will continue to drive the cost of tags into the range of fractions of pennies. Tags made of polymer semiconductors, instead of silicon, are currently developed by several companies, aimed at creating very inexpensive printable tags similar to barcode labels.

According to whether the tags are powered or not, we have three types of RFID systems: active, passive, and semi-active. **Active RFID tags** have their own power supply, although limited. These tags have similarities to WLAN access points but with limited duty cycles to conserve energy. Active RFID tags can be designed to contain sensors that measure temperature, humidity, vibrations, or radiation. Therefore, they can be used to provide a network of radio-based sensors and actuators in industrial automation and control networks along with supply chain tracking and logistics.

Passive RFID tags do not have their own power supplies. The energy required for reception and transmission is provided by the reader's electromagnetic field. The tags are activated when they move within the reader's proximity. The induced current in the antenna is used to power up the tag and backscatter or modulate the reader's signal to transmit the information. These tags are low cost technology items, generally very small, the size of a coin or postage stamp, with no power consumption. There is no data processing involved. RFID tags do not initiate transmission or relay each other's traffic. However, a tag chip can contain non-volatile EEPROM to store data. Therefore, the implementation of RFID systems requires RFID readers and a change in the existent business processes. **Semi-active RFID tags** have a very small battery supply that allows an increased range of action compared to passive tags. The life of these tags is limited.

A very small tag, developed by Hitachi, called **μ-chip**, measures 0.15×0.15 mm with a thickness of 7.5 μm. It can transmit a 128 bit ID number that is hard-coded into the chip during the manufacturing process. A RFID that costs a dime can store a complete description of an object using the **Electronic Product Code** (EPC), an **Extensible Markup Language** (XML)-based standard for describing physical objects. The information stored in active and semi-active tags is usually larger than in passive tags.

The Auto-ID Laboratory at Fundan University, Shanghai, China has been working to develop RFID tags that integrate the RFID antenna with the silicon chip. This approach poses a major challenge, knowing that the chipset is powered by the energy provided by the RFID reader. This energy is just a tiny fraction of the energy provided to regular passive tags, given the overall size of the tag, less than a square millimeter.

RFID tags are largely used in **RFID access cards** (buildings/tickets) and RFID transponders attached to windshields for toll reading. The list of applications will be described in subsequent sections. RFID-based wireless tags have found an unusual implementation in personal RFID chip implants. Applied Digital's company subsidiary, VeriChip, has received the US Food and Drug Administration's approval to sell chips for human implantation. As of 2007, there were several thousand live human implants. The information carried by these chips is primarily health-related, with specific warnings and contraindications. This can be critical for unconscious, non-verbal, or mentally ill people. These chips are expected to last 10 years and they are not susceptible to MRI scanning. There is no evidence that they can be deactivated by magnetic energy.

9.4 RFID Standards

The great promise of RFID technology cannot be achieved without international standardization. The standardization starts with spectrum harmonization to assure cross business and cross country RFID applications. For example, in Europe, there is less UHF bandwidth available compared with the United States, which might tend to give an unfair competitive advantage to American companies. There is also a need to develop a common, standardized, way of coding products and the content of information carried.

Given the nature of RFID short proximity wireless communication, there is no global governing body to control frequency allocation and RFID applications. Each country or community of countries can set their own rules such as those emanating from the Federal Communications Commission (FCC) [40] or the European Telecommunications Standard Institute (ETSI) and the European RFID Forum [41].

The first international standard adopted for RFID was the **Electronic Product Code** (EPC), initiated by ISO. EPC standardization has followed some of the concepts adopted in the Universal Product Code (UPC) which is used in barcode labeling. EPC was followed by EPCglobal network standardization. The name of the second standard is the same as the internationally recognized standard body, a consortium of retail and consumer goods industries. This group adopted the Class 1 Generation 2 (C1G2) specifications. In one word, combining the "world of items" with the "world of RFID data" and to exercise this complex correspondence across the Internet. The standard provides interface specifications for communications between RFID readers and RFID tags [42]. Characteristics of the C1G2 interface include:

- The spectrum for RFID is in the ultra-high frequency (UHF) range of 860–960 MHz;
- Support for spread spectrum technologies which allow selective communications with multiple card readers at different frequencies within the accepted spectrum range;
- The read range for UHF readers is 10–20 feet in the open air;

- It allows speeds of 1500 tag-reads per second in the USA and 600 in Europe. (Current tags are read at 100–300 reads per second, while most of the consumer-goods manufacturing/pharmaceuticals lines move cases or bottles at a rate of 200–400 units per second);
- There are three modes of operation for tag readers: single-interrogator (allows one reader within a 1 km (0.62 mile) radius); multiple interrogators (allows 10 readers within a 1 km radius), and dense interrogators (allows 50 or more readers within a 1 km radius);
- The standard covers essential chip, reader, label, printer, and software technical specifications to allow building RFID infrastructures;
- There is a royalty-free framework in building products by independent vendors;
- The standard has already been adopted in Europe, North America and parts of Asia.

The centerpiece of the EPCglobal network initiative is the creation of the **Object Name Service** (ONS), where each product is assigned a unique identifier. Currently, the ONS directory service is operated by VeriSign, an American company that also provides the Internet DNS directory. There is a generally accepted opinion that a centralized ONS service, i.e., a European and Asian hierarchical service, will be necessary to assure future independence in RFID applications.

The most important RFID standard specifications are listed below:

- ISO 18000 series of standards (Part 1, Part 2 (below 135 KHz), Part 3 (13.56 MHz), Part 4 (2.45 GHz), Part 5 (5.8 GHz), Part 6 (860–930 MHz), Part 7 (433 MHz) [43];
- ISO 18000–6C (EPCglobal UHF Class 1 Generation 2);
- ETSI Recommendations EN 300 220 and EN 302 208;
- ETSI Recommendations EN 302 065 (UWB communications);
- ISO 14443-based International Civil Aviation Organization (ICAO) 9303 standards for RFID passports;
- ISO 15693, HF RFID (13.56 MHz) for non-contact smart payments;
- ISO 11784 and 11785 RFID tags for animals;
- ISO 14443A-3 Mifare family of standards;
- ISO 14443A-4 TCL family of standards;
- Tag standards: ISO15693, Philips Icode, EPC, TI Tag-It, Legic, DESfire, TagSys (C210/220/240/270), MIFARE ISO 14443-A/B, Hitag 1 & 2, ISO FDXB, ISO 1800 6A/6B/6C.

There are also specifications provided by national organizations for specific applications such as ANSI 371 Part 1/2/3 for beacon locating tags, AAR S918 for railcar identification in the USA, and IEEE Rubee for monitoring industrial applications.

9.5 RFID Applications

One of the major drivers in RFID technology implementation in the USA was Wal-Mart who mandated, starting in 2005, that all large incoming shipments have attached RFID tags. It was mandatory for the 100 biggest suppliers and is applicable only to cartons and pallets. No individual items on shelves will have RFID tags in the near future. Similar requests are coming from Target and the US Government.

While there is no longer discussion about the future of RFID, there is still a debate of how to implement it, and what kind of industries and applications can benefit most from these technologies. Perusing current implementations and future plans to introduce RFID technologies in current operations, as expressed by industry leaders, we found numerous examples of RFID applications:

- **Retail Industry**: Tracking the movement of labeled goods, a flexible continuation of bar coding, identification of individual items or batches/pallets as they are supplied, shelved, and ultimately sold. This includes protection against pirated, counterfeit products;
- **Health Care and Pharmaceutical Industry**: Many aspects of health care from the labeling of medications, to follow up on operations and treatment, and the logistics of providing timely provisioning of medical supplies;
- **Consumer Applications**: One of the selling points of RFID is to ease many aspects of every day life, from transportation, banking, shopping, entertainment, to the intuitive, easy, exchange of information between digital/computing devices;
- **Automotive industry**: The production of any car implies tens of thousands of individual components. They are aggregated into ensembles representing dash-boards, engines, transmission and steering systems, wheels, doors, and additional functional and security components. Assembling all these components requires a thorough tracking of individual components and a sequential addition of them during the manufacturing process;
- **Aviation Industry**: All the arguments presented for the automotive industry are valid for aviation where individual parts that constitute a plane are numbered in millions. RFID is applicable to the maintenance of airplanes, scheduled movement of parts used for maintenance, and overall security;
- **Global Transportation**: These applications serve the tracking process and monitoring operations related to the movements of goods shipped in pallets and containers as part of the global international trade using multi-modal transportation that includes ships, trains, trucks, barges, and airplanes.

The success of NFSN, especially RFID-based, is highly dependent on the information system infrastructure. The easiest part is the creation and reading of tags as articles move or are moved as part of normal flows of goods and people. Therefore, there is a need for systems that can collaborate and instantly transmit this information whenever it is needed. For example, reading a RFID-based passport should be correlated with a database that provides additional information to fully authenticate the passport holder and display possible restrictions put on his/her movement in and out the country. All these operations need to be performed in accordance with nationwide regulations or special requirements inspired by the security policies of a particular country.

Each application is different and may require use of a particular RFID technology, meaning a certain frequency band, type of tag, and supporting infrastructure. The variability of applications is determined by the variety of processes and operations that should

Table 9.2 RFID Classes and Typical RFID Applications

RFID Class	Frequency Range	Reading Range	Data Rates	Tag Reading	Typical Application
Low Frequency	125 KHz 134 KHz	<0.5 m <1.5 ft	Slow	Near wet Metal surface	Access Control Animal Tracking POS Applications
High Frequency	13.56 MHz	1 m 3 ft	Average	Near wet Metal surface	Access Control Smart Cards Smart Shelves
Ultra-High Frequency	860 MHz– 930 MHz	3 m 10 ft	Faster	–	Packet Tracking Luggage Tracking Toll Applications
Microwave Frequency	2.4 GHz– 5.8 GHz	1 m 3 ft	Very Fast	Only dry surface	Supply Chain Tracking Toll Applications

be tracked, environmental conditions surrounding the tags (air, liquid, metal), range of action, and the use of different RFID tag and reader types. A high level comparison of the main RFID technologies is presented in Table 9.2.

Currently, the most common implementations and applications are based on HF RFID using the 13.56 MHz band.

9.6 RFID Security Aspects

The wide spread implementation of RFID technology poses serious security concerns. First, there is a possibility of illegal reading, tracking, and even disabling RFID tags. To protect this information, data transmission can be encrypted and use of security public keys can be made. In this way, secret information contained in the tag can be transmitted only to authorized RFID readers using challenge-response protocols. Jamming signals can be used to prevent identification from outside sources. Legislation can be used to prevent the unauthorized use of consumer personal information or credit card data that is generated in conjunction with RFID-tagged consumer products acquisition. Since RFID tags could remain functional when they are physically destroyed, they are susceptible to being read in the presence of readers used for surveillance or eavesdropping. Such abuse can also come from retail organizations, eager to augment the protection of goods against shoplifting or to collect consumer shopping pattern information through RFID spy chips.

There are security issues concerning human implanted RFID chips since spam generated by these chips through anonymous tracking, analog to spyware and adware infecting our computers, are an abuse on individual privacy. Beth Israel Deaconess Medical Center in Boston and Harvard Medical School are among those perfecting the RFID

Chip implantation operation so that it will not cause any side effects, pain or change in muscle functions. Currently the data on these chips is not encrypted and so the readers are networked in a closed web site to eliminate the possibility of unwanted dissemination of information.

For those concerned about security issues in the use of RFID tags, there are methods available to shield personal information. For example, one way to protect passport information is to shield the RFID card or passport in aluminum foil, thus creating a Faraday cage. This method works well for HF tags in the 13.56 MHz band. For UHF tags, an antistatic plastic bag does a similar job.

Another security concern comes from the vulnerability of RFID tags that have been destroyed or disabled. For example, the tags can be shielded as described above and rendered non-functional. Other methods include using sharp devices to cut or disable the visible antenna, using powerful light to disable RFID tags, or inducing high currents by placing the tags in the presence of strong electromagnetic media such as microwave ovens.

9.7 RFID Vendors, Products, and System Integrators

The prospects of very strong growth in wireless RFID tagging and its applications, has attracted a large number of companies involved in research, development, and production of RFID components as well as the implementation of business infrastructures based on RFID technologies.

Vendors of RFID Tags (active/passive transponders) and Smart Cards

- **Alien Technology**: C1G2 ALN 9540, 9529, 9562, 9554, 09534 Tags, (www.alientechnology.com);
- **Baltech AG**: CCT Smart cards and Smart labels, (www.baltech.com);
- **Checkpoint Systems Inc.**: Security Systems, (www.checkpt.com);
- **Dynamic Systems GmbH**: HF/UHF Hard Tag on Metal, (www.dynamic-systems.de);
- **Hitachi Ltd**: μ-chip tags, (www.hitachi.com);
- **Huber+Suhner AG**: UHF RFID antennas, (www.hubersuhner.de);
- **Identa**: IDECard Identifications System, (www.identa.com);
- **Microsensys**: RFID miniature transponders/sensors, (www.microsensys.com);
- **Rako Security-Label**: RFID inlays Clinch Process, selfadhesive labels, hangtags, hardtags, (www.rako-security-label.de);
- **Sato Labelling Solutions**: UHF RFID C1G2 tags, FlaTag Solution, (www.sato.com);
- **Savi Technology Inc.**: Asset, Data Rich, Sensor Tags, USA, (www.savi.com);
- **Siemens AG**: RF 300, Rewritable SmartLabels, (www.automation.siemens.com);
- **TagStar Systems**: smart inlays for labels/tickets/smart cards, (www.tagstar-systems.de);
- **Tagsys**: Rigid Tags (nano-size, disk, thermo, small & large), Fexible Tags (370 L, AK C1G2), (www.tagsys.net);

- **Texas Instruments**: Tag-it (circular, rectangular, square, miniature) HF/LF Tags, (www.ti.com);
- **VeriChip**: RFID chips for humans, (www.verychip.com).

Vendors of RFID Reader/Writers (mobile and stationary)
- **Alien Technologies**: ALR-9900, ALR-9800, ALR-9650, ALR-8800 Tag Readers, (www.alientechnology.com);
- **Baltech AG**: ID-Engine Series, ACCESS series, (www.baltech.de);
- **Brooks Automation**: LF60, LF70, LF80, HF70, RFID Readers, (www.brooks-rfid.com);
- **Deister Electronics**: Ident & Automation, Security & Safety, (www.deister.com);
- **Mobile Data Processing**: DAT500 RFID Reader (combines RFID reader with WiFi, IrDA, and GSM/GPRS handheld device, (www.4p-online.com);
- **Psion Teklogix**: Workabout Pro G2 family of products (handheld device that combines WIFI, Bluetooth, GSM/GPRS and a multi-band RFID reader/barcode reader, (www.psionteklogix.com);
- **Savi Technology**: SMR-650 Mobile, Handheld Reader Kit, (www.savi.com);
- **Siemens AG**: 300,000 read/write units worldwide, (www.automation.siemens.);
- **Symbol Technologies (now Motorola)**: XR440/480 Fixed Reader, MC 9090 Handheld Reader, RD5000 Mobile Reader, (www.symbol.com);
- **Tagsys**: Short-Range Medio S0002/s003, Mid-Range Medio P101, P013, L-P101, Long-Range Medio L100/L200, TR-L100/200, L-L100, (www.tagsys.net);
- **Texas Instruments**: HF/LF RFID AFE, 2000 Micro Reader, (www.ti.com);
- **TopShop Finland**: RFID mobile phones, RFID readers, (www.topshop.com).

RFID Software/Infrastructure, Systems Integration
- **7iD Technologies**: Batch and Item Tracking, AidBand, (www.7id.com);
- **ADT Sensormatic GmbH**: Performance Test Lab, (www.adt-deutschland.com);
- **Brooks Automation**: RFID Hardware/Systems Integration, (www.brooks.net);
- **GS1**: EPCglobal standards development and implementation, (www.gs1-germany.de);
- **IBM Corporation**: Secure Trade Lane, TREC, EPCIS, (www.ibm.com/us);
- **Rf-iT Solutions**: RFID Operating System You-R@OPEN, (www.rf-it-solutions.com);
- **Savi Technology**: SmartChain Software Applications, (www.savi.com);
- **Siemens AG**: RFID systems MOBY E, MOBY, MOBY U, SIMATIC RF300 and RF600 for Transparent Supply Chains, (www.automation.siemens.com);
- **Salomon Automation**: WAMAS logistic software/management, (www.salomon.de);
- **Tagsys**: e-Xecute RFID Methodology and Program, (www.tagsys.net);
- **Ubisense AG**: RFID/UWB Real-time Location Systems, (www.ubisense.net).

9.8 Near-Field Communications

In the past few years, **Near-Field Communications** (NFC) has become a distinct NFSN field, aimed at providing very short range wireless communications for consumer electronics devices, primarily for handheld mobile phones. As an expression of technical and commercial interest in this field, an industry consortium was organized in 2005, the **Near Field Communications Forum** (NFC Forum) [44]. Among the initial founders were Philips Semiconductors (now NXP Semiconductors), Nokia, and Sony Electronics. As of March 2007, NFC had more than 110 members. Given the importance of this field, the GSM Applications (GSMA) group has joined the technical work by proposing a NFC Mobile Initiative. The declared goal of the NFC Forum, a non-profit organization, representing all major mobile telecom service providers, is to support standardization and implementation of NFC technology. The first NFC developers' summit was held in April 2007 in Monaco. An additional consortium, Store Logistics and Payments, was formed to explore the benefits of NFC technology in the context of European countries, given the specificity of regional and country-by-country regulations.

The major NFC applications envisioned are:

- Banking (smart wallet and banking card alternative);
- Mobile commerce (purchasing, smart commercial transactions);
- Access to transportation systems (charging of tickets and ticket replacements);
- Access to entertainment (ticket reservations);
- Travel (electronic tickets, vouchers);
- Access to digital information (books, magazines, brochures);
- Peer-to-Peer transactions (exchange of files, digital pictures, business cards);
- Interoperability with wireless technologies such as GSM/GPRS, WLAN, WPAN (Bluetooth, UWB).

The main technical specifications of NFC technology:

- Operations in the HF unlicensed band of 13.56 MHz, shared with RFID;
- Range of action from 0 to 20 cm;
- Data transmission rates of: 106 Kbps, 212 Kbps, and 424 Kbps.

NFC devices, like RFIDs, can be classified into two categories: **passive** and **active**. In passive NFC devices, touching or slow motion of the initiator will create an electromagnetic field that supplies the tiny energy needed by the target device to receive the signal. The target device will modulate the signal to transmit the desired information. Both initiator and target active devices have their own power supply that generates the coupling inductive field for short proximity communications. Among the most common NFC tags are: **Topaz** (mandated by NFC Forum), **Jewel** (for mass transit ticketing system), and **io** (for global mobile handset and consumer electronics devices).

NFC implementation is supported by two sets of standard specifications developed by ISO/IEC in cooperation with ECMA:

- ISO/IEC 18092/ECMA-340 "Near Field Communication and Interface and Protocol 1" (NFCIP-1);
- ISO/IEC 21481/ECMA-352 "Near Field Communication and Interface and Protocol 2" (NFCIP-2).

The standards include specifications for modulation types, transfer speeds, coding schemes, protocol frame format of the air interface, initialization methods, and collision control schemes.

Several trials are taking place as we write this book. In one of them, Cingular Wireless (now AT&T) is working with financial institutions such as Citigroup and MasterCard Worldwide and mobile phone manufacturer Nokia (model 6131) to develop a NFC service called MasterCardPayPass. The Nokia handset is a multi-mode handset. It allows credit card payment using the handheld device by touching a special NFC reader. Most global phone operators and mobile phone manufacturers are currently involved in one way or another in projects related to NFC use in consumer applications. It is expected that, in 2012, 20% of the mobile phones worldwide will have NFC capabilities.

Major NFC Vendors and Innovators

- **Arygon Technologies**: ARYGON GSM PDA AGPA 13.56 MHZ NFC and MIFARE RFID, (www.arygon.com);
- **Bundesdruckerei International Services**: Maurer 5000 Passport Identification System, (www. bundesdruckerei.com);
- **Inside Contactless**: System on Chips (SoC) NFC plus other wireless technologies, (www.insidecontactless.com);
- **Nokia Corporation**: Nokia 6131, 3220, NFC combined with GSM/GPRS and Bluetooth, (www.nokia.com);
- **NXP Semiconductors**: (formerly Philips Semiconductors), Common NFC projects with Sony Electronics, (www.nxp.com);
- **Innovision Research & Technology**: Topaz NFC SoC family of tags, (www.innovision-group.com);
- **Narian Technologies**, NFC platform and applications, NFC integrator, (www.nariantechnologies.com);
- **Sirit Inc.**: SoC for Kyocera NFC handsets for Cellular South, (www.sirit.com);
- **Tracient Technologies**: NFC Sliver to initiate Bluetooth, (ww.trecient.com);
- **ViVOtech**; NFC Payment and Promotion Solutions, (www.vivotech.com).

9.9 Advantages and Disadvantages of RFID and NFC Technologies

Based on the number of working applications and successful trials, both **Radio Frequency Identification** (RFID) and **Near-Field Communications** (NFC) are promising wireless technologies. The great resemblance between these two technologies justifies the presentation of their relative advantages and disadvantages together:

- They address short range and very short range (distance) needs of wireless links;
- Easy and quick implementations;
- Cost effective solutions, helped by reduced needs of processing and memory;
- RFID is based on the general concept of bar coding tagging technology;
- RFID is applicable in a large array of industries based on supply chains;
- RFID provides higher productivity and reduced costs in tracking goods;

- RFID provides a good anti-counterfeit mechanism;
- RFID and NFC are less vulnerable to eavesdropping and interference;
- RFID and NFC can be implemented as no power, or limited power technologies;
- Both are based on international standards and globally accepted specifications;
- Active RFID tags can be combined with sensor networks;
- NFC provides peer-to-peer wireless communications, easily applicable by cellular mobile handsets and their user-centric applications;
- Extensive vendor and market acceptance.

There are limitations and disadvantages in using the RFID and NFC technologies:

- The cost of the RFID/NFC tags is not yet on par with the cost of bar code labels;
- The very large number of standards and specifications requires harmonization;
- Thc varicty of tags, opcrating bands, and possibly different national/regional regulations pose interoperability problems;
- Complex readers may be needed to support the large variety of tags;
- Transmission rates are low, unfit for video and multimedia applications;
- There are concerns about the ability of IT infrastructures to handle full-blown global tracking implementations.

9.10 Ultra Wide Band Network Architecture

Ultra-Wide Band (UWB) pulse radio technology, as its name implies, provides transmission by using a much wider frequency band (minimum 500 MHz or 25% of the center frequency) in comparison with the other systems described earlier. Thus, a signal centered at 5GHz will require 1.25 GHz wide bandwidth. However, given the very short non-sinusoidal pulse signals, with the energy spread across a very large bandwidth, UWB signals do not interfere with other signals using the same spectrum.

UWB promises short range, less than 3m, and data rates up to 480 Mbps at very low power consumption. Thus, UWB becomes a very competitive solution to WPAN Bluetooth/ZigBee technologies. Currently, UWB is approved in the USA for operation in the 3.1–10.6 GHz unlicensed spectrum at a maximum of one microwatt. This low power level minimizes possible interference with Wi-Fi, GPS, and cell phones operating in the same spectrum. Currently, the primary UWB application is **Wireless Universal Serial Bus** (WUSB) which resembles the 480 Mbps interfaces linking PCs to peripherals. UWB is also known as baseband pulse radio, impulse radio, and impulse/ground penetrating radar technology.

UWB or WUSB networking implies a two-tier architecture that consists of a transmitting segment and a receiving segment, both communicating at data rates from 100 Mbps up to 480 Mbps. A high-level UWB architecture is presented in Figure 9.2.

- The **transmitting segment** consists of a USB memory stick connected to a Multi-Band OFDM Alliance (MBOA) Field Programmable Gate Array (FPGA) Device Wireless

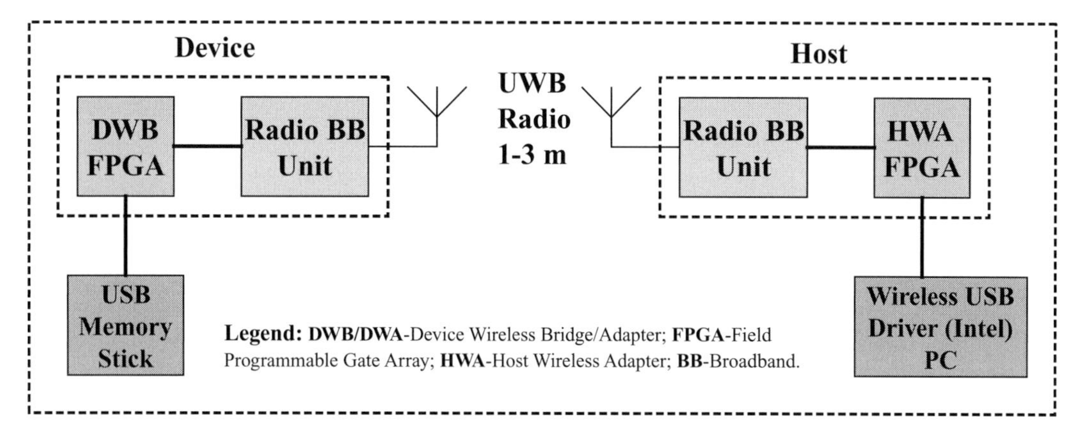

Figure 9.2 Ultra-Wide Band Architecture

Bridge (DWB) MAC adapter (Phillips) that feeds the Radio Broadband Unit (Staccato Communications) transmitter.

- The **receiving segment** consists of a RF BB receiver (Staccato Communications) and a Host Wireless (MAC) Adapter (HWA) (Phillips) that feeds a Wireless USB (WUSB) driver (Intel).

The main characteristics of UWB technology are:

- The UWB pulse: Non-sinusoidal signals that can share the spectrum with sinusoidal signals without causing interference;
- UWB transmitted power: Spread across the spectrum with energy concentrated in any part of the spectrum;
- UWB antennas: Non-resonant, current-based so they are very small and can be directly driven by CMOS;
- UWB transceivers: Require low power so they can be integrated into CMOS without reactive components;
- The UWB spread spectrum: Provides better protection against interception and eavesdropping as well as excellent multipath immunity for indoor environments;
- UWB uses a larger bandwidth: Has a good penetrating capability and operates well within buildings, forests, and urban areas;
- Requires IEEE 802.15.3a standardization of Physical Layer and MAC Sublayer.

In February 2004, the 802.15 Working Study Group, working specifically on UWB standard (802.15.3a), was unable to decide between two competing proposals from Intel and Motorola/Freescale. The group split, with the Intel & Texas Instruments faction forming a new group called **Multiband OFDM Alliance** (MBOA) [45]. The group developed its own proposal and presented it to the IEEE for approval. The counter proposal, **Direct Sequence UWB** (DSUWB), is still part of the **UWB Forum** specifications (www.uwbforum.org). This split will certainly delay UWB standardization and implementation.

In the long run, WPAN Bluetooth and ZigBee wireless technologies might be replaced by the UWB radio technology which promises data rates up to 480 Mbps at 2 m distance. UWB will allow fast media file transfer between media servers and mobile devices, or between video media servers and HDTV flat screens. Products have been demonstrated since 2006. However, the technology will take around 5 years to mature.

The Wireless USB Promoter Group, together with the 1394 Trade Association, collaborated with the WiMedia Alliance to complete the Wireless Media Convergence Architecture (WiMCA) for the coexistence of wireless USB and Wireless 1394 (FireWire). Among the current UWB players are: Intel, Texas Instruments, Staccato Communications, Freescale (Motorola's subsidiary), Abocom, Fujitsu, NEC Electronics, NXP Semiconductors, Belkin, Icron, Gefen, and Pulse-LINK.

9.11 Advantages and Disadvantages of UWB Technology

Ultra-Wide Band (UWB) was touted as, and still is, one of the most promising wireless technologies to provide communications from very short to medium range distances. Developed initially for the military, it quickly found markets as possible replacements of existent technologies such as NFC, Bluetooth, and even the 802.11x. Advantages and benefits of using the UWB technology include:

- Very high, scalable data rates;
- Low power requirements;
- Easy and quick device-based implementations;
- Cost effective solutions, helped by the simplicity of the electronics involved;
- No interference with other wireless sources because of the pulse wave nature of the signals, power is spread over a wide spectrum, and different frequency bands can be used;
- Good penetration inside buildings and urban areas with multi-path immunity;
- It provides precise ranging, or distance measurement, and is therefore useful for location identification and radar types of applications.

The limitations and disadvantages in using UWB derive from regulatory and organizational aspects:

- Lack of agreement among major players regarding unique modulation and access methods with risks for future interoperability;
- FCC current mandate to use UWB only in a short range; less than 10 meters;
- Lack of commonality in worldwide spectrum allocation; and
- Lack of internationally recognized standards.

With all its shortcomings, UWB has begun to have a growing presence and impact on the market through one major application, the Wireless USB (WUSB) [46].

9.12 Wireless USB

We mentioned earlier that Wireless USB (WUSB) has a great potential given the size of the current USB market with over two billion USB-based devices sold worldwide. Many of these devices are wireless/mobile candidates with the same advantage of USB, i.e., fast connectivity at low cost.

Wireless USB employs the **Orthogonal Frequency Division Multiplexing** (OFDM) schema with an aggregate of 128 sub-carriers, each using a 528 MHz band. No hubs are present while a WUSB can logically connect to a maximum of 127 WUSB devices. The maximum data rate of WUSB over-the-air, 640 Mbps, is coded for error correction at 55, 80, 110, 160, 200, 320, 400, and 480 Mbps information rates. Pulse repetition rate is low, between 1 and 200 megapulses per second, for radar and imaging systems. In communications systems, the pulse repetition is typically between 1 and 2 gigapulses per second.

Wireless USB (WUSB) port applications (taken over by USB Forum [47]) based on UWB radio technology, are expected to be used in three classes of configurations [46]. Use of UWB technology eliminates the possibility of interference with 802.11x WLANs that operate in the same 2.4 GHz band of spectrum.

- **Host Configuration**: A WUSB defines a **Host Wire Adapter** (HWA), essentially a bridge from a wired USB connection to a WUSB host connection. A HWA, as a USB memory stick, can plug into a wired USB host port on a PC, standalone powered USB hub, or a USB monitor or as an ExpressCard USB 2.0 interface for notebook PCs;
- **Device Configuration**: A WUSB defines a **Device Wire Adapter** (DWA), providing a bridge from a wireless USB device to a wireless USB host port which plugs into a peripheral's USB device port such as DMA, Cardbus, PCI, SD IO interface;
- **Dual Role Device Configuration**: Capable of either host or device connectivity. This allows a peripheral (e.g., printer) to connect to either a host PC to print files or another peripheral (digital camera) to print photos.

Certified WUSB products following the USB Forum certification program, started shipping in the first half of 2007. The initial products were WUSB dongles and WUSB hubs that plug into existing USB ports. The expected transfer data rate will be around 100 Mbps. Since WUSB allows only one connection between a host and device, a numeric pairing is required. This means a WUSB device should be plugged into a traditional USB port to configure, before the wireless connection is set up.

Part IV

Fixed Wireless Cellular Mobile Networks Convergence and Integration

10 Fixed-Mobile Convergence Overview

10.1 Why Convergence?

In the previous chapters we presented and analyzed all major forms of wireless communications and divided them into two major groups: The world of cellular mobile networks as represented by three generations of networks with the corresponding technologies, and the other, the world of so called "fixed wireless" communications grouping WLAN, WPAN, WiMAX, and Near-Field Sensor Networks. In the pursuit of the ultimate goal of communications "**anytime, anywhere, any technology**" there is a need for interoperability between the cellular mobile radio networks of the latest generations and any form of fixed wireless networks, i.e., **Fixed-Mobile Convergence (FMC)**, the subject of this book. From the user's perspective, convergence means use of the same mobile cell phone or mobile handset across any type of wireless network and transmission of digital information at the highest available data rate, all at the lowest possible cost.

This chapter will provide an overview of the fixed-mobile convergence concept. We will address the terminology, architectural components, interfaces, protocols, the overall requirements, and the technical forums that have been created to advance fixed-mobile convergence concepts. We will also introduce two major solutions that address this convergence at the standards level. Convergence is also one major aspect of the grand schema of designing the **New Generation of Wireless Networks**; along with associated architectures, applications, and services. The actual pair solutions of convergence between cellular mobile networks and individual fixed wireless networks will be presented in subsequent chapters.

The high-level characteristics of this integration/convergence are:

- A meshed wireless infrastructure as the conduit for voice, data, and video communications;
- Increased broadband capacity to customers;
- Wide variety of multimedia applications and services;
- Distributed intelligence and management;
- Easy installation, provisioning, operation, and maintenance;
- End-user service configuration capability.

The main issues associated with the integration/convergence of wireless networks are:

- Standardization of handover performance criteria and interoperability mechanisms;
- End-to-end guaranteed Quality of Service;
- End-to-end latency to reduce the effects of redundant headers, buffering, routing;
- Discarding and retransmission of packets;
- Use of unlicensed transmission spectra that are prone to interference;
- Need to support large, complex networks and application infrastructures.

10.2 Convergence Explained

The notion of "**convergence**" or "**integration**" in wireless communications has matured along with the development and implementation of new networking technologies and applications. The wireless "convergence/integration" goal is to design a transparent communication path across heterogeneous wireless networks, regardless of radio access technology used. Fixed-mobile convergence implies the following fundamental technical aspects:

- It assumes a **multi-technology** networking environment, **multi-service operations, multiple service providers**, and **interworking** at the **physical** (different radio access), **data link**, and **network layers** (different protocols).
- Use of **a single mobile terminal** across different networks and radio access interfaces, ideally with no limits in area covered, mobility, and radio conditions, while maintaining an acceptable level of security and quality of services.
- "Convergence/integration" is done by adopting a unique network layer protocol, the **IP-based Internet Protocol**, to support a wide range of **voice, data**, and **video communications services**.
- Interoperability mechanisms between those mobile and fixed wireless networks that use different access interfaces. These mechanisms include the **Initial User Assignment** (IUA) based on optimal network selection (well defined criteria) and the transparent **Inter-System Handover** (ISH).
- **Convergence** at the level **of customer services**, including bundled Wireless WANs, PANs, LANs, MANs, and NFC with **Internet access**, all with packaged pricing, consolidated into one bill, if necessary.
- Well defined **performance criteria** and **system level parameters** that trigger the handover process between heterogeneous wireless networks, or handovers within networks that use the same technology and protocols.
- Use of internationally recognized standard solutions for convergence. Included should be criteria regarding spectrums used, radio access methods, and other relevant technical specifications.

In a more subtle way, the notion of **convergence/integration** in wireless communication implies continuous support for the newest generations of cellular networks (UMTS, HSDPA/HSUPA) and new developments in fixed wireless networking (NFC, UWB, ZigBee, etc.).

10.3 Fixed-Mobile Convergence History

There were several developments in wireless networking technology that can be considered as milestones in the advancement of WLAN/WPAN and GSM/GPRS convergence/integration.

- 1997: **Ericsson** provides the **first dual-mode DECT/GSM handset** implementation;
- 1997: **IEEE** issues the first set of **802.11a** standards;
- 1999: **IEEE** issues the **802.11b** standards;
- Late 1990s: Wireless networks, primarily those designed for voice communication, started **adding** on large-scale **data** processing and transmission capabilities (2G, 2.5G);
- Late 1999: **British Telecom** provides the first **GSM/DECT convergent service** (not successful because of required manual switching between networks);
- 2003: **IEEE** issues the **802.11g** standards;
- Early 2004: The **Unlicensed Mobile Access** (UMA) alliance is formed to develop the WLAN/WPAN and mobile convergence architecture and specifications;
- Late 2004: The Wi-Fi Alliance initiates work on Wi-Fi/Cellular convergence;
- 2005: **UMA specs** are adopted and incorporated in 3GPP2 Release 6 specifications;
- 2005: IEEE issues 802.11e, dealing with WLAN QOS aspects;
- 2005: **British Telecom (BT)** and **South Korea Telecom** started providing fixed-mobile convergence services (Bluetooth/GSM and Wi-Fi/GSM);
- 2006: The **IP-based Multimedia Subsystem** (IMS) is adopted as the long-term strategic architecture of next generation mobile networks. IMS includes convergence of fixed-mobile networks.

10.4 A High Level Wireless Convergence Concept

Wireless convergence is a multidimensional concept that combines **IT infrastructures, networks, applications, user interfaces**, and **management** aimed at supporting voice, data, and multimedia over IP-based wireless networks. A high-level depiction of FMC overall construct, in the form of a pyramid, is presented in Figure 10.1.

At the base of the convergence pyramid are physical network components of both fixed and mobile networks. The next level consists of the networks themselves, as collections of components with well-defined topological layouts and coverage domains such as WLAN, WPAN, WMAN, and cell-based mobile WWANs. The next level is represented by the applications that provide support for a variety of services such as voice, data, and multimedia services. Access to all these systems requires user interfaces, where the centerpiece of FMC is the mobile phone. The whole convergent network and services should be managed to provide the expected QOS as agreed upon in SLAs.

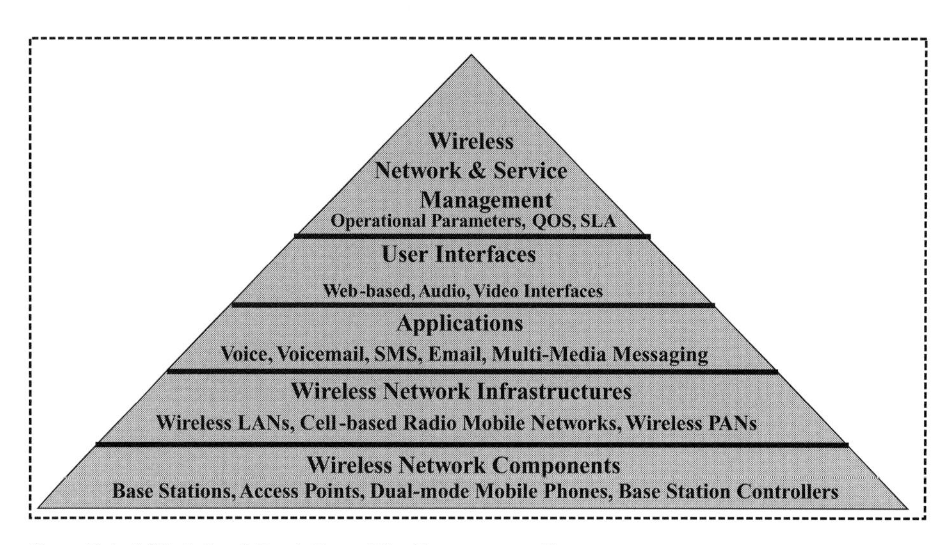

Figure 10.1 A High-level Depiction of the Convergence Concept

10.5 Fixed-Mobile Convergent Network Architecture

A more intuitive depiction of the FMC network architecture is in the form of the chain of networks involved in convergence. This is shown in Figure 10.2.

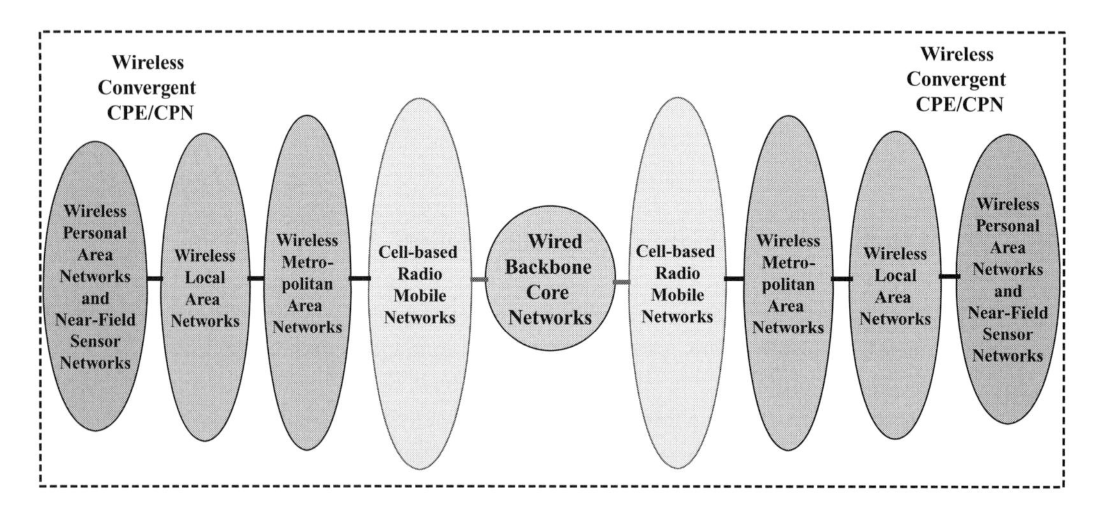

Figure 10.2 Fixed-Mobile Convergent Network Architecture

At the highest level, cell-based mobile networks convergence takes place at the Customer Premises Equipment/Networks (CPE/CPN) represented by Wireless Personal Area Networks (WPAN), Wireless Near-Field Sensor Networks, and Wireless Local Area Networks (WLAN). Although WLANs are CPE/CPN they are presented as separate networks because Mesh WLANs can also provide metropolitan area coverage. Convergence also takes place in the Wireless Metropolitan Area Networks (WMAN) represented by broadband wireless access networks such as WiMAX. At the center of this symmetric diagram is the core backbone network, mostly based on wired (fiber optic and copper) networks.

10.6 Fixed-Mobile Convergent Network Components

Each of the wireless network areas, CPE/CPN, metro, cellular mobile and wired core backbone, is supported by a variety of competing networking technologies and a multitude of network components. This short chapter gives a general indication of the major convergent network components that stand behind the architectural boxes. An asymmetric layout of FMC network components is presented in Figure 10.3. The acronyms used in this diagram were spelled out in Chapter 1, Section 1.6.

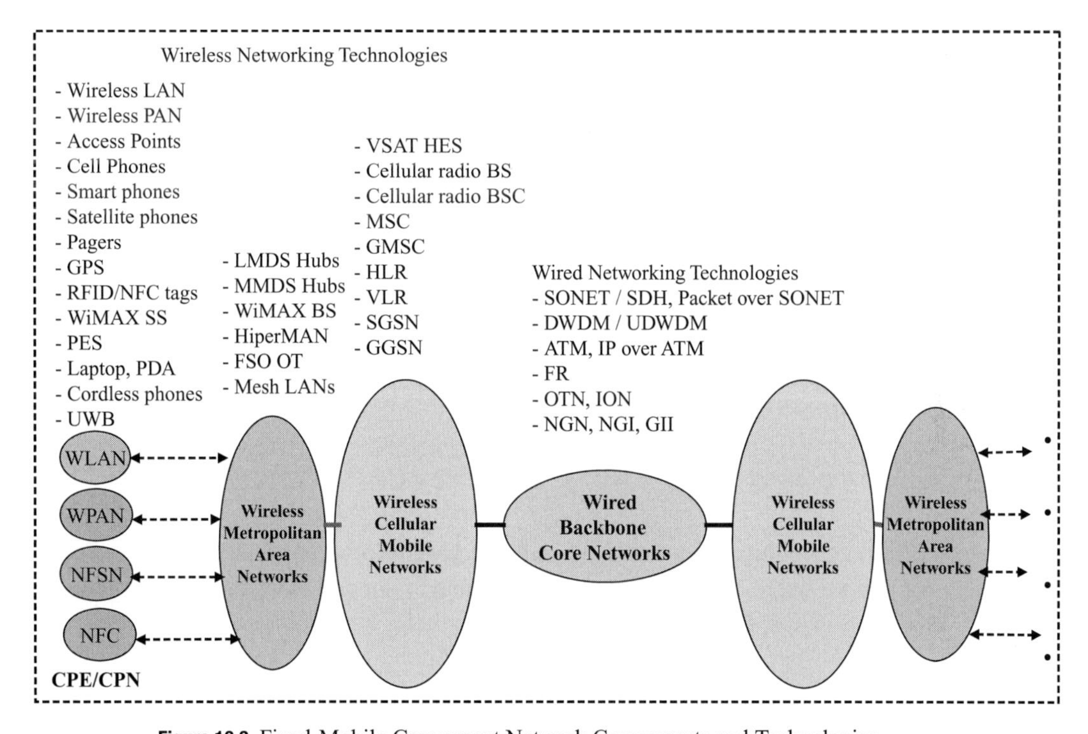

Figure 10.3 Fixed-Mobile Convergent Network Components and Technologies

Starting from the right, the wired core backbone networks are realized as a mixture of various technologies, mostly fiber optics-based. The core cell-based radio mobile network infrastructure consists of components supporting primarily voice services and data services. Wireless metropolitan area networks are supported by a variety of networks, such as WiMAX-based, FSO, and VSAT, all serving specific applications. Other major sets of components belong in the wireless CPE/CPN class that includes WLANs, WPANs, and Near-Field Sensor Networks.

10.7 Fixed-Mobile Convergent Network Interfaces and Protocols

Each of the wireless network constituent technologies is accompanied by a set of individual protocols and communications stacks that may be limited to a wireless network area or can be used across the whole network. An asymmetric layout of major FMC network interfaces and standard protocols for radio access is presented in Figure 10.4. Most acronyms used in this diagram are related to the wireless networks that were introduced in Chapter 1 where we described the various fixed and mobile wireless networks. Once again, we will use the FMC architecture presented in section 10.4.

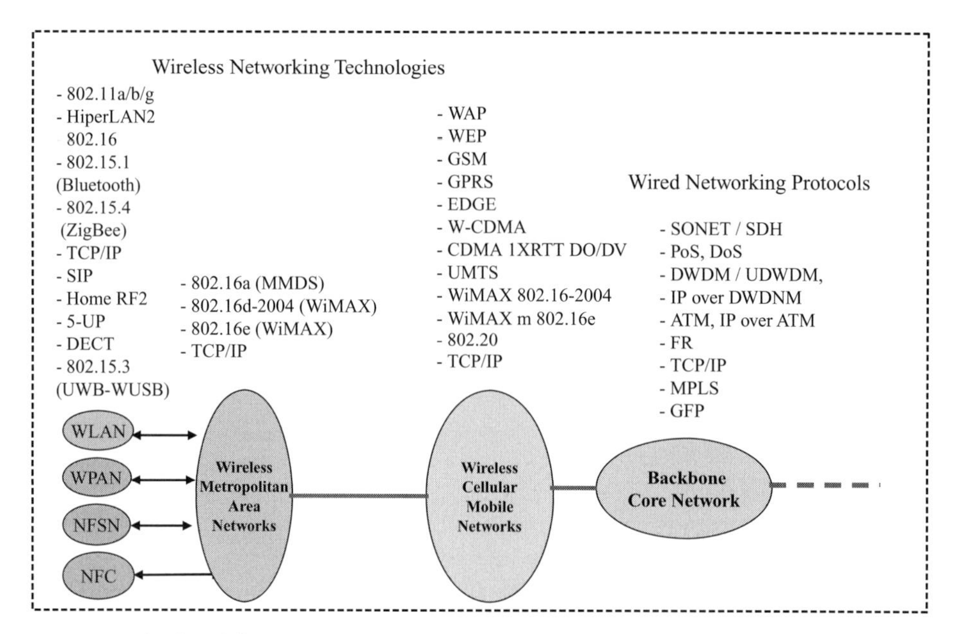

Figure 10.4 Fixed-Mobile Convergent Network Interfaces and Protocols

Starting from the right are the wired core backbone networks, each using different fiber optics-based technologies. Each technology is supported by a set of international standards. Since the subject of this book is related to wireless networks, we do not enumerate these standards. The core cell-based radio mobile network infrastructure is also supported by a set of international standards dealing with applications, security, data services and technologies related to the latest generations. Wireless metropolitan area networks are supported by the WiMAX technology for fixed and nomadic/mobile users (IEEE 802.16d and 802.16e). Another major set of standards is related to the wireless CPE/CPN class that includes WLANs, WPANs, and Near-Field Sensor Networks.

10.8 Drivers of Fixed-Mobile Convergence

Convergent wireless communications provide critical means to transmit voice, data, and multimedia applications, using Internet Protocol (IP), across heterogeneous wireless networks. The need for this convergence is driven by applications that require the highest available bandwidth at lowest possible cost. Quality of Service is expected to be comparable to that provided by conventional wired voice-based telecommunications networks. It is also expected that convergent services be provided transparently across all wireless networks, regardless of a user's location and the network used. Typical high-level FMC applications are:

- Voice communications;
- Voice over IP (VoIP) communications;
- Voice mail, unified messaging, Interactive Voice Response (IVR);
- Voice and data over converged WLANs and UMTS networks;
- Standard Internet access and applications (Web access, file transfer);
- Multimedia applications (MP3, web camera, videoconferencing).

In addition, some fixed wireless networks might originate applications/services specific to those networks that can be beneficial if extended across the mobile networks. For example, wireless broadband access and metropolitan broadband via fixed and mobile WiMAX or Mesh WLANs can provide support for additional services targeting small, medium, or large multi-tenant units, and small and medium-size businesses such as:

- LAN-to-LAN, campus and buildings interconnection;
- Broadband access for multimedia communications;
- Backhaul channel for mobile wireless and wired communications;
- Disaster recovery using WiMAX as a backhaul channel;
- Metropolitan wireless networking.

10.9 Convergence Functional Requirements

At a high level, the following functional requirements can be associated with fixed-mobile convergent networks and services:

- Support for voice communications across heterogeneous wireless networks;
- Support for multimedia services: voice, data, and video, in a wide array of applications;
- Support for total location, mobility, and service transparency;
- Separation of transport, control, signaling, and management functions;
- Network and service management capabilities, including customer network management;
- Soft vertical handoff (across different networks) and horizontal handoff (same network);
- Security of wireless communications and security of management operations;
- Support for end-to-end quality of service and associated service level agreements;
- Interoperability and transparency across multiple service providers.

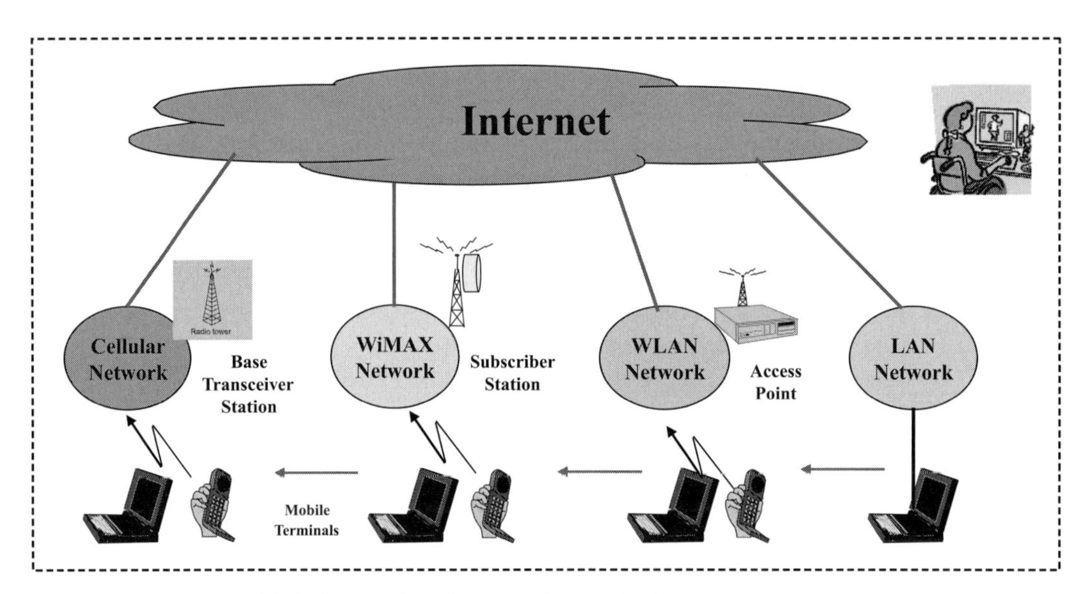

Figure 10.5 Media Independent Handover Service Architecture

10.10 Media Independent Handover Services (IEEE 802.21)

The handover mechanism is a basic function of cellular mobile networks that provides user mobility. Handover occurs when users move and change their locations. Therefore, transparent handover, independent of media, is a key requirement of fixed wireless cellular mobile convergent networks. There are handovers that occur within single-provider networks or between different service providers. In general terms, we call this type of handover, **horizontal handover**.

Similar handovers take place within mesh networks that are built based on IEEE 802.11r and 802.16e standards (for example the Wi-Fi and WiMAX technologies). **Vertical handover**, or **Media Independent Handover** (MIH), is a service provided between heterogeneous networks such as fixed wireless and cellular mobile networks. IEEE 802.21 [48] is an emerging standard written to specify the mechanisms and algorithms that will support vertical handover across any type of wireless networks. These mechanisms could be applied to handovers between wired and wireless networks. One example is transparent communications from LAN to WLAN to WiMAX to cellular mobile networks. A high-level architecture of MIH is shown in Figure 10.5.

MIH service provides seamless handover of mobile terminals that are initially attached to LANs. Based on handover algorithms and triggers, the link to the Internet will be maintained as the connection is established through a WLAN, then through a WiMAX. Ultimately, when the mobile terminal is out of the range of any fixed wireless network, the connection will be continued through the cellular mobile network. A high-level protocol architecture that supports MIH is shown in Figure 10.6.

Handover functions are performed by Layers 2 and 3 of the MIH protocol stack and are independent of the network hardware and protocols that are used in a LAN, WLAN,

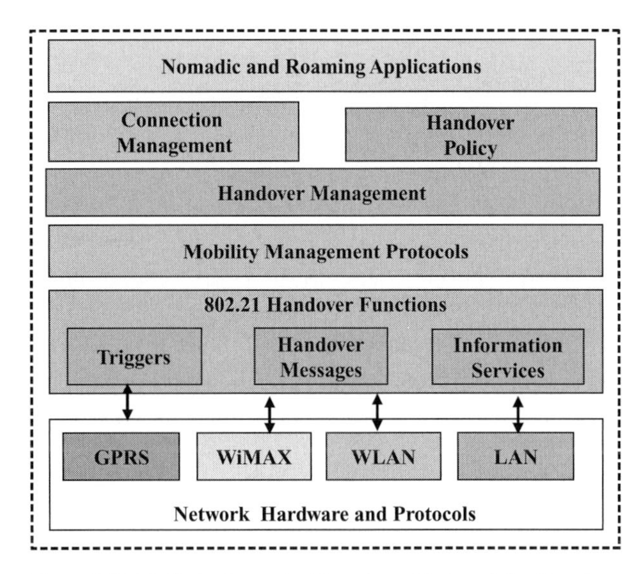

Figure 10.6 Media Independent Handover Protocol Architecture

WiMAX, or GPRS cellular mobile network. Smart triggers and message handovers will be used. The upper layers of the stack deal with mobility management and handover management according to a handover policy. The sublayers use standardized protocols and services that are based on IETF specifications.

Major requirements for generalized MIH services include:

- Transparent roaming across WLANs (802.11 a/b/g/n), WPANs (Bluetooth), and cellular mobile (GSM/CDMA) networks;
- Transparent roaming across LANs and WLANs;
- Initial network assignment and subsequent handovers based on performance (signal strengths and bandwidth availability) as well as on overall cost;
- Allow multiple technologies, vendors, and users;
- Compatibility with the existent 802.xx family of standards;
- Support for data, voice, video, and multimedia communications;
- Assurance of adequate QOS while roaming across various type of networks;
- Ability to manage the MIH using the standard SNMP protocol;
- Low power handover operations in multiple networks;
- Maintain the security features provided by other wireless standards.

MIH assumes three distinct operations related to handover:

- Initiation (when network discovery, selection, and negotiation takes place);
- Preparation (when a new link is set up and Layer 2 and 3 links are provided);
- Execution (when the connection is transferred and packets are received).

The idea of providing a universal MIH service was first offered in 2003 and a MIH Working Group was established in 2004. The process of calling for and selecting proposals led to a letter ballot in early 2006. MIH specifications are expected to be ratified at the end of 2007, with products deployed in 2009.

10.11 Fixed-Mobile Convergent Networking Solutions

The convergence between **fixed wireless** systems ranging from WPAN, WLAN, WiMAX, Near-Field Sensor Networks, and **GSM/CDMA cellular mobile** networks requires integration at all levels, from **access segment** to **packet switch core network**. Several convergence steps are envisioned on this road and will be analyzed in the following chapters. There are three aspects of this integration that we will discuss here:

Integration of Packet Switch core networks infrastructures serving fixed wireless networks and mobile cellular networks. This is an on-going process transparent to the users. It will be achieved as a feature of the mobile IP-based data offerings of 2.5G, and 3G services, ultimately using the Internet infrastructure.

Integration at the access level of both fixed and mobile wireless networks using digital technology such as Voice over IP (VoIP) over WLAN and use of a Base Station Controller (BSC) emulation box responsible for transparent handover between fixed and mobile wireless networks. Originally, the convergence solutions in this scenario were collectively known as **Unlicensed Mobile Access** (UMA) architecture specifications. They were generated by the UMA consortium. A variation of this step is a scenario that places convergence control into Mobile Switching Center (MSC) components, thus minimizing overall costs. Later, once it was adopted by 3GGP in April 2005, UMA's name was changed to **Generic Access Network** (GAN).

Integration at all levels by promoting an all IP-based network from access to the core network. This means that both data and voice will be carried over a common IP-based network, once the phase-out of circuit switch technologies that serve PSTN and GSM is completed. Convergence solutions in this scenario are collectively known as the **IP-based Multimedia Subsystem** (IMS) architecture and interfaces, as generated by the 3G Partnership Project (3GPP) Release 6 specifications.

10.12 Fixed-Mobile Convergent Networking Forums

In the previous section we indicated two major approaches for standardization of FMC, UMA and IMS. Each of these grand solutions is supported by international consortiums or forums that have among their members cellular mobile operators, fixed wireless service providers, manufacturers of wireless networks components, systems integrators, testing and certification organizations, and even traditional wired telecommunications operators. Three major forums dominate the work on fixed-mobile convergence:

- **Unlicensed Mobile Access/Generic Access Network Consortium** (UMAC);
- **IP-based Multimedia Subsystem Alliance** (IMS Alliance);
- **Fixed-Mobile Convergence Alliance** (FMC Alliance).

Given the fact that several areas of wireless networking are involved in fixed-mobile convergence, there are many other organizations that are involved in and have a direct impact on the definition, standardization, and implementation of FMC:

- **Wi-Fi Alliance** (WLAN);
- **WiMAX Forum** (WMAN);
- **Bluetooth SIG** (WPAN);
- **ZigBee Alliance** (WPAN);
- **Near-Field Communications Forum** (NFC Forum) (NFSN);
- **Institute of Electrical and Electronics Engineers** (IEEE);
- **European Telecommunications Standards Institute's Telecom & Internet Converged Services and Protocols for Advanced Networks** (ETSI TISPAN; Group with input into ITU-T's NGN work);
- **3rd Generation Partnership Projects** (3GPP and 3GPP2);
- **Open Mobile Alliance** (OMA);
- **Ultra Wideband Forum** (UWB Forum);
- **Session Initiation Protocol Forum** (SIP Forum);
- **Internet Engineering Task Force** (IETF);
- **Seamless Converged Communications Across Network** (SCCAN);
- **Broadband Wireless Internet Forum** (BWIF), (www.bwif.org);
- **Orthogonal Frequency Division Multiplexing** (OFDM) **Forum**, (www.ofdm-forum.com);
- **Wireless Communications Association International** (WCAI), (www.wcai.org);
- **Voice over IP Forum** (VoIP Forum).

11 Wireless LAN Cellular Mobile Convergence

11.1 WLAN Convergent Network Architecture

In the previous chapter, we defined and described Fixed-Mobile Convergence (FMC) as a multidimensional concept that implies convergence of terminals (cellular handsets and computing devices), networks (NFSN, WPAN, WLAN, WMAN, and cellular mobile WWAN), and applications (data, voice, video, and multimedia). The primary focus of this book is on network convergence. A simplified high-level depiction of FMC at the network level is shown in Figure 11.1.

In this diagram we see that the most compelling convergence/integration cases are between cellular mobile networks and any of the wireless contained networks. The prevalent case, which introduced the notion of FMC to the market, is between WLAN (Wi-Fi) networks and cellular mobile networks. This convergence is attractive because WLANs are included within the wide area coverage of mobile networks but provide higher throughput. Another convergence case is between WPANs and WLANs, where the higher data rate and larger area coverage of WLANs can extend the functionality of WPAN technology such as Bluetooth and Near-Field Sensor Networks such as NFC and RFID. This chapter will focus on convergence between cellular mobile and WLANs.

11.2 WLAN Convergent Applications

In Chapter 6 we listed the functionality built into WLAN networks without indicating those applications that are facilitated by convergence of WLAN with other fixed or mobile wired and wireless networks. A short list of WLAN convergent applications follows. Some of these applications will be detailed when we analyze WLAN convergent case studies in subsequent sections of this chapter.

- WLAN and cellular mobile convergence (dual-mode handsets and connectivity across cellular mobile networks);
- WLAN and RFID networks convergence;
- WLAN and Wireless Sensor Networks convergence (test and measurement instrumentation, alarm and surveillance systems);
- WLAN and BlackBerry convergence;
- WLAN and iPod convergence;

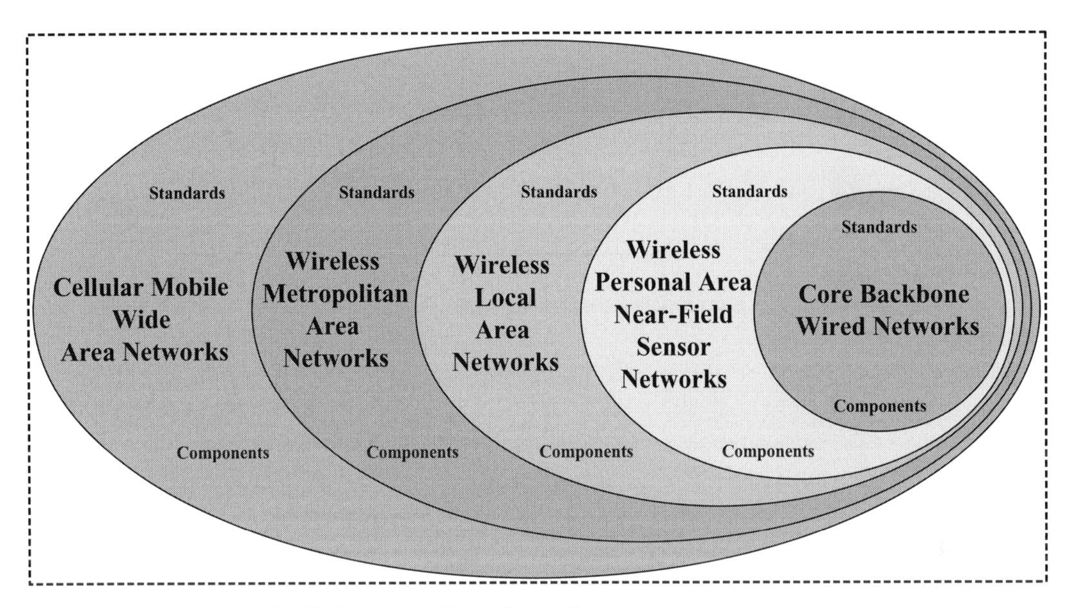

Figure 11.1 WLAN Convergent Network Architecture

- WLAN as an alternative for WPAN;
- WLAN Mesh as an alternative for WMAN; and
- WLANs as hotspots.

11.3 802.11n–based WLAN Implementation Case Study

An example of early implementation of 802.11n Draft 2.0 standard for very high speed WLANs is that proposed by the Morrisville State College, New York State, USA. IBM was selected as the project manager and system integrator [49]. The 802.11n-based architecture is shown in Figure 11.2.

In the final stage, the WLAN network will consist of 900 Meru Networks AP 300 Access Points, a new class of equipment designed to support 802.11n in combination with legacy 802.11 a/b/g standards. The AP 300s are MIMO-based 2×2 or 3×3 devices with up to six antennas, three used as transmitters and three for receivers. The APs are powered through the Power-over-Ethernet (PoE) connections to the wired LANs.

Over 2000 students and college staff spread over 45 buildings on the campus will use the WLANs with an expected user level performance of 130 Mbps. To achieve this performance, Meru Networks adds several high-end Air Traffic Controllers, in this case the MC 5000. These controllers follow the design of an earlier family of controllers, the MC 500, MC 1000, and MC 3000, which were designed to support a smaller number of APs as required by small offices, branch offices, and large scale enterprises. The Wireless Meru System is connected to the wired network through Gigabit Ethernet switches. The controller itself is connected to the wired LAN through a Gigabit port.

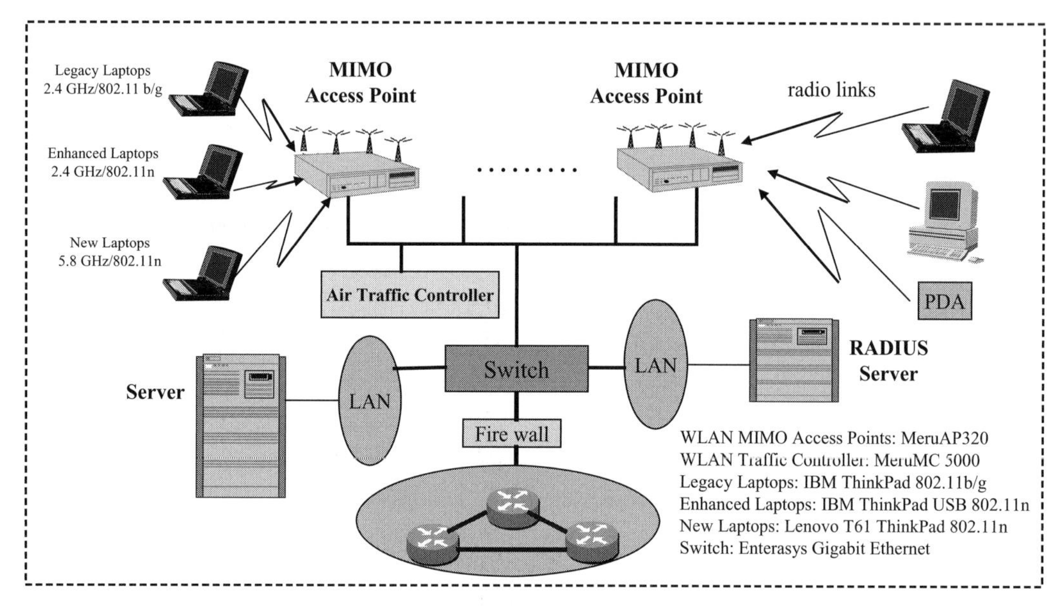

Figure 11.2 IEEE 802.11n-based WLAN Architecture

The radio interface traffic controllers are Meru Networks-specific products that allow control of channel allocation of individual APs. They use a patented scheme of coordination and timing that results in higher density of users per access point. This control allows detection of traffic types and allocation of bandwidth for time sensitive voice packets. Automatic detection of flows generated through SIP, H.323, Cisco SCCP, and SpectralLink SVP are included. The innovation includes the creation of virtual cells as coverage zone of multiple APs that allow transparent handoffs and automatic load balancing among clients assuring consistent and reliable throughput for voice and data communications [50].

Included in the unlicensed 2.4 GHz band, there are three large non-interfering 20 MHz channels that can be used separately or one 20 MHz channel and one 40 MHz channel. This is compatible with 802.11b/g channelization. Included in the unlicensed 5 GHz band that is used by the 802.11a APs (5.15–5.25, 5.25–5.35, and 5.75–5.85 GHz), there are twelve available non-overlapping 20 MHz channels or six non-overlapping 40 MHz channels. It is expected that regulatory changes will release some of the 5.47–5.75 GHz frequency bands for use by unlicensed WLANs.

Among the customers of the new WLANs are users and terminals ranging from legacy IBM ThinkPad with 802.11b/g 2.4 GHz cards to enhanced ThinkPad laptops having 802.11n support in the form of 802.11n 5 GHz USB dongles. Another candidate is Lenovo T61 ThinkPad with integrated 802.11a/b/g/n chipsets working in the less crowded 5.8 GHz unlicensed frequency band. During the testing phase, downloading and uploading large files from the network to stations connected to the WLAN showed dramatic increase in data rates compared with IEEE 802.11 a/b/g WLANs. The best results were achieved with laptops using plug-in 5 GHz 802.11n cards. Throughput was three times faster than on laptops using 2.4 GHz USB 802.11n dongles and ten times faster than legacy 802.11b/g equipped laptops [51].

11.4 Convergent WLAN Cellular Mobile Network Architecture

We have noted here that the ultimate goal of fixed-mobile convergence, having as its centerpiece the ubiquitous mobile handset or cell phone, is to provide transparent communications across any type of wireless network. More precisely, to select and use the wireless network that provides the highest performance at the lower cost. The goal of having a single device comes with the advantage of having one telephone number, one voice mailbox, and one telephone/address book. To have the full picture, we must add to these important features the mobility and roaming capabilities. Those having a phone at the office, another phone at home, and one or two cell phones can appreciate the ultimate goal of convergence, i.e., one handset and transparent communications across all wireless networks.

Perhaps the best example of convergence is in the integration of two prevalent wireless technologies that both have solid supporting infrastructures. One technology is cellular mobile with an estimated three billion handsets in use by roughly half of the global population. Then, there are the Wireless LANs, present in most organizations and businesses, in 300,000 worldwide hotspots, and in many home networks. What are needed are dual-mode handsets and an infrastructure that recognizes these terminals. The convergent system will transparently handover the call or data transfer from one network to another network, according to the circumstances, i.e., depending on the proximity of the networks to each other. When the signal fades because one of the parties is moving out from the coverage area, the call will be automatically transferred from the WLAN network to the cellular mobile network.

Convergence between WLANs and mobile GSM/CDMA networks requires interoperability between the networks i.e., initial user access and inter-system handover. A major advantage of the core mobile network is that connectivity within the mobile network and with the wired world remains unchanged. Portable or nomadic dual-mode handsets and terminals are the mobile agents. An extended functionality is required from the Gateway GPRS Support Node. The WLAN and mobile GSM/CDMA convergent network architecture is shown in Figure 11.3.

The mechanism is simple. When the user wants to place a call, and is outside the range of the WLAN AP, the call will be handled by the mobile network by accessing the BTS and engaging the BSC, MSC, and GMSC, connecting the called party across the PSTN. When the mobile terminal is active on a call, and moves into the WLAN coverage area, the Wi-Fi enabled terminal will connect with the closest AP. Through proper signaling, an intelligent BSC will provide a soft handover to the WLAN AP that is considered a cell in the extended mobile network. A Voice over IP (VoIP) path will be established across the WLAN and packet switched data network, and the call will continue through proper signaling. Similarly, when the user moves outside of the WLAN coverage area during a call, the call will be transferred from the WLAN to the mobile network.

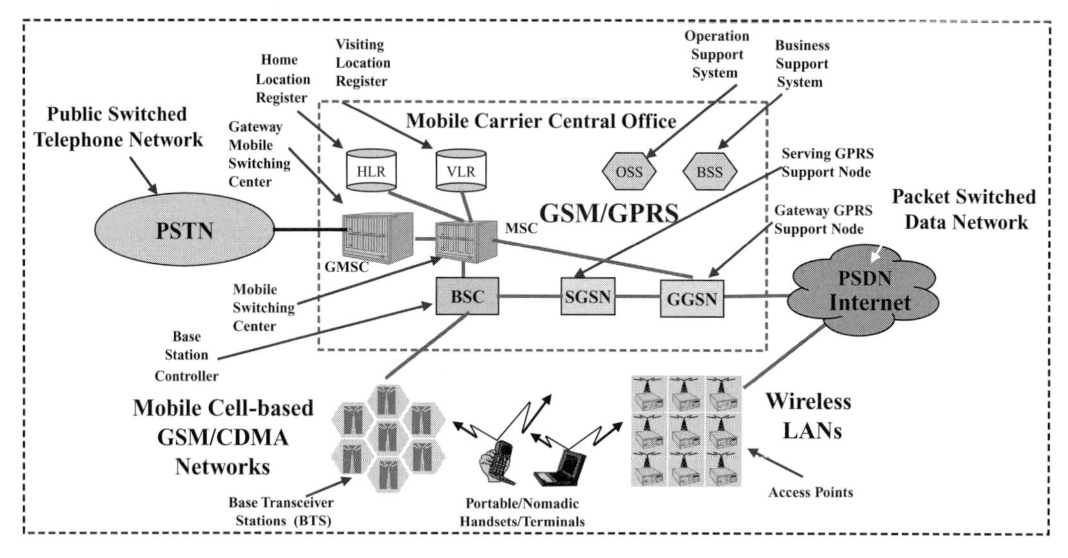

Figure 11.3 Wireless LAN and Mobile GSM/CDMA Networks Convergence

It is evident that specific functionality should be added to the current BSC that can play the role of moderator between the two networks. Also, an agreed upon algorithm and mechanism is necessary to determine initial user assignment based on optimal network selection (well defined criteria) and the transparent inter-system handover. A more detailed description of the handover from WLANs to mobile networks and vice versa will be given in Chapter 15. There, we will discuss the UMA-based standardized solution for Wi-Fi cellular mobile convergence.

11.5 Dual Mode WLAN Mobile Convergent Handsets

The tremendous success of WLANs in large enterprises, small businesses, and home networking is well recognized. Building on this success, Wi-Fi enabled laptops and even desktops are standard offerings today. These developments were focused on data communications and, to a lesser extent, on VoIP. Inclusion in the Wi-Fi development of mobile handsets, with focus on voice communications, is a more recent phenomenon. There are several reasons. One is the reluctance of mobile service providers, who perceive the use of WLAN networks as a threat to the revenue generated by the exclusive use of mobile operators' networks. This is the case when data is transmitted using WLANs connected to the Internet. The same is true when voice is transmitted, even occasionally, over WLANs in the form of VoIP. Another reason was the lack of standard dual-mode Wi-Fi cellular mobile (GSM, CDMA, W-CDMA) handsets. The third major reason was the lack of infrastructure to support this convergence on the solid base of internationally accepted standards. However, under the pressure of the market, things are changing. This is reflected in the fact that all major mobile phone vendors are bringing convergent mobile phone models to the market. In this context, ABI Research predicts that in 2009 we will have over 100 million cell phones with Wi-Fi capabilities (300 million in 2011) and convergent services will be offered by most of the mobile carriers. Issues, among them battery life, will remain and need to be solved. Currently, handset power consumption working in a WLAN is 2–3 times higher than power consumption of the same handset working in a regular cell-based mobile network. A short sample of dual-mode cellular and Wi-Fi convergent handsets follows:

Dual Mode Handsets (DMH) GSM/CDMA and Wi-Fi

- **D-Link**: V-CLICK, dual-mode tri-band GSM-Wi-Fi, (www.dlink.com);
- **Fujitsu-Siemens**: Pocket Loox T-800, UMTS, 802.11 b/g, Bluetooth 2.0, (www.fujitsu-siemens.com);
- **Hewlett Packard**: iPAQ hw6900 PDA phone with Wi-Fi, GPS, Bluetooth, (www.hp.com);
- **High Tech Computer** (HTC) Corporation: P3600, 3G PDA HSDPA and Wi-Fi, Bluetooth 2.0, (www.htc.com);
- **Kyocera Wireless**: CDMA Wi-Fi BREW DMH based on Voice Call Continuity (VCC) specifications, (www.kyocera-wireless.com);
- **Nokia Corporation**: E90 DMH quad-band GSM/WCDMA 2100 and Wi-Fi b/g, (www.nokia.com);
- **Motorola Corporation**: A910 UMA-based GSM/GPRS and Wi-Fi 802.11b/g, (www.motorola.com);
- **Paragon Wireless**: Hipi 2200, tri-band dual-mode GSM Wi-Fi b, MobileIGNITE-based, (www.paragonwireless.com);
- **Pirelli Broadband Solutions**: Dual Mode Handset GSM-Wi-Fi b/g DP-L10, (www.pirellibroadband.com);
- **Research in Motion** (RIM): Blackberry 8220 GSM/CDMA/iDEN and Wi-Fi, (www.rim.com);
- **Samsung**: SGH T709 GSM and Wi-Fi, (www.samsung.com);
- **Tiger Netcom**: WP000, tri band GSM and Wi-Fi b/g, (www.tigernet.com).

To project the evolution of handsets, we reproduce the transcript of United States Patent 6763226, the futuristic "multifunction, interstellar device" World-Wide-Walkie-Talkie [52]:

"A high speed multifunction interstellar wireless computer/instant messenger communicator, Personal Digital Assistant (PDA), coupled with a resilient, robust, VoIP data network and internet server method, deploying multiple wireless networks and protocols such as Voice Over IP, GPRS, WAP, Bluetooth, PCS, I-Mode, comprising a high speed Intel Pentium 4 Mobile.TM. or compatible Processor, to formulate a internet gateway system (99) and network bridge (150) for establishing instant low cost, real time global communications to the Public Switched Telephone Network via the internet (54). A PUSH-TO-TALK-WORLDWIDE button (21) instantly initiates global bisynchronous communications, or videoconferencing sessions. Fax, VideoMail, and unified messaging services are immediately available. GPS and mass memory provides global navigational tracking and data storage. Internet users, telephones, and cellular/satellite phone users can intercommunicate with the invention via VoIP/IM services. The invention provides uniformed global wireless communications, eliminates traditional long distance costs, and operates anywhere on earth", http://www.freepatentsonline.com/6763226.html.

11.6 Siemens WLAN Cellular Mobile Convergent Network Case Study

Siemens's HiPath MobileConnect is a Fixed-Mobile Convergent (FMC) solution for enterprises. It is based on Dual-Mode Handsets (DMH) that combine GSM cellular mobile and Wi-Fi capabilities to support transparently both voice and data communications in both networks. The Siemens WLAN cellular mobile convergent network architecture is shown in Figure 11.4.

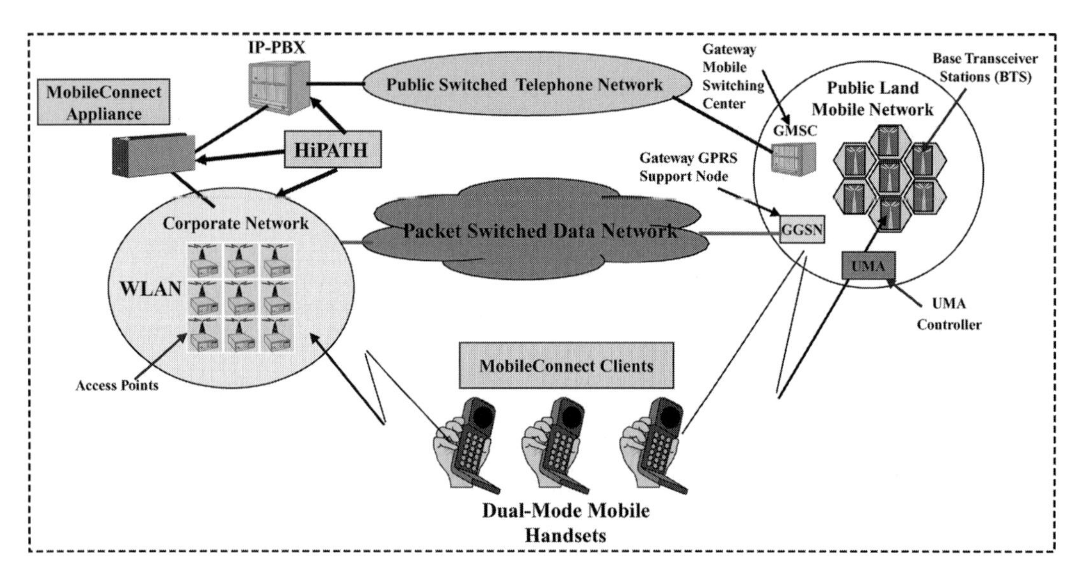

Figure 11.4 Siemens WLAN Cellular Fixed-Mobile Convergence Architecture

The HiPath MobileConnect solution consists of HiPath dual-mode handsets that run the **MobileConnect Client Session Initiation Protocol** (SIP)-based software. This is in addition to standard support for GSM/CDMA/W-CDMA mobile communications and Wi-Fi IEEE 802.11 a/b/g specifications. The enterprise **MobileConnect Appliance** provides centralized coordination by assigning clients use of corporate WLANs and SIP-based Private Branch Exchange (PBX) facilities (details about SIP are provided in Chapter 16, entirely dedicated to this subject). One number and one voicemail are allocated to each client who is then free to roam between enterprise WLANs and standard cellular mobile networks with the benefits of improved productivity and reachability. This functionality can be extended to WLAN Wi-Fi based hot spots and home networks. This convenience is important because most users are often located in working areas, in hotspots, or connected to home area networks. The advantage of using the SIP and VoIP is that both voice and data communications are handled similarly. The advantage of using an IP-based PBX is that it can handle both wired and wireless calls, based or not based on VoIP.

When a roaming DMH client comes into the proximity of enterprise WLANs, they are automatically registered with the MobileConnect Appliance as a SIP client on the enterprise PBX. From this moment, a VoIP session could be established and all the calls initiated or received by the user will be routed and handled by the corporate PBX.

Many of the IP-based PBX features such as call waiting, call forwarding, caller ID, and 3-way conferencing will be available to the user in addition to the convenience of having one telephone number. The MobileConnect Client software can run on many dual-mode devices. Two operating systems are supported: Windows Mobile 5.0 and Symbian 9.1. When a session is established a supporting display will help the user to make calls and to take advantage of standard PBX features.

When the user moves out of the corporate WLAN coverage area, the MobileConnect Appliance senses this and will automatically handover the call to the cellular network through the PSTN network. All calls received from the outside will be checked through the MobileConnect Appliance and if the user is not local or not on the WLAN at that moment, the calls will be redirected through the PBX to the outside public wired or wireless network.

The HiPath MobileConnect Appliance comes in three models, depending on the maximum number of users and the maximum number of concurrent sessions supported: MobileConnect 50, MobileConnect 250, and MobileConnect 1500. The aggregate throughput of these three models is 400 Mbps, 2.2 Gbps, and 6 Gbps, respectively. The MobileConnect Appliance also includes firewall type inspections to make sure that only authorized clients are using the corporate network.

WLANs can be built using the HiPath Wireless family of products. These include HiPath Wireless dual band (a/b/g) Access Points, centralized WLAN controllers, and HiPath management systems. Configuration, fault, performance, and accounting management modules, as part of HiPath 4000 and 8000 systems, use collection agents that allow traffic monitoring, hardware diagnostics, and real-time management. HiPath QOS Management is a distinct application used to monitor the VoIP components of the HiPath network to assure proper QOS.

The HiPath Siemens MobileConnect FMC solution is particularly suited for enterprises where there are multiple parallel operations with employees moving around a great deal of time such as retail warehouses/malls and industrial plants. It helps reduce the total cost of operations by providing a single infrastructure that can transparently handle any type of call and by selecting the lowest cost and the highest data rate network facility (WLAN), whenever possible.

The MobileConnect FMC solution is part of Siemens's overall vision of open communications across wired and wireless networks. An important piece of this solution is the HiPath family of convergence platforms that are used to support voice, data, and video across any type of network. Three models of softswitches, HiPath 3000, HiPath 4000, and HiPath 8000 are optimized to support a range of users from small businesses (1000 users or less) to very large carrier-grade configurations (100,000 users or more). They can handle a mix of packet switched and circuit switched environments. More details about softswitches are provided in Chapter 16.

11.7 WLAN Mesh Networks

Traditional WLANs are based on **single-hop communications** with the Access Points as focal points and a single connection to the wired WAN. Among the major shortcomings of this architecture are the limited coverage area and the poor performance when the number of users and loading increases. Wireless mesh networks, such as the new WLAN developments of the past few years, offer solutions to these limitations.

Wireless Mesh Networks (WMN) are multiple-hop WLANs that provide transparent handover over multiple Access Points, combining access and routing functions. The advantages of mesh WLANs are:

- **Spatial reuse** by connecting the AP to the wired network through multiple paths/nodes;
- **Robustness** by eliminating the single point of failure when there is only one AP;
- **Higher bandwidth** by transmitting data over multiple channels; this results in less interference, lighter loading, and lower signal strength needed;
- **User mobility** across a larger area network.

A typical wireless mesh network is presented in Figure 11.5. It consists of numerous interconnected **mesh routers** nodes, which are WLAN access points capable of handing over communication established with one of the **mesh clients**. Some of the mesh routers, the **mesh gateways**, provide connectivity to public networks such as the Internet. Other mesh routers may provide connectivity to other type of networks such as WiMAX and WPANs. In this case, by employing more complex functions, they are called **mesh bridge-gateways**. An important feature of WMN is the multi-hop wireless connectivity between mesh routers and backhaul connectivity to a packet switch network.

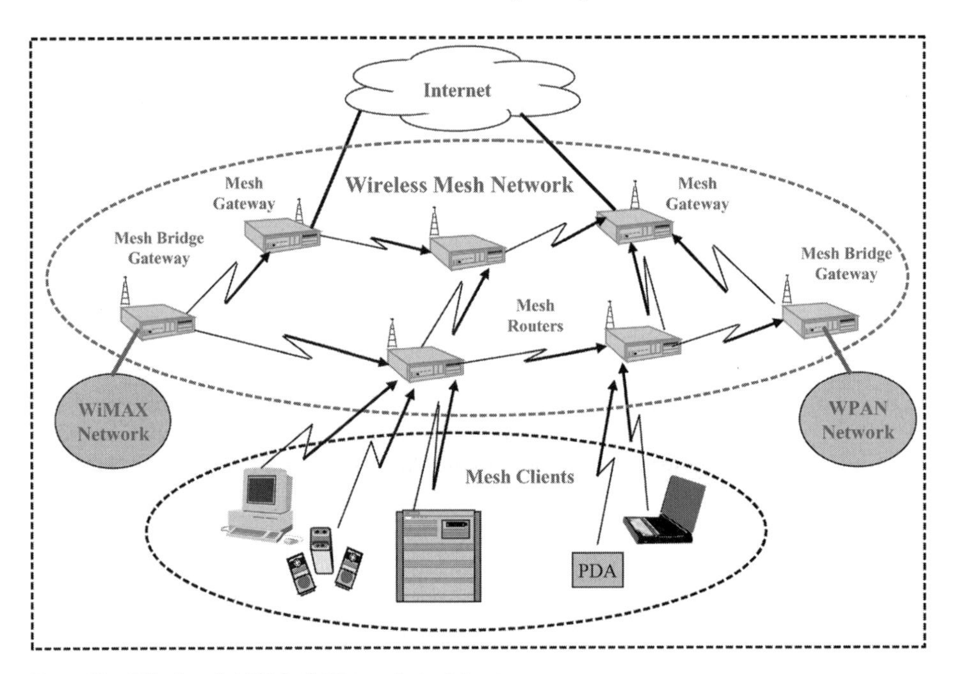

Figure 11.5 Wireless LAN Mesh Network Architecture

Mesh networks can be implemented to support "always-on, anytime, city-wide, metro Wi-Fi networks". They can also be used for one-time temporary exhibitions, for entertainment, and for emergency situations. From this point of view, there are similarities between ad-hoc mobile networks and mesh WLAN networks. However, mesh networks are different from **hot-spots**, which are mainly congregations of APs in defined coverage areas with limited or no wireless connectivity between APs.

Metropolitan Wi-Fi networks provide carpet/blanket contiguous wireless coverage over several square mile areas with backhaul connectivity to the Internet. Grown out of traditional indoor Wi-Fi networks, mesh networks are challenged by important technical issues. One is the need to provide powerful mesh gateway routers capable of handling the growing number of users and the traffic created by broadband applications such as streaming audio and video, VoIP, and, in the future, HDTV. Another issue is interconnecting APs to relay packets through backhaul wireless connections to and from mesh gateways.

Multiple gateways are required for redundancy and to minimize the number of hops. Traffic should be controlled and congestion should be avoided through careful designs that take into account factors such as environmental conditions (natural obstacles, buildings), the radio equipment used (antenna type, RF signal strength), and the distance between mesh nodes. A combination of regular APs working in the crowded 2.4 GHz band, and APs working in the 5 GHz band, can help, the latter serving as mesh gateway nodes. There are also techniques that can overcome the limitations born from the original WLAN 802.11 standards that are based on CSMA/CA multiple access. Such a solution is Kiyon's TDMA MAC, a different way to handle packets that increases the amount of usable bandwidth. It does this by basing transmission of packets on slot availability rather than node availability.

Several projects in the USA are in different stages of network design and implementation such as metro Wi-Fi in Philadelphia, San Francisco, Boston, Cupertino, Santa Clara, and Corpus Christi, to name a few. Opposition from major carriers such as Verizon, on the basis that government should not compete against private businesses, melted away when a telecommunications act, approved by the US Senate Commerce Committee, affirmed that states cannot prohibit their own municipalities from offering broadband services, particularly since the intent of metro Wi-Fi projects is to attract downtown businesses and to alleviate the growing digital divide among the metropolitan populations.

In Philadelphia, a test case in the implementation of Wi-Fi based mesh networks is underway. Initially, a 15 square mile area, out of a total of 135 square miles, will be used as the testing ground. The primary contractor of this project is Earthlink, who will be responsible for implementing 450 mesh nodes, the gateway nodes, and the cluster nodes. Tropos Networks mesh nodes and Dragonwave gateways are partners in this project, Alvarion and Motorola are also involved. The target hop count of 3 or fewer requires installation of one mesh gateway in the center of each square mile with clusters of 8 adjacent mesh nodes.

Some current WLAN mesh network vendors and service providers include:

- **Alvarion**: Backhaul Technology, Breeze Access Family;
- **BelAir Networks**: BelAir Wireless Mesh, BelAir BelView NMS;
- **DragonWave**: AirPair Flex Wireless Platform (Canada);
- **Earth Link**: Service Provider, primary contractor of Pittsburgh's metro Wi-Fi;
- **EnGenius**: ECB 3220 wireless bridge for Internet connectivity;

- **Motorola**: HotZone Duo, Mesh Manager, Canopy;
- **Nortel Networks**: Wireless Mesh Gateway7250, Bridge 7230, AP 7215, 7220;
- **SkyPilot Networks**: Boston WiFi Hot Zones, Mesh Gateways, Sky Extender;
- **Strix Systems**: Access/One Network OWS, Tempe Arizona;
- **US Wireless Online**: Operator of Pittsburgh's metro Wi-Fi;
- **Tropos Networks**: Architecture, mesh routers, analysis and control tools.

11.8 Metropolitan Mesh WLAN Convergent Network Case Study

One of the technical solutions that provide full infrastructure for mesh networking, and the supporting management, is Motorola's **Canopy HotZone Duo** [53]. Canopy is an umbrella broadband wireless public network infrastructure aligned with Motorola's vision of broadband solutions called **MOTOwi4**. Motowi4 capabilities include fixed wireless broadband, WiMAX, mesh networking, and broadband over power line solutions. The Canopy HotZone Duo architectural diagram for a city-wide WLAN Mesh network is shown in Figure 11.6.

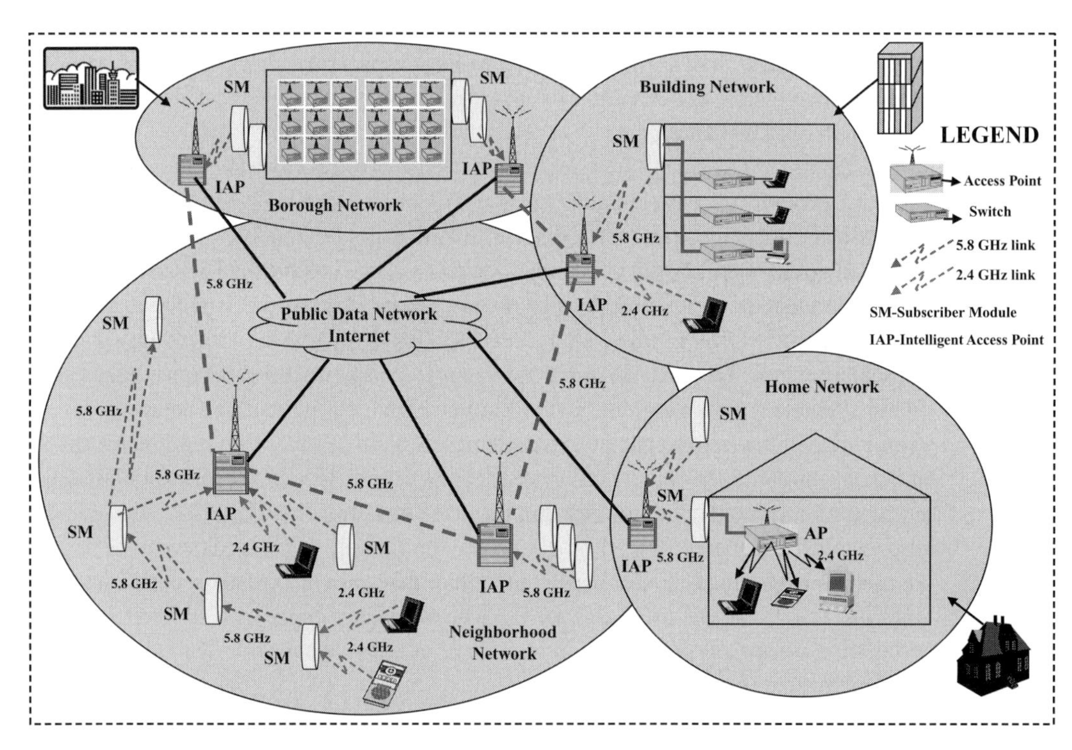

Figure 11.6 Motorola Canopy HotZone Duo Mesh Network Architecture

The network infrastructure includes three layers: access, concentration, and backhaul. The **access layer** consists of **Subscriber Modules (SM)** which are outdoor WLAN transceivers working in the ISM 2.4 GHz unlicensed spectrum band. These modules connect via Ethernet cable with home or building networks that consist of wired and wireless devices connected to the LAN. Wired devices can be desktops, laptops, residential gateways, printers, or scanners. Wireless devices can be WLAN-enabled laptops, wireless routers, or short range wireless access points connected to the local Ethernet.

The **concentration layer** consists of interconnected **Canopy Advantage Access Points** with a wide range of operation. These APs can be organized as clusters of 1 to 6 Access Points, communicating with a large number of subscriber modules. Typically, there are about 200 subscriber modules attached to one cluster access point. The access points communicate among themselves using the U-NII 5.8 GHz unlicensed spectrum band. Hence, one part of the name of this solution, HotZone Duo, i.e., a dual-radio operation. This partition will release SMs from the burden of handling both bands of frequencies. The **Canopy Advantage Cluster Management Module** (CMM), attached to the cluster of access points, provides the functions of wireless controllers, important components of second generation WLAN Systems. CMM controls the access of multiple subscriber modules to the access points. It uses a point-to-multipoint protocol.

The concentration layer can be divided in three "zone points", leading to the second part of the name of this solution, HotZone. Outdoor weatherized, rugged zone points can be mounted on buildings or utility poles and street lights. Nomadic zone points can be mounted on vehicles that support user mobility. We can also have indoor zone points which are wireless routers with access point functions. Remote users found outside access point coverage areas are connected through point-to-point wireless bridges.

The **backhaul layer** provides connectivity to the public network using high capacity wired Ethernet links to a router-based network. Overall coverage at 2.4 GHz can reach a radius of 15 miles, while the 5.4 GHz band links between APs can operate within a radius of 10 miles. The APs, weighting less than 5 pounds, attached to traffic lights, utility poles, billboards, and buildings, can be safely and easily deployed in about 15 minutes.

The HotZone Duo solution is accompanied by the **MeshConnex** routing mechanism and the **MeshManager**. The MeshConnex minimizes the number of hops while delivering the information with a minimum of overhead. The Wireless **Adaptive Routing Protocol** (WARP) is a scalable mesh routing algorithm for municipal-sized applications. The **Prizm MeshManager** is a centralized Element Management System that provides over-the-air provisioning, fault, performance, configuration, and security management as well as software updates. Open management standards such as SNMP version 1 and 2 are used. For security, Canopy systems use either the Data Encryption Standards (DES) or, when specifically required, the Advanced Encryption Systems (AES). The latter is based on 128-bit encryption key that is believed impossible to crack.

11.9 Wi-Fi and BlackBerry Convergence

BlackBerry wireless handheld devices from Research in Motion (RIM) (the first models were launched in 1999) are good examples of convergent terminals, networks and applications. RIM's initial focus on email in the cellular mobile environments has gradually changed. As of 2007, BlackBerry models support data, voice, and numerous consumer electronics features and applications. The primary carriers of data and voice applications are GSM/GPRS/EDGE/UMTS based networks. BlackBerry also works well on iDEN (Sprint/Nextel) and various CDMA generations of handsets from CDMA 2000 to CDMA1xRTT to CDMA 2000-EV-DO.

Major applications supported are: Internet access, push email (real-time delivery of email to the handheld device), text messaging (SMS), instant messaging (IM), web browsing, mobile telephony, and push-to-talk (two way communications) as well as all the PDA functions such as address list, schedule, and list-to-do. In addition, consumer electronics features such as MP3 player, GPS receiver, speaker phones, USB port, and microSD flash cards are supported.

Targeting business customers, BlackBerry delivers seamless email access to roaming users by building a client server architecture supported on the enterprise side by the BlackBerry Enterprise Server (BES). This server provides the push email service and the security features by encrypting data and voice communications. Triple DES (Digital Encryption System) was used in the early models, later augmented with Advanced Encryption System (AES) functions.

Success of the BlackBerry led to its embrace by all major mobile operators from the USA (Sprint/Nextel, AT&T, Verizon Wireless, T-Mobile), and later, in Europe (Vodaphone, T-Mobile, Orange, Telefonica Moviles) and Asia (NTT DoCoMo and Hong Kong Smartone). There were over 10 million users in 2007. Consequently, regular cell phones manufactured by several vendors such as Nokia, Motorola, Samsung, LG, HTC, Siemens and Sony Electronics have included BlackBerry clients in their handsets to support the popular push email applications.

Support for Bluetooth Version 2.0+EDR and WLAN 802.11 Wi-Fi makes BlackBerry a convergent terminal. Cellular mobile and Wi-Fi convergence is supported through the Unlicensed Mobile Access (UMA), also called Generic Mobile Access (GAN), standard. Implementation of the functions and protocols in the standard makes possible increases in the available data rate when the BlackBerry operates in enterprise WLANs, Wi-Fi hot-spots, and municipal WLAN-based mesh networks.

11.10 Wi-Fi and iPhone Convergence

iPhones from Apple Computers are another example of convergent terminals, networks and applications. iPhone represents a new stage in the evolution of a very successful device, the iPod, along with other developments such as iPod Classic, the iPod nano, and the accessories iTunes and iPod shuffle. iPhones support data, voice, video, and multimedia applications as well as many consumer electronics features. The primary carriers of data and voice applications are GSM/EDGE/UMTS based networks, in all four bands (850, 900, 1800, and 1900 MHz), and Wi-Fi 802.11b/g WLAN networks. Another important iPhone feature is its support for Bluetooth 2.0+EDR.

Major applications supported by the iPod are: mobile voice communication, voice mail, ringtones, text messaging (SMS), Internet over GSM/EDGE, Internet over Wi-Fi, and browsing using the Safari browser. In addition, the iPod provides support for audio and video entertainment consumer electronics features such as video/music store and playback with stereo earphones with built-in microphones. A 2.0 megapixel digital camera, PC or MAC connectivity with the USB 2.0 port, and an 8GB flash drive completes the picture. iPhones use Apple's OS X operating system. Bluetooth provides connectivity to the headset and other Bluetooth-enabled devices [54].

Targeting individual customers/consumers, iPhone provides seamless email access to roaming users by using the GSM EDGE technology and by switching back and forth to Wi-Fi WLANs when the iPhone is in the proximity of a hot spot. Mail access to most POP3 and IMAP servers, access to Google maps on a 3.5 inch wide touch display, Visual Voice Mail, widget applications such as stock reports, and weather information, are other valuable features.

What makes the iPhone a convergent terminal is its support for GSM cellular mobile, Bluetooth version 2.0+EDR, and WLAN 802.11b/g Wi-Fi, in line with fixed-mobile convergence developments. Convergence allows much higher data rates when the iPhone operates in enterprise WLANs, Wi-Fi hot-spots, and municipal WLAN-based mesh networks. The initial iPhone business target was to have 10 million users by 2010. The first one million iPhones were sold in just 74 days after it went on sale. This success follows the steps of iPod used today by 90 million worldwide customers.

The iPod was launched with exclusive support in the USA from AT&T, O2 in United Kingdom, Orange in France, and T-Mobile International in Germany. AT&T was selected since it had the highest number of subscribers in the USA (67 million), there was very good geographical coverage (as a result of mergers of Cingular, AT&T Wireless, and Bell South), there was support for GSM radio access with a worldwide presence, and network operation and management is robust. To understand the complexity involved when a new convergent phone is introduced into a mobile network, we can look at the preparation done by AT&T before the official launch of the iPhone [55]. Based on the initial assessments of data needs in dense populated areas, 20,000 new base stations were added to increase the overall cellular mobile networks capacity. This was done as part of the consolidation of the network and elimination of old 2G TDMA based systems. Consolidation was also applied to the six Network Operation Centers (NOCs) that will be concentrated in one spot, Atlanta (Georgia), with a disaster recovery alternate in Redmond, Washington.

An inconvenience of iPhone, typical for mobile phones used in the USA, is its exclusive use in AT&T mobile networks. It was just a matter of time until the software enabled locks were cracked and the iPhone was unlocked by hackers. The same independent spirit led many users to download additional applications, circumventing the recommended use of Safari, iPhone's native browser. Another major issue is the low bit rate of the data services based on the GSM/EDGE technology, a step backward, in the era of 3G and 3.5G mobile GSM HSDPA/HSUPA developments.

11.11 Siemens HiPath WLAN Network Management Solution

Siemens's networking vision has found expression in the HiPath Communications Systems. HiPath is a collection of communications solutions and services that address both

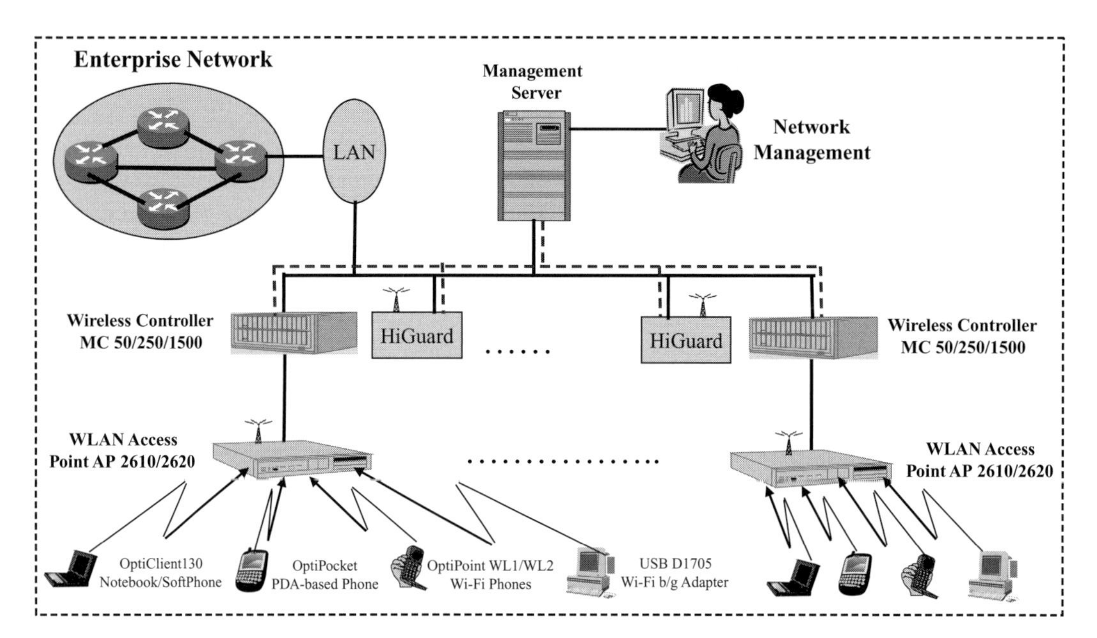

Figure 11.7 Siemens HiPath Wireless Network Architecture

wired and wireless worlds. The HiPath portfolio covers applications, client devices, management aspects, and security solutions. HiPath Wireless Solutions address specific aspects of wireless communications with focus on WLAN networks and their management. As previously mentioned, WLANs have evolved from simple data communications bridges between wireless devices and wired Ethernets or the Internet to become standalone networks that carry both data and voice and support complex functions. The HiPath Wireless Management Platform [56] is one component of the HiPath Wireless Network architecture. This is shown in Figure 11.7.

The HiPath Wireless Network topology is based on a centralized hierarchical architecture, where the management server monitors up to ten Wireless Controllers. These management servers run several real-time communication applications to collect information related to their status and the status of APs found in their domain. Three models of wireless controllers are used: C10 that can control up to 30 APs and a total of 512 users; C100 with 75APs and 1536 users, and the high-end C1000 that supports up to 200 APs and 4096 users. Other important management functions, related to RF interfaces to the users, are performed by the APs.

The users in this diagram are represented by various wireless devices such as: OptiClients 130 running on Notebook combined with a softphone, OptiPacket that provides similar functions for PDA-based Wi-Fi enabled devices, OptiPoint WL1 or WL2 Wi-Fi b/g smart phones, and desktops or laptops equipped with USB D1705 Wi-Fi b/g adapters. Two types of APs are used in HiPath solution; the AP2610 for indoor WLANs and AP2620 for outdoor WLANs.

The management platform also supports HiGuard, an advanced security service, which allows monitoring and reporting of any unauthorized intrusion in the WLAN coverage areas. The HiGuard software module can be loaded on dedicated APs or on regular APs with the corresponding additions. The HiGuard-based APs serve as sensors to scan the air surrounding the wireless network.

Worldwide implementations such as the Metro Toronto Convention Center, Canada, and the Municipal Clinic from Gorlitz, Germany are based on Siemens Wireless Management Solutions. For example, the Toronto Convention Center is the host of 700 events every year. Giant floor exhibitions require quick set up of wireless networks to support both data and voice communications for thousands of sophisticated users. Security and management capabilities are paramount for this shared network. Good scalability and versatility is required in these implementations.

11.12 QOS in WLAN Cellular Mobile Convergent Networks

In Section 11.4 of this chapter, we provided a high-level WLAN cellular mobile convergent network architecture with emphasis on the components that allow interworking of these networks. Two mechanisms were identified as key for the interoperability: initial user assignment and inter-system handover. However, since WLANs were initially designed to carry only data and cellular mobile networks were primarily intended for voice traffic, the differences in requirements for QOS must be taken into account in the convergent networks. Further, WLANs and cellular mobile are domains that usually are deployed and serviced by different operators. This is the context in which we have to assure transparent communications across convergent networks while maintaining a satisfactory end-to-end performance for all services. To explain how the QOS function is performed in convergent networks, we will use a high-level architecture, as shown in Figure 11.8.

We selected a loosely coupled architecture [57] where Gateways GPRS Support Nodes (GGSN) and routers/gateways connect the mobile networks and WLANs to the Internet. There is no direct relationship between cellular mobile networks and WLANs other than they physically overlap. The cellular mobile networks provide total coverage of an area while the WLANs, as represented here, have an island-like presence (enterprise networks, hot spots, and multi-tenant units).

The meaning of QOS in a WLAN cellular mobile convergent network is shown in Figure 11.9.

From the QOS analysis provided in Chapter 3, we know that the transmission of voice signals requires real-time communication and is very sensitive to delays and jitter (variations in delay). A small ratio of packet loss for digitized voice is admissible but a constant rate and fixed bandwidth are required. On the other hand, data communications is not sensitive to delay and jitter but is intolerant to loss of data. Data communications can be bursty with a variable bandwidth required, depending on the size of files transferred and the level of interactivity between the entities that exchange files.

The QOS of a convergent network will include the WLAN and the core mobile network as well as the QOS of the public network, the Internet. Service providers of the networks are responsible for the QOS in their respective sections. End-to-end QOS may

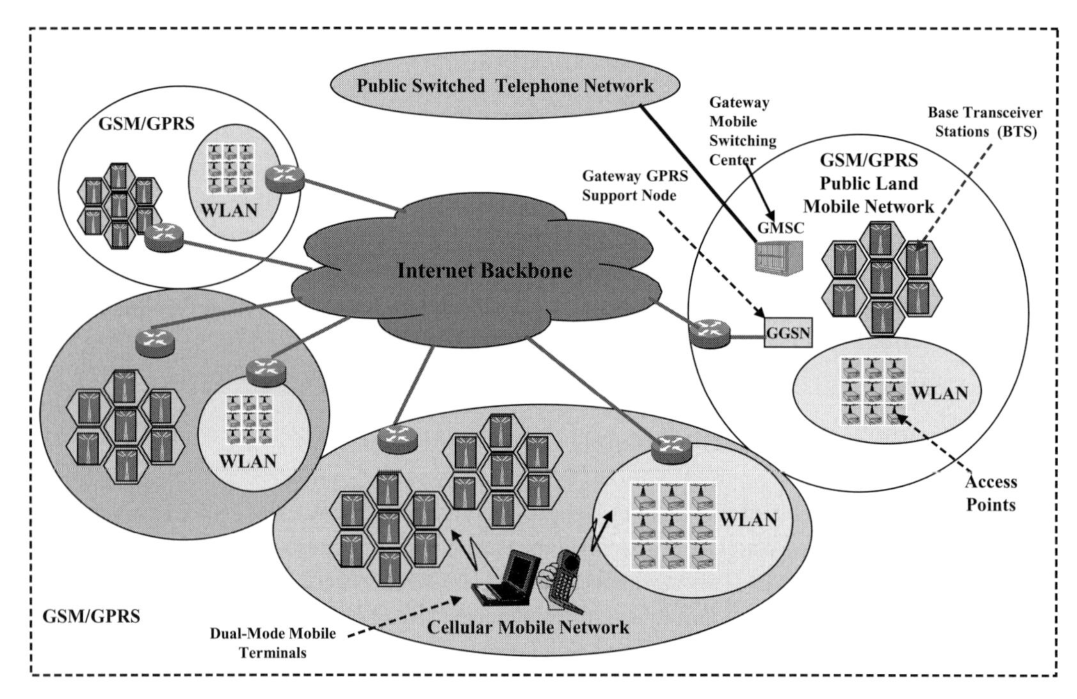

Figure 11.8 QOS in a WLAN Cellular Mobile Convergent Architecture

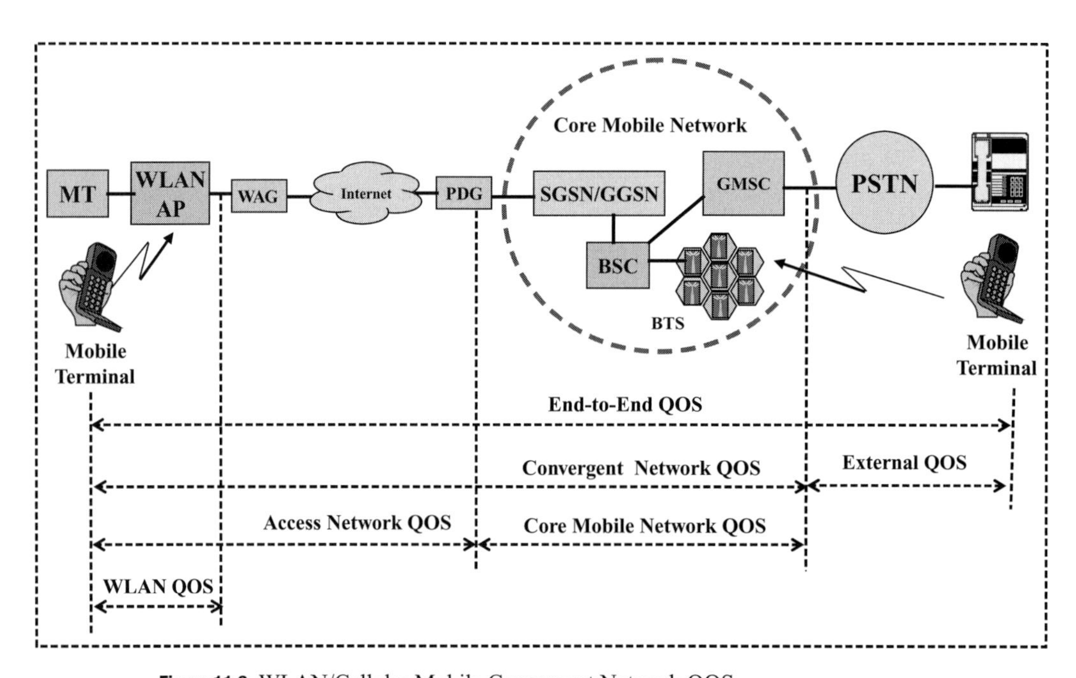

Figure 11.9 WLAN/Cellular Mobile Convergent Network QOS

include the PSTN QOS in cases where a call originates in the wired public telephone network.

Three grand schemes will be used to assure QOS in convergent networks: Integrated/Differentiated Services (Int/Diff Services), Multi-Protocol Label Switching (MPLS), and Policy-based Management (PBM). Chapter 18 will be dedicated to analyzing the methods used to assure adequate QOS in fixed-mobile convergent networks.

12 Wireless PAN Cellular Mobile Convergence

12.1 Bluetooth Networking

In Chapter 7 we introduced the WPAN infrastructure and Bluetooth, a technology that supports short range communications and home networking. Initially, Bluetooth's primary function was "cable replacement", i.e., elimination of short-range wired-based communications. Therefore, most of the Bluetooth applications presented were based on point-to-point and point-to-multipoint wireless links. However, the applicability of Bluetooth has gradually extended, covering various aspects of office automation, industrial process automation, and even mobile ad-hoc networking. Many of these extensions include Mobile-to-Mobile (Mo2Mo) networking, beyond the traditional clustering within piconets. An essential condition of Bluetooth networking is its ability to work in extended ranges, up to 1 km, using omnidirectional antennas. Another condition is low power consumption while maintaining a high data rate. The Bluetooth networking architecture is shown in Figure 12.1.

Bluetooth networks consist of multiple **piconets** interconnected into larger **scatternets**, providing a community area network of multiple groups of users. A piconet consists of a single **Master Station** (MS) and several **Bluetooth** (B)-enabled devices. Sophisticated home networks can be organized as scatternets where each piconet represents an ad-hoc grouping of Bluetooth-enabled devices with common functionality such as environmental automation, audio/visual systems, and computing systems.

Within a piconet, the master station establishes connections with all the Bluetooth devices, including a smart cell phone that can provide connectivity to a mobile service provider and the Internet. Within scatternets, Bluetooth connections are established between master stations to allow multi-hop routing. In this case, additional characteristics such as range, power, routing, and Quality of Service become important. Among them, energy saving is a critical issue because of the low-power consumption requirements of some Bluetooth devices. That is the case in industrial processes where power cabling is not possible and the life of piconets and scatternets depend on the longevity of batteries.

The configuration of scatternets depends on the number of piconets and the internal configuration of overlapping piconets. It is expected that in the near future every electronic device manufactured will be Bluetooth-enabled. Each device will have an electronic address that reflects the supported Bluetooth profiles and the types of device. Examples of Bluetooth-enabled devices include: computers (laptop, PDA, pocket PC), computer accessories (keyboard, mouse, joystick), peripherals (printer, scanner), home

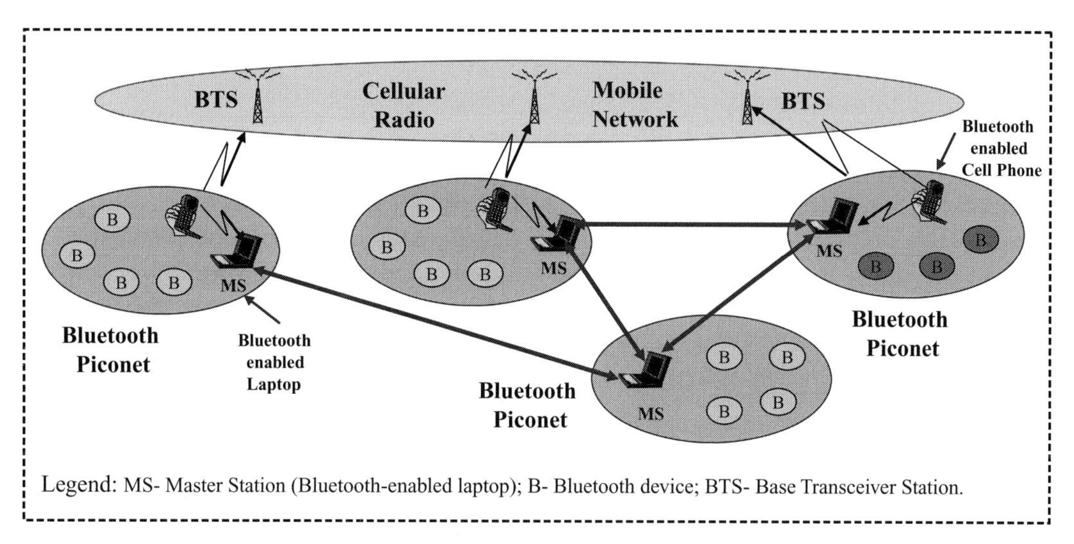

Figure 12.1 Bluetooth Networking Architecture

theater components (TV, radio, projector, DVD and MP3 players), voice telephony (cordless phones, smart phones, cellular handsets), and home environmental and surveillance systems (HVAC, home appliances, video camera). All of these devices can automatically self-organize in individual piconets around a pre-programmed Bluetooth-enabled master station. This is done as part of the normal process of discovery that starts when the devices are powered up.

The configuration of scatternets depends on the algorithm established for communication between piconets. This is not standardized, and is one of the reasons that Bluetooth-based networking is still in its infancy. The confinement of a Bluetooth device to a particular piconet and its master station can be changed so that some devices, such as PDAs, can interact with multiple master stations. For example, they can communicate with desktops or laptops belonging to a different computer piconet to synchronize an address list and in the same time can act as a remote device to monitor the home theater devices or surveillance systems.

12.2 Bluetooth Convergent Applications

A short list of Bluetooth convergent applications follows. Some of these applications will be discussed in more detail when we will analyze some Bluetooth convergence case studies in later sections of this chapter. There were more than one billion Bluetooth units shipped in 2007, supported by hundreds of applications. The units sold are the best witness of the pervasiveness of this technology beyond anything envisioned by its inventors and the founders of the Bluetooth SIG. The convergent applications are:

- Bluetooth and cellular mobile convergence (head set extensions and connectivity across cellular mobile networks);
- Bluetooth and WLAN networks convergence;

- Bluetooth and Wireless Sensor Networks convergence (test and measurement instrumentation, alarm and surveillance systems);
- Bluetooth and Ultra Wide Band;
- Bluetooth and RFID networks convergence;
- Bluetooth and Consumer Electronics convergence (audio/visual, photography, cordless telephony, video gaming controllers, digital file/music transfer);
- Bluetooth and computer world convergence (peripherals, keyboard, mouse, joystick, speakers, USB dongles, serial and RS232 adapters, and PDAs wireless connectivity);
- Bluetooth and automotive networks convergence (moving parts and moving platforms wireless monitoring, data logging, and communication); and
- Bluetooth and customer location/navigation (Bluetooth GPS, Bluetooth watches).

Automatic discovery, synchronization, and communications with other Bluetooth-enabled devices of the largest variety are measures of Bluetooth's versatility. For example: synchronization of calendars/schedules between desktop/laptop computers and PDAs, connectivity between phone jacks and phones, hands free car kits that sense the presence of mobile phones in pockets or briefcases, and connectivity with other phones supporting Push-to-Talk (walkie-talkie applications). Other examples: instant postcard (still pictures taken with a Bluetooth-enabled digital camera automatically sent over the Internet via a Bluetooth-enabled mobile phone); sharing businesses cards and meeting notes, transferring voice messages from PCs to mobile phones, and LAN connectivity. Another popular application is the portable, automatic email sending once a place with Internet connectivity is reached.

12.3 Multi-mode Bluetooth Mobile Convergent Terminals

Since the mobile Bluetooth version 2.0 and IEEE 802.15.1 standards were published, hundreds of vendors have designed Bluetooth devices for a large variety of applications. Among them, the highest level of sophistication was achieved by multi-mode convergent mobile phones and pocket PCs, well on the road to become the focal points of future mobile communications.

Bluetooth technology offers three types of devices, based on power output. Class 1 devices can have up to 100 mW of power, allowing them, with a standard antenna, to operate in a range of 10 to 100 meters. Class 2 devices operate at about 2.5 mW. With a standard antenna their range is about 15 to 30 meters. Class 3 devices, at about 1 mW, operate in a range of 5 to 10 meters.

All major manufacturers have incorporated various levels of Bluetooth profiles and classes into their cell phones: Nokia, Motorola, Samsung, LG, Sony Ericsson, Nextel, Sanyo, Blackberry, and Kyocera. The major drivers of these implementations were the hands-free headsets and car sets. Bluetooth headsets allow phone conversations to continue hands-free without any manual configuration or initiation and, of course, not using any cable for such connectivity. Mass production of the headsets was boosted by the decision of many national governments to impose heavy fines and penalties for those driving while talking on mobile phones.

The major characteristics of Bluetooth headsets are: quality of receiving and transmitted audio, form factors including weight, comfort, ergonomics, battery life, and, evidently, the cost. When we refer to ergonomics, we have to take into account how comfortable the sets are to use for long durations and how easy it is to access the control buttons. Another qualifying aspect is the Bluetooth class, i.e. the normal range of operation.

An example of the complexity of multi-mode convergent cell phones is the latest Motorola tri-band GSM/GPRS model, the MPx300 running on the Microsoft Pocket PC Windows Mobile 2003 SE (second edition) operating system. It includes embedded Bluetooth 2.0 and 802.11b Wi-Fi connectivity. Additional features include a QWERTY keyboard, 1.2 Megapixel camera, USB port, 1 GB memory, and large dual-hinge color display. And, there is compatibility with Pocket PC applications such as ActiveSync to access Microsoft Outlook email, SMS, and Multi-Media Messaging Service (MMS), and the ability to send photos and ring tones. Web-browser, a Windows Media player, and support for WAP 2.0 are also included. Another interesting feature is the interchangeable SIM card that allows worldwide operation with any mobile carrier.

Another example of multi-mode convergent terminals is the HP iPAQ615 PocketPC. This device combines a quad-band GSM/GPRS mobile phone with Bluetooth class1 and Wi-Fi 802.11b wireless technologies. This allows a user to seamlessly roam across mobile networks, private WLANs and public Wi-Fi hotspots, and connect with any other Bluetooth device that is within 30 feet. It supports the Bluetooth GSM navigation system and a quasi Push email feature, courtesy of T-Mobile. Push email is realized through a T-Mobile Web service that peruses POP3 and Outlook Web Access inboxes looking for new messages to be delivered to your account. iPAQ615 also employs a snap-on QWERTY thumb keyboard for easy typing, a built-in 640×480 mega pixel digital camera with the ability to create your own slide shows. A separate application, HP Image Transfer allows you to move your images from iPAQ to a laptop computer. It comes with the Windows Mobile 2003 operating system and employs all the common PDA features.

12.4 Convergent Bluetooth Cellular Mobile Network Architecture

In Section 12.3, we presented several multi-mode handsets as examples of the convergence between Bluetooth wireless technology and cellular mobile technology. There, the embedded Bluetooth technology served as a cable replacement to hands-free Bluetooth headsets or car sets, ubiquitous extensions of mobile phones.

Bluetooth cellular mobile convergence can be extended to the convergence between Bluetooth networks and cellular networks. This is shown in Figure 12.2. The primary Bluetooth function, as the facilitator that allows devices to talk to other devices, or to accessories, remains the same. However, the application is larger, allowing mobile handsets to transparently communicate across either Bluetooth or cellular radio networks.

Bluetooth-enabled handsets/cell phones can communicate over Bluetooth-based WPAN **piconets** and over Internet-based broadband networks. Peer-to-peer Bluetooth

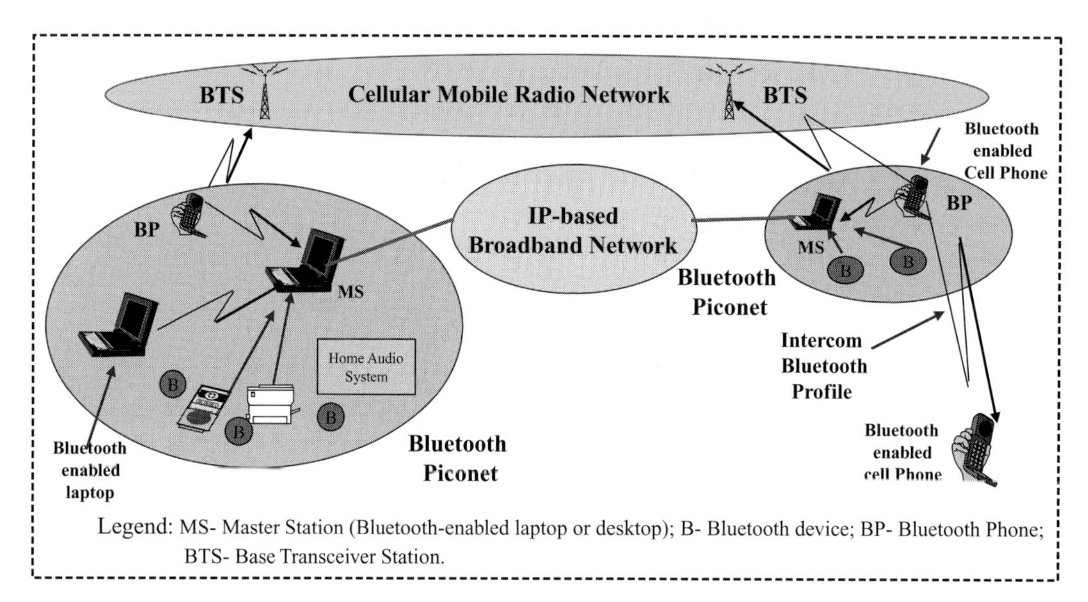

Figure 12.2 Bluetooth Cellular Mobile Convergent Network Architecture

connectivity can be provided within the same piconet between laptops and desktops and between handsets provided their support for intercom profile.

Bluetooth-enabled cell phones behave like regular cell phones when outside of the Bluetooth piconet range, usually less than 30 meters. When the mobile phone is in the range of piconet operation, it can connect with the master station of that Bluetooth cell and gain access to a broadband network. In this way, a connection can be established with a peer Bluetooth-enabled mobile phone that is in the range of another Bluetooth cell and has a connection established with its master station. This alternative makes sense when higher capacity is required than that offered by the regular mobile radio access. Conversely, when the broadband access network is not available, the Bluetooth-enabled phone can be used as a relay to establish a connection between the master station and the GSM/GPRS cellular mobile network.

Direct peer-to-peer connectivity between two Bluetooth-enabled mobile phones can be set-up using the "intercom" Bluetooth profile. Communication range is limited by the Bluetooth class adopted, and is about 150 meters in an environment relatively free of major obstacles. Provided that mobile phones are augmented with hands-free Bluetooth headsets, this can be useful for people biking or motorcycling because it gives them a hands-free "walkie-talkie" type of application.

Similarly, Bluetooth technology allows peer-to-peer communication between Bluetooth-enabled desktops and laptops or between two laptops using one of the profiles that allows file transfer (data, images, or music). This functionality can be extended to PDAs and other Bluetooth-enabled devices that are found in the same piconet or in a neighboring piconet.

Bluetooth-enabled cell phone functionality can also be augmented with cordless telephony features that allow Bluetooth-based connection between the mobile phone playing the role of a regular plain old telephone and a wired base station connected to the PSTN network via ISDN or a standard dial-up line.

12.5 Bluetooth and GSM Health Care Convergent Networks Case Studies

Health Care is one of the social enterprises and businesses that involves a large number of customers/patients, doctors, pharmacies, and other medical and administrative staff. Continuous interactions between patients and doctors, monitoring and logging of medical information, and prompt interventions are all necessary aspects of providing good medical care. Elderly care is one of the activities of an overall health care system that requires daily visits to the elderly and to outpatients requiring medical attention. The Health Department of Copenhagen Municipality, in collaboration with BLIP systems, provides a case study of Bluetooth implementations in a large synchronization network that we will examine here [58].

The Elderly Care Support System network will serve over 20,000 customers with the support of 3,000 municipal workers, each having a handheld Bluetooth-enabled PDA. Every morning, each active worker synchronizes his or her PDA, downloading patient related information along with some necessary operational data. Each evening, through a network of 400 Bluetooth Access Points (BAP) aggregated in seven city centers, the PDAs are synchronized, with the system downloading current information about daily visits. The main components of the synchronization network are shown in Figure 12.3.

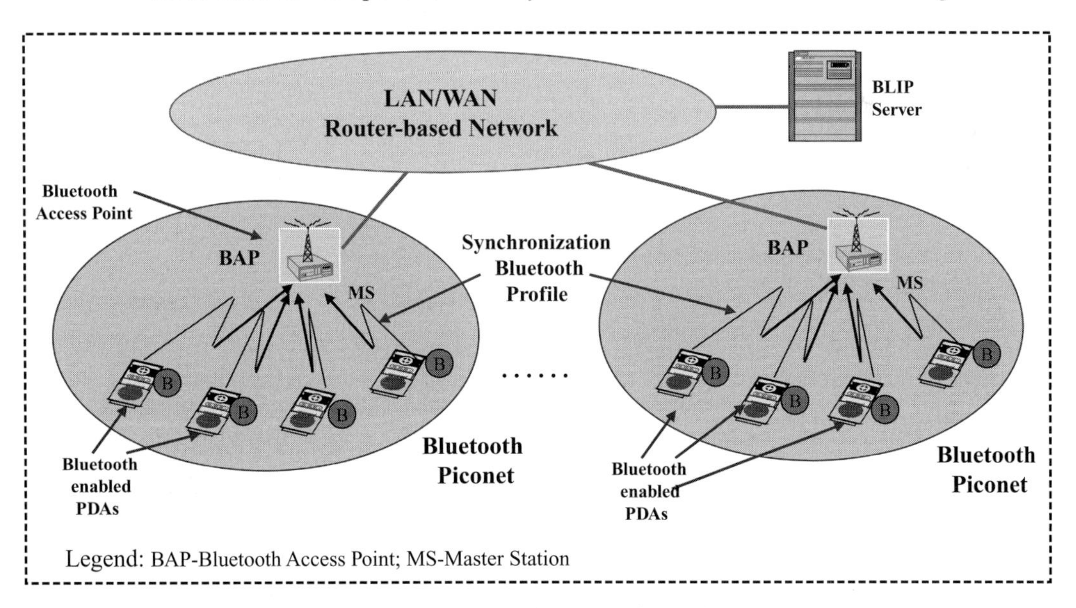

Figure 12.3 Bluetooth-based Synchronization Network Architecture

Each BAP acts as master station of one of the Bluetooth piconets, serving up to seven Bluetooth devices. Usually, four PDAs are synchronized through an access point. Each BAP is connected through a router-based network to a central BLIP Server. This server is responsible for monitoring the access points, pushing information to Bluetooth synchronized PDAs, collecting and processing the information, and storing the information in databases.

The Bluetooth-based synchronization system replaces a written, paper-based, time consuming process, where the information requested and collected was provided manually. In a transitional phase, the PDAs were synchronized using a wired Ethernet-based network. The selection of a Bluetooth-based solution was based on the advantages of using a wireless solution with low power consumption. An additional deciding factor was the high level of automation that is built-in the Bluetooth synchronization process.

Another case study of Bluetooth application in the health care industry is offered by the Norwegian Center for Telemedicine, as implemented at the University Hospital in Tromso. The network supports automatic collection of blood glucose levels from diabetic patients and transmission of this information to the hospital data center, physicians, and parents, if necessary. A high level architecture of this network is shown in Figure 12.4.

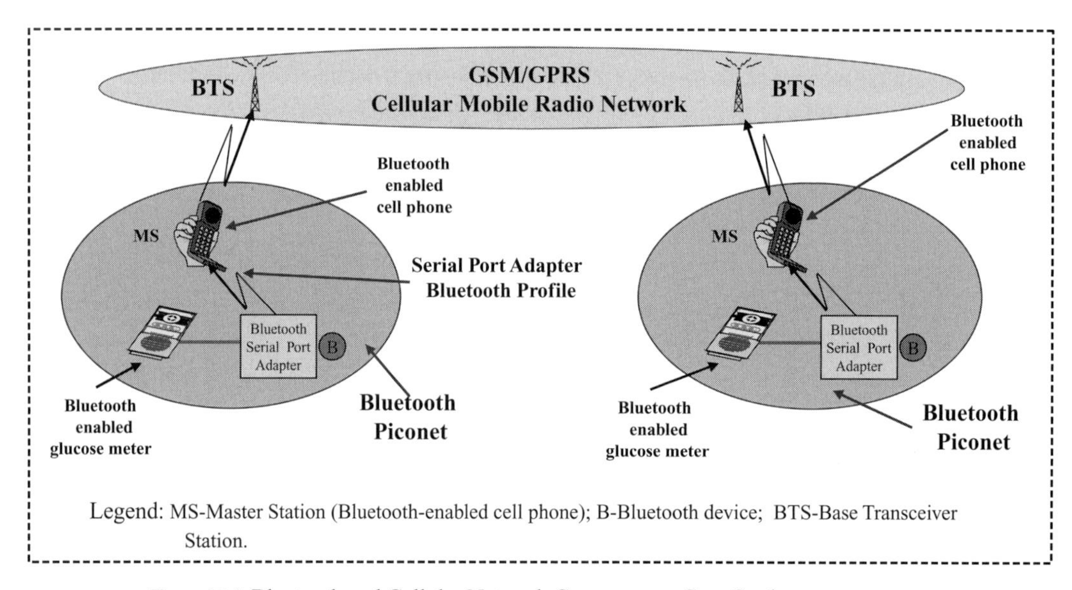

Figure 12.4 Bluetooth and Cellular Network Convergence Case Study

Each blood glucose meter is connected to a small Bluetooth-enabled serial port adapter (RS 232 cable replacement) that is connected to a Bluetooth-enabled mobile phone. The phone acts as the master station of the Bluetooth piconet. A special application on the mobile phone incorporates the test results into an SMS message that is sent to the hospital data center via the GSM/GPRS global cellular mobile network. ConnectBlue AB, Sweden, provides the Bluetooth network and applications [59].

12.6 ZigBee Networking Standards

The first set of ZigBee, IEEE 802.15.4 standards, dealing with Physical Layer and MAC Sub-Layer, was adopted in 2004. However, the large scale implementations envisioned

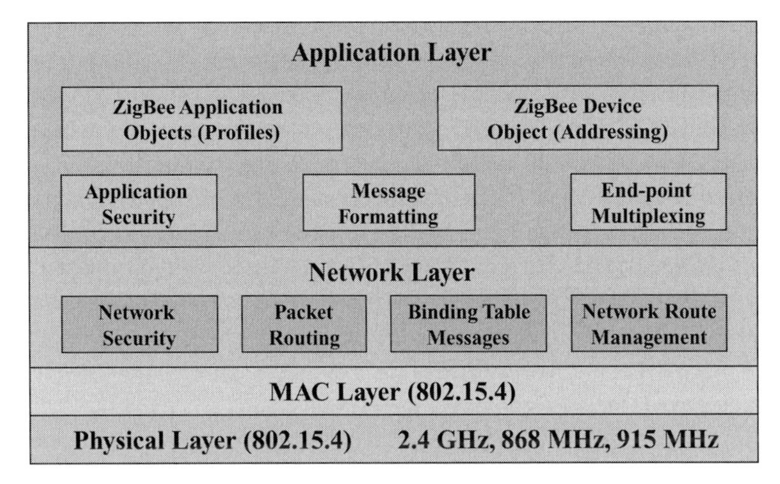

Figure 12.5 ZigBee Protocol Stack Architecture

for this technology have been delayed because of a lack of common understanding in designing mesh networks. These networks require appropriate routing protocols and a mechanism to efficiently couple application profiles with the lower layers of the ZigBee protocol stack. Many of the shortcomings of ZigBee version 1, 2004, have been addressed in an enhanced version called **ZigBee Professional**, scheduled to be ratified in 2007 [60].

The high-level ZigBee protocol architecture is shown in Figure 12.5. The protocol stack, as presented in this diagram, based roughly on the OSI model, bundles communication functions into four layers: Physical, Data Link, Network and Application. We have already analyzed the **Physical Layer** and the Data Link **MAC Sub-Layer** in greater detail in Chapter 7, Section 7.11. Here, we will examine details of the Network Layer and Application Layer that are relevant to networking in various mesh configurations (star, tree, and clusters). This stack should be supported by all the ZigBee network components, from end nodes, to routers, to ZigBee coordinators.

Four major modules provide the functionality supported at the **Network Layer**: packet routing, multi-hop routing management, security, and message binding. Binding is a process that allows pairing of compatible end units to support automation. For example interaction between end units that support switching a light on and off or keeping a door open or closed. The ZigBee coordinator keeps the master look-up binding table to simplify the firmware of end units. The binding table is abstracted using Key-Value Pairs (KVPs). The end units know only the value provided by the coordinator in a Get Request and Send Reply exchange of messages [61]. Since the exchange of information takes place between nodes supporting a combination of point-to-point, point-to-multipoint, and multipoint-to-multipoint links, architected in several levels of cluster and tree nodes, a message broker is required to facilitate the packet routing.

The **Application Layer** consists of two sub-layers. The bottom modules are part of the Application Support Layer and sit between the Application Layer and the Network

Layer. They provide "Presentation Layer" type of functions such as message formatting, end-point multiplexing, and additional security functions.

The Application Layer includes ZigBee Applications Objects (ZAO) and ZigBee Device Objects (ZDO). The applications objects run on the end-points and are software applications that allow devices to perform the desired function. A ZigBee product includes application objects according to a profile. Multiple application profiles can be supported on the same device, hence the need of proper addressing. ZigBee Device Objects provide device management, and these are the objects called when the device is discovered and binding is requested. ZDOs are resident on all ZigBee nodes.

Addressing comes into play when a device requests an association as part of the discovery process. In the 2004 standards [62], a tree structure was adopted where addresses were allocated as the device joins the tree. However, this schema is not scalable in large deployments because of the risk of reaching the maximum allowed number of devices on a particular branch. Therefore, in the latest version of Professional ZigBee, 2007, a random allocation of addresses was adopted.

Another line of improvement of the current ZigBee standard was to allow multiple routing algorithms to be used in mesh configurations. The **table-based algorithm** establishes routes from gateway nodes to each node (routers and end nodes). This is applicable for networks with fairly static devices. The other **any-to-any algorithm** involves any set of devices that must communicate across the network. Since some end-devices do not support the same data rate in both directions of communication, another improvement was made to properly identify these asymmetric links.

To perform a particular task, each type of ZigBee application requires a series of KVP values defined and stored by the coordinator. These KPVs are grouped together and form a profile for a specific application. This allows development of applications that can interoperate with ZigBee products from various vendors. For example, the Home Control – Lighting profile includes clusters for turning lights on and off, and for setting dimming levels. Similar profiles can be developed for home and building automation (lighting, HVAC, surveillance) and industrial automation. The ZigBee Alliance tracks and adopts public profiles to encourage development of compatible ZigBee devices.

To better understand the ZigBee technology, we can look at the interactions that take place between ZigBee network components when a new device is installed and attempts to join the network. The registration process starts automatically when the new device sends a beacon request, i.e. a very short signal to make its presence known but is still saving power. This request is intercepted by ZigBee routers and by the ZigBee coordinator, which are always "on", listening to the network and responding in kind by issuing their beacons. The beacons are "wake up" calls to which the end nodes are required to respond. The device will select a beacon and will send an association request to join the network to the router or coordinator. The binding router or coordinator will respond, sending an association response to indicate acceptance or rejection of the new device. Because of its limited capacity to handle new devices, the payload in the association response will show full acceptance from the router or coordinator, or partial acceptance from the router but not from the coordinator. Once the new device is accepted it will issue an acknowledgment. There are different types of acknowledgements depending on the type of control or data messages.

12.7 ZigBee Convergent Applications

In Chapter 7 we listed some applications provided with ZigBee technology but did not separate out those applications that are facilitated by convergence of ZigBee with other fixed or mobile wired and wireless networks. A short list of ZigBee convergent applications follows. Some of these applications will be detailed when we analyze ZigBee convergent case studies in later sections of this chapter.

- ZigBee and cellular mobile GSM/CDMA networks convergence;
- ZigBee and WLAN Wi-Fi networks convergence;
- ZigBee and RFID networks convergence;
- ZigBee and Wireless Sensor Networks convergence (test and measurement instrumentation, alarm and surveillance systems);
- Home automation applications (digital home, green home, appliances);
- Building Automation applications (utilities, surveillance);
- Industrial Automation (plant floor machineries, wireless mobile robots);
- City-wide Utility Management (electrical power, water, gas);
- Health Care (remote telemedicine, remote patient medical monitoring); and
- Tracking Applications (transportation means location, presence).

12.8 ZigBee-based Electrical Power Management Case Study

The advantages of ZigBee technology – low data rate radio communications, ultra low power, and little firmware resource needed, positions this technology for use in those home, building, and industrial automation applications that do not have high memory, data storage, or special security needs. Examples are automatic meter readers, home/building security systems, light switches, smoke detectors, surveillance systems, and a variety of HVAC sensors. Another application is electrical power management based on ZigBee technology [63].

Several cities in Sweden such as Goteborg and Stockholm, and utility companies in California, Texas, and Florida have started projects to automate real-time monitoring and reading of electrical meters. The peak power used as well as the energy consumed can be measured and the customer billed accordingly. Because it eliminates non-real-time, manual readings by agents and it can go as far as shutting down power in case of disasters or imminence of power black-outs, this automation is an important cost saving solution. The advantages of ZigBee-based networking are augmented by the self-configuration and self-healing capabilities of this technology. A high-level architecture of ZigBee-based metering infrastructure is presented in Figure 12.6.

Each power meter represents a ZigBee node. The power metering network is essentially a cluster of ZigBee end-nodes, repeaters, and ZigBee routers with one ZigBee

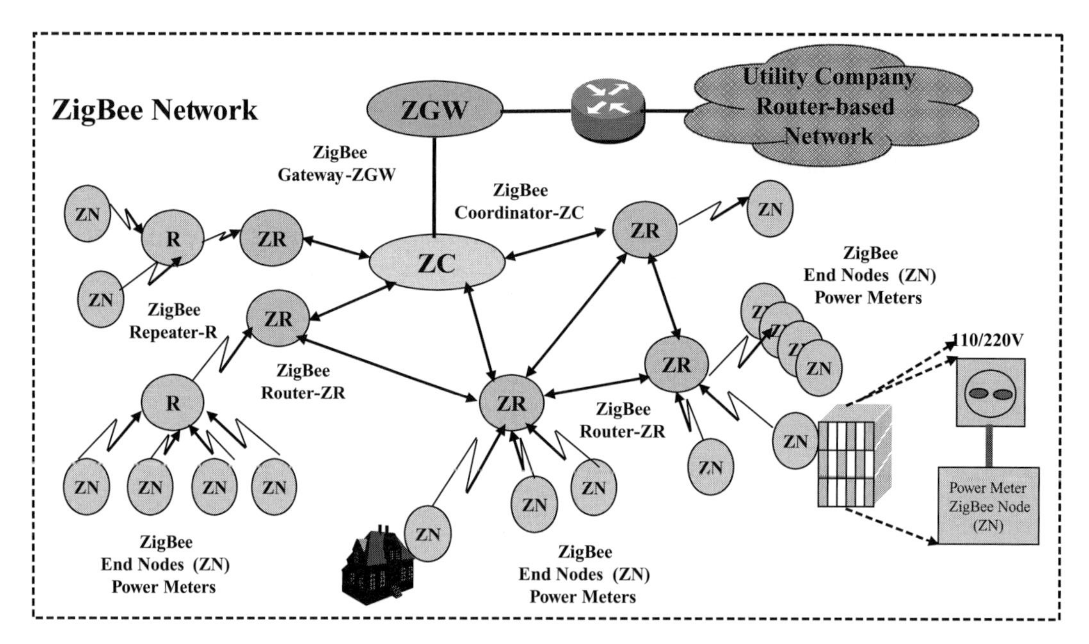

Figure 12.6 Networking Case Study: ZigBee-based Electrical Power Management

coordinator. It provides, through a ZigBee gateway, a link to the utility LAN/WAN router-based data network. The mesh architecture follows the geographical dispersion of myriads of ZigBee nodes and consists of point-to-point and point-to-multipoint connections. Two way communications between ZigBee nodes and the electrical power management systems provides both monitoring and control capabilities.

The applicability of ZigBee sensor-based networks can be extended to the realm of water and gas systems monitoring. Another group of applications are those related to health care or telemedicine, providing remote medical patient monitoring. An example of such an application is provided by Orange mobile service provider and is based on the convergence of ZigBee and GSM/GPRS mobile networks. This application envisions sensor-based monitoring of medical patients including a "panic call" feature in case of emergency and has similarities with Bluetooth-based health care solutions.

12.9 ZigBee, Wi-Fi, GSM Convergence Case Study

Convergence of WPAN, WLAN, and cellular mobile is a challenge since each of these networks has its own wireless technology, access methods, modulation techniques, and level of mobility. The first step towards convergence is to use the cellular mobile handset to work transparently across cellular mobile, WLAN and WPAN networks. Another aspect is to use the networks as a conduit of information, given the limited range of action of native WPAN technologies compared with WLANs and cellular mobile. An interesting case study is the convergence of ZigBee, Wi-Fi, and GSM/GPRS networks in providing home and building automation. This is shown in Figure 12.7

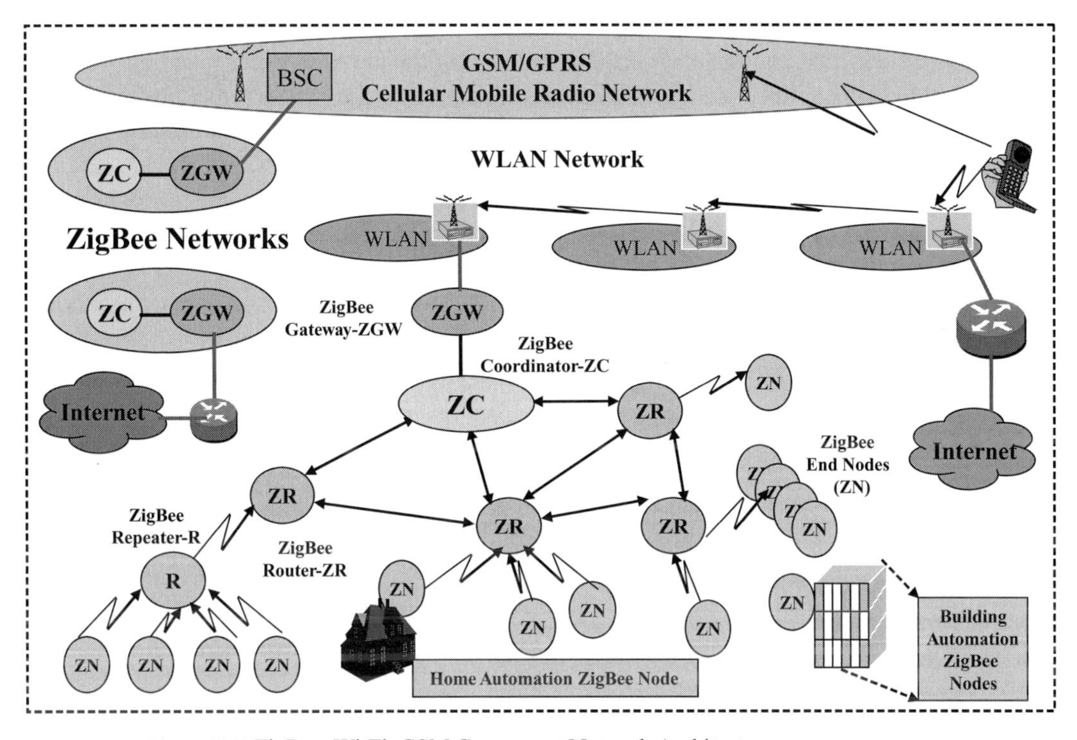

Figure 12.7 ZigBee, Wi-Fi, GSM Convergent Network Architecture

The underlying sensor network is provided by ZigBee end-nodes devices that monitor public utility domains such as electrical, gas, and water distribution. The primary function of these devices is to provide measurement values that can be used for billing purposes. The same devices can have a degree of control by sending alarms when critical conditions are sensed and by shutting down the supply if the conditions call for it.

These ZigBee nodes are connected through repeaters and routers to a ZigBee coordinator. In turn, the coordinator has access through gateways to a higher order network such as municipal or city-wide WLAN mesh network. This approach takes advantage of public mesh network with its higher transmission capacity and broadband connectivity to the public Internet.

In cases where a WLAN infrastructure is not in place, there are two possible scenarios. First is a direct broadband connection of the ZigBee Gateway node to the Internet through an existent or created facility. Second is a direct connection of ZigBee to a Base Station Controller (BSC), taking advantage of the GPRS/EDGE/WCDMA data capabilities of mobile networks.

Two major applications were designed based on ZigBee and Wi-Fi convergence. One is that promoted by LG Electronics, a major manufacturer of mobile networks. The other is from Orange, a major mobile service provider in Europe. Ember Corporation, a leader in ZigBee chipset technology, designed a ZigBee System on Chip that is used by both applications.

The LG Electronics solution is built around the HomeNet networking system. This system provides home automation by monitoring and controlling home devices and incoming utilities via a central communications hub facilitated by a mobile operator. A similar solution is provided by Orange for remote medical monitoring of patients. There, home patients and nursing home residents wear sensors that provide location notifications. Additional information can be transmitted using simple, non-intrusive methods of measuring blood glucose levels. Panic call features are also included in the system.

12.10 ZigBee and Wi-Fi Coexistence and Interference

A major impetus in the development of fixed wireless technologies was the use of unlicensed frequency bands of the spectrum. These bands are known as Industrial Scientific Medical (ISM) bands and are shared by several technologies in the realm of WLAN and WPAN. The most commonly used band is the 2.4 GHz worldwide band, and it is shared by three technologies: WLAN Wi-Fi IEEE 802.11b/g, WPAN Bluetooth IEEE 802.15.1, and WPAN and Sensor Networks ZigBee-based IEEE 802.15.4. These technologies have overlapping short-range coverage from 1m to 100m, immediately raising the question of possible interference when they are used in applications related to home, building and industrial automation. The same 2.4 GHz band is also shared by electrical appliances such as microwave ovens.

Initially, the technologies mentioned above were designed to carry data, but the scope has been widened to also include voice communications. The topology of these networks has also changed from point-to-point directional wireless links to complex multi-hop wireless mesh configurations. Therefore, there is a need to satisfy both voice and data communications QOS requirements. We need to note that each of these technologies use slightly different techniques for network access, spectrum bandwidth allocation, and channelization. And, there are other terms of differentiation such as transmitting power and the duration of transmission that can cause compatibility concerns. For example, the power emitted by a WLAN Access Point is 30 times higher than that of a ZigBee node so there is a real possibility of signal interference. And, the duration of transmission varies widely since WLAN APs send signals with a high duty cycle while ZigBee nodes transmit short-lived beacons, as described earlier. Bluetooth is less vulnerable when working in shorter proximity, helped by the frequency hopping mechanism used in RF access links.

Laboratory test measurements performed in real and simulated environments, opposing 802 11b WLANs to IEEE 802.15.4 ZigBee, have confirmed the concerns regarding the potential interference and the magnitude of this interference [64]. However, the same tests showed ways of avoiding these situations by careful allocation of working channels. Standard commercial equipment was used in these tests.

To explore the degree of interference, a worst case scenario was chosen: A WLAN 802.11b working at maximum admissible power, at maximum possible data rates, and for a prolonged period of time was operated in a proximity of 2 meters with ZigBee devices sharing the same spectrum. Two overlapping channels were selected: channel 6, 2,437 GHz, for the WLAN and Channel 6, 2,440 GHz for the WPAN. The results were conclusive. 90% of the 802.15.4 ZigBee frames were destroyed in the presence of the more energetic 802.11 frames. In short, the interference can be devastating for normal communications and should be avoided.

The solution lies in controlling the transmission pattern of WLAN 802.11b to allow short idle periods between frames; enough to send the very short control frames or even data in the 802.15.4 WPAN. This will require adaptive protocols. Another solution might be a frequency plan that allows use of WLAN frequencies distant from WPAN channels (for example use of WLAN channel 4, centered around 2.427 GHz) that have no interfering effect on WPAN frames. Another solution could be the selection of two WPAN channels, for example 15 and 20, that fall between two neighboring WLAN channels. In addition, there are two other WPAN channels, 25 and 26, at the edge of the spectrum that are not affected by WLAN frames.

12.11 AirBee ZigBee Network Management System

Network management is different, and more complex, when developing solutions for ZigBee. Here, networks consist of clusters of tens, hundreds, if not thousands of ZigBee end nodes attached to micro devices, and, there are very stringent limitations in power consumption and processing power. Therefore, the scale and the topology of ZigBee implementations require management capabilities. High-level requirements are the almost universal ones: the need to discover, to interact, and provide a minimal level of monitoring and controlling capability in a centralized, secure fashion. An example of a ZigBee network management system is shown in Figure 12.8.

AirBee's flagship product, **ZigBee Network Management System** (ZNMS), allows control and monitoring of ZigBee devices, end nodes, and routers, networked in a star, cluster, or in a mesh configuration. Each device employs an **AirBee ZigBee Agent** and an **AirBee ZigBee protocol stack**. An optional superframe structure with beacons for time synchronization and Guaranteed Time Slot (GTS) capability for high priority communications might be included. Up to 65,000 devices can be networked in various topological configurations.

AirBee-ZNMS™ software is a centralized service and network management system. It delivers sophisticated, standards-based management and control of ZigBee mesh networks and proprietary mesh sensor networks. The software is also used to manage gateways interconnecting ZigBee and IP networks. It is the industry's first fully distributed Linux/Unix/Windows-based platform and multi-service manager, providing a comprehensive management solution and enabling integration and interoperability with ZigBee devices.

AirBee-ZAgents run on managed network devices. The AirBee-ZAgents listen for requests from the manager and respond with the appropriate information. AirBee-ZAgents also generate SNMP-based "traps" to identify the nature of any faulty condition that might occur. The browser runs on the client displaying HTML pages. A variety of

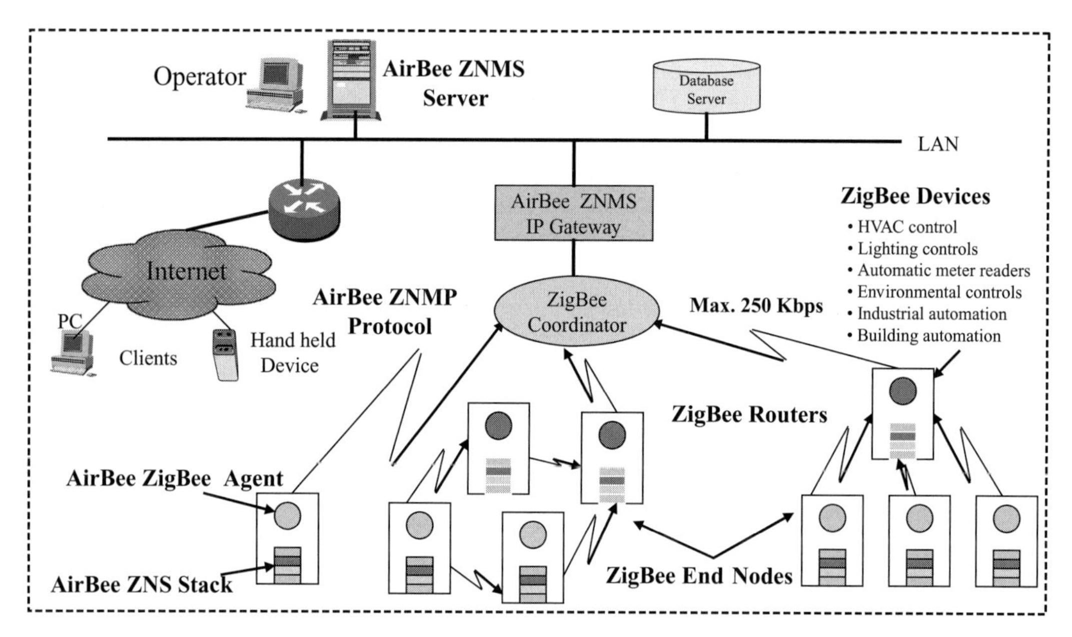

Figure 12.8 AirBee ZigBee Network Management System

clients, including Web browsers and wireless handsets, can access the application layer via the Internet. There are many choices for selecting controllers as provided by Texas Instruments, Atmel, Zilog, Motorola. The main characteristics of these controllers are:

- Choice of any IEEE 802.15.4 compliant radio, including Chipcon, Motorola, Atmel, and ZMD;
- Choice of real-time Operating Systems (OS) including AirBee OS, TinyOS, μCOS, and PalmOS;
- Customizable to the smallest possible code footprint with lowest possible memory (RAM).

At the Network Layer, ZigBee specifies how the sensor network forms (for example a mesh configuration that heals itself, grows, and routes messages). For security, 802.15.4 specifies 128-bit Advanced Encryption Standard (AES) encryption but says nothing regarding how to manage encryption keys. ZigBee Version 1.0 specifies key management.

AirBee-ZProfiles contain libraries that enable system integrators and device manufactures to create wireless applications. AirBee-ZProfiles support ZigBee device features such as binding, device discovery, service discovery, and dynamic configurations.

AirBee-ZNMS™ is a Java based, web-enabled technology. AirBee-ZNMS™ uses the proprietary AirBee-ZNMP protocol, a variant of the standard SNMP protocol tailored to suit the requirements of managing WPAN networks. The AirBee-ZAgent, similar to the SNMP agent, resides in ZigBee devices and communicates with the server using the AirBee-ZNMP protocol.

12.12 WPAN PLC-based Management System

Management of PLC-based mixed PAN and WPAN environments is eased by the fact that all PLC devices are tied to the power distribution lines, so power consumption is not an issue. The same is true concerning the form factor, memory size, and processing power of managed devices and management agents attached to them. An example of a PLC-based network management architecture is shown in Figure 12.9.

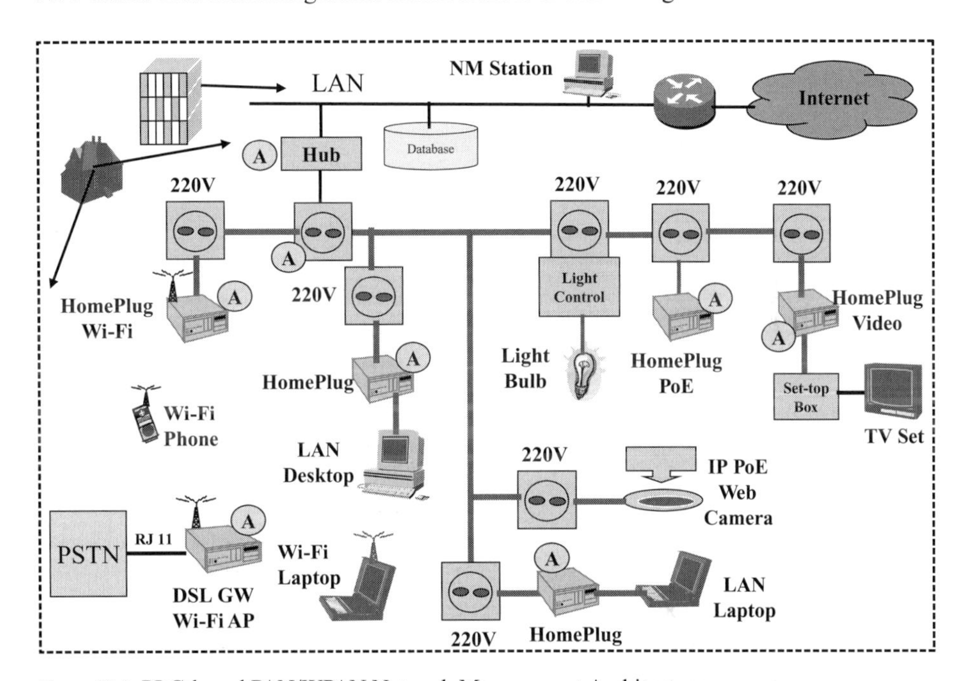

Figure 12.9 PLC-based PAN/WPAN Network Management Architecture

Network management means that an SNMP agent, providing advanced monitoring and controlling capabilities, can easily be implemented along with each Homeplug device. The need for management is even greater when PLC systems provide services to multi-tenant units or hospitality suites. A common alternative to SNMP-based network management is basic Web-based management that is suitable for networks with less than a hundred Homeplugs.

A network management station plays the SNMP Manager role. To manage the PLC network, each Homeplug device contains an SNMP Agent. Standard MIB, RMON, and enterprise-specific MIBs, developed by each equipment manufacturer, allow both monitoring and controlling of each individual device. The control function of management architecture is important when active components such as heaters and light switches are part of the network.

An example of a network management product designed to manage devices that use electrical power lines and coaxial cables is the SpiDMonitor Remote Control Application from SPiDCOM Technologies. The monitor supports SNMP-based interfaces and operates on Windows and Linux platforms. SpiDCOM devices are equipped with SNMP agents. The enterprise-specific MIB provides insight and access to internal physical parameters such as modulation schemes and pilot frequencies that allow the powerline devices to be notched on any of 896 carriers to minimize interference with other radio sources. Bit Error Rate (BER) and Signal to Noise Ratio (SNR) values are collected and graphically displayed. The management product is not limited to SPiDCOM devices but is usable by other PLC developers. The product is incorporated in AvistoTelecom company's NemSIS Configurator Lite management system, which is used to manage larger PLC networks.

12.13 The Concept of Femtocell

Another advanced concept in home network automation is the **femtocell**. In this case the users, whether home, small and medium sized businesses, or large enterprises, will have their own Base Station Transceiver (BTS), the femtocell [65]. Each location will then represent a small micro cell access point in the constellation of a myriad of cells. The high-level functionality built into femtocells will resemble that of a mobile BTS. However, one major difference will be the radiating power of femtocells.

Femtocells will be relatively small indoor boxes accessed through a mobile phone in house or in the proximity of the house. We are not talking about big, expensive RF base stations. The same convergent handset that works in a standard mobile network is handed over to the femtocell once the user is in the transmission range. In this case, there is no need for dual-mode convergent handsets or cell phones. The femtocell acts as a mobile base station and is connected to the public data network using the broadband access connectivity of that location. One example of a femtocell is Samsung's Ubicell, currently employed by Sprint/Nextel as part of an initial test of this concept. The service called Airave is being trialed in Denver and Indianapolis areas. The service that lets users make unlimited calls from home is priced at $30 per month per family. This price comes on top of the subscriber's regular mobile plan.

To further their use, a not-for-profit **Femtocell Forum** was formed. The initial founders included the Airvana, NetGear, Tatara Systems, ip Access, Ubiquisys, and picoChip companies. The forum's mission is to advance the development and adoption of femtocell products and services as part of a collaborative effort of mobile operators, telecoms hardware vendors, and software vendors.

ABI Research analysts predict 70 million femtocell installations by 2012, with 140 million users. Another forecast from In Stat predicts 40.6 million installations by 2012 and 103 million users. Strategy Analytics predicts, for the same period, 88.4 million users, mostly concentrated in Europe and North America.

Both cellular mobile operators and WiMAX technology promoters have a great interest in this technology because femtocells will address indoor coverage, a weak point of WiMAX technology. Sprint/Nextel, T-Mobile, Vodaphone, Softbank, Telefonica, O2, and France Telecom have issued requests for proposals to infrastructure vendors. The list of potential vendors and integrators of femtocells includes: Alcatel-Lucent, Sony Ericsson, Huawei, Motorola, Nokia Siemens, and Samsung. PicoChip UK, a company specializing in providing arrays of fully programmable elements for signal processing, is one of the leading chipmakers in the effort to make the femtocell a ubiquitous consumer electronic fixture.

13 Wireless MAN Cellular Mobile Convergence

13.1 WiMAX Mobile Convergent Applications

In Chapter 8 we listed the applications carried over WiMAX networks but didn't separate out those that are facilitated by convergence of WiMAX with other fixed or mobile wired and wireless networks. A short list of WiMAX convergent applications follows. Some of these applications will be detailed when we analyze WiMAX convergent case studies in subsequent sections of this chapter.

- WiMAX and cellular mobile convergence (extension of reach and connectivity of cellular mobile networks to remote, rural areas);
- WiMAX and WLAN networks convergence;
- WiMAX and WLAN metro mesh networks convergence;
- WiMAX, WLAN, and WPAN networks convergence;
- Backhaul connectivity for cellular mobile networks and traffic;
- Backhaul channel for wired Internet Service Providers;
- Disaster recovery using WiMAX links as wireless channel alternatives;
- WiMAX and EPON (as a backhaul link) networks convergence;
- WiMAX-based broadband location/navigation and customer search; and
- Internet Protocol Television (IPTV) and High Definition Television (HDTV).

13.2 WiMAX and Internet Protocol Television

IPTV, one of the applications listed above, is considered the killer application in both the fixed wired public Internet and cellular mobile networks. WiMAX, as a broadband wireless access network in metropolitan areas, is a good candidate to support IPTV. This support will include video on demand, live content, and multicast video communications, either as managed (paid TV) or unmanaged service (You Tube, for example).

WiMAX technology is strengthened by the degree of mobility conferred by the newly adopted IEEE 802.16e standard. Also, WiMAX, as a point-to-multipoint technology, has built-in mechanisms for allocation of bandwidth on demand (reservation-based) and QOS management based on four QOS classes and associated parameters. This was presented earlier in Table 8.3, Chapter 8. QOS management specifically addresses the time sensitive video applications. In addition to these technical aspects, there are the economics of WiMAX implementations that do not require an existent infrastructure or stringent direct line-of-sight communication.

This support for IPTV comes with drawbacks because of the diversity of natural fading conditions in WiMAX transmissions. Use of WiMAX is also limited by the introduction of HDTV since much higher bandwidth is required by this application compared to existing ones. When associated with multiple TV sets and multiple PCs, bandwidth demand can be as high as 20 Mbps. That bandwidth should be compared with the highest bandwidth delivered by a WiMAX BS, which is around 60–70 Mbps, in good conditions.

The sources of IPTV video content are Internet Service Providers (ISPs) specialized servers. The ISPs are independent organizations, or major telecom or cable operators that own the core network or the access network. In most cases, video signals are encoded at the source using video coders. In the process, the signals are compressed with technologies such as MPEG2 or MPEG4 and encrypted to protect the content. A Set-Top-Box (STB) is required on customer premises to decode/decompress the video signal. Use of services and the billing aspects of video delivery are controlled in this way.

13.3 Multi-mode WiMAX Mobile Convergent Terminals

The WiMAX Forum has published a Certified Product Registry that lists products that are based on IEEE 802.16-2004 (fixed WiMAX) standards. In the second half of 2007, several laboratories were set up to begin certification of products based on the mobile WiMAX 802.16e-2005 specifications.

Certified WiMAX Products

- **Airspan Networks**: MacroMAX BS, MicroMAX SoC BSR BS, EasyST SS, (www.airspan.com);
- **Alvarion**: BreezeMAX SI SS, BMAX PRO-S CPE SS, MACRO MODULAR BS, MICRO MODULAR BS, (www.alvarion.com);
- **Aperto Networks**: PacketMAX SS, Packet MAX 3000, PacketMAX 5000 BS, ProST SS subscribers units, (www.apertonet.com);
- **Axxcelera Broadband Wireless**: ExcelMAX BS, ExcelMAX CPE SS, (www.axxcelera.com);
- **E.T Industries**: Appolo BS, Appolo SS, Appolo MAX-SU, (www.etiworld.com);
- **Nokia/Siemens**: WayMAX@Advantage BS, (www.nokia.com);
- **Proxim Wireless**: Tsunami MP 16 3500 BS (3.5 GHz), (www.proxim.com);
- **Redline Communications**: RedMAX BS, RedMAX SS, (www.redlinecom.com);
- **Selex Communications**: YSEMAX BS and SS, (www.selex-comms.com);
- **Sequans Communications**: SQN 1010-RD (FDD) SS, SQN 2010 BS, (www.sequans.com);
- **Siemens AG**: Gigaset SE461 WiMAX SS, (www.siemens-mobile.com);
- **SR Telecom**: SSU5000 SS, SYMMETRY BS, (www.srtelecom.com);
- **Telsima**: StarMAX 2140-3.G SS, StarMAX 4120-3.5G BS, (www.telsima.com);

- **WaveSat** Inc.: miniMAX 3.5G (TDD) SS, miniMAX 3.5G (FDD), (www.wavesat.com).

WiMAX products are based on WiMAX processors, most of them System on Chips (SoC). SoCs can be used for a variety of subscriber terminals such as PCMCIA cards, Compact Flashes, USBs, and WiMAX Subscriber Stations as well as Base Station Channel Cards, Systems on Board, Mini Base Stations, and Pico Base Stations.

WiMAX Systems on Chip
- **Atmel Corporation**: WiMAX chipset, (www.atmel.com);
- **Beceem Fierce Wireless**: First 802.16e chipset, (www.fiercewireless.com);
- **COMSYS Communications**: COMAX processor, (www.comsysmobile.com);
- **Denali Software**: BCS200 chipset for WiMAX, (www.denali.com);
- **Fujitsu**: WiMAX MB87M3400 Forum Wave 2 SoC, (www.fujitsu.com);
- **Intel**: Centrino Mobile-based platform, WiMAX/Wi-Fi chipset, (www.intel.com);
- **Next Wave**: NW100 WiMAX series of chipsets, (www.nextwave.com);
- **Runcom Technologies**: SoC and BS System-On-Board, (www.runcom.com);
- **Sequans Communications**: WiMAX chipsets Wave1 & 2, (www.sequans.com);
- **TeleCIS Wireless**: TCW1620 (fixed), TCW2720 (mobile), (www.telecis.com);
- **Telsima**: StarMAX 2140–3.G SS, StarMAX 4120–3.5G BS, (www.telsima.com).

WiMAX PCMCIA Vendors
- **AWB Networks**: AWB PC100 WiMAX PCMCIA, (www.awbnetworks.com);
- **Accton**: WM 8911BE/2.3G, WM8021/2.5G, 8931BE/3.5G, (www.accton.com);
- **Broadband Wireless**: PCMCIA cards for WiMAX, (www.bbwexchanage.com);
- **Cameo Communication**: WiMAx MIMO PCMCIA, (www.cameo.com);
- **Intel**: WiMAX PCMCIA card, (www.intel.com);
- **Motorola Corp**.: (NextNet Wireless) PCMCIA WiMAX, (www.motorola.com);
- **Navini Networks**: PCMCIA WiMAX 802.16e, (www.navinini.com);
- **Sequans Communications**: WiMAX Wave 2 Express card, (www.sequans.com);
- **Telsima**: StarMAX 3200 PCMCIA for laptops and PDAs, (www.telsima.com);
- **WaveSat Inc**.: U Mobile Access Card, (www.wavesat.com);
- **ZyXel Communications**: MAX-100 PCMCIA, (www.zyxel.com).

The maturity of WiMAX technology is also measured by the sophistication of testing systems on chip, reference designs, and full-blown WiMAX products:

List of WiMAX Testing Products and Testing Organizations
- **Agilent Technologies**: N4010A Wireless Connectivity Test Set; WiMAX Optimization Solution for 802.16.e, (www.agilent.com);
- **Azimuth Systems**: WiMAX Tester, (www.azymuth.com);
- **Innowireless**: Wibro Mobile WiMAX SS Tester, (www.innowireless.com);
- **Seasolve Software**: SeaMAX Analyzer/Generator, (www.seasolve.com);
- **Spirent Communications**: SR5500 WiMAX tester, (www.spirentcom.com);
- **Tektronix**: K127-G35 Protocol Analyzer, (www.tektronix.com);
- **AT4 Wireless Inc**.: Cetecom Spain test lab, (www.at4wireless.com);

- **Rohde and Schwartz**: 1MA97 WiMAX signal generator and analyzer, (www.rohde-schwarz.com).

A good example of a convergent WiMAX terminal is the Samsung SPH-P9000. It is a PDA-like device that supports wireless communications and Internet access via mobile WiMAX. It includes multimedia features such as Direct Mobile Broadcast (DMB), photo camera, and interactive games. SPH-P9000 has a Qwerty foldable keyboard and a 30 GB hard drive to store and play music, multimedia files, and video downloads. Nicknamed Mobile Intelligent Terminal (MIT), it incorporates CDMA mobile communications and is considered the first convergent WiMAX mobile phone terminal.

13.4 Convergent WiMAX Cellular Mobile Network Architecture

From the beginning of the WiMAX development, the viability of this technology was linked to the convergence of WiMAX and cellular mobile networks. This convergence included a functional connectivity between the two networks, where WiMAX acts as a long-range complementary arm that brings remote, rural areas into the fold of cellular mobile coverage. A typical WiMAX cellular mobile convergent architecture that supports both fixed and mobile wireless versions is shown in Figure 13.1.

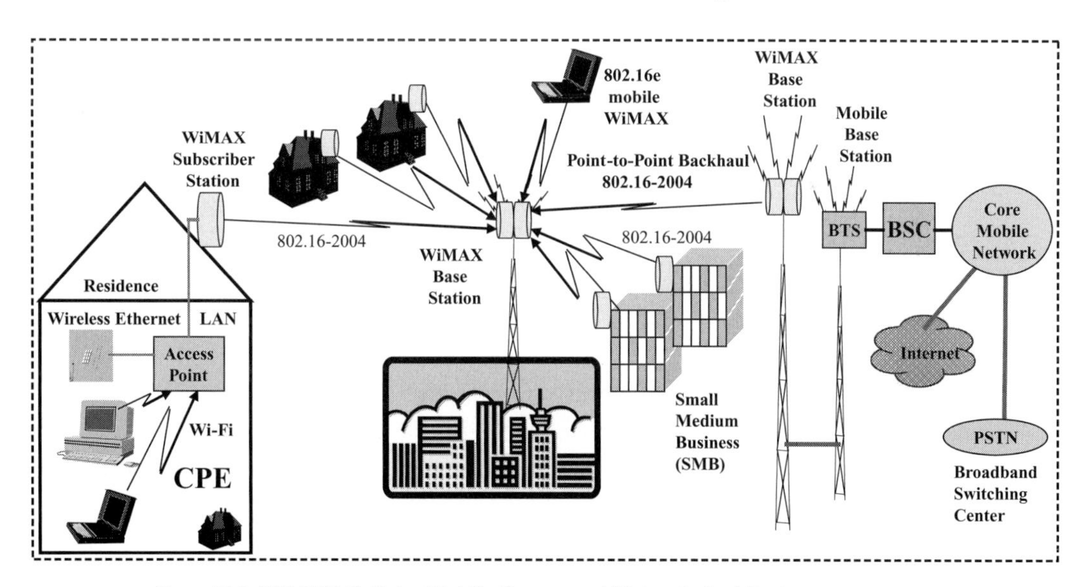

Figure 13.1 WiMAX Cellular Mobile Convergent Network Architecture

Fixed and mobile WiMAX links connect WiMAX CPE Subscriber Stations (SSs) to WiMAX Base Stations (BSs) through air interfaces based on the WiMAX 802.16-2004 and 802.16e-2005 standards. Both indoor and outdoor SS units are supported. A wireless point-to-point backhaul connection to another BS uses one of the WiMAX air interfaces. The WiMAX BS can be collocated with one of the cellular mobile BTSs/BSCs to provide direct connectivity between the two networks.

Another scenario of WiMAX cellular radio fixed-mobile convergence is for WiMAX BS to act in a dual role. This means, it functions as the WiMAX BS in its relationship with the constellation of SSs supported, and, plays the role of a BTS and the radio cell when associated with the wide area mobile cellular networks. This solution assumes that the functional capability of the cellular network does not impede WiMAX's performance as a high-speed broadband wireless solution. The cellular mobile network provides connectivity to a broadband switching center and from there through cable or fiber links to public Internet and the public telephone networks.

This convergence is questionable since WiMAX and 3G/B3G cellular mobile networks are competitors, with the same goals of providing broadband services to customers. An interesting case of promoting this convergence comes from Sprint/Nextel, one of the major cellular/mobile wireless carriers in the USA, and holder of major chunks of the 2.5 GHz WiMAX frequency band. Sprint/Nextel is the sole carrier betting on WiMAX cellular convergence to the point that it considers WiMAX mobile as the centerpiece and holder of future promises of "4G" mobile networks. This convergence is exemplified by the Sprint/Nextel's ultra-optimistic plan to provide, as early as 2008, WiMAX broadband commercial services in the USA in major geographical and metropolitan areas. The areas not covered by WiMAX, will be covered by 3G EV-DO Revision A cellular mobile upgrades. It is planned to reach 100 million plus potential subscribers by 2010.

This plan follows an earlier attempt by Sprint/Nextel to align all major cable operators in the so called Pivot Joint Venture (Pivot JV) behind its approach. The major thrust of this alignment with cable operators such as Comcast, Cox Communications, and Time Warner Cable, was to develop and share convergent wireless 3G and 4G technologies under the banner of fixed-mobile convergence. In this case, "fixed" means the wired cable world and "mobile" is essentially the cellular radio world and the new addition, the mobile WiMAX. This should be clearly understood, since the FMC term, as introduced in this book, is used differently by different organizations and people.

13.5 WiMAX and Ultra Wide Band Convergence

IEEE 802.16e mobile WiMAX has paved the way to bring terminals such as laptops, mobile handsets, and even PDAs into metropolitan area coverage. The drivers of this development are the high-speed broadband services provided over WiMAX/WMAN technology. As presented in Chapter 9 of this book, a separate development, Ultra Wide Band (UWB) technology, provides very high data rates over relatively short distances. The most common UWB application is the Wireless USB (WUSB), allowing communication at close proximity at speeds up to 480 Mbps.

In the USA, the spectrum allocation for WiMAX (2.4 GHz, 2.5 GHz, 5.8 GHz) and UWB (3.168 to 4.752 GHz) do not overlap, so WiMAX and UWB can coexist in the same area. However, overlapping frequency allocation of WiMAX (3.5 GHz band) and UWB (3.168 to 4.752 GHz) in other countries (Europe and Asia) is a major issue. Therefore, global compatibility of WiMAX and UWB will require a cognitive radio

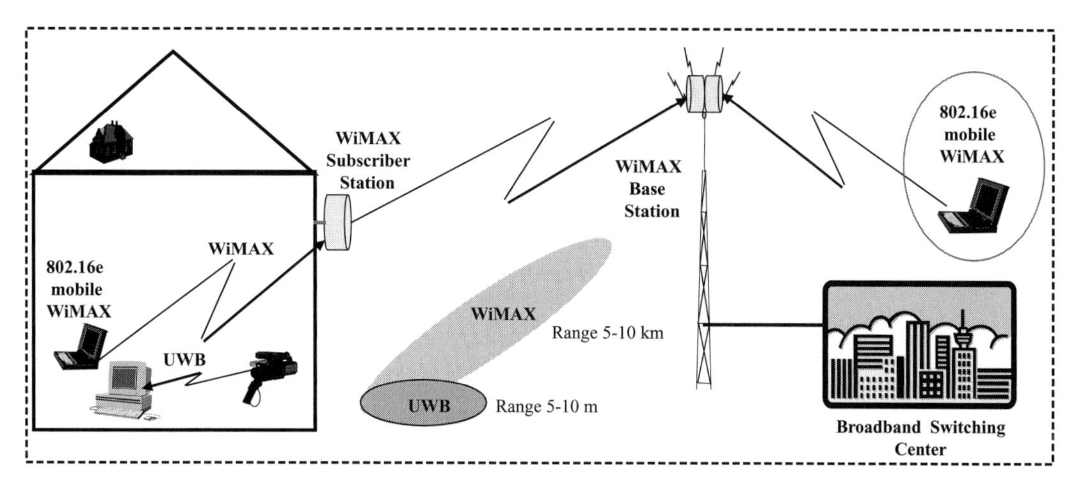

Figure 13.2 WiMAX and Ultra Wide Band Convergent Network Architecture

Detect and Avoid (DAA) mechanism for collocated users. This situation is depicted in Figure 13.2 [66].

Convergence of these two types of technologies is achieved at the level of mobile terminals such as laptops that are equipped with both WiMAX and UWB capabilities. This means that the laptop can act as a standalone WiMAX Subscriber Station (SS) and a WUSB host in its communications with other WUSB equipped devices such as digital cameras/video cameras and storage devices that store multimedia and music files. Mass implementation of WiMAX and UWB WUSB on laptops is expected in 2008.

Devices such as Wireless USBs, acting as secondary users, must detect mobile WiMAX terminals such as laptops or computers, that are acting as primary devices. The DAA mechanism is part of a larger technique/technology known as cognitive radio. It allows the use, by secondary users, of a primary user's licensed spectrum, when it is idle or not used at all. The secondary users should implement a DAA mechanism to avoid disruptions and to make the two classes of users compatible. Such a mechanism should be implemented when overlapping occurs between WiMAX and UWB, making convergence possible and the two technologies complementary.

13.6 WiMAX and Fixed Wired EPON Convergence

The material presented so far has been confined to fixed-mobile convergence in the world of wireless communications, the subject of this book. However, convergence can also take place between various wired and wireless communications networks. One such example is the integration of WiMAX and Ethernet Passive Optical Network (EPON).

EPON technology evolved from the Passive Optical Network (PON) by adopting the main characteristics of Ethernet access methods and transmission protocols. EPON provides symmetric downstream and upstream transmission bandwidths of 1GHz. The commonality between these two technologies is that they both provide broadband access

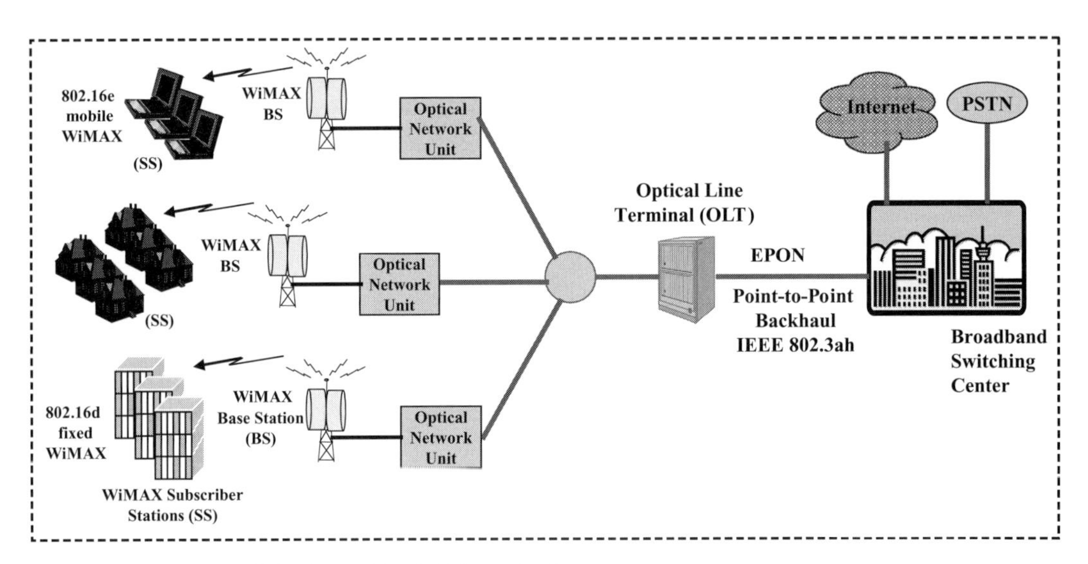

Figure 13.3 WiMAX and EPON Networks Convergence

solutions. The interesting part of this integration is that EPON takes the role of a backhaul networking technology to connect dispersed WiMAX Base Stations (BSs) [67].

One of the applications of WiMAX technology is to provide wireless broadband access to dispersed users in rural and remote areas. However despite all the advancements in optical fiber-based technologies, it is still very expensive to provide Fiber (optic) To the Home (FTTH). It is even less cost-effective if subscribers are geographically dispersed over large areas. Thus, the economic solution is to provide broadband wireless access through WiMAX to remote and dispersed customers. In this case, EPON is used as a long backhaul technology to connect the integrated network to a public switching center. The convergent WiMAX/EPON architecture is shown in Figure 13.3.

In this configuration, the Optical Line Terminal (OLT) maximum bandwidth is split to support a maximum of 16 Optical Network Units (ONUs), each with approximately 60 Mbps bandwidth. This matches the maximum bandwidth supported by WiMAX technology, around 70 Mbps, both downstream and upstream.

The OLT does ONU bandwidth allocation using a request/grant mechanism. All the control messages and management messages are Ethernet MAC frames. The downstream messages (from the OLT to the ONU) are tagged using a Logical Link ID to select the proper ONU and the associated WiMAX BS. Upstream messages (from the ONU to the OLT) are aggregated and sent during time slots granted by the OLT. The ONU and WiMAX BS can be distinct units with their own protocol stack or can be combined in one physical unit. Because the users may move across different WiMAX BS boundaries, these hybrid units should provide a handover mechanism. The handover mechanism will help facilitate WiMAX mobility.

EPON technology, adopted in 2004, is specified in the IEEE 802.3ah standard. Commercial products designed and manufactured according to these specifications support applications such as analog video telephony, high-speed Internet access, videoconference, gaming, and video broadcast (downstream). An EPON standard counterpart is known as ITU-T G.984 or Gigabit Passive Optical Network (GPON). Currently, one of the IEEE Task Forces is working on specifications for the 10GB EPON (GEPON) technology, expected to become the IEEE 802.3av standard in 2009. GPON serves as the backhaul technology for the next generation of wireless networks such as those defined in IEEE 802.11n standards. GPON provides up to 100 Mbps bandwidth for each mobile terminal [68].

13.7 WiMAX, Wi-Fi, and RFID Convergence Case Study

Broadcasting of ad-hoc sporting events that take place in environments with no communications infrastructure is the stage where WiMAX, in tandem with other wireless technologies, can demonstrate its versatility. A good example is live broadcasting of the Triathlon World Championship ("Ironman") competition. The races take place in rugged terrains where the athletes have to swim 2.4 miles, bike 112 miles, and run a full 26.2 mile marathon course, all in one day.

To provide a live high-quality video broadcast and web-cast of these events, beyond the starting and finishing line, communications facilities must be dispersed along the racing course with the ability to transmit data, video image footages, voice, and sound. Real-time data is critical in tracking each athlete's progress along the route and all the information must be relayed to a data center with its associated databases. At the core of this network infrastructure is the WiMAX long-reach, wireless broadband technology. Convergence of WiMAX with Wi-Fi WLANs and RFID WPANs, is shown in Figure13.4 [69].

Airspan Networks' AS3030 WiMAX Base Station (BS) and a constellation of relaying WiMAX Subscriber Stations (SS) provide the 8 Mbps conduit required by high-quality video streams. The video transmission sources are several Sony IP TV fixed video cameras capturing video footages along the racing course and images shot by moving motorcycles and helicopters. A wired backhaul link provides connectivity to a broadband switching center and from there to the Internet and PSTN.

Several Wi-Fi hot-spots, equipped with Cisco AirNet Access Points, are installed along the racing course. Equipment at these points includes Wi-Fi-based Dell notebooks, dual-mode Wi-Fi and Bluetooth handsets, Wi-Fi and Bluetooth PDAs, and Dell's Axim handheld devices. These devices can also download or provide additional information since the hot-spots are connected via WiMAX SS to the WiMAX BS and from there to the public network.

Several RFID timing mats equipped with Intel's Xscale RFID readers with Wi-Fi capabilities are used along the course to monitor each athlete's progress. Since the athletes are equipped with ankle bracelets containing RFID tags, exact timing information is transmitted via the Wi-Fi network and WiMAX connectivity to the data center. In case of emergency situations, the RFID tags also provide instant access to medical data and contact information for a particular athlete.

All this information from the race: live footage, pre-event interviews, real-time commentaries, and database information containing tracking time information is edited in

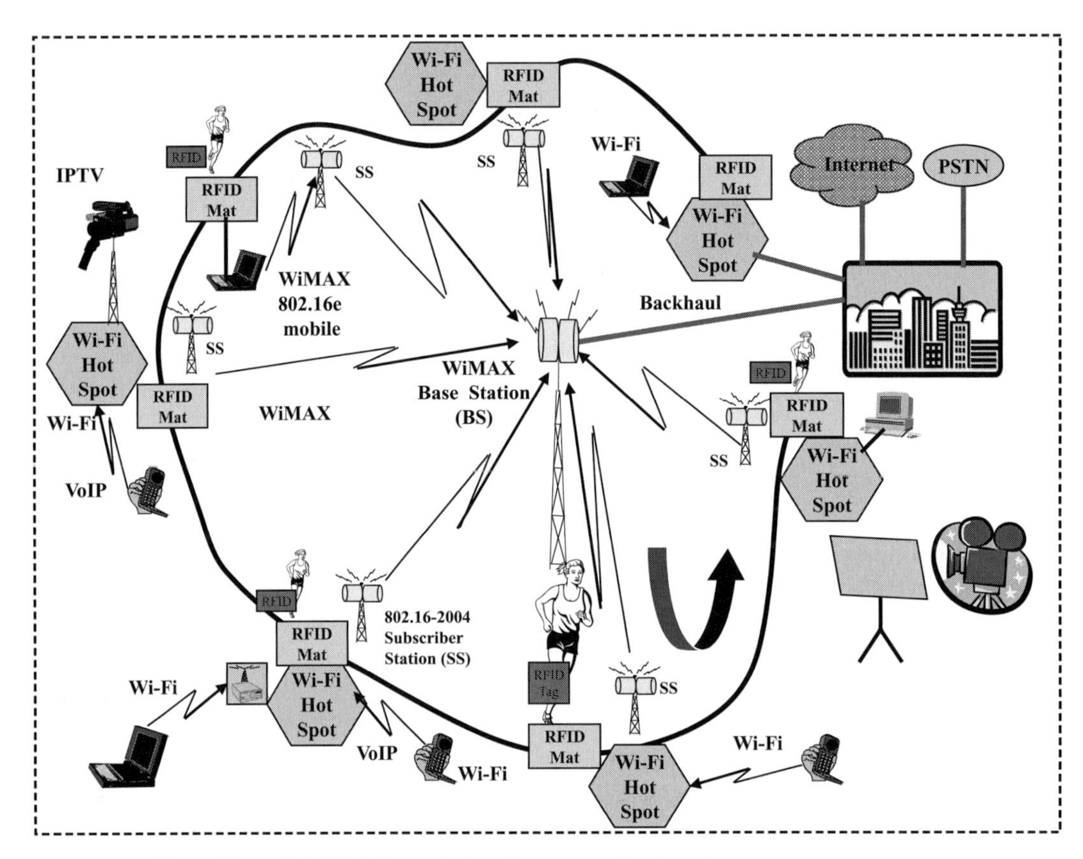

Figure 13.4 WiMAX, Wi-FI, and RFID Convergent Network

the studio-like data centers. It then becomes the source of a live web-cast that runs on the **ironmanlive.com** site and feeds video data servers belonging to major broadcast operators.

This case study demonstrates the capabilities of WiMAX technology in providing channels for broadband video, VoIP, and high-speed data for ad-hoc, short-term events such as sporting events and other events that take place over large geographical areas. It shows the applicability of WiMAX in rural, suburban, or remote areas without requirements for cable or DSL high-speed Internet connectivity, and the technical applicability of integrating several fixed and mobile wireless convergent technologies.

13.8 WiMAX and Metro Mesh Convergence Case Study

The wireless convergent network architecture presented in Chapter 10 shows that WiMAX, as the representative WMAN technology, borders the cellular mobile WWAN world on one hand and the WLAN/WLAN Mesh on the other. Both adjacent networks are, in a way, competing technologies in a given metropolitan area. This is evident in terms of coverage. Cellular mobile networks natively have been designed with full metropolitan coverage in mind. Besides this competition, there is room for

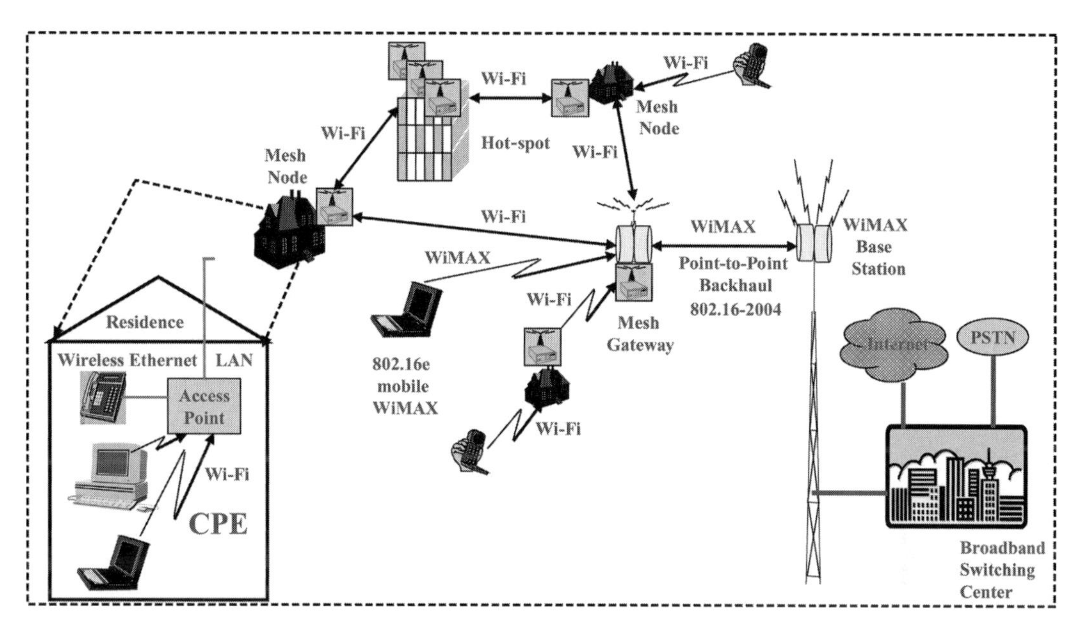

Figure 13.5 WiMAX and Wi-Fi Mesh Convergent Network Architecture

complementary designs since WiMAX has unique features regarding operational distance, mobility, and bandwidth.

An example of convergent wireless technology promoting the best features of both WiMAX and WLAN is offered by Hopling Technologies, Netherlands, in collaboration with Fujitsu Microelectronics America [70]. This network uses a relay-oriented architecture that connects individual WLANs, WLAN hot-spots, and WLAN hot zones to provide broadband services for metropolitan areas. The role of WiMAX is to provide broadband, long-range, backhaul connectivity of metropolitan users to traditional, public networks or wired and wireless carriers. The HopWARE WiMAX and Metro Mesh network is shown in Figure 13.5.

This architecture consists of Wi-Fi indoor and outdoor Access Points and mesh routers. Individual APs are part of home WLANs while congregations of APs are part of business WLAN or hot-spots/hot zones. Hopling Xnet Mark II mesh nodes are interconnected to relay information across the metropolitan network. One or two mesh gateways, exemplified by the Xnet Raptor II hybrid nodes, include WiMAX Subscriber Station functionality that provides long-range connectivity outside of the metropolitan coverage area.

The Hopling metro mesh architecture is applicable to large or small metro areas. For smaller areas, two micro WiMAX base stations are planned: the HopMAX 4600 and HopMAX 6600 products. The silicon foundations of HopMAX WiMAX products are provided by Fujitsu Microelectronics System on Chip (SoC). Their design is based on IEEE 802.16 and 802.16e standards.

The HopWare convergent network architecture works in the 900 MHz, 2.4 GHz, 2.5–2.6 GHz (Europe), 3.3–3.5 GHz, 4.9 GHz, and 5 GHz bands. The layered protocol stack used in the metro mesh solutions contains two protocol additions: the Hopling Mesh Protocol and the Hopling Discovery Protocol. These protocols run on top of standard

802.11b/g and WiMAX protocol stacks and provide traffic relaying capabilities between mesh nodes, self-configuration, self-healing, and routing (there is no single point of failure).

The primary function of metro mesh networks is to provide roaming capabilities within metropolitan networks. Additional functions are: bandwidth control, automatic channel selection, and RADIUS-based security services. The mesh metro solutions can be deployed as private municipal networks to provide support for Wi-Fi and WiMAX broadband services, last-mile access, metropolitan video surveillance, and mesh backhaul connectivity.

13.9 QOS in Fixed-Mobile WiMAX Convergence

Interest in nation-wide implementation of WiMAX has been shown by the government of South Korea and South Korea Telecom, one of the early WiMAX service providers. Their designs are based on WiBro, a fixed metropolitan WiMAX alternative and service. It operates in the 3.6 MHz band. Ten service classes have been proposed in conjunction with WiMAX implementation:

WiMAX Basic Services

- **Class 1**: Wireless Internet.
- **Class 2**: Wireless Internet and Voice over IP.

WiMAX Enhanced Services for Personal Multimedia

- **Class 3**: Enriched Telephony Service (WiMAX modem and video phone).
- **Class 4**: Portable data service with rich content (PCMCIA WiMAX card).
- **Class 5**: Wireless triple play for mobile users (video streaming, unicast/multicast).

WiMAX Enterprise Services

- **Class 6**: Wireless office services (wireless PBX, multimedia conferences).
- **Class 7**: Push-to-Talk service for the field worker (portable mobile).
- **Class 8**: Tourist Portal Service (city-wide WiMAX coverage).
- **Class 9**: Remote monitoring (home or buildings through digital camera).
- **Class 10**: National events support (networking services for VIP).

13.10 WiMAX Mobile Convergence Service Providers

Unfortunately, the slow pace in the development of standards has left room for manufacturers, vendors, and service providers to claim that their products are WiMAX conformant, with no proof of certification. These claims will be verified when the WiMAX Forum's certification program starts in mid 2007 with focus on mobile WiMAX specifications. This certification, as mentioned in Chapter 8, will begin with the SoC that will drive WiMAX devices, in the so-called Wave 1 and Wave 2 certification processes.

A good place to validate products will be Taiwan, where 95% of the Wi-Fi equipment is manufactured. The government's involvement, as expressed in the published guide

"Taiwan's WiMAX Acceleration Program", is backed by a large government subsidy. In addition, the 2.5 GHz spectral band was released to facilitate the development of chipsets and corresponding WiMAX products.

Another development regarding testing and certification is the decision of the WiMAX Forum's Applications Working Group and the Application Business Task Group to set up two proof-of-concept laboratories. One of the groups will be at the University of California, Berkley, USA, and the other at the University of Taipei, Taiwan. Focus will be on comparative technology testing to prove the performance of WiMAX against other mobile wireless broadband technologies. This will include end-to-end testing of WiMAX applications in real-world environments. Traditional interoperability testing will be left to independent organizations.

Sprint/Nextel has decided to set up a WiMAX test lab at Herndon, Virginia. Emphasis there will be on "certification" of products based on the 802.16e standard and development of WiMAX profiles. It is known that the standards often do not provide clear-cut definitions and range of parameters for all components. For example, there are several spectral bands opened to WiMAX technologies, some licensed, others unlicensed. Also, there are different sizes of channel allocations (5 MHz, 7.5 MHz, 10 MHz or 20 MHz), and different ways of duplex communication (TDD, FDD). In addition, there are different requirements from country to country regarding adoption of WiMAX technology, use of smart MIMO antennas, and QOS requirements.

There is some fear that Sprint/Nextel and its supporters such as Motorola, Intel, Nokia, and Samsung, will end up designing a custom-built WiMAX, thereby controlling the certification of interoperable products from the position of major players. These players were already given the task of building WiMAX services in major USA markets such as Chicago, Washington DC, and Dallas/Austin. Usually, the natural consensus around a new technology comes after debates, customer collaboration, and interoperability testing. In this situation such a dialogue may be cut short to the benefit of a few players.

13.11 WiMAX as an Alternative to 4G Cellular Mobile

Since Sprint/Nextel's announcement to set up a testing program, the debate around the elusive 4G mobile networks and services has intensified. The testing program is related to the plan to pursue WiMAX as the stepping stone technology for the next generation of mobile networks. Sprint/Nextel architects, with the backing of close partners such as Motorola, Samsung, Intel, and Nokia, went even further, calling WiMAX the ultimate 4G technology.

Since it was published as a final standard in early 2006, the WiMAX Forum has taken an aggressive approach to set up certification processes, certification labs, and proof-of-concept labs. This has forced virtually every major country to take a position in releasing spectrum to support this technology. The WiMAX Spectrum Owners Alliance (WiSOA) is another driver organization helping with education, commercialization, and deployment of the WiMAX spectrum. As of 2007, the WiMAX Forum listed over 350 current projects, testing trials, and implementations of that technology. In this context,

Table 13.1 Comparison of WiMAX with other 4G Frontrunners

Technology	Standard	Targeted Use	Radio Access	Downlink	Uplink	Notes
WiMAX	IEEE 802.16e	Mobile Internet	MIMO-SOFDMA	70 Mb	70 Mb	10 Mb at 10 km
HIPERMAN	ETSI-BRAN	Mobile Internet	OFDMA	56.9 Mb	56.9 Mb	
WiBro	WiBro	Mobile Internet	OFDMA	50 Mb	50 Mb	50 Mb at 1 km
iBurst	802.20	Mobile Internet	HC-SDMA	64 Mb	64 Mb	60 Mb at 3 km
W–CDMA	UMTS HSDPA/HSUPA	Mobile Phone	CDMA/FDD	14.4 Mb	5.76 Mb	Current rates 1–2 Mb
3GSM	UMTS-TDD	Mobile Internet	CDMA/TDD	16 Mb	16 Mb	
4GSM	LTE UMTS	Mobile Internet	OFDMA/MIMO SC-FDMA	>100 Mb	>50 Mb	Under development
CDMA 2000	EV-DO 1× Rev.A	Mobile Internet	CDMA	3.1 Mb	1.8 Mb	
CDMA 2000	EV-DO 1× Rev.B	Mobile Internet	CDMA/FDD	4.9 Mb × N	1.8 Mb × N	Not deployed

Note: N is the number of 1.25 MHz chunks of spectrum used.

we show in Table 13.1 a high-level comparison of WiMAX with the other frontrunners contending for 4G leadership.

One of the goals set up by ITU-T for the 4G generation of mobile networks was to achieve maximum data rates of 100 Mbps for mobile terminals and 1 Gbps for fixed-nomadic terminals. These rates can be achieved using smart, adaptive MIMO antennas. Technologies that may provide the capabilities to achieve these rates are the ITU-T UMTS Long-Term Evolution (LTE) and WiMAX 802.16e.

WiMAX's standing as a promotor of 4G was reinforced by the demonstrations in South Korea, in 2006. The demos show that from the standpoint of mobility and data rates delivered, WiMAX 802.16e technology is in the class of what is expected from mobile 4G networks. For example, one of the early promoters, Samsung Electronics, demonstrated WiMAX mobile communications at 100 Mbps on a moving bus with simultaneous streaming of 32 HDTV channels on a single carrier in stationary mode. Using 8x8 MIMO antennas, in stationary mode, the aggregate WiMAX delivered high-speed data at 3.5 Gbps.

The preliminary results of mobile WiMAX implementations, based upon the 802.16e-2005 standard, has led IMT-2000 to accept the IP-OFDMA access method as an approved technology. That helped WiMAX acceptance by regulatory authorities and operators in many countries to release and use the cellular spectrum for WiMAX-based applications. It is expected that future development of this technology, WiMAX (IEEE 802.16m), will be proposed for the IMT-Advanced 4G architecture.

13.12 The Upcoming Auction of the 700 MHz Spectrum

The development of new technology and the success of wide coverage services are dependent on the available spectrum. This is a national issue that can have worldwide implications. The upcoming auction of the 700 MHz band spectrum, scheduled for

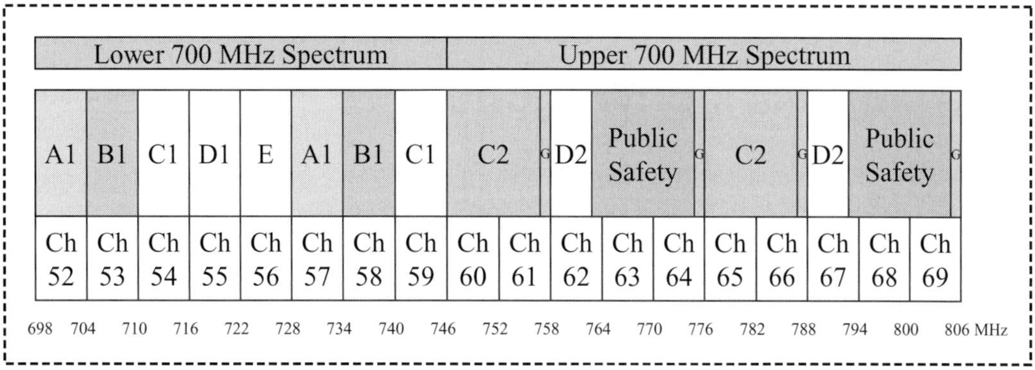

Figure 13.6 FCC 700 MHz Auction Frequency Allocations

January 2008, is one such example [71]. This will have an impact on several technologies that include cellular mobile and WiMAX. The attraction is great because working in this band provides higher indoor penetration of RF signals and longer range communications. A similar auction, named Advanced Wireless Services (AWS), took place in 2006 to sell licenses for the 1700 MHz and 2100 MHz spectrum bands.

The 700 MHz auction covers 62 MHz of spectrum in the 698–806 MHz bands. It is shared with portions of spectrum set aside for Public Safety [72]. The layout of spectrum blocks and channel allocations is shown in Figure 13.6.

In the Lower 700 MHz Spectrum there are two 12 MHz blocks open for auction, the A1 and B1 (two 6 MHz paired bands each). They will be auctioned as 176 licenses, and 734 licenses, respectively. The most appealing block in the Upper 700 MHz Spectrum, C2, is 22 MHz wide (two 11 MHz paired bands). It will be auctioned as 12 regional licenses or as a block in a nation-wide license. This single offering is of great interest for incumbent wireless service providers or new entrants who want to build nation-wide wireless networks. Among those interested is the Google Company, provided that use of the spectrum will allow them the open access they requested. The notion of open access, in their view, assumes that any device and any application can be used on the licensed spectrum and any Internet Service Provider could connect to this network. Another condition would be the right to resell the spectrum to third parties as partners in building a nation-wide network.

14 Wireless Sensor Networks Cellular Mobile Convergence

14.1 RFID Technology Development

Radio Frequency Identification (RFID) is a technical evolution of the bar code system. The goals are similar, i.e., to identify goods, items, and objects in general, using labels and dedicated label readers. Both bar code systems and RFID employ wireless technologies [73]. Despite these similarities, there are some differences. This is shown in Table 14.1.

The differences are in the areas of convenience (line-of-sight), environmental requirements, and, above all, in their capacity to identify individual objects. Bar coding is capable of identifying classes of similar objects but not individual items. And, here is the major advantage of RFID; it is an excellent candidate to provide tracking systems for individual items throughout the supply chain.

In this context, the supply chain is a staged suite of processes. It starts with the procurement of raw materials or basic components, followed by manufacturing or assembling goods, quality control, distribution, and, finally, the retail processes. The needs and the advantages of labeling have been known since the time of assembly lines, and systems resembling wireless identification were applied decades ago. But only in the current stage of mobile communications do we have a technology capable of identifying and tracking items in real-time.

At the heart of RFID technology labeling systems is the Electronic Product Code (EPC). It resembles the Universal Product Code (UPC) used in bar code systems. The EPC coding system was conceived by the Auto-ID Center, an adjunct group of professionals linked to the Department of Mechanical Engineering at the Massachusetts Institute of Technology. After EPC gained global recognition and standardization, the work of advancing the implementation of RFID was taken over by the EPCglobal organization, and the Auto-ID group was disbanded.

The EPCglobal code structure can range from 64 bits to 256 bits. Most of the implementations use the 96 bit type of coding, organized into four distinct fields [74]. The EPCglobal code structure is shown in Figure 14.1.

The Header field is 8 bits long and indicates the type of code, and, implicitly, the length of code. For example, a hexadecimal 01 indicates a type 1 code with a length of 96 bits. The **EPC Manager** Field is 28 bits long and indicates the manufacturer of the product the EPC tags are attached to.

Table 14.1 Bar Code System and RFID Technologies Comparison

Bar Code System	RFID
Requires direct line-of-sight between the reader and the bar code label	Allows reading at different angles and even through some materials
Requires clean, undusted label surfaces	It can work in diverse environmental conditions
It is a mature technology but with no potential for future advancement and networking	New RFID chip design and packaging provides extension into various fields
The UPC can identify only classes of objects but not individual items	The EPC code allows identification of up to 2^{96} unique objects
It can not be used for tracking items in the supply chain tracking systems	It can track items in real-time as they move throughout the supply chain
It is based on worldwide standards with recognized benefits despite limitations	It is a relatively new technology but with a great potential in many fields

Header bits 0-7	EPC Manager bits 8-35	Object Class bits 36-59	Serial Number bits 60-96

Figure 14.1 RFID EPCglobal Code Structure

The **Object Class** field is 24 bits long and indicates the class of product of the item. It is similar to the Stock Keeping Unit (SKU) that is assigned and tracked by each merchant in an area of operation. The **Serial Number** field is 36 bits long and uniquely itemizes 2^{36} individual objects belonging to the same object class.

A critical aspect of RFID technology is linked to the level of sophistication built into individual tags. Earlier, we classified tags based on their power source as either passive, or semi-active, or active. However there are other characteristics that differentiate tags. Shown in Table 14.2 [73] are the five different classes recognized by EPCglobal.

Table 14.2 RFID Tags Classification

Tag Class	Name	Memory	Power Source	Applications
Class 0	EAS	None	Passive	Anti-theft
Class 1	EPC	Read-only	Any	Identification
Class 2	EPC	Read & Write	Any	Data Logging
Class 3	Sensor Tags	Read & Write	Semi-active/Active	Sensor Networks
Class 4	Smart Dust	Read & Write	Active	Ad-hoc Networks

Class 0 tags have no memory; therefore they do not contain any data beyond their shown, "car plate"-like ID. They are considered to be Electronic Article Surveillance (EAS) types of tags. They are used only to protect items against theft and are activated when the tag passes through gated antennas. The ID is factory programmed. There are also **Class 0 Plus** types of tags that are UHF read/write tags.

Class 1 tags are EPCglobal standardized tags used for identification of pallets, boxes, and containers. They can be either active or passive. One example is the Write Once Read Only (WORM) tag. The ID is either factory or user programmed. **Class 1 Generation 2 (C1G2)** is the most common UHF tag used for supply chain automation based on RFID technology.

Class 2 tags are also standardized EPCglobal-based tags that have a read/write capability, either semi-active or active. They perform local data collection of environmental or motion data from attached sensors and retain the information in memory for later acquisition initiated by a RFID reader.

Class 3 tags are either semi-active or active with read/write capability. These tags form the basis of sensor networks, detailed later in this chapter.

Class 4 tags, code name "smart dust", are full blown read/write active tags with their own ability to transmit, discover, receive, and communicate with similar tags without any reader. They are like the active nodes of an ad-hoc network.

14.2 RFID Tag Standards and Code Structures

The development of RFID technology was driven by two directions of standardization. One line of work was done under the ISO/IEC processes and the other was driven by the EPCglobal consortium. The ISO/IEC focus was on technology, while the consortium emphasized the practical applicability of RFID. These standards mainly deal with the physical RF air interface. However, there is a great deal of confusion regarding interoperability between these two solutions. This is because of some differences in the two directions in standardization, the use of possibly different frequency domains, and how the types of tags are classified. In addition to these aspects, there are some potentially important differences in regional and country-based regulations regarding the use of RFID tags. The most important RFID ISO standards are:

- ISO 18000 Part 1: Generic Parameters for Air Interface Communications for Globally Accepted Frequencies;
- ISO 18000 Part 2: Parameters for Air Interface Communications below 135 KHz;
- ISO 18000 Part 3: Parameters for Air Interface Communications at 13.56 MHz;
- ISO 18000 Part 4: Parameters for Air Interface Communications at 2.45 GHz;
- ISO 18000 Part 5: Parameters for Air Interface Communications at 5.8 GHz;
- ISO 18000 Part 6: Parameters for Air Interface Communications at 860–930 MHz; and
- ISO 18000 Part 7: Parameters for Air Interface Communications at 433 MHz.

ISO 18000-1 provides the RFID reference architecture for item management and defines the parameters that characterize the air interface. The specific parameters for the RFID air interface working in various frequency bands are detailed in Parts 2, 3, 4 and 6. ISO 18000-5 was withdrawn because of lack of global acceptance. ISO 18000 Part 6A provides specifications for a basic one-time programmable tag with only 96 bits of user data. The tag can be read at rates up to 200 tags per second. The tag requires only two

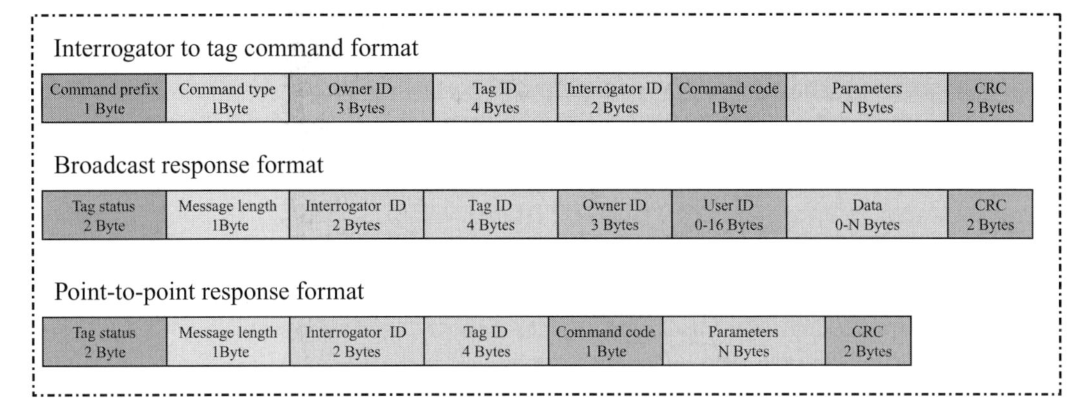

Figure 14.2 UHF RFID ISO 18000 Part 7 Code Structure

commands. It is based on pulse interval encoding. ISO 18000 Part 6B uses bi-phase modulation and Manchester encoding.

Two particular standards are of great interest. One is ISO/IEC 18000 Part 6C for UHF (860–930 MHz) passive tags which has been harmonized with the EPCglobal Class 1 Generation 2 (C1G2) specifications. These specs provide definitions of data structures and a complex set of commands. The other standard is ISO/IEC 18000 Part 7 dealing with VHF (433 MHz) active tags. These two standards are important because they indicate two distinct trends in tag design [75]. One trend, followed by the ISO 18000 Part 6C design, is based on integrated circuit architecture with Complex Instruction Set Computers (CISC)-based processors. The other trend, followed by the ISO 18000 Part 7 is based on integrated circuit architecture with Reduced Instruction Set Computers (RISC)-based processors. In short, the tradeoff is between using few commands with complex code structures as opposed to many commands with simpler code structures. Nevertheless, simpler code structures consume less power, a critical element in the design of passive tags. An example of reader or interrogator commands and tag responses, conforming to ISO 18000 Part 7 standards, is shown in Figure 14.2.

The command format sent by the RFID reader to the tag contains a command prefix and type, each of them one byte long. The command type indicates a broadcast or a point-to-point type. The next fields provide IDs about owner, reader, and the tag addressed, given the fact that many tags may be in reach of the reader in the same time. Broadcast types of commands do not contain tag IDs. The owner IDs are used to differentiate classes of tags. They are programmed into tags during manufacture. The command code, one byte long, indicates the name of the commands performed. For example, collection of data, reading or writing into tag's memory, status information request, or commanding the tag to go into the sleep mode. The command structure ends with a CRC field that allows verification of the message received.

Tag responses can have two formats depending on the type of command format. The broadcast response follows a command format where the command code is a collection type of command. This allows the collection of information from a group of tags. Broadcast responses contain status information and an indication of message length.

To link the command with the response from a specific tag, the reader ID, tag ID, and owner ID are sent back in the response frame. Specific user ID and data fields contain the information requested. The point-to-point response has similar format to the broadcast format, except for the user ID field. To associate the command and the corresponding response, the point-to-point code includes the command code field.

14.3 RFID Tag Evaluation Criteria

The following criteria and characteristics should be used when evaluating RFID tags and selecting tags for a particular application:

- Type of tag (passive, active, semi-active);
- Usage of tags (read/write or read only);
- Reusability of tags (one time use or reusable with updated information);
- Frequency (LF, HF, VHF, UHF);
- Supporting standards (ISO, EPC, ECMA);
- Regional and country regulations (USA, Europe, Asia);
- Class of tag (1, 2, 3 as defined in the previous section);
- Memory size (the amount of information stored);
- Size and form factor (to fit the application);
- Complexity (standalone or sensor linked tag);
- Range of operation (the maximum distance between reader and tag);
- Power supply (battery, rechargeable battery, AC/DC);
- Battery type (for small form active tags);
- Environmental requirements (temperature, dust, vibration);
- Ruggedness (capability of working in harsh conditions);
- Tag movement in the field (requires readers of certain speed);
- Tag orientation (antenna polarization);
- Anti-collision (ability to support multiple tags in the vicinity);
- Closeness of tags (density of tags in space);
- Tag security (encrypted data);
- Reader support (support for most types of tags);
- Packaging type (paper, PVC, electronic module).

14.4 RFID Reader Evaluation Criteria

RFID readers should be evaluated on the basis of the following criteria and characteristics they possess.

- Multiband RFID support (LF, HF, UHF classes as described earlier);
- Multitag RFID reading and writing capability (ISO15693, Philips Icode, EPC, TI Tag-It, TagSys (C210/220/240/270), MIFARE ISO 14443-A and B, Hitag 1 & 2, ISO FDXB, ISO 1800 6A/6B/6C;

- Range of action (cm, feet, meter);
- Integrated reader and encoder capabilities (tag read/write);
- Processor type and memory size;
- User display (LCD color and brightness levels);
- Integration with computing devices (e.g., laptop, printer);
- Serial/parallel networking capability (Ethernet PoE, TCP/IP, serial UIDSB, serial RS-232, parallel Centronics IEEE 1284);
- Multimode capabilities (Wi-Fi, Bluetooth, WPAN, GSM/GPRS WWAN);
- Integration with IrDA barcode reader capability;
- Operating system supported (Microsoft Windows CE, Windows Mobile);
- Antenna options (indoor, outdoor, directional, omni-directional);
- Size and weight;
- Environmental specifications;
- Ruggedness;
- Power supply (main rechargeable battery);
- Management capability (SNMP, MIB II, Enterprise MIB);
- Technical assistance (on-line assistance, preventive maintenance, Mean-Time-to Repair);
- Documentation and training;
- Expertise in RFID software implementation for large businesses;
- Compatibility with UHF, EPC Generation 2, and ISO 18000-6C;
- Compatibility with the legacy infrastructure.

14.5 RFID and Cellular Mobile Networks Convergence

RFID, as a short-range RF identification technology, implies a RFID Reader or interrogator and a RFID Tag or transponder. Details of RFID two-way communication across air-interfaces were presented in previous sections of this chapter and in Chapter 9. The tags are attached to individual items, aggregate items such as shelves, packages, pallets, and major items such as vehicles, trucks, or containers.

Information collected as the result of interactions between readers and tags should be processed by a local host computer or by remote computers supporting specific RFID applications. In this case, convergence between the RFID network and cellular mobile networks is needed. A high-level architecture of a RFID and cellular mobile convergent network is shown in Figure 14.3.

This convergence has two aspects. First is the convergence of mobile terminals. This means a dual-mode or a multi-mode handset that combines a mobile RFID reader with a mobile phone. In most cases, this is an RFID reader that incorporates extended cellular radio mobile capabilities. An example of such a terminal is one in the Psion Teklogix Workabout Pro G2 family of products. This product combines a RFID reader with a GSM/GPRS phone that has Wi-Fi and Bluetooth capabilities; all in the same device. The second aspect is convergence of RFID and cellular networks. This convergence will require the data transmission capability of mobile networks (for example

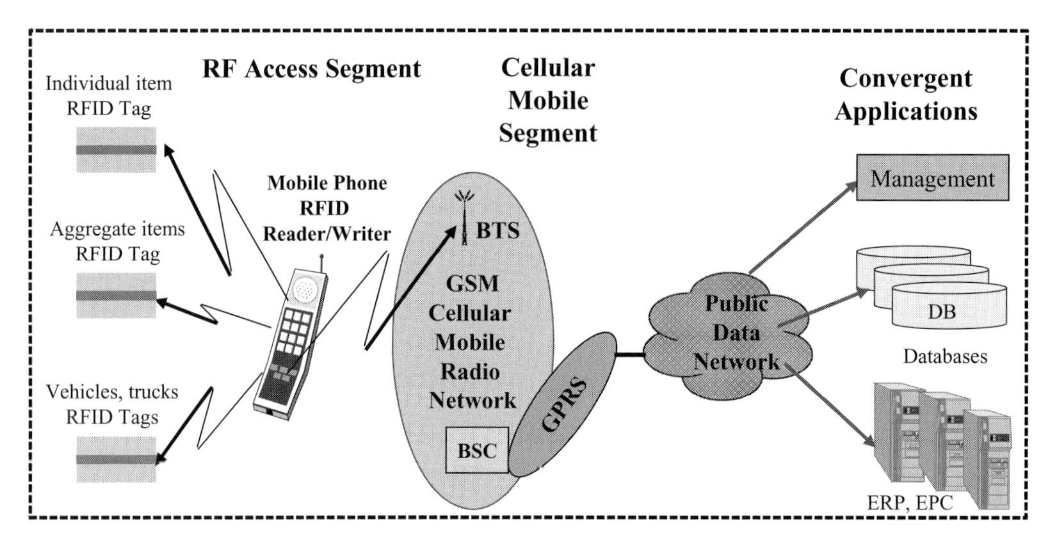

Figure 14.3 RFID and Cellular Mobile Convergent Network Architecture

the GSM/GPRS/EDGE) to communicate information that results from interrogation of enterprise specific convergent applications.

The convergence includes writing operations, executed through the same convergent RFID reader/writer terminal, and information being downloaded and uploaded through the mobile network.

14.6 RFID-based Health Care Services Case Study

An example of an RFID-based convergent network is provided by Alvin Systems' **RFID and Wireless Software Platform for Health Care** [76]. This is a platform designed for patient and asset identification and tracking that integrates with hospital information systems. An RFID-based Health Care Network is shown in Figure 14.4.

The major functions supported by this system are:

- Real-time patient identification based on "Smart RFID Wristbands";
- Medical assets identification and tracking;
- Specimen collection/identification and matching specimens with patients;
- Temperature monitoring for laboratory items, specimens, and medicine.

The process of identification and tracking is based on RFID technology. All the patients, items, and assets are tagged with passive and active tags. The readers consist of RFID-enabled Pocket PCs or Tablet PCs. Upon admission to the hospital, each patient is issued a unique ID with an RFID wristband passive read/writable tag that contains the patient's information and medical record number. A special RFID tag printer provides the encoded information on the wristband along with some human-readable information. When doctors, nurses and authorized hospital personnel wave the reader over a patient's RFID wristband it will provide positive identification of the patient. The medical information

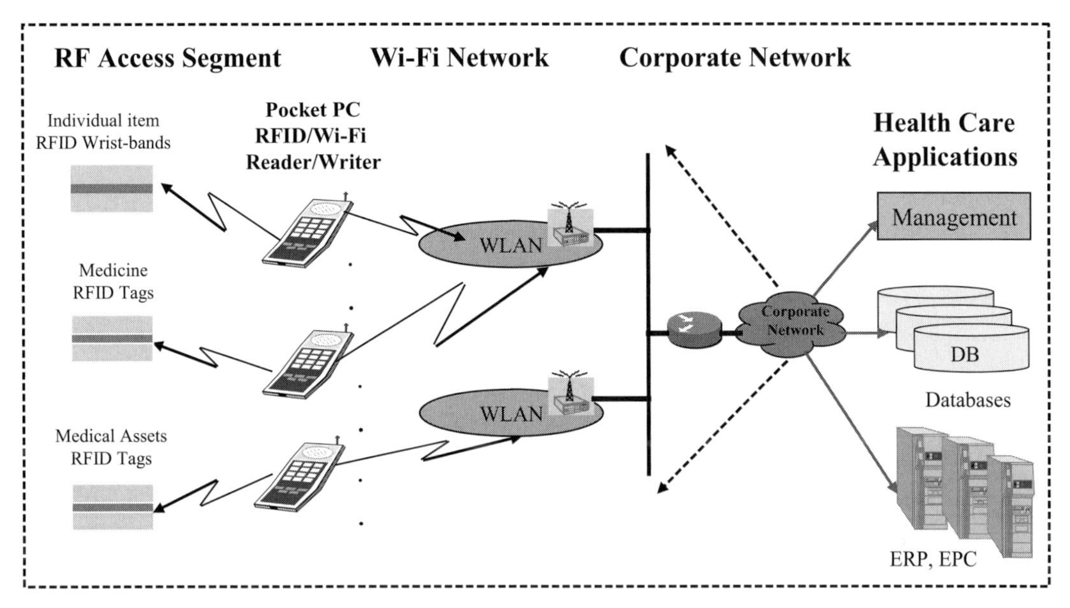

Figure 14.4 RFID-based Health Care Service Network Architecture

about the patient, and past monitored information from a central database, can now be transmitted over the corporate network.

Near real-time acquisition of information about patients allows accurate health care interventions and prevents mix-ups and medication errors. This is critical in emergency situations. Reading patient information and medications can be done when the lights are off or when the patient sleeps. Newly acquired information from patient monitoring (blood pressure, pulse, temperature) can be entered on the spot into the system and is immediately transmitted, processed, and stored in the central database. If the network is not available, the information is stored on a dual-mode Pocket PC and downloaded later via Wi-Fi or wired network. The WLAN Wi-Fi network consists of a constellation of Access Points that covers the hospital buildings with wired connectivity to the internal corporate network and applications.

Different tags are used for asset identification and tracking. Large items such as wheelchairs, moveable beds, intensive care units, and mobile electronic and surgical equipments can be tagged and monitored for their location, usage, and inventory management. This item management can be extended to cover staff and surgical garments, sheets, bed covers, and any items that require cyclical replacement.

Errors can be prevented by affixing RFID tags with patient information to vials containing blood samples and to any other plastic containers. Further, individual test tubes can be tracked down in a laboratory environment by waving the RFID reader in the general physical proximity. Critical sensitive samples or pharmaceuticals can be monitored for storage or control of temperature using paper-thin, reusable, low cost RFID sensor tags. These sensors measure temperature values and record these values into the tag's memory. Similar RFID tags can be associated with the movement and administration of medication destined to a certain patient, including shelf and stock management of medicines.

The network consists of an RFID segment that provides identification and tracking functions, and a Wi-Fi network that provides wireless connectivity to the corporate network and health care applications.

14.7 IBM's Secure Trade Lane RFID-based Case Study

Key to the success of sensor network technologies such as RFID, NFC, and UWB is the network infrastructure supporting these systems and the management capability of this infrastructure. The network should be capable of handling the increasing flow of data related to sensor networks, especially RFID technologies. First, there is a local flow of data when thousands of tags are read in a relatively short time during the auto ID process. Second, this information is transmitted through a global network for processing. Third, the pertinent information needs to be stored in multiple distributed databases. Through these phases, various parties such as suppliers, customs, and multinational companies will be involved as part of the global supply chain. This process will allow interested parties to track any items via land, sea, or air.

An interesting application of near-field sensor networking that serves as a good example of sensor networks and their management is provided by **IBM's Secure Trade Lane** (STL) solution [77]. A high level architecture of the STL application is shown in Figure 14.5.

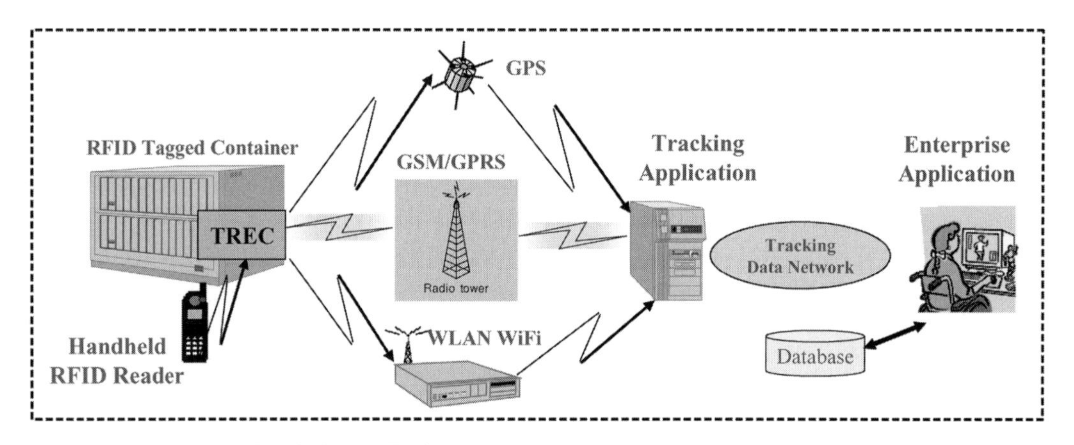

Figure 14.5 IBM's Secure Trade Lane

This application addresses the management of a complex tracking process and the monitoring operations related to the movement of goods shipped overseas in ocean-going containers, a major component of goods exchanged in global international trade. This large scale movement implies that cooperation is required between suppliers, shippers, logistics companies, and importers. Currently, only 2–4% of containers are inspected by customs when reaching the final destination; primarily because of the lack of supporting technologies. The STL solution is applicable to multi-modal transportation and can include beside ships, trains, trucks, barges, and airplanes.

The key component of STL is the **Tamper Resistant Embedded Controller** (TREC) developed by IBM's Research Laboratory in Zurich, Switzerland. This controller integrates the collection and processing of data from multiple sensors and processors, acting as a complex wireless communication device and data storage element. It is a miniature security keeper of the container on which it is attached. The built-in sensors detect door openings, light, temperature, humidity, acceleration, and above all global positioning. The data collected from sensors is filtered, aggregated and periodically sent to the **tracking application**. In this way, real-time tracking of the container takes place, backed by the complete history of movements and condition of the container. The movement can be related to a planned itinerary so any change is detected and can launch an alert to the logistic companies or security organizations if the situation warrants it.

As the container is moved along the planned route, a TREC can be accessed via Bluetooth or RFID handheld devices to enter the initial configuration or to read the stored data by authorized personnel. Some information can be read with regular barcode readers. Key to the application is, while the container is in a warehouse, incorporation in the TREC of a Global Positioning System (GPS), GSM/GPRS terrestrial mobile communications, and a Wi-Fi wireless communications with WLAN access points. Communications are controlled by a tracking application that is part of a complex tracking data network, enterprise applications, and data storage systems. Critical for this application is that the TREC batteries have enough capacity to last for months or even years.

The tracking application provides communications and continuous monitoring of multiple TREC devices. Its management functions include configuration, fault, and security. The information collected is filtered, aggregated, and correlated with that provided by the enterprise application regarding route planning, order management, and accounting. It is sent to relevant parties through any standard data communication network. The tracking application is also responsible for passing on any alarm information regarding tampering, incidents, or delays in shipment.

STL is based on EPCglobal networking concepts and specifications that apply to the collection of information through RFID readers and passing on of that information to the tracking application, and to warehouse management systems. Since multiple parties are involved in the process of tracking shipments, data will be stored at multiple points to secure privacy and integrity. STL gives owners full control of the data accessed by and disseminated to relevant parties. STL is implemented as an open, multivendor platform based on **Service Oriented Architecture** (SOA). This architecture includes distinct shipment monitoring, information, security, and efficiency service modules. It is backed by IBM's WebSphere, the DB2 Relational Database Management System (RDBMS), and the Tivoli Management Applications software.

The complex relationship between IBM's WebSphere RFID Information Center and the EPCglobal standard architecture is shown in Figure 14.6 [78].

At the core of the infrastructure is the Electronic Product Code Information System (EPCIS), a complex tracking application accessed through operator consoles. It is capable of interacting with multiple WebSphere RFID Premises Servers, the infrastructure

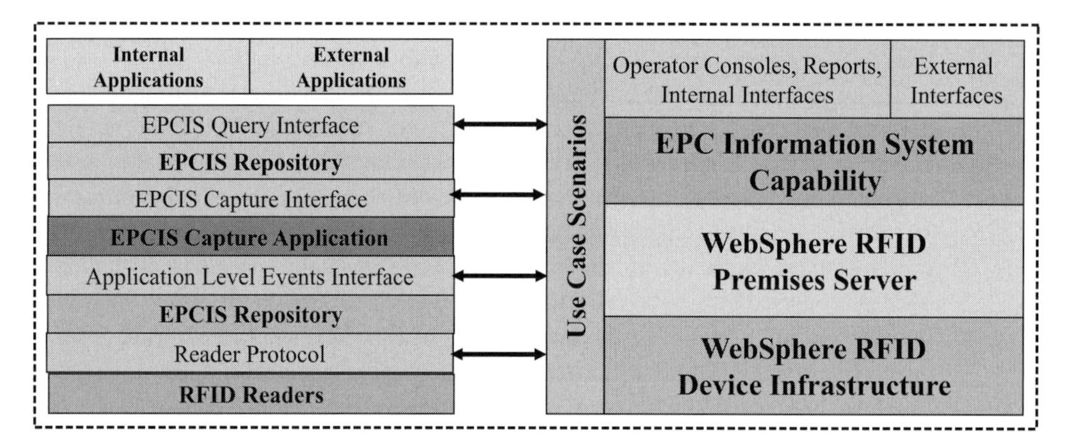

Figure 14.6 EPCglobal Standard Architecture and IBM's EPCIS

component connected to the RFID reader, here called WebSphere RFID Device Infrastructure. This provides local processing of information collected through the reader. EPCIS also provides external interfaces to data tracking networks and relevant databases. EPCIS also sends reports regarding the information collected through RFID readers.

RFID readers are state-of-the-art devices capable of collecting information from active RFID tags and communications with WebSphere RFID Premises Servers. The readers also have the capability to filter events. All communications capabilities are done in accordance with EPCglobal standard specifications that include reader protocols and interfaces.

14.8 NFC Networking, Standards, and Applications

There is a strong resemblance between RFID and NFC technologies. Both address short range communications, but NFC only operates in very short ranges, only a few centimeters. Another difference is the fact that a NFC-enabled terminal provides both reading and tag capabilities in the same device. That opens the door for direct peer-to-peer communications between NFC-enabled devices. All these operations are performed by "touching" or "holding close" the two NFC-enabled devices or a device to a NFC tag. The NFC Forum mandates four tag types to support interoperability between NFC tags and NFC readers/writers:

Type 1 tag is based on the ISO14443A standard. Tags are read and re-write capable; users can configure the tag to become read-only. Memory size is 96 bytes, expandable to 2 KB. The data rate is up to 106 Kbps.

Type 2 tag is also based on ISO14443A and resembles type 1 tags. These tags are read and re-write capable and users can configure the tags to become read-only. However, the available memory is limited to 48 bytes, expandable to 2 KB. The data rate is up to 106 Kbps. Type 1 and 2 tags with small memories are suitable for low cost, single use, discardable tags.

Type 3 tags are based on the Japanese Industrial Standard (JIS) X 6319-4 standard, also known as FeliCa. Tags are pre-configured by the manufacturer to be either read and re-writable, or read-only. The memory available is variable and can reach 1 megabyte per tag. The maximum data rate is 212 Kbps or 424 Kbps.

Type 4 tag is based on ISO14443A and ISO14443B standards. These tags are also pre-configured by the manufacturer to be either re-writable or read-only. Memory availability is variable, up to 32 KB per service. Maximum communication speed is 424 Kbps.

In June 2006, the NFC Forum published specifications for Data Exchange Formats (DEF) and Record Type Definitions (RTD) for smart posters (NFC-based embedded tags), text content, and Internet resources for reading applications. This is in addition to ISO 14443A/B standards, dealing with tags, and the ISO 18092/21481 standard, dealing with NFC air interface and protocols (ECMA 340/352 equivalent).

Examples of type 1 and 2 tags are the "Topaz" family of NFC read/write tags. They measure less than a quarter of a square millimeter and are manufactured by Innovision Research and Technology [79]. These tags have a memory capacity of 96 bytes organized in 12 blocks of 8 bytes each. A special block of 7 bytes is used as a unique ID-field. This field, programmed and locked at the time of manufacturing, provides an authentication and anti-cloning mechanism. Another block of 6 bytes is a one-time programmable field and can be used as a single use token. A block of 10 bytes is reserved for text type of information such as a long URL address. Incorporated into the NFC tag, it can be used by the reader; a NFC-enabled cell phone, for example.

Convenience in the use of NFC technology is experienced by the user through the gentle "touch" or close proximity swing of the cell phone to establish a "contactless" link and to initiate a transaction. However, success will be determined by the authentication time over the network as part of the payment related processes. The reader-tag communication is just a fraction of the overall transaction time. That shows how critical the NFC supporting network infrastructure is.

14.9 NFC and Cellular Mobile Networks Convergence

NFC, as a very short-range communication technology, is defined by the NFC Reader and a NFC Tag RF segment. The details of two-way communications across the very short air-interface were presented in previous sections of this chapter and in Chapter 9. NFC capabilities can be incorporated in mobile phones, PDAs and Pocket PCs. In the simplest implementations, these devices are just active tags. In more complex implementations these devices can play the role of NFC readers combined with communication capabilities such as Wi-Fi, and Bluetooth. The information collected as a result of interactions between the reader and the tag is processed by a local host computer or by remote computers found across public or private data networks. A high-level convergent architecture of NFC and a cellular mobile network is shown in Figure 14.7.

A NFC-enabled handset can communicate with a NFC reader and with other NFC-enabled devices such as handsets, laptops, various tags, and NFC-enabled electronic keys

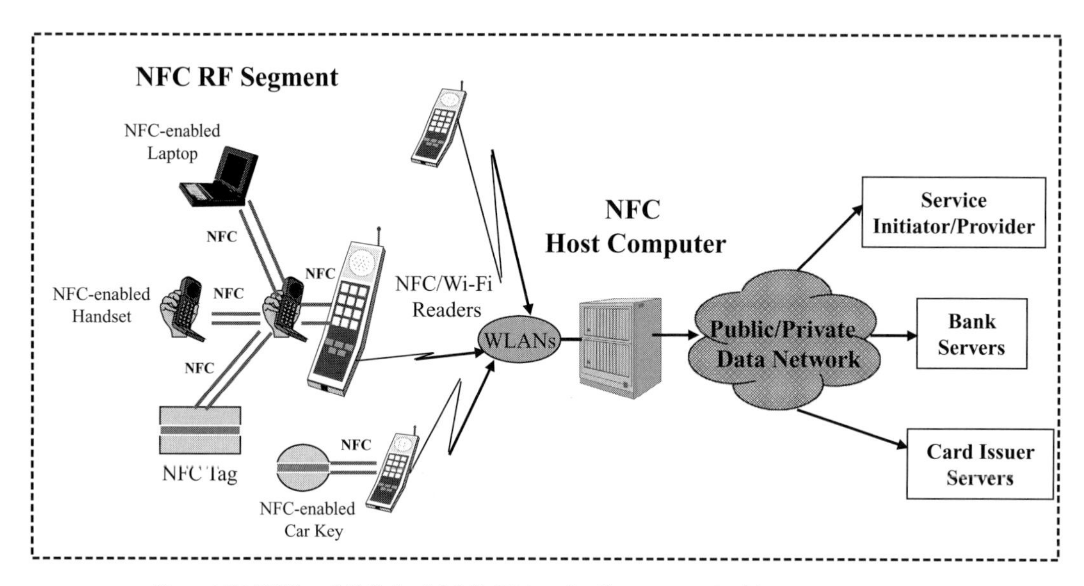

Figure 14.7 NFC and Cellular Mobile Networks Convergent Architecture

(car, office, house, and hotel room keys, for example). Various applications with strict security requirements such as identity documents verification (passports) or mobile commerce (payments) are better served with NFC technology rather than RFID or Bluetooth because of the very short range of operation.

NFC and mobile networks convergence has two components. First is the convergence of mobile terminals. This implies a dual-mode or multi-mode handset that combines a NFC-enabled device with a mobile phone and Wi-Fi. The second aspect is the convergence of NFC and Wi-Fi or cellular networks. This means the ability to use the data transmission capability of WLANs or cellular mobile networks to communicate collected information through NFC readers to specific convergent applications. The applications run on servers belonging to service providers, banks, and card issuers. Therefore, the spread of NFC technology, beyond peer-to-peer near-field communications (exchanging files, photos, video clips), will be highly dependent on building the supporting infrastructure. This will require collaboration between network service providers, banks, and credit card issuers.

14.10 NFC Payment and Promotion Case Study

Several Asian countries, notably Japan, Korea, and China, have already implemented the NFC technology for several years now. However, the use of non-standard NFC air interface protocol, Sony's FeliCa, remains a major issue outside Asia. At the same time, in Europe, and in the USA, several trials of NFC-based technology have been conducted, involving major mobile operators, mobile NFC-enabled handset manufacturers, banks, and credit card service providers. A major trial was performed in Atlanta (Georgia) in 2005/2006 involving AT&T (formerly Cingular Wireless), Nokia, Visa, and Chase Bank.

Over 200 regular fans of Atlanta's Thrashers hockey team and the Hawks basketball teams were given Nokia 3220 mobile NFC-enabled cell phones to be used to gain access to the games, to purchase items at the arena's convenience stores, and to exchange files and pictures among themselves [80]. In other words, in addition to regular mobile phone functions, the cell phones were used as replacements for credit cards to authorize access and to pay for services. NFC tags were incorporated into posters, allowing the users, by near touching the tags, to download web-based Universal Resource Locator (URL) addresses. Internet access and a web browser accompanied each phone. Based on the URL selected, the users could access the web pages provided by the mobile operator, AT&T. The content consisted of photos, video clips, and players' information, all uniquely designed for the participants in this test. The participants must have had an active Visa account through the Chase bank to use the cell phones and to make credit card payments.

This test was preceded by a study commissioned by Visa and NXP (formerly Philips Semiconductors) that involved NFC operations performed with a Nokia NFC-enabled cell phone in Philips' laboratory environment, in Atlanta. It included NFC-based purchase of food and services in a virtual NFC kiosk, downloading the URL information regarding a movie store from the NFC tags, purchasing a movie to be watched through a NFC-enabled set-top-box, and purchasing a ticket at one of the sporting events through a NFC-enabled poster [80]. This test highlighted the usability of this technology and above all the extraordinary convenience in performing contactless, touch-based, purchase of items and services, all based on rapid transfer of information between tags and the NFC cell phones.

The VIVOtech Company provided the software application that runs on the phone's main processor. This NFC chip-based software controls the exchange of data between the cell phones and the NFC-tagged posters. Another application that runs on the phone's Smart MX microchip enables mobile phones to transmit encrypted payment data to the NFC readers linked to Point-of-Sale (PoS) terminals located in the arenas, which VIVOtech was also supplying.

A similar deployment having again the VIVOtech Company as system integrator and NFC software applications provider, started in 2007 in Taiwan. The payment and promotion solution for the Master PayPass service involved the Taipei Fubon Bank and the Taiwan Mobile operator.

The results of this trail were upbeat; users were amazed by the versatility of using cell phones as credit cards. In this context, ABI Research Company predicted, in 2006, quite optimistically, that in 2010 50% of all mobile phones manufactured will contain NFC capabilities. Since then, a correction was made, the new estimate calling only for 20% of the cell phones delivered by 2012 to have NFC capability. One explanation of this was the lack of uniquely standardized RF NFC interfaces and lack of applications programming interfaces. Another explanation was that American, and especially European mobile operators, are still burdened by the huge cost of licensing spectrum for 3G and B3G operations. In the meantime, in Asia, over 100 million users continue to benefit from the convenience brought by the NFC technology.

14.11 UWB-based Wireless USB Products

Despite a split that occurred in the UWB Forum between two competing groups of players, the market has agreed that the time has arrived to focus on marketable products based one UWB technology. There are more than 2 billion wired USB-based consumer electronics devices on the market. Therefore, the idea of providing Wireless USBs (W-USB) to eliminate cable in the process of high-speed transfer of huge amounts of data is very appealing for users and businesses alike. This can be achieved by using the UWB technology.

The task of standardizing W-USB was taken over by the USB Implementation Forum (USB-IF), a non-profit organization formed to promote the development and implementation of USB technology. In a short span, a W-USB Promoter Group, supported by major players such as Agere Systems, HP, Intel, Microsoft, NEC, NXP, and Samsung, was formed. These companies have defined the core specifications for certified W-USB products. These specifications were adopted by USB-IF and the first UWB-based W-USB products are expected on the market in 2007. Examples are: W-USB dongles that plug into PCs' existing USB ports, W-USB hubs, PCI Wireless USB adaptors, and PC embedded W-USB, the last group providing the highest data rates.

W-USB is based on WiMedia Multi-band Orthogonal Frequency Division Modulation (MB-OFDM), an Ultra Wide Band access method developed by the WiMedia Alliance. It allows point-to-point connectivity between devices and a Wireless USB hub. W-USB specifications are backward compatible with wired USB 2.0 devices. They allow connectivity of up to 127 devices and deliver data rates up to 480 Mbps at 3 meters and 110 Mbps at 10 m range. For security, Certified Wireless USB devices ask for connections that use the 128 bit AES encryption. A short list of certified W-USB products follows:

- **D-Link**: UWB USB Hub and Adapter, (www.d-link.com);
- **Belkin**: Cable Free USB Hub, (www.belkin.com);
- **Dell Computers**: 1720 Notebook, (www.dell.com);
- **IOGEAR**: W-USB Hub and Adapter Kit, (www.iogear.com);
- **Lenovo**: Think Pad T61/T61p, (www.lenovo.com);
- **ICRON**: Extreme USB family, (www.icron.com);
- **Gefen**: Wireless USB Extender, (www.gefen.com).

Use of W-USB devices in the USA will not interfere with most WLANs and WPANs operating in the unlicensed 2.4 GHz band because UWB band allocation is between 3 GHz and 10 GHz. In Europe, where lower frequency bands are used, UWB-based operations will require the implementation of Detect and Avoid (DAA) techniques. These techniques, in the class of "cognitive software", allow detection of used channels. To avoid interference the devices using unlicensed bands will automatically switch channels.

14.12 Mobile Ad-hoc and Wireless Sensor Networks

It has been more than 10 years since the advent of wireless LANs, when the idea of building Mobile Ad-hoc Networks (MANET) was launched. Vehicular, military

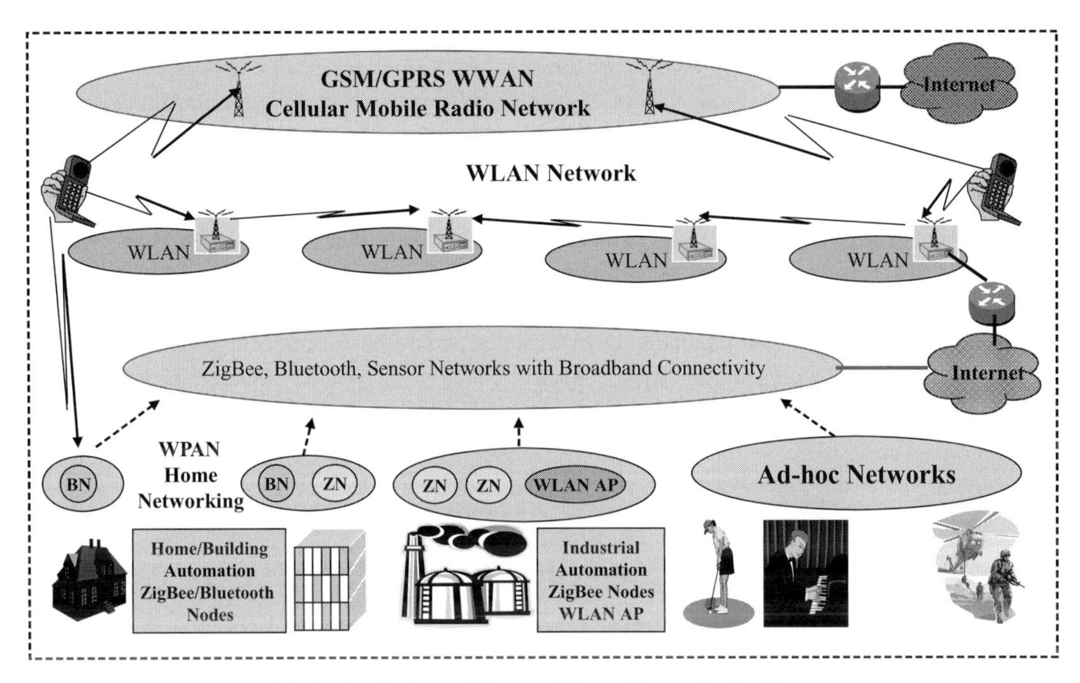

Figure 14.8 High-level Wireless Ad-hoc and Sensor Networks Architecture

battlefield, transportation systems, performing arts and sport events, and disaster recovery ad-hoc networks were among the drivers of these developments. A major tenet of MANET architecture has been a fully distributed network. IETF took interest in these trends by coalescing the ideas and the work done under the IETF MANET Working Group. Mobile ad-hoc networks can be achieved through combinations of WLANs and WPANs, and clusters of WLAN Mesh and WPANs in areas that sometimes can be called Community Area Networks. Despite the numerous studies and proposals, the overall architecture was not embraced by the industry because of lack of management capabilities [81].

However, a great deal of the research work, done especially in academia, has found applicability in the practical aspects of design and implementation of wireless sensor networks. In this class of networks there is a reduced emphasis on the mobility features of the terminals. The important aspects are those related to minimal power consumption of the terminals/sensors and effective algorithms for handling the handover in a multi-hop network environment. Other important aspects are higher data rates within the network and between the base stations/routers, as well as connectivity with the public network, the Internet. Several solutions were proposed and implemented such as mesh networks based on Wi-Fi technologies, and wireless sensor networks based on ZigBee and even Bluetooth technologies. Broadband access to communities and to automated industrial processes are among major applications. A high-level architectural diagram of mobile ad-hoc and sensor wireless networks is shown in Figure 14.8.

Important aspects of these sensor networks are the level of mobility allowed for the base stations (fixed or moderately mobile) and the clustering of mobile terminals around base stations. This should be correlated with the major requirements of sensor networks, low power consumption and high longevity of the mobile sensors. This aspect, generally known as energy management, involves battery power, transmission power and overall systems power management. The design of terminal hardware (specialized sensors, laptops, PDAs, Pocket PCs) that includes CPU, memory, display, and wireless card adapters, is a critical element in saving power [82].

The devices that are part of mobile ad-hoc networks can be classified into two groups: low data rate devices and high data rate devices. Low data rate devices can exchange data at 50–200 Kbps. They are typically sensors and actuators used for monitoring and controlling processes such as environmental and power control measurements. In many instances, these devices do not have their own power sources. Therefore, they rely on batteries with low power consumption and high longevity requirements. These devices may not have standardized interfaces and they might be based on proprietary protocols. High data rate devices, capable of exchanging data at rates of 2 to 5 Mbps include computer terminals such as laptops, PDAs and the latest generations of mobile phones. Generally, these devices have their own power and they communicate using standard interfaces and protocols.

Part V

Fixed Wireless Cellular Mobile Networks Convergence: Standardized Networking Solutions

15 UMA-based Fixed Wireless and Cellular Mobile Networking Solutions and Products

15.1 What is Unlicensed Mobile Access or GAN?

Unlicensed Mobile Access (UMA) is an architecture and a set of specifications that provides convergence of **cell-based mobile radio GSM/CDMA** networks with IP-based fixed wireless access networks. Examples of fixed wireless networks include IEEE 802.11 a/b/g WLANs, IEEE 802.16 WMANs (WiMAX), IEEE 802.20 Ultra Wideband, IEEE 802.15.1 Bluetooth WPANs, and Near-Field Sensor Networks. UMA subscribers are provided with total location, mobility, and service transparency.

UMA development was initiated in January 2004 by a group of mobile service infrastructure providers and mobile handset manufacturers: Alcatel, AT&T Wireless, British Telecom, Cingular, Ericsson AB, Kineto Wireless, Motorola, Nokia, Nortel Networks, O2, Roger Wireless, Siemens AG, Sony Ericsson, and T-Mobile. By mid 2005, key components of the UMA specifications, dealing with user, architecture, protocols, and conformance testing aspects were developed. These specifications were submitted to the 3G Partnership Project (3GPP), as part of a work item called "Generic Access to A/Gb Interfaces". Since then, these specifications, also known as TS 43.318, have been approved and incorporated in 3GPP Release 6 documentation. Consequently, the UMA initiators have decided to disband the independent working group and to continue their effort under the umbrella of the 3GPP organization. Since its adoption by the 3GPP, the UMA name has been changed to **Generic Access Network** (GAN). However, the old name is still used in technical and marketing circles, so for this reason, we will continue to use it in this book.

UMA specifications allow development of high performance voice, data, and multimedia services over WLANs that are located in homes, offices, hot-spots, or metropolitan areas. UMA uses dual-mode, tri-mode, or multi-mode mobile phones for communications over unlicensed spectrum technologies such as IEEE 802.11x WLANs, Bluetooth WPANs, and WiMAX.

15.2 UMA-based Network Architecture

Unlicensed Mobile Access Networks (UMANs) are parallel radio access architectures that interface with the existing terrestrial mobile public network components. UMAN

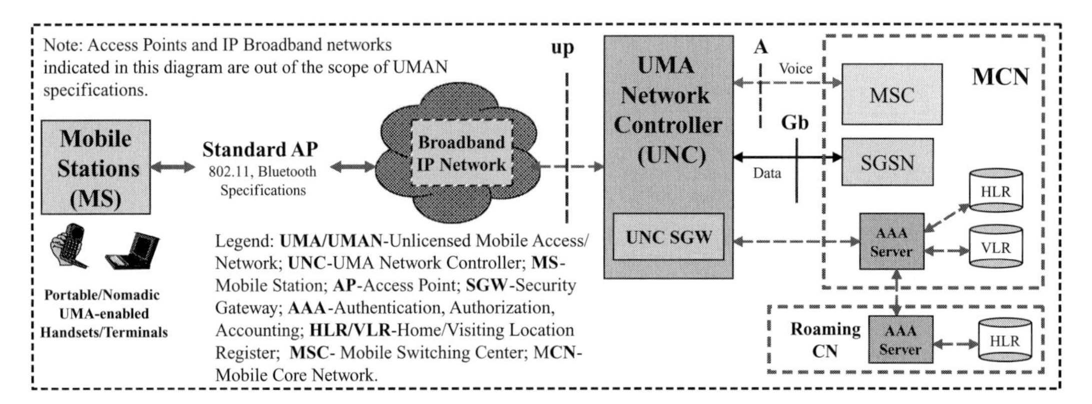

Figure 15.1 High-level UMA-based Network Architecture

allows transparent handover to and from **GSM/GPRS/EDGE Radio Access Networks** (GERAN) and **Wireless Local/Personal Area Networks** (WLANs and WPANs). A specialized **UMA Network Controller** (UNC) or **Generic Access Network Controller** (GANC) is needed. This controller has similar functionality and connectivity as the Base Station Controllers (BSCs) in core mobile networks. Since the land-based mobile core network remains unchanged, it is possible to deliver full service and operational transparency across fixed-mobile networks. Standard Access Points and broadband IP-based networks are used for communications. The high-level UMA-based network architecture is shown in Figure 15.1.

Two components make up the new UMAN network: the dual mode **Mobile Stations** (MS), generally in the form of mobile handsets and portable terminals, and the **UMA Network Controller** (UNC). UNCs are collocated with mobile core network components and their coverage depends on the area designated to support WLAN/GSM convergent services. UNCs support discovery, registration, intra-system handover to other UNCs and inter-system handover between WLANs/WPANs and the **GSM/GPRS EDGE Radio Access Network** (GERAN).

In this diagram, the public land-based Mobile Core Network (MCN) is represented by the Mobile Switching Center (MSC), Home/Visiting Location Registers (HLR/VLR), and the Authentication, Authorization and Accounting (AAA) server. All these components provide well-known roaming and security features in traditional cellular mobile networks. Voice services are provided through the **"A" interface** while data services are provided through the **"Gb" interface**, via the SGSN.

15.3 UMA/GAN-based Networking Overview

UMA/GAN is a fixed-mobile convergent network and a technology that allows transparent access and roaming between cellular mobile wide area networks and fixed wireless local and personal area networks. This is accomplished by using dual-mode or multi-mode mobile phones. The use of a single handset and a single phone number are key features of UMA/GAN technology. This convergence is applicable to those second and third generations of mobile networks that are based on GSM/GPRS and UMTS. These architectures were shown in figures in the previous sections. The change in name of Unlicensed Mobile Access to Generic Access Network is very much related to the ambiguous connotation of the word "unlicensed" which has sometimes been construed as illegal. In any event, all the technical considerations regarding UMA remain the same with the GAN.

The landscape of fixed-mobile convergence, using dual-mode or multi-mode handsets, is by far more fragmented than the one suggested in the introduction to this chapter. In addition to UMA/GAN, there are three other proposals and implementations: MobileIGNITE Alliance (Integrated Go-to-Market Network IP Telephony Experience), SCCAN (Seamless Converged Communications Accross Networks), and GSM Interworking Profile (GIP).

MobileIGNITE is a SIP-based signaling and convergence technology promoted by BridgePort Networks. Collaborating with them are: Kyocera Wireless, Bingo Wireless, and several SIP and wireless infrastructure vendors. So far, however, there has been no support from the larger mobile carriers. This technology uses the VeriSign software, a pilot implementation that has Net2Phone as an operator. More details are provided in Section 16.14.

SCCAN was initially promoted by Avaya, Proxim, and Motorola, later joined by Chantry Networks (now Siemens), Colubris Networks, Meru Networks, and 2Wire. SCCAN extends PBX functionality to mobile employees in and out of the enterprise. It is focused on Voice over WLAN and mobility management between WLANs and cellular networks. It was tested as part of the Syracuse University Real-World Laboratory.

GSM Interworking Profile (GIP) or simple **Interworking Profile** (IWP) is a follow-up on the attempt to converge DECT cordless phones with GSM mobile phones. It is based on DECT standards, implementing DECT/GIP. It requires an infrastructure to connect DECT base-stations supporting DECT/GIP to GSM networks.

15.4 UMA-based WLAN and GSM/CDMA Convergent Networking Solution

A good example of the convergence of networks having the **UMA Network Controller** (UNC) component as the core, and based on the UMA architecture and specifications, is that between WLANs and mobile GSM/CDMA. This is shown in Figure 15.2.

The core mobile network provides interconnectivity within the mobile world and connectivity to the wired world PSTN and to the Internet through a GSM/GPRS data-oriented network. The mobile agents/clients are UMA-enabled handsets or terminals that support both WLANs (802.11x or HiperLAN2) and WPANs (Bluetooth or ZigBee).

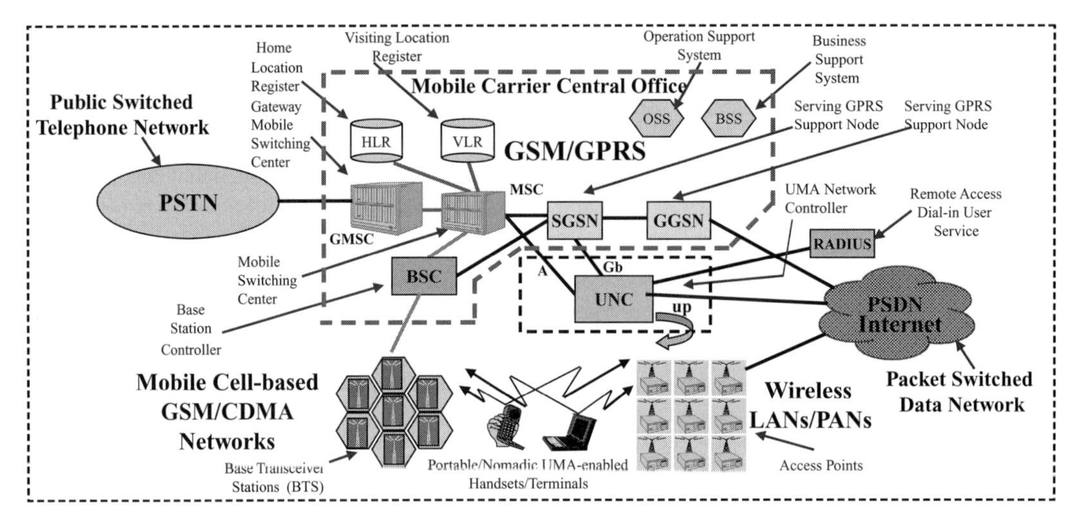

Figure 15.2 UMA-based WLAN and GSM/CDMA Convergent Network Architecture

15.5 Advantages and Disadvantages of the UMA/GAN Technology

The advantages of implementing UMA/GAN derive from the following aspects:

- Use of unlicensed portions of the frequency spectrum for convergent fixed wireless networks, such those used in WLANs (IEEE 802.11x) and WPANs (IEEE 802.15x);
- Use of existent GSM/GPRS and EDGE/UMTS network infrastructure with no essential changes; only the addition, for convergence, of the UNC/GANC component to support the UMA/GAN architecture;
- Leveraging on existent cellular mobile voice, data, and multimedia services and applications;
- Extension of cellular mobile physical coverage by adding cheap arrays of WLAN access points and Bluetooth-enabled masters;
- Providing high transmission bandwidth at lower cost, usually a fraction of the cost of pure mobile networks; this will serve to reduce loads on congested mobile networks;
- Use of common, dual-mode or multi-mode GSM/CDMA handsets instead of two different phones: a Wi-Fi phone and a regular cellular mobile phone;
- Bundling mobile services with WLAN services in traditional hot-spots (such as the case with T-Mobile);
- Use of VoIP services and Internet network and service infrastructure, both at no cost or lower costs, in comparison with traditional mobile networks data services.

There are some disadvantages that come with the implementation of UMA/GAN:

- Use of more expensive, complex, heavier handsets;
- The relatively high power requirements for WLANs may drain batteries faster, a critical feature for mobile users;

- The current UMA/GAN implementations are focused on 2G and 2.5G mobile networks convergence, so future generations of mobile networks will require enhancements, yet to be standardized;
- It is not clear yet what the charges, and resulting impact on the mobile service revenues, will be for UMA/GAN types of services.

15.6 UMA/GAN Standard Specifications

UMA/GAN specifications, prior to their adoption by 3GPP in June 2005, were developed by the **Unlicensed Mobile Access Consortium** (UMAC). The UMA/GAN 3GPP specifications are contained in the **TS 43.318** and **TS 44.318** technical standards. These specifications, known as Release R1.0x UMA, consist of the following distinct parts:

- UMA Stage 1 Specification (**User Perspective**), last update September 2004;
- UMA Stage 2 Specification (**Architecture**), last update May 2005;
- UMA Stage 3 Specification (**Protocols**), last update May 2005;
- UMA Conformance Specification (**Testing**), last update, June 2005.

There are other standard activities that contribute to the development and implementation of FMC such as:

- ETSI TISPAN, (www.etsi.org);
- 3GPP2 IMS Consortium, (www.3gpp.org);
- IETF SIP Working Group, (www.ietf.org); and
- CTIA & Wi-Fi Alliance, common dual-mode handsets testing, (www.ctia.org), (www.wi-fi.org).

It is planned to have the UMA network access specifications compatible with the IP-based Multimedia Subsystem (IMS) recommendations developed by the 3G Partnership Project. IMS will provide convergence of GSM/GPRS/UMTS/CDMA networks with any wireless and wired network. UMA is viewed as an FMC access technology where IMS is a long-term service-oriented technology. Additional technical and implementation-based information can be found at the (http://www.umatechnology.org) website.

15.7 UMAN UNC Design Requirements and Functionalities

UNC is the core component of the UMA network architecture. As noted above, UNC communicates with the mobile core network through the **"A" interface** for circuit switched voice services, and with SGSN, through the **'Gb" interface**, for packet data services. Connectivity with multi-mode terminals is through the **"up" interface** that relays user and signaling information. The primary functions of the UNC are:

- Provides discovery, registration, and redirection of services to allow the mobile stations to connect to the appropriate UNCs (provisioning, default, current);

- Provides secure, private, communications over open IP networks between each mobile station and the service provider core network. UNC contains a security gateway that implements a secure IP interface towards each mobile station, using IPsec tunnels. The tunnel assures data integrity and confidentiality while authentication is provided through the RADIUS interface to an AAA server;
- Relays core network control signaling for higher layer stations and GSM/GPRS;
- Sets up and tears down UMAN bearer connections for circuit and packet services;
- Transcodes the voice bearer from Voice-over-IP transport to voice-over-circuit transport towards the conventional PCM-based A interface;
- Emulates paging, handover, and similar radio access procedures for UMAN mobile access; and
- Provides standards-compliant A and Gb interfaces with appropriate physical, signaling, and bearer interfaces.

15.8 UMAN UNC Discovery and Registration

UMAN discovery and registration allows UMA-enabled mobile stations to access UMA services, assuming there is UMA/UNC support throughout a larger coverage area. The discovery and registration process is depicted in Figure 15.3.

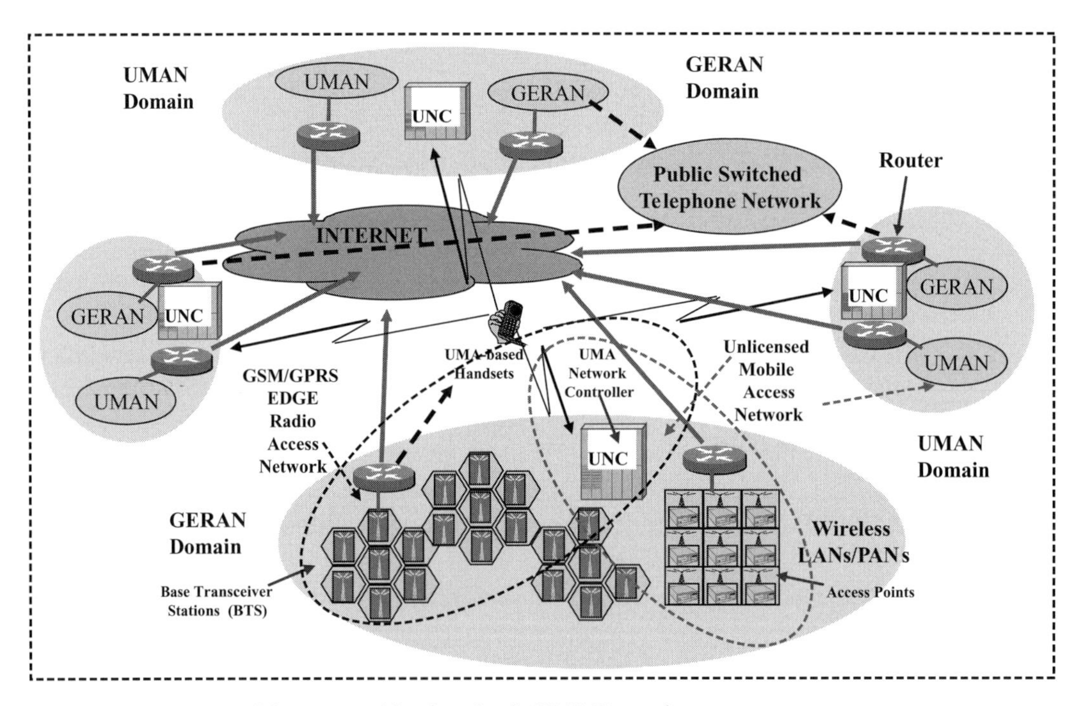

Figure 15.3 Discovery and Registration in UMA Networks

Regular voice services, including roaming, are provided through MSCs that have connectivity to the PSTN network. Regular data services are provided through

GSM/GPRS EDGE networks that contain GGSNs, not shown in the diagram. Interspersed within the global cellular domains, there are islands of WLAN/WPAN Access Points with direct connectivity to the Internet through a router-based network. Each UMA/UNC provides data and voice services handoff of UMA-enabled MS between two domains. One domain is the Unlicensed Mobile Access Network (UMAN), backed by fixed wireless access points. The other is the GSM/GPRS EDGE Radio Access Network (GERAN), backed by a GPRS EDGE data network. In this diagram, several combinations of GERAN and UMAN domains are presented as islands. In reality, mobile cell-based coverage is quasi-total for a given area. UMA coverage depends on the presence of UNCs. Although there is one physical UNC, it logically belongs to both the UMAN and GERAN functional domains.

In the process of registration and discovery, we can distinguish three functional types of UNCs: **provisioning, default, and serving**. When subscribing for UMA services, each MS, based on its original location, is assigned a provisioning UNC. This is one aspect of registration. The actual discovery process starts when the MS connects to the provisioning UNC whose IP-address is provided through an Internet Domain Name System (DNS) lookup.

Based on location information and subscription to UMA services, the provisioning UNC will assign a default UNC to the MS. This information is stored in the MS for future use. To obtain the UMA services, the MS registers with the default UNC. Once the MS is registered, the default UNC will become the serving UNC for that particular session, assuming that the MS continues to stay in the location covered by the default UNC.

When a MS moves outside the default UNC coverage area, the UNC will determine the appropriate serving UNC based on the MS's location. The MS will once again register with the new serving UNC and will store this new information for future use, thereby avoiding the redirection process. So the calls can be sent and received through the mobile network, the MS information is sent through the UMAN domain to the core network. Each time the MS connects with a new serving UNC, discovery and registration messages are protected and authenticated by using an IPsec tunnel.

15.9 Securing the UMA "up" Interface

As shown in Figures 15.1 and 15.2, the "**up**" interface is established between the UMA-enabled mobile station and the UNC over the air to the AP, over the public IP-based Internet or private packet switched data network, then to the mobile core network. This path is not secured. To provide the necessary security (data integrity and confidentiality), the "up" interface creates an **IPSec tunnel** between the mobile station and the UNC [83]. The UNC MS performs authentication and encryption of all communications on that link. The IPsec tunnel will handle GSM/GPRS signaling and any voice, data, or multimedia packet flows between the MS and UNC. Initial configuration of the UMA-enabled handsets includes service set IDs, user names, and security keys, all stored in a profile on the handsets along with home/hot-spots profiles [84].

Voice traffic is carried over the Voice over IP (VoIP) protocol stack. The User Datagram Protocol (UDP) is used in the Transport Layer in combination with the Real-Time Protocol (RTP). This stack is the same as that used on fixed line networks. Therefore, the VoIP protocol stack can interface with compatible fixed-line media gateways.

15.10 GERAN to UMAN UNC Handover Operation

To understand how UMA provides transparent handoff/handover between GERAN and UMAN, we refer to Figure 15.3. To begin, assume that the MS is in a voice call originated in, and was handled by, a regular GSM network. The voice traffic is carried through a mobile network BTS, and the tandem between BSC/MSC assures mobility, roaming, and connectivity to the PSTN.

When the UMA-enabled dual-mode handsets come into proximity of a UNC, the UNC starts the discovery and registration process described in the previous section. Once registered, the serving UNC will provide system information to the registered MS to allow connectivity to an Access Point of the UMAN (WLAN or WPAN). Based on the MS measurements and its reporting to UNC, GERAN will treat the UMAN access point as a regular cell with adequate signal strength and quality.

If all conditions are met, based on established algorithms, the GERAN will ask the serving UNC to initiate the handover operation using standard GSM signaling procedures. To assure the necessary resources are available, the UNC will inform the core mobile network about the coming change. A Voice over IP path will be established between the MS and the serving UNC and the MS software client will switch from access over GSM to access over WLAN/WPAN. The new connection over the "up interface", as supported by the UMA-enabled WLAN/WPAN APs, is established before breaking the GSM connection to assure a soft handoff. Any traffic, voice or data, will be channeled over this AP as long as the MS stays in its proximity.

15.11 UMAN to GERAN UNC Handover Operation

To understand the UMAN-to-GERAN UNC handover, consider a mobile station that already has a voice call or data transmission in progress over the UMAN. This means the UMA-based MS has gone through the discovery and registration process and a serving UNC is supervising the ongoing call. Nevertheless, the handover from UMAN to GERAN is always initiated by the multi-mode handset or mobile station. This decision is taken by the MS based on continuous monitoring of WLAN signal strength, characteristics of the voice channel, and status information coming from the serving UNC. When such a decision is taken, the MS sends a "handover required" message to the UNC and indicates the GSM cell to be used for handover.

At this time, the serving UNC, using common signaling, informs the core network about the request for handover, indicating the GSM cell that should be used to take over the call. Using standard GSM handover signaling, the core network will request that the target GSM cell allocate the resources required for handover. To assure a soft handoff, the call is handled as an inter-BSC operation while the UMA client in the MS switches from the WLAN/WPAN to the GSM network. Subsequent voice calls and data transmitted from the mobile station will be sent over the GERAN.

For new voice calls initiated by the UMA-enabled MS, requests for services will be sent by the MS to the GSM network using standard connection management procedures. Signaling messages using the "up" interface are relayed to the serving UNC and forwarded to the MSC and GWMSC, responsible for preparing the PSTN to take the call if necessary.

When the UMA MS requires a GPRS connection, the MS activates a UMA Radio Link Control (URLC) transport channel to the serving UNC over the "up" interface. This channel will transport packets to the serving UNC that then forwards them to the SGSN over the UMA Gb interface. In a similar fashion, GPRS packets coming from the mobile core network are passed by the SGSN to the serving UNC and from there to the MS over the active URLC transport channel. This channel is established when there is a data flow and terminated when the flow of packets ends.

15.12 UMAN Signaling Protocol for Voice Communications

UMA Network Controllers (UNCs) communicate with **Mobile Stations** (MSs) through "**user plane**" or "**up**" interfaces. UNCs use the IP-based protocol layer architecture over standard Access Points and Broadband IP networks. They maintain communication with MSs and relay signal call control information from **GSM** networks, more precisely from one of the **Mobile Switching Centers**, which are part of the core public mobile land-based network. UMAN signaling for voice communications is shown in Figure 15.4.

The signaling protocol is based on the SS#7 architecture. This consists of Message Transfer Parts 1, 2, and 3 (MTP1, MTP2, and MTP3), for the first three layers of the protocol stack, and the Signaling Connection Control Part (SCCP) for the Transport Layer. Signaling protocols in mobile networks include additional fields for the upper layers that contain information related to the applications and services supported as part of the fixed-mobile convergence. These fields are: Basic Station Subsystem Management Applications Part (BSSMAP), Mobility Management (MM), and Call Control/Short Message Service (CC/SMS).

15.13 UMAN Signaling Protocol for Data Communications

When transmitting data across convergent networks, a UNC communicates with one of the **Serving GSM Supporting Nodes** (SGSN) through a "**Gb**" interface over IP-based standard Access Points and Broadband IP networks. UNCs maintain communications with MSs by transparently relaying data to/from the **GSM/GPRS** or **GSM EDGE Radio Access Network** (GERAN). These are subnetworks of public land-based mobile

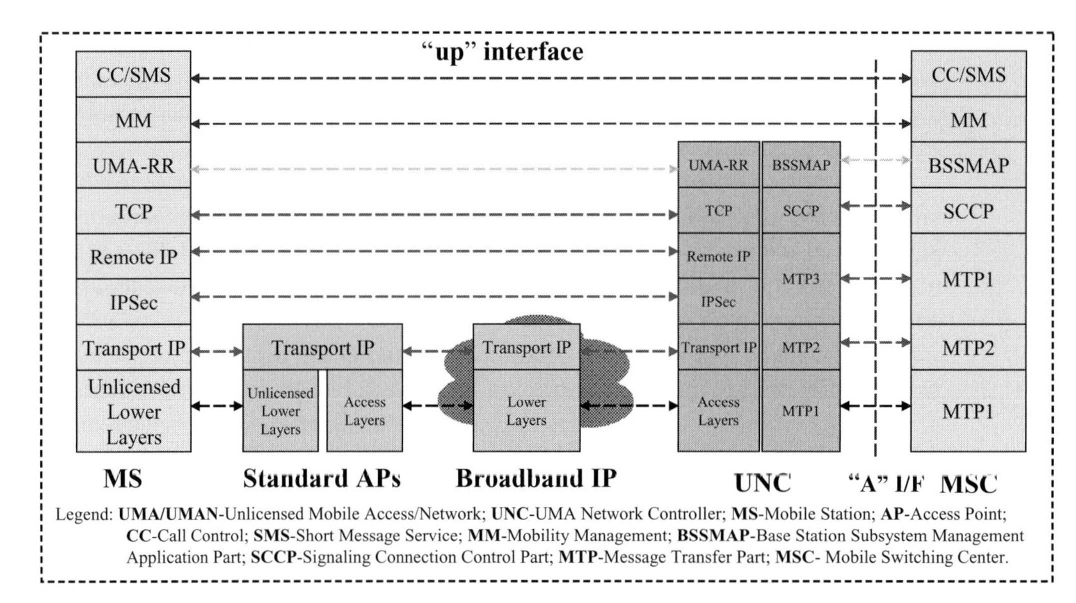

Figure 15.4 UMAN Signaling Protocol Architecture for Voice Communication

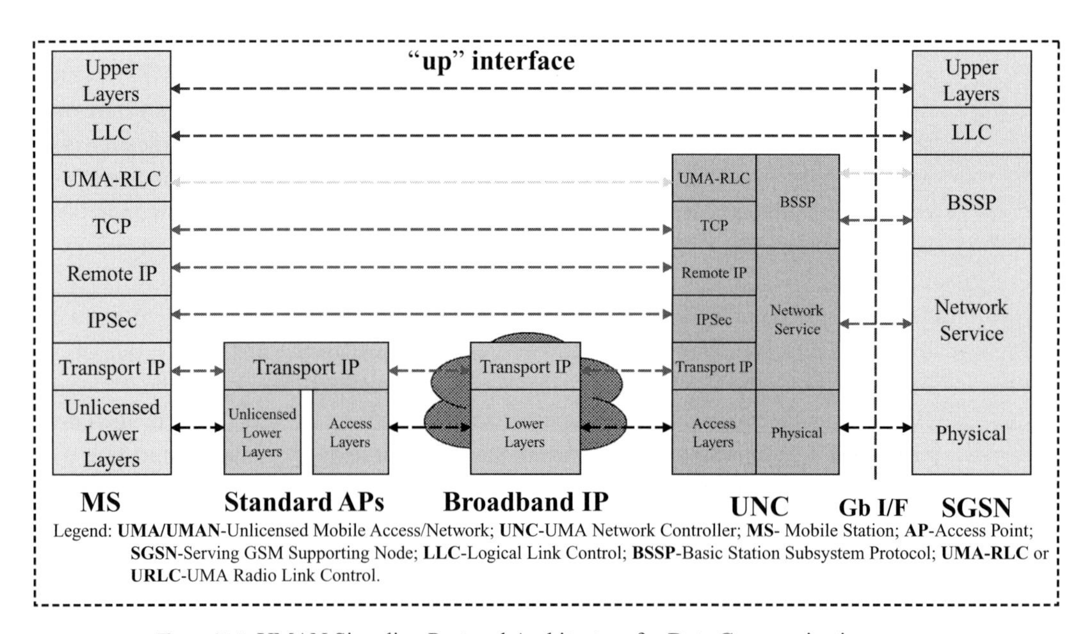

Figure 15.5 UMAN Signaling Protocol Architecture for Data Communications

networks that support data services. UMAN signaling for data communications is shown in Figure 15.5.

Transmission of data across fixed-mobile convergent networks is done by encapsulating or tunneling the TCP/IP stack used in APs and broadband IP-based networks into the Data Link Layer of the protocol stack used between the Serving GSM Supporting

Node (SGSN) and Mobile Stations. The encapsulated stack includes the Basic Station Subsystem Protocol (BSSP) sublayer, specific to the mobile environments. On the top of the SGSN-MS Data Link Layer is the standard Logical Link Control (LLC) sublayer followed by the network and upper layers of the SGSN protocol stack.

The "up" interface allows transparent communication of GPRS LLC PDUs for signaling and data transmission between a UMA-enabled MS and a SGSN. This means that all GSM/GPRS services available in a GERAN BTS, such as SMS, MMS, WAP, and IN/CAMEL will be available to the MS with no additional modifications required to the core mobile network. Change appears only in the GPRS Radio Link Control (RLC) protocol stack where GPRS RLC is replaced with the UMA-RLC (URLC) protocol. As in a GERAN base station, the UNC, with functions similar to BSC, terminates the UMA-RLC protocol. All the packets are forwarded through the Gb interface to the SGSN using the Basic Station Subsystem Protocol (BSSP) messaging system [84].

15.14 UMAN Mobile Station Lower Layers Protocols

UMA/GAN Mobile Stations (MS) act either as mobile handsets or as terminals that support, in addition to the standard GSM/CDMA protocol stacks, the lower layer protocols for WLANs (IEEE 802.11× and ETSI HiperLAN2) and WPANs (802.15.1 Bluetooth). Hence, the names of dual-mode, tri-mode, or multi-mode handsets. This architecture allows communications with standards based Access Points or Master Bluetooth. The lower layers protocol stack for UMA/GAN mobile stations is shown in Figure 15.6.

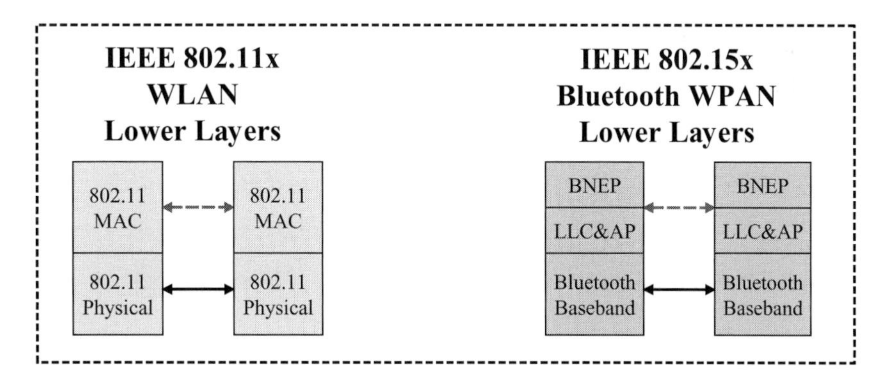

Figure 15.6 UMA/GAN Mobile Stations Lower Layers Protocol Stack

Dual mode WLAN/GSM mobile stations have four modes of operation:

- GERAN only; The MS is used only as a GSM/CDMA mobile phone;
- UMAN/GAN only; The MS is used only as a Wi-Fi smart phone;
- GERAN preferred; The MS connects to the GSM/CDMA network, as the first choice, provided signal strength is acceptable; and
- Wi-Fi preferred; The MS connects to the WLAN network, as the first choice, provided the signal strength is acceptable.

Similar modes of operations can be established for convergence between Bluetooth and GSM/CDMA. In that case, the tri-mode MS will interact with the designed Bluetooth master, hosted by a home desktop or laptop computer. The Bluetooth lower layers protocol stack consists of Bluetooth Network Encapsulation Protocol (BNEP) on top of the Logical Link Control and Adaptation Protocol (LLC & AP or L2CAP) on top of the Bluetooth radio baseband Physical Layer. If the underlying network is an Ethernet, the BNEP header replaces the Ethernet header. The Ethernet payload remains the same. The BNEP header and the original payload are encapsulated by the L2CAP header. Maximum BNEP payload is 1691 bytes, where 191 bytes are reserved for the BNEP header [85]. The Ethernet payload can vary from 0 to 1500 bytes.

15.15 UMA-based FMC Solutions, Products, and Services

Incorporation of UMA/GAN specifications into the 3GPP technical standards opened up opportunities for manufacturers and mobile operators alike to provide convergent products and services. By bundling Wi-Fi calls with their mobile networks, carriers can enhance coverage and presence in the relatively under-penetrated residential sector. It is expected to see, in the near future, use of Dual-Mode Handsets (DMH) not only in WLAN hot-spots but in enterprise WLANs as well. Despite the advantages listed earlier, major operators in the USA, with the exception of T-Mobile, still fear they will lose revenue by offering convergent services. The introduction of convergent features in the latest mobile phones models, including the 2007 BlackBerry and iPhone editions, will certainly change this situation.

Companies such as **ABI Research** predict an increase of DMH to reach 50 million units in 2009, 100 million in 2010, and 256 million in 2012. **In-Stat's** forecast calls for 66 million DMHs in operation by 2009. Similar predictions based on VoIP, advancements that go along with FMC, are from **Frost & Sullivan**. DMHs costing less than $100 are key for this penetration. Lists of current UMA-based products and services follow:

UMA-based chipsets

- **Airify Communications Inc** and **Helic S.A.**: Chipset for Cellular-WLAN convergence, (www.airify.com);
- **Infineon TechnologiesAG**: BlueMoon Unicellular chipset for Bluetooth Cellular FMC, (www.infineon.com);
- **Texas Instruments**: Embedded UMA for Instant Wi-Fi/GSM handoffs, (www.ti.com).

UMA-based Dual Mode Handsets (DMH)

- **Calypso Wireless**: DMH C1250i Wi-Fi/GSM-GPRS, (www.calypso.com);
- **HP**: iPAQ 510 DMH (based on Kineto Wireless technology), (www.hp.com);
- **Kyocera**: DMH Slider Remix CDMA/Wi-Fi, (www.kyocera-wireless.com);
- **LG Electronics**: Mobile Communications, UMA-enabled tri-band cellular and Wi-Fi LG CL400 mobile phones, (www.lge.com);
- **Motorola Corporation**: A910 GSM/Wi-Fi/Bluetooth, (www.motorola.com);

- **Motorola Corporation**: DMH CN650, Avaya Seamless Communication Solutions (Avaya, Motorola, Proxim), (www.motorola.com);
- **NEC**: DMH N900iL 3G/WLAN UNIVERGE, (www.nec.com);
- **Nokia**: Nokia N80 and Nokia 6136, UMA-based Wi-Fi/GSM, (www.nokia.com);
- **Nokia Corporation**: DMH Nokia 9500 Communicator, GSM/GPRS EDGE and Wi-Fi tri-band, DMH, (www.nokia.com);
- **Research in Motion** (RIM): Blackberry 8220 GSM/CDMA/iDEN and Wi-Fi, (www.rim.com);
- **Samsung Corporation**: P200 SGH-T709, GSM/GPRS, EDGE/Wi-Fi, (www.samsung.com).

UMA/GAN Systems Integrators

- **Alcatel/Lucent**: NGN-UMA Architecture, Standalone UMA Architecture, Alcatel 1430 Home Subscriber Server, Alcatel 1300 OMC-CN, (www.alcatel-lucent.com);
- **Aruba Wireless Networks**: WLAN AP for FMC using NTT DMH, (www.aruba.com);
- **BridgePort Networks**: NomadicONE Network Convergence Gateway software for wireless carriers and MVNOs, (www.bridgeport-networks.com);
- **Ericsson**: Mobile @Home SupportNode, Mome Base Station Controller, Security Gateway, (www.ericsson.com);
- **Kineto Wireless Inc.**: Mobile over Wireless LAN (UMA-based MoWLAN), UMA-compliant handset client software, UMA development tools, (www.kinetowireless.com);
- **Meru Networks**: Cellular WLAN and Meru System Director Version 3, (www.merunetworks.com);
- **NXP**: Nexperia Wi-Fi Cellular Systems Solution 7210, (www.nxp.com).

UMA-based Services and Implementations

- **British Telecom**: Bluephone FMC Project, "BT Fusion", UMA-based Bluetooth/GSM service and Wi-Fi/GSM service, (www.bt.com);
- **Cincinnati Bell**: CB Home Run, based on tri-mode Nokia 6086 headsets, (www.cincinnatibell.com);
- **Embarq**: (Sprint USA), SMART Connect Plus Service using STARCOM PC6200 dual mode handsets (WiFi/CDMA), (www.embarq.com);
- **Korea Telecom**: FMC Service covering Wi-Fi hot-spots and residential WLANs, (www.koreatelecom.com);
- **NTT**: DoCoMo DMH 3G/Wi-Fi, (www.ntt.com);
- **Osaka Gas**: FMC 3G/Wi-Fi using Meru Networks APs and NEC N900iL DMH, (www.osakagas.co.jp);
- **Orange**: "unik/unique", UMA WLAN/GSM based service, (www.orange.com);
- **Saunalahti**: (Finland MSP), Kineto Wireless Technologies and Nokia handsets;
- **Telia Sonera**: (Denmark) "Home Free", UMA-based Wi-Fi/GSM service, (www.teliasonera.com);
- **Telecom Italia**: "Unica", UMA-based WLAN/GSM service, (www.telecomitalia.com);

- **T-MobileUS**: "HotSpot@Home", UMA-based Wi-Fi/GSM/GPRS EDGE and WCDMA 3G, 7000 hot-spots USA and Europe, (www.t-mobile.com).

UMA-based Convergence Testing

- **Agilent Technologies**: UMA testing, (www.agilent.com);
- **Anritsu Corporation and Setcom**: Anritsu MD8470A Signaling Tester and Setcom UMA/GAN Application Testing, (www.anritsu.com, www.setcom.eu);
- **CTIA & Wi-Fi Alliance**: DMH testing, (www.ctia.org), (www.wi-fi.org);
- **Rohde & Schwartz**: UMA testing, (www.rohde-schwartz.com).

15.16 UMA-based Nokia Dual-mode 6301 Handset

Wi-Fi Alliance data indicates that almost 90% (82 out of 92) of the Wi-Fi certified phones introduced since 2004 function in dual or multi modes. In addition to Voice over WLANs, these phones will also support one of the cellular mobile access technologies that will be used in standard handsets that operate outside of WLAN coverage. We listed earlier several UMA-based or "UMA-like" dual-mode handsets from major manufacturers such as Nokia, Motorola, Samsung, RIM Blackberry, and Apple Computers. One example of UMA-based handset/cell phones are Nokia's series 6301. Orange, one of the mobile carriers, is expected to be the first operator to offer these cell phones in Europe for its Unik/Unique service. Nokia has already tested previous UMA-based models 6136 in a well publicized field trial in Oulu, Finland.

The main characteristics of these phones are [86]:

- GSM tri-band support 900/1800/1900 GHz;
- Wi-Fi 802.11 b/g support;
- Form factor (93g weight and 13.1 mm thick);
- 2-inch QVGA 320×240 screen;
- Mini USB/PC synchronization;
- Internal user memory in-box microSD card of 30MB and 128MB, expandable to 4GB;
- Voice dialing, voice commands and voice recording;
- Digital camera, 2 Megapixels;
- MP3 player, FM radio;
- Music and video streaming;
- Bluetooth 2.0-enabled; Headset BH-208 or BH-602;
- Integrated hands-free speaker;
- Push to Talk;
- GPRS class 10, 53.6 Kbps and EGPRS class 10, 236.8 Kbps data services;
- Talk time up to 3.5 hours (UMA up to 3 hours);
- Standby time up to 14 days (UMA up to 100 hours).

16 Session Initiation Protocol

16.1 What is the Session Initiation Protocol-SIP?

The **Session Initiation Protocol** (SIP) is an Internet application layer protocol designed to control voice calls or multimedia session setups. SIP is a family of IETF-recommendations and draft standards that provide signaling specifications for Internet conferencing, VoIP, event notifications, and instant messaging over IP-based networks.

SIP-based signaling serves as multimedia call control between calling parties, and is used to set up calling sessions such as audio/video conferences, peer-to-peer communications, or to just indicate the caller's presence. SIP works in tandem with another application layer protocol, **Session Description Protocol** (SDP), hence the tandem abbreviation of SIP/SDP.

As indicated in Figure 16.1, there are three major components that make up a SIP network/system: Endpoints **SIP User Agents** (SIP-UA), the callers and beneficiaries of services; **SIP Proxies** (SIP-P), acting as relays of SIP messages; and **SIP Registrars** (SIP-R), acting as SIP directories.

As indicated with broken lines, communications between SIP UA and SIP Proxies as well as between SIP Proxies is limited to signaling. The actual transport service is delivered by specialized Applications Servers (AS) such as VoIP Servers, Presence Servers or Push-to-Talk over Cellular (PoC) Servers. The user information, data or digitized voice/video multimedia, is carried in separate frames and packets, using different transport protocols.

16.2 SIP System Architecture

A high-level SIP-based functional model is indicated in Figure 16.2.

Communications between **SIP User Agents** may use a series of SIP proxies that relay SIP/SDP call setup messages. The actual traffic flow, indicated with solid line, takes a different route and involves the **Real-Time Transport Protocol** (RTP) and **Real-Time Control Protocol** (RTCP). SIP proxies may be part of the same SIP domain or different SIP domains. Call setup or session establishment follows an exchange of multiple messages between SIP User Agents end-points and SIP Proxies. This is shown in Figure 16.3.

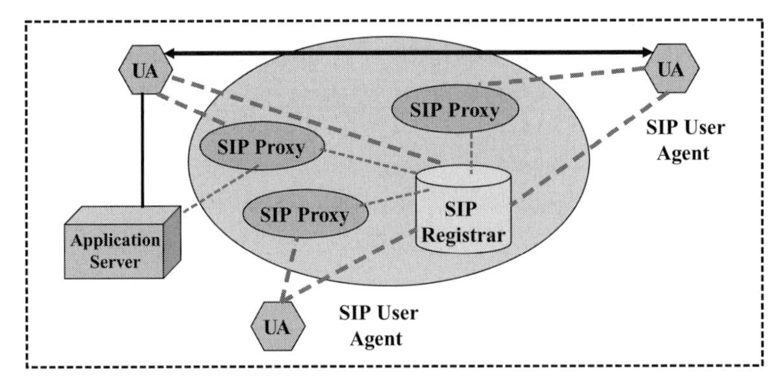

Figure 16.1 Session Initiation Protocol (SIP) System Components

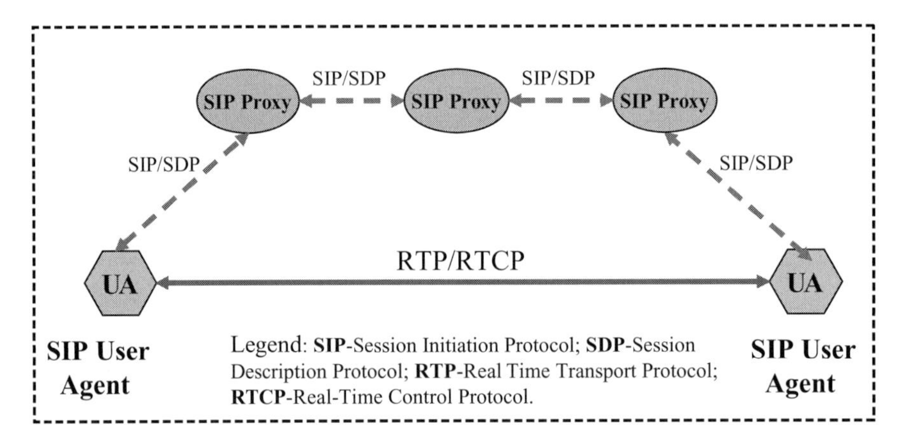

Figure 16.2 SIP Basic Functional Model

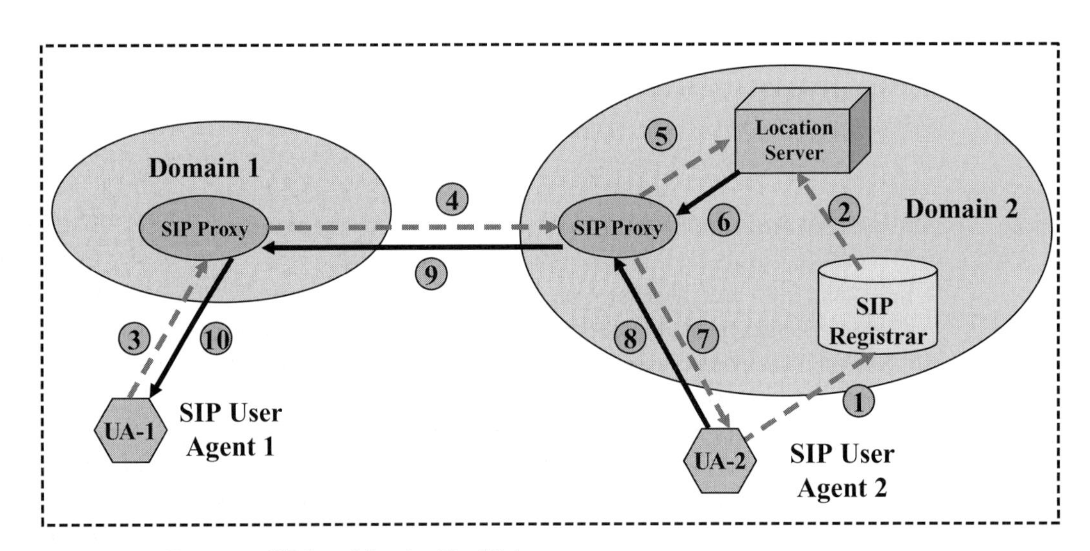

Figure 16.3 SIP-based Session Establishment

One of the services proposed as a VoIP application is Peer-to-Peer (P2P) communication based on SIP/SDP signaling. This is an extension of traditional VoIP architecture, and is based on SIP client-server communications architectures [87]. Each User Agent is identified by a **Universal Resource Identifier** (URI) that has a format similar to the Universal Resource Locator (URL) used in web services.

To establish a communications session between SIP User Agent 1 (UA1), the initiator of the call, and SIP User Agent 2 (UA2), the called party, the following steps are performed:

1. UA2 must register with its local SIP Registrar;
2. UA2 information (SIP URI and the current IP address) is stored in the Location Server;
3. When UA1 (already registered) wants to establish an instance of communication with UA2 it sends an INVITE message to its SIP Proxy 1, which belongs to SIP Domain1;
4. After a DNS lookup in its Location Server (not shown in this diagram), the request is relayed to SIP Proxy 2, which is found in SIP Domain 2;
5. SIP Proxy 2 searches for UA2 in the Location Server;
6. The Location Server acknowledges UA2 and its URI;
7. This information is passed as an acknowledgement to UA2;
8. The information can be exchanged starting from UA2;
9. The information is relayed between SIP Proxy 2 and SIP Proxy 1;
10. Finally, the conduit between UA2 and UA1 is ready to support the session.

In many instances, the SIP Registrar, the Locations Server, and the SIP Proxy belonging to a SIP domain are implemented in the same physical unit, a generic SIP Proxy.

16.3 SIP Overview

SIP is a message-based client-server signaling protocol, similar to HTTP and SMTP. It is used to initiate interactive communication sessions between end-users to support voice, video, data, chats, and games. The main SIP features are:

- SIP transparently supports name mapping and redirection services, allowing the implementation of ISDN and Intelligent Network (IN) telephony subscriber services. These facilities also enable personal mobility;
- Capabilities to support establishment, maintenance, modification, and termination of multimedia communications. These capabilities are:
  - **User location**: Determination of the end system to be used for communications;
  - **User capabilities**: Determination of media and media parameters to be used;
  - **User availability**: Determination of the willingness of the called party to engage in communications;
  - **Call setup**: "Ringing", establishment of call parameters at both called and calling party;
  - **Call handling**: Transfer and termination of calls.

SIP, as an Application Layer protocol, runs on top of any of the three transport protocols: **Transport Control Protocol** (TCP), **User Datagram Protocol** (UDP), and **Stream Control Transmission Protocol** (SCTP). TCP is a reliable, connection-oriented transport protocol. UDP is a connectionless, best-effort transport protocol. Both protocols were introduced in Section 2.6, Chapter 2 of this book dealing with the Internet TCP/IP-based communication protocols stack.

SCTP is a connection-oriented transport protocol that was designed specifically to support SIP. It uses message or datagram formats, combining the strengths of TCP and UDP. These features make it relatively easy to parse and interpret signaling messages, at least in comparison to TCP which is a stream-oriented protocol. Parsing the stream to extract the signaling message requires application layer framing. SCTP was developed by the IETF SIGTRAN Working Group and it is also known as the SIGTRAN signaling protocol.

Unlike other known application layer protocols such as HTTP, SIP may use UDP. When sent over TCP or UDP, multiple SIP transactions can be carried in a single TCP connection, UDP datagram, or SCTP association. UDP and SCTP datagrams, including all headers, should not be larger than the path Maximum Transmission Unit (MTU), where it is known, or 1500 bytes if the MTU is unknown.

SIP, despite its market success (it was selected as the IP-based Multimedia Subsystem, IMS, signaling protocol), is not the only signaling protocol used. There is a whole world of voice communications using SS#7. And, there is H.323, the "legacy" signaling protocol for Internet voice and video conferencing. There are also special-purpose signaling protocols such as the Media Gateway Control Protocol (MGCP), which is used in IMS. And, there are many proprietary signaling protocols like Avaya's H.323 variation, Cisco's Skinny, and Nortel's Unistim, used in VoIP enterprise systems.

Further, many of the successful applications promoted by Google, Skype, MySpace, and YouTube, related to instant messaging, peer-to-peer communications, and streaming audio video, do not use SIP. For example, Asterisk uses IAX, GoogleTalk uses XMPP, and Skype uses still another proprietary signaling protocol called SIP for Instant Messaging and Presence Leveraging Extensions (SIMPLE). To be fair, we have to mention that the popular application AOL Instant Messaging (AIM) and Microsoft Messenger and Yahoo Messenger, use SIP. Reasons why SIP extensions and SIMPLE are not used more extensively are their increasing complexity and the eternal desire of major players to dominate the market with proprietary protocols, and the real fact that there are other signaling protocols more suitable for particular applications.

Callers and called parties are identified by SIP addresses. When making a SIP call, a caller first locates the appropriate server and then sends a SIP request. The most common SIP operation is the invitation. Instead of directly reaching the intended called party, a SIP request may be redirected or may trigger a chain of new SIP requests by proxies. The "objects" addressed by SIP are users at hosts, identified by a SIP URL. This SIP **Universal Resource Identifier** (URI) takes a form similar to a mailto or telnet, i.e., user@host. The user part is a name or a telephone number. The host part is either a domain name or a numeric network address. A user's SIP address can be obtained out-of-band, learned via existing media agents, included in some mailers' message headers, or recorded during previous invitation interactions. In many cases, a user's SIP URI can be implied from their email address.

16.4 SIP-based Message Exchange

To understand the SIP-based message exchange, we will consider a very common scenario, where the calling party, a SIP User Agent 1 (UA1) using a SIP-based mobile handset, tries to reach User Agent 2 (UA2). UA2 is equipped with a SIP-enabled fixed phone. As presented earlier, in Section 16.2, the session establishment requires a SIP Proxy Server.

To establish a session between endpoints, UA1 and UA2 must negotiate, beforehand, the session parameters (binding) using the Session Description Protocol (SDP). Typical transmitted information is network address, port number (where the information requested should be sent), and media format to be used. This offer is delivered to the called agent, UA2, which answers with its own SDP. After this exchange, UA1 and UA2 can continue the process of establishing the session using SIP-based specific signaling messages. This is indicated in Figure 16.4.

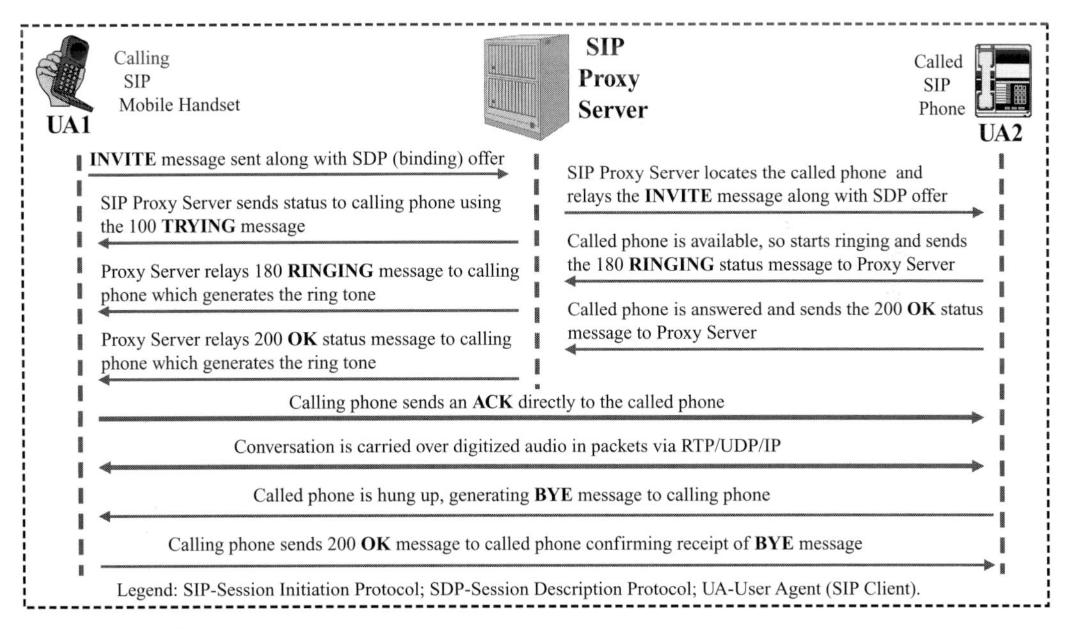

Figure 16.4 SIP-based Signaling Messages in VoIP

As the result of the INVITE message sent by calling party UA1, the SIP Proxy will locate the called phone and relay the INVITE message to UA2. In the meantime, the SIP Proxy will send the "100 TRYING" message to UA1. Once the invitation is accepted along with binding information, UA2 will send the "180 RINGING" message to the Proxy Server. The message is relayed to UA1, and will generate the ring tone. The acceptance of UA2 comes with the "200 OK" message, indicating UA2's readiness to participate in the multimedia session. Finally, UA1 sends an "ACK" message to acknowledge the UA2 response and the acceptance that will go directly to the called phone.

The conversation is carried between UA1 and UA2 using either the RTP or UDP as Transport Layer protocols over IP. At the end of the conversation, the called phone will generate the "BYE" message to the calling phone, followed by a "200 OK" message from UA1 to UA2, confirming termination of conversation.

16.5 SIP Message Format, Fields, and Options

A SIP message is either a Request from a client to a server, or a Response from a server to a client. Both Request and Response messages use the basic HTTP 1.1 format of RFC 2822, even though the syntax differs in character set and other specifics. Both types of messages consist of a start-line, one or more header fields, an empty line indicating the end of the header fields, and an optional message-body. SIP header fields are similar to HTTP header fields in both syntax and semantics. In particular, SIP header fields follow the definitions of syntax for the message-header and the rules for extending header fields over multiple lines. SIP message format is shown in Figure 16.5.

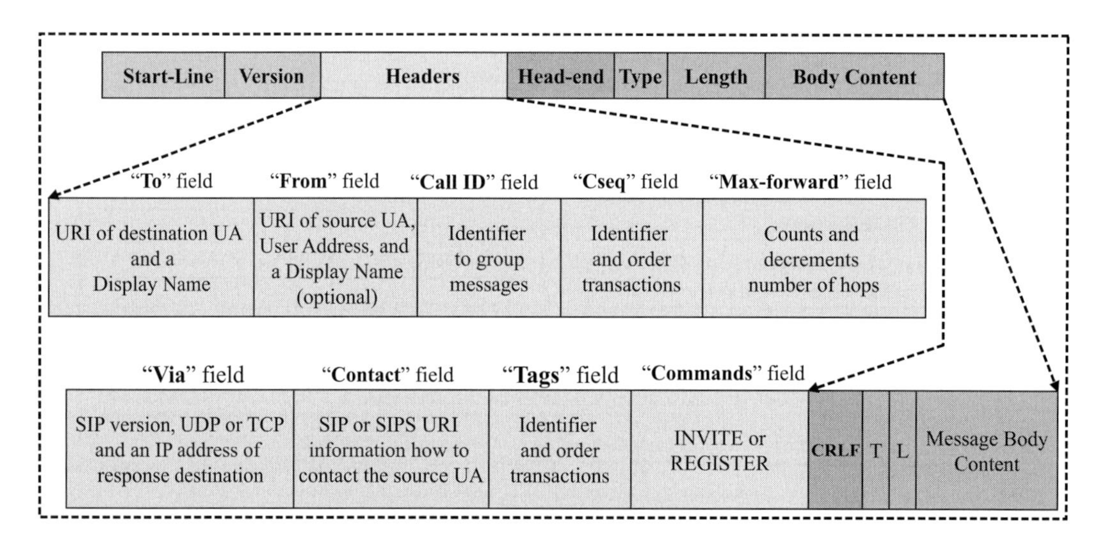

Figure 16.5 SIP Message Format, Fields, and Options

- The "**To**" header field specifies the desired "logical" recipient of the request, or the address-of-record of the user or resource that is the target of this request. This may or may not be the ultimate recipient of the request. The "To" header field may contain a SIP or SIPS URI, but it may also make use of other URI schemes. All SIP implementations must support the SIP URI scheme. Any implementation that supports TLS must support the SIPS URI scheme. The "To" header field allows for a display name;
- The "**From**" header field indicates the logical identity of the initiator of the request, possibly the user's address-of-record. Like the "To" header field, it contains a URI and

optionally a display name. It is used by SIP elements to determine which processing rules to apply to a request (for example, automatic call rejection). The "From" header field allows for a display name;

- The "**Call-ID**" header field acts as a unique identifier to group together a series of messages. It must be the same for all requests and responses sent by either UA in a dialog. It should be the same in each registration from a UA;
- The "**Cseq**" header field serves as a way to identify and order transactions. It consists of a sequence number and a method. The method must match that of the request. For non-REGISTER requests outside of a dialog, the sequence number value is arbitrary. The sequence number value must be expressible as a 32-bit unsigned integer and must be less than 2^{31}. As long as it follows the above guidelines, a client may use any mechanism to select "CSeq" header field values;
- The "**Max-Forwards**" header field serves to limit the number of hops a request can transit on the way to its destination. It is an integer that is decremented by one at each hop. If the Max-Forwards value reaches 0 before the request reaches its destination, it will be rejected with a "483" (Too Many Hops) error response;
- The "**Via**" header field indicates the transport used for the transaction and identifies the location where the response is to be sent. A "Via" header field value is added only after the transport that will be used to reach the next hop has been selected;
- The "**Contact**" header field provides a SIP or SIPS URI that can be used to contact that specific instance of the UA for subsequent requests. The Contact header field must be present and contain exactly one SIP or SIPS URI in any request that can result in the establishment of a dialog. For the methods defined in this specification, only the INVITE request is valid. For these requests, the scope of the "Contact" is global. That is, the "Contact" header field value contains the URI at which the UA would like to receive requests, and this URI must be valid even if used in subsequent requests outside of any dialogs;
- The "**option tags**" listed must refer only to extensions defined in standards-track RFCs. This is to prevent servers from insisting that clients implement non-standard, vendor-defined features to receive services. Extensions defined by experimental and informational RFCs are explicitly excluded from usage with the supported header field in a request, since they too are often used to document vendor-defined extensions.

NOTE: The "From" field of the response must equal the "From" header field of the request. The Call-ID header field of the response must equal the Call-ID header field of the request. The "CSeq" header field of the response must equal the "CSeq" field of the request. The "Via" header field values in the response MUST equal the "Via" header field values in the request and must maintain the same ordering. A SIP message is either a request from a client to a server, or a response from a server to a client. Both messages use the generic-message format of RFC 2822 [88] for transferring entities (the body of the message). Both types of messages consist of a start-line, one or more header fields (also known as "headers"), an empty line (i.e., a line with nothing preceding the carriage-return line-feed-CRLF) indicating the end of the header fields, and an optional message-body. To avoid confusion with similar-named headers in HTTP, we refer to the headers describing the message body as entity headers. The initial SIP standards include security mechanisms implemented at the Transport layer. They include basic Transport Layer Security (TLS) that provides encrypted SIP call control, Digest Authentication (DA) that uses the MD5 algorithm to verify the sender's identity, and Secure RTP (sRTP) that provides encrypted voice streams.

16.6 SIP IETF Standards and Extensions

SIP was initially developed within the IETF **Multiparty Multimedia Session Control** (MMUSIC) working group and published as RFC 2543 in September 1999 (now obsolete). Subsequently, the original specifications were extended both in scope (applications supported) and the volume of documentation (now approaching a thousand pages). Current SIP development is now done in the **IETF SIP Working Group**. The recommended standard specifications are included in the following documents:

- **Session Initiation Protocol**, RFC 3261;
- **Reliability of Provisional Responses in Session Initiation Protocol**, RFC 3262;
- **Session Initiation Protocol Locating SIP Servers**, RFC 3263;
- **An Offer/Answer Model with Session Description Protocol**, RFC 3264;
- **Session Initiation Protocol Specific Event Notification**, RFC 3265;
- **Session Description Protocol**, RFC 2327;
- **Session Initiation Protocol Extension for Instant Messaging**, RFC 3428; and
- **A Model for Presence and Instant Messaging Presence Protocol Requirements**, RFC 2778/RFC 2779.

Additional IETF Requests for Comments regarding protocol specifications used in SIP systems are:

- **Real-Time Transport Protocol** (RTP), RFC 3550;
- **Real-Time Control Protocol** (RTCP), RFC 3605;
- **Session Announcement Protocol** (SAP), RFC 2974;
- **Resource Reservation Protocol** (RSVP), RFC 2205;
- **Real-Time Streaming Protocol** (RTSP), RFC 2326;
- **Stream Control Transmission Protocol** (SCTP), RFC 2960/RFC 3309; and
- **E 164 (telephone) Numbering (mapping with DNS)** (ENUM), RFC 2916.

In addition to the IETF groups dealing directly with the initial design of SIP-based specifications, there are other organizations that provide contributions to the development and implementation of SIP, such as:

- **3GPP IMS Consortium**: WCDMA-based cellular mobile, (www.3gpp.org);
- **3GPP2 IMS Consortium**: CDMA2000-based cellular mobile, now called MultiMedia Domain (MMD), (www.3gpp2.org);

- **European Computer Manufacturers Association**: CTI calls, (www.ecma-international.org);
- **International Telecommunications Union-Telecommunication Sector**: H.248 for media control, Q1912, signaling SIP/ISUP conversion, (www.itu-t.int).
- **European Telecommunications Standards Institute**: Tiphon project, SIP-H.323 interworking, (www.etsi.org);
- **IETF SIGTRAN Working Group**: SCTP transport layer, (www.ietf.com);
- **SIP Center**: SIP testing, (www.sipcenter.com).

16.7 Advantages and Disadvantages of SIP

The advantages of implementing SIP-based systems derive from the following aspects:

- SIP is based on open, public IETF Internet standard specifications;
- SIP is designed to run on public or private IP-based networks and their suite of protocols;
- SIP can run on several transport layer protocols: TCP, UDP, and SCTP, as suited for particular applications;
- SIP provides support for voice, video, instant messaging and presence, as well as various forms of multimedia communications;
- SIP was embraced as the signaling protocol of choice by both wireless and wired telecommunications operators;
- SIP was adopted as the signaling protocol for IP-based Multimedia Subsystem (IMS), the next generation convergent network;
- SIP has numerous extensions tailored for specific architectures and applications;
- SIP is backed by a SIP Center that provides guidance for implementations and a development environment for SIP-based applications;
- SIP is conceived as a signaling platform to allow development of new applications by third parties using open standard APIs;
- SIP design is flexible because it is based on functional modules (Proxies, Location Server, Registrar) with few restrictions regarding their aggregation and implementation in specialized servers;
- There is a large pool of vendors supporting various SIP technology components: SIP phones, SIP servers, SIP gateways, and SIP development platforms.

Disadvantages of SIP implementations include:

- It is a complex system, with numerous functional extensions that can create implementations and interoperability problems;
- Many SIP-related drafts are still not finished; there are revisions with no backward compatibility, omissions exist, and there is no consistency in feature implementations;
- SIP is confronted with a huge legacy of voice communication systems that still use the SS#7 signaling protocol suite and other proprietary protocols;

- There is no proven evidence regarding economies of scale when implementing the SIP in new systems or replacing legacy networks;
- New applications promoted by Google, YouTube, and Skype do not use SIP: instead they use their own proprietary signaling protocols;
- A lack of SIP implementation profiles and best practice case studies;
- A lack of comprehensive security features (the basic Transport Layer Security-TLS, Digest Authentication-DA, and Secure RTP are not adequate); and
- A lack of built-in management and provisioning capabilities.

16.8 SIP Applications

Use of the SIP/SDP suite of protocols by a large variety of applications is a proof of its validity and benefits. This is the reason that SIP was selected as IMS signaling protocol to help integration and convergence of wired and wireless networks and delivery of new multimedia applications. Third party SIP applications development is facilitated by the existence of vendor independent, publicly available standard specifications and development tools. These applications can serve carriers, medium and large enterprises, and individual users by extending its use from the traditional client-server environment to peer-to-peer communication. The SIP range of applications includes the following:

- **VoIP** (most of the basic features)
- **Instant messaging and presence**
- **Videoconferencing**
- **Multimedia conferencing**
- **Audio and video streaming**
- **Event subscription and notification**
- **Peer-to-peer communications**
- **Web collaboration, Web publishing**
- **Group chatting**
- **Desktop sharing**
- **Click to call**
- **Buddy lists**
- **Unified messaging**
- **Contact Center** (CTI, ACD, IVR, Call Data Recording)
- **Unified messaging**
- **Speech-based access**
- **Announcements**
- **Whiteboard**
- **Video and content sharing**
- **Web services**

SIP can also initiate multi-party calls using a Multipoint Control Unit (MCU) or a fully meshed interconnection instead of multicast communications. Internet telephony gateways that connect Public Switched Telephone Network (PSTN) parties can also use SIP to set up calls between them.

16.9 ITU-T H.323 Signaling Protocols

To understand the advantages of using SIP as the call control mechanism, we will present a few technical details about the H.323 standards that preceded the SIP and are still in use by all major carriers. The **H.323 ITU-T family of standards** provides specifications for audio, video, and data communications over IP-based networks, including public Internet and private Intranets. Multimedia communications covers non-guaranteed bandwidth packet switched networks and includes IPX and TCP/IP over Ethernet, Fast Ethernet, and Token Ring. There are three main H.323 versions:

- H.323 v.1 (1996), H.323 v.2 (1998);
- H.323 v.3, v4, v5 (1999, 200/2003) that includes main text, Annexes A to G, K, M.1, M.2, and Appendices I-V
- Appendices J, L, M3, P, Q, and R.

H.323 is part of a larger spectrum of standards produced by ITU-T that deals with transmission of non-telephone signals in PSTNs: H.320 (narrowband ISDN), H.321 (broadband ISDN, ATM, LAN), H.322 (guaranteed bandwidth), H.323 (non-guaranteed bandwidth), and H.324 (PSTN, POTS analog systems). H.323 compliant devices are hardware and operating systems independent; they include a variety of terminals and special components:

- **IP-enabled telephone sets**;
- **Video-enabled PCs**;
- **Cable TV set-top boxes**;
- **Multiple Point Control** (MCU): units for three party conferencing;
- **Gateways**: hardware and software for translation between H.323 devices and other non-H.323 terminals i.e., between packet and circuit-switched networks;
- **Session Controllers** and **Session Border Controllers** (SBC): Hardware and software for translation between H.323 and SIP plus security features;
- **Gatekeepers**: Server-like hardware and software for call control services (admission, directory) to registered end-points; PBX/ACD like functions.

H.323 standards also provide support for special features such as: multipoint connectivity (multiple end-points) and multicast (in multipoint conferences bandwidth management). H.323 terminals and associated applications communicate on both the reliable and best-effort transport layer protocols, i.e., Internet Transmission Control Protocol (TCP) and User Datagram Protocol (UDP). Only one Network Layer protocol is accepted, the Internet Protocol (IP). H.323 standards, although known mainly for voice and videoconferencing, are comprehensive in scope, covering all aspects of non-telephony terminals interactions and adaptation with various multimedia Physical, Data Link, and Network Layer protocols. H.323 standards coverage is shown in Figure 16.6.

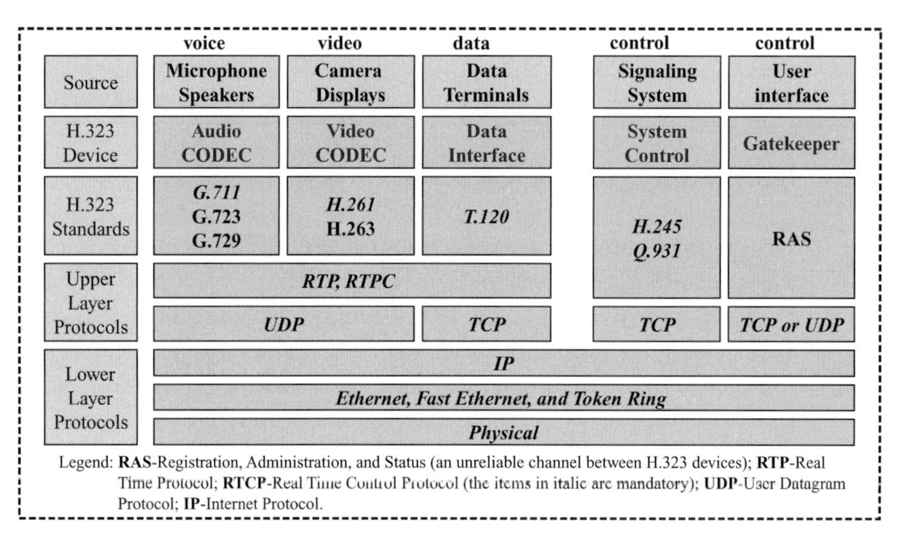

Figure 16.6 H.323 Family of Standards and Protocol Stacks

All the mandatory standards are indicated in bold italic (G.711 voice, H.261 video, T.120 data, H.245, and Q.931 for signaling/call control), running on top of TCP/UDP and IP. Any standard Data Link Layer and Physical Layer are accepted. In this context, it is useful to make a comparison between H.323 legacy systems and SIP as presented in Table 16.1.

The table presents a comparison of the main features of the H.323 family of standards and the IETF family of standards. Both are aimed at providing call control/signaling for non-telephony devices. Abbreviations used include: **H.245**-Control messages and communications procedures for opening/closing channels; **RAS**-Registration, Administration, and Status (an unreliable channel between H.323 devices); **Q.931**-ITU-T ISDN call signaling and setup standards; **SS7 ISUP**-Signaling System 7 Integrated Services User Part call control (Q.761-Q764); **ASN.1**-Abstract Syntax Notation One; **PER**-Packet Encoding Rules; and **RTSP**-Real-Time Streaming Protocol. The table does not reflect the expansion of SIP-based documentation as the number of pages and headers used in the SIP approaches in complexity that of the H.323 family of standards.

16.10 Signaling Gateways/Session Controllers

In a world still dominated by PSTN legacy signaling systems, such as SS#7, Signaling Gateways (SGWs) will be required at any points of connectivity to and from PSTNs. SGWs have a special role in VoIP networks where H.323, the first choice of signaling, must be converted to PSTN's SS#7. The same is true for newer VoIP systems that support SIP and any softswitch that must be compatible with systems such as MGCP. In addition, there is a world of VoIP systems designed by major players that use either proprietary

Table 16.1 ITU-T H.323 and IETF SIP Signaling/Call Control Comparison

Features/Standards	IETF SIP V1/V2/V3	ITU-T H.323 V1/V2/V3/ V4/V5
Basic set of standards	RFC 2543 (x) RFC 3261	H.3xx, G. 7xx H.2xx, T.1xx
Standards status	1999/2002/2004	1996/1998/1999/2000/2003
Additional signaling standards	RFC 2719 MGCP	H.245, RAS, Q.931 SS7 ISUP
Protocol stack	RTP/TCP/UDP	RTP/TCP
Operator-assisted transfer	Yes	No
Blind transfer	Yes	Yes
Holding	Yes through SDP	Not yet
Multicast conferencing, Bridged conferencing	Yes, Yes	Yes, Yes
Call forward, Call park, Directed call pickup	Yes, Yes, Yes	Yes, Yes, Yes
Encoding rules and their standards	English HTTP and RTSP	Binary ASN.1 and PER
Complexity	128 pages, 37 headers	736 pages, hundreds of message types

Legend: H.245-Control messages and communications procedures for opening/closing channels; **RAS**-Registration, Administration, and Status (an unreliable channel between H.323 devices); **Q.931**-ITU-T ISDN call signaling and setup standards; **SS7 ISUP**-Signaling System 7 Integrated Services User Part call control (Q.761-Q764); **ASN.1**-Abstract Syntax Notation One; **PER**-Packet Encoding Rules; **RTSP**- Real-time Streaming Protocol.

extensions of standardized signaling protocols such as H.323, or their own proprietary protocols.

Among the SGWs, **Session Controllers** are special signaling gateways that provide interoperability between two VoIP systems that use different signaling protocols and call setup messages (H.323 and SIP/SDP). The actual traffic flow takes a different route and uses the Real-Time Transport Protocol (RTP) and Real-Time Control Protocol (RTCP) [89]. The architecture of the session controller is shown in Figure 16.7.

Users on the left side, using either an IP Phone, IP software on the desktop, or a regular POTS phone connected to an IP PBX, call a SIP-based Unified Messaging system to access their voice mail. Since the IP PBX supports the H.323 signaling protocol, a call setup message will be sent from the IP PBX across the session controller signaling gateway. This gateway translates the H.323 format into SIP INVITE format and forwards the request to the Unified Messaging server. The server will respond, acknowledging the binding and its readiness to start the data session using the same signaling path.

Traffic is passed directly between the Unified Messaging Server and IP PBX using the RTP/RTCP protocol. Depending on the security policies adopted the traffic may pass a firewall integrated with the Session Controller. In this case, the Session Controller is an Integrated Access Device (IAD) that contains the functionalities of a H.323 gatekeeper and a SIP Proxy.

Both sides are assisted by additional protocols and message exchanges that provide the binding between the endpoints. On the SIP side is the Session Description Protocol

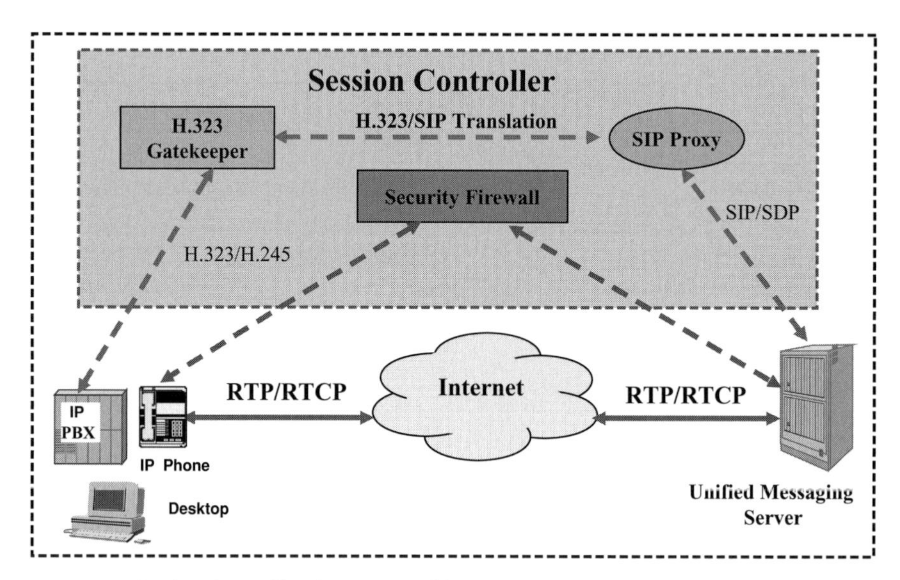

Figure 16.7 Session Controller Gateway Architecture

(SDP) and on the H.323 side is H.245. Mandatory features of both SIP and H.323 are supported.

In the same class of signaling gateways, there is a group of devices called **Session Border Controllers** (SBCs). They are used in SIP networking to provide connectivity over SIP domains which are separated by networks that do not support SIP. These gateway functions are not limited to SIP domains and most SBCs provide SIP-H.323 interworking, a very common occurrence in the real world. SBC devices also provide security features such as authentication, Transport Layer Security (TLS), Denial of Service (DoS) protection, secure RTP (SRTP), security firewalls with Network Address Translation (NAT), and encryption.

16.11 SIP-based Interoperability in Wireless Networks

IETF SIP standards specify signaling interfaces for multimedia products and the corresponding applications from multiple vendors in wired and wireless networks. A high level architecture of SIP-based system integration with wireless networks is shown in Figure 16.8.

This high level diagram shows the placement of a SIP-based system that provides convergence between cellular mobile networks such as UMTS-based 3G mobile networks and Wireless LANs. The SIP platform is presented as one domain that incorporates SIP Proxies, Registrars, and a Location Directory. In practice, all these components are part of the same box, and multiple SIP domains might be required for larger networks. In case there is a lack of SIP-support at the level of UMTS or WLAN, signaling gateways are needed.

Similar to the generic SIP-based architecture, presented in Section 16.2 of this chapter, the User Agents are incorporated in mobile terminals, phones or laptops, providing

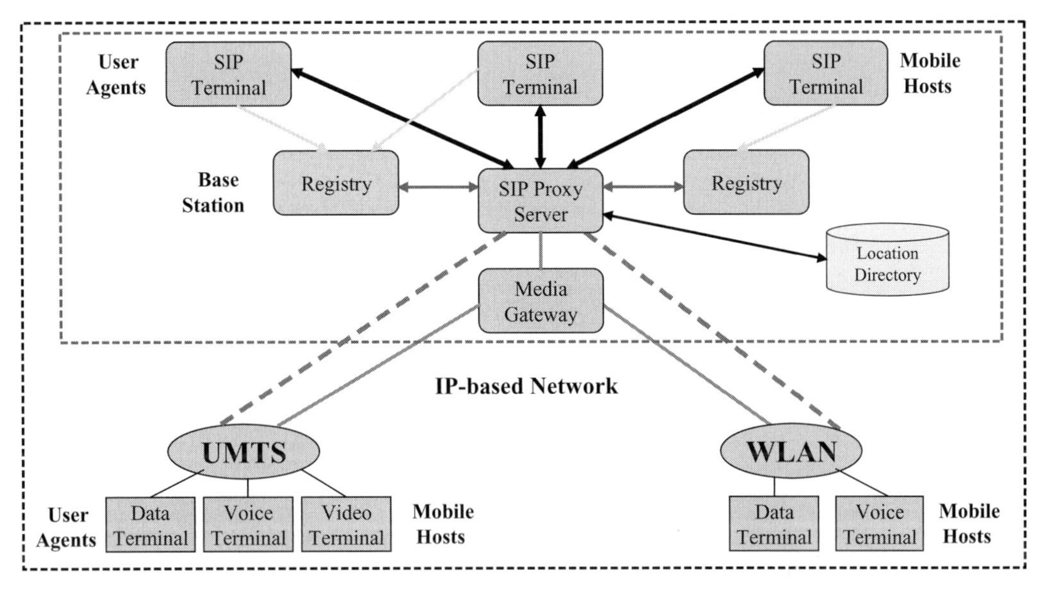

Figure 16.8 SIP-based Mobile Hosts Interoperability in Wireless Networks

full support for the SIP protocol stack. The traffic will be carried across the Media Gateway.

16.12 SIP-based VoIP Network Architecture

One of the major applications of SIP-based systems is VoIP, and within VoIP the softswitch, the essential signaling component that assures convergence between circuit-switched and packet switched networks. An example of the use of SIP-based systems in VoIP network architectures is shown in Figure 16.9.

The SIP system consists of SIP-based User Agents that act as clients of a SIP Proxy Server. The SIP terminals are SIP Phones that communicate with the Proxy Server through a LAN. Other IP-Phones communicate directly with the branch office IP-PBX that acts as a SIP User Agent. The IP-PBX that acts on calls from POTS phones, performing the necessary digitization and coding. The SIP Proxy Server relays signaling messages to the softswitch.

The softswitch controls the entire call setup process. It has four major functions:

- Passing signaling messages to the PSTN via the signaling gateway and handling calls;
- Controlling the Media Gateway via the Media Gateway Control Protocol (MGCP);
- Controlling application servers for new services; and
- Providing registration and accounting information.

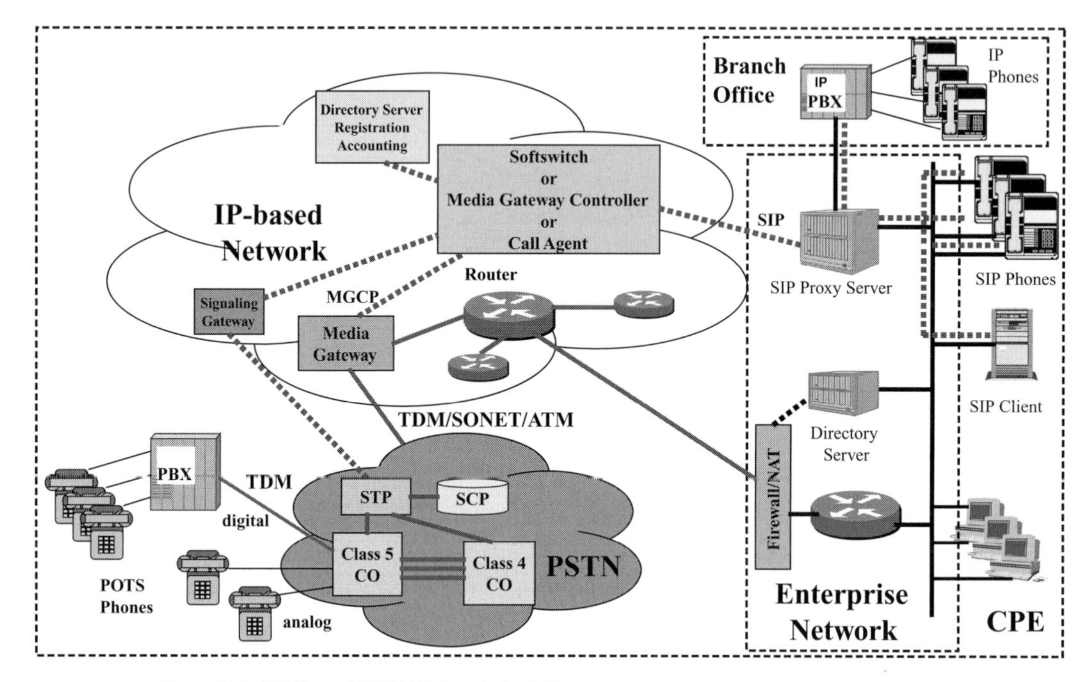

Figure 16.9 SIP-based VoIP Network Architecture

In this position, the softswitch acts as a central office switch and is touted as a possible replacement for the traditional Class 4 and even Class 5 digital switches. More details concerning the softswitch will be given in Chapter 17. Within the USA switching hierarchy, classes of offices are used to rank **Central Offices** (COs). Class 5 offices are part of the **local exchange network** while the **toll network** consists of class 4 or higher level switching offices. Signaling is handled on separate out-of-band networks using redundant Signal Transfer Points (STPs).

Working traffic is carried over the LAN, an IP-based router network, and the **Media Gateway** (MG) using either TCP or UDP as Transport Layer protocols. The MG converts IP telephony messages, and packetized voice traffic, into GR 303 messages used by the class 5 switch. This allows the switch to deliver all the features normally provided to analog and ISDN phones to IP Centrex users. The richness of softswitch features, including SIP features, determines how well it performs in comparison to a central office switch.

16.13 SIP-based Solutions, Products, and Services

As presented here, the balance between SIP's advantages and limitations, presented in the previous sections, is in the favor of SIP as a widely accepted signaling protocol. This is proven by the number of vendors and products that are already incorporating the SIP specifications. The list of SIP-based products includes SIP-based phones/terminals, SIP gateways, SIP call controllers, and SIP-based development tools.

SIP Phones

- **3Com**: 31xx family of IP Phones (wired, cordless, wireless), (www.3com.com);
- **Avaya**: 4600 series of IP Phones, (www.avaya.com);
- **Cisco**: 3900 series Unified SIP phones, (www.cisco.com);
- **Dialexia**: Dial-Com Pro (2.1.0.5) SIP softphone, (www.dialexia.com);
- **Mitel**: Mitel 5212 IP Phone (SIP and Mitel IP MiNET based), (www.mitel.com);
- **Nortel Networks**: i2050 SoftPhone Software, (www.nortelnetworks.com);
- **Polycom**: Sound Station VTX 1000 Speaker Phone, (www.polycom.com);
- **Siemens**: optiClient 130 SIP wireless phones, (www.siemens.com);
- **Swyx**: SwyxPhone L400 and L500 series IP phones, (www.swyx.com).

SIP Gateways

- **3Com**: 3 Com VoIP Gateway, (www.3com.com);
- **Adtran**: NetVanta 7100 Embedded SIP Gateway, (www.adtran.com);
- **Cisco Systems**: Cisco SIP Gateway, (www.cisco.com);
- **Firsthand Technologies**: SIPquest Solutions, (www.firsthandtech.com);
- **Interactive Intelligence**: Interaction SIP Proxy, Gateway, (www.inin.com);
- **Mitel**: Mitel 3300mWireless Gateway, (www.mitel.com);
- **MultiTech**: MultiVoIP SIP Gateways, GSMCDMA Cellular Gateways, (www.multitech.com);
- **Siemens**, QuesCom 400 IP **SIP** proxy, (www.siemens.com);
- **Swyx**: Swyxware Embedded SIP Gateway, (www.swyx.com).

IP/PBX with SIP Support/Call Controller

- **3Com**:VCX, (www.3com.com);
- **Adran**: NetVanta 7100, (www.adtran.com);
- **Avaya**: SIP Enablement Services, (www.avaya.com);
- **Cisco Systems**: Unified Call Manager, Call Manager Express, (www.cisco.com);
- **Dialexia**: Dial-Office IP PBX, (www.dialexia.com);
- **Digium**: Asterisk Business Edition, Open Source PBX, (www.digium.com);
- **Mera Systems**: Mera IP PBX, Mera VoIP Transit Softswitch, (www.mera-systems.com);
- **Mitel**: 3300 ICP, (www.mitel.com);
- **NEC-Philips**: UNIVERGE NEAX IPX, (www.nec-philips.com);
- **Nortel Networks**: Multimedia Communications Systems 5100, (www.nortelenetworks.com);
- **Pingtel**: SIPxNano, (www.pingtel.com);
- **Siemens**: HiPath 8000 softswitch, (www.siemenscommunications.com);
- **Swyx**: Swyxware IP PBX, (www.swyx.com);
- **Veraz Networks**: Softswitch-Control Switch, (www.veraznetworks.com).

Open SIP Application Platform

- **3Com**, 3Com Convergence Application Suite, (www.3com.com);
- **Adomo**: Mobile Unified Communications, (www.adomo.com);
- **Antepo (now Adobe)**, OPN Open Presence Network, (www.antepo.com);

- **Avaya**, Meeting Exchange, (www.avaya.com);
- **BridgePort Networks**: NOMADIC ONE Network Convergence Gateway, (www.bridgeport-networks.com);
- **Cisco Systems**, Unified Presence Server and Personal Communicator, (www.cisco.com);
- **Continuous Computing**: Trillium SIP software C source code implementation of the SIP protocol stack, (www.ccpu.com);
- **Dialexia**, Dial-Office DXO, Dial-Gate Softswitch, (www.dialexia.com);
- **Digium**: Asterisk Business Edition, Open Source PBX, (www.digium.com);
- **Genesys Lab**: Contact Center SIP Server, (www.genesys.lab);
- **Interactive Intelligence**: Customer Interaction Center, (www.inin.com);
- **IP Unity**, Mereon 6000, Mereon 3000 Media Servers, (www.ipunity.com);
- **Mera Systems**, Mera SIPrise, (www.mera-systems.com);
- **Microsoft**, Microsoft Office Live Communicator Server, (www.microsoft.com);
- **NEC-Philips**: UNIVERGE SV7000 IP Communications Server, (www.nec-philips.com);
- **Nortel**: Multimedia Communications Server 5200, (www.nortelenetworks.com);
- **Personeta**: TappS NSC Network Service Controller, (www.personeta.com);
- **Pingtel**: SIPxchange Enterprise Communications Server and Call Manager, (www.pingtel.com);
- **Siemens**: HiPath OpenScape, (www.siemenscommunications.com);
- **Swyx**: SwyxIt! Now, (www.swyx.com).

16.14 SIP-based Signaling in the MobileIGNITE Architecture

MobileIGNITE (Integrated Go-to-Market Network IP Telephony Experience) is a SIP-based signaling technology platform, and a set of specifications to promote fixed-mobile convergence. The technology was proposed by BridgePort Networks in early 2005 and promoted through the **MobileIGNITE Alliance**, a consortium of over 50 members. Founding members were: BridgePort Networks, AirSpace Inc, IBM, and VeriSign. Among the current members are: Acme Packet, Apertio, Aruba Networks, BroadSoft, Colubris Networks, FirstHand Technologies, Kyocera Wireless, Meru Networks, Net2Phone, Paragon Wireless, Sylantro, Tekelec, and WorldCell.

One example of application of this technology is in the establishment of transparent communications between Microsoft Office Communicator client software and mobile phone users. For this capability, the most important features supported are: call handling using a SIP-based signaling setup, voice messaging, and presence/location. This solution can be extended by integrating the Microsoft Live Communications Server with an enterprise PBX. The convergence is facilitated by the BridgePort Networks' NOMADIC ONE Network Convergence Gateway (NCG) platform.

The key MobileIGNITE specification is Release Version 1.0 "Functional Specification for FMC Handover", developed by the Handover Interoperability Group. Implementation

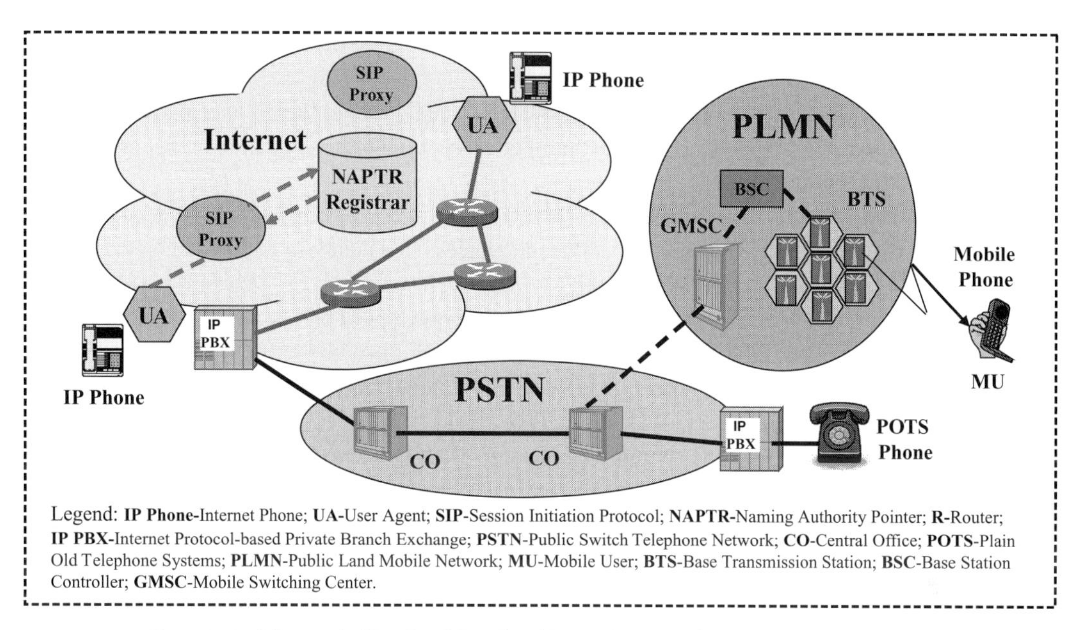

Figure 16.10 Electronic Number Mapping (ENUM) Network Architecture

of the specification supports WLAN Wi-Fi (fixed) and GSM/UMTS cellular radio (mobile) networks convergence using SIP-based dual-mode handsets. This specification conforms with the 3GPP Release 99 and Release 4 reference architectures as well as with the IMS reference architectures specified in 3GPP's Release 5, 6, and 7. [90]. The current 3GPP2 Release 7 draft specs include the Voice Call Continuity (VCC) standard along with protocol implementation agreements and test cases necessary for formal interoperability certification.

16.15 Electronic Number Mapping (ENUM)

The implementation of VoIP in both wired and wireless mobile communications is tightly linked to the use of SIP for signaling (setup calls) and RTP/RTCP for digitized voice packets transfer across heterogeneous network environments. Identification of users to each of the two major networks, the Internet and the Public Switched Telephone Network, (PSTN) is quite different since the PSTN uses dialed telephone numbers according to ITU-T Recommendations E.164 whereas the Internet uses IP addresses and the Domain Name System (DNS). Therefore, there is a need for translation between telephone numbers and Internet addresses. A similar architectural context is created by VoIP over a Public Land Mobile Network (PLMN). At a high level, this mapping is provided by **Electronic Number Mapping (ENUM)**. ENUM contains a set of specifications, protocols, and data formats that allows this translation to take place. A simplified view of the ENUM process in the context of a VoIP network architecture is shown in Figure 16.10.

Three components are necessary to provide this translation:

- The Uniform Resource Identifiers (URI): Used on the Internet; as explained below;
- The E.164: Personal telephone number associated with the personal URI, used on the PSTN network; and
- The Naming Authority Pointer (NAPTR): To record call preferences; accessible via the personal URI. The protocol used to access this database is the SIP.

When a call is initiated through a SIP-based phone, Smart phone, or IP phone, the calling party will use the E.164 telephone number format (area code followed by seven digits). The User Agent (UA) will convert the E.164 number into a URI. This URI is used in the ENUM gateway DNS to look up and extract the NAPTR record containing preferences regarding call forwarding and termination. The preferences may include another VoIP address (for calls terminated on the Internet) or a mobile/fixed telephone number for calls terminated on the PSTN or a Public Land Mobile Network (PLMN). This mapping can be extended to support instant messaging, voice mail systems, fax, web addresses, and email addresses. A similar process takes place when the call is initiated on the PSTN and the caller is represented by an ENUM address. After the look up process, the terminal application sets up a communication link, and the call is routed.

The ENUM number is the dialup number without any spaces or dashes between numbers and, for the Internet, the suffix "e.164" is added. On the IP side of the network, the ENUM represents the phone number and the suffix "e164.arpa" [91].

IETF's ENUM Working Group has specified services that support the mapping of telephone numbers into URIs. The Application Unique String for the ENUM application is the E.164 telephone number with the dashes removed. The First Well Known Rule is to remove all characters from the telephone number and then use the entire number as the first Key. For example, the phone number "201-555-1212", represented as an E.164 number, would be "+1-201-555-1212". Converted to the Key it would be "12015551212".

To convert the first Key into a form valid for the ENUM database, periods should be inserted between digits. The entire Key is inverted and "e164.arpa" is appended to the end. The above telephone number would then read "2.1.2.1.5.5.5.102.1.e164.arpa.". This domain-name is then used to retrieve Rewrite Rules as NAPTR records.

Internet IP-DNS address translations to PSTN phone numbers has been a subject of standardization since the early days of the Internet. With the advent of VoIP these standards were reviewed and augmented. The relevant ENUM standards are:

- RFC 2168 "Resolution of Uniform Resource Identifiers (URI) using the Domain Name System (DNS)", was made obsolete by RFC 3401-3404;
- RFC 2915 "The Naming Authority Pointer (NAPTR) DNS Resources Record" (obsoleted by RFC 3401–3404);
- RFC 3401 "Dynamic Delegation Discovery System (DDDS): The Comprehensive DDDS";
- RFC 3402 "Dynamic Delegation Discovery System (DDDS): The Algorithm";

- RFC 3403 "Dynamic Delegation Discovery System (DDDS): The Domain Name System (DNS) Database";
- RFC 3404 "Dynamic Delegation Discovery System (DDDS): The Uniform Resource Identifiers (URI) Resolution Application"; and
- RFC 3761 "The E.164 to Uniform Resource Identifiers (URI) Dynamic Delegation Discovery System (DDDS) Application (ENUM)".

There are some issues regarding the use of ENUM and SIP:

- Currently, the use of SIP phones is limited to enterprises where SIP clients have subscribed to the same services and a compatible directory (e.g., Microsoft).
- Calls outside the enterprise must be converted from Internet packets into PSTN accepted frames, otherwise they will be stopped at the firewall servers and NAT servers. Firewalls have not been designed to support SIP's dynamically assigned port numbers and real-time traffic. Companies such as Check Point, Cisco, Microsoft, Ridgeway, and Vonage use proprietary solutions. Encapsulation of SIP's RTP/UDP into HTTP (port 80) and HTTPS (port 433) is needed. However, HTTP and SMTP SIP are sent in the clear as plain text, so a protocol analyzer can see the messages in ASCII format.
- The ability to interwork with existing legacy systems that operate with older signaling protocols such as MGCP, H.323, SIGTRAN, and PSTN SS7.

17 IMS-based Fixed Wireless and Cellular Mobile Networking Solutions and Products

17.1 What is the IP-based Multimedia Subsystem?

The **IP-based Multimedia Subsystem (IMS)** is the **Next Generation Network** (NGN) architecture, set of components, and interface specifications that allow convergence of wired and wireless networks. Convergence/integration of fixed and mobile wireless networks, the subject of this book, is part of the broader convergence just defined. The IMS convergent network will emerge from an Internet Protocol (IP)-based network infrastructure and a common service platform, allowing the development of a large array of telecommunications and multimedia applications. IMS is a user/operator-centric architectural framework that shifts much of its intelligence to the network periphery. Users of IMS services will benefit because of full mobility and service transparency across all networks.

IMS development was initiated in 1999 by a group of leading mobile service providers in conjunction with the promotion of future generations of mobile networks. This work was taken over by the **3G Partnership Project** (3GPP) and presented for the first time in 3GPP Release 5 (3GPP R5) specifications. Release 5 was augmented by the addition, in Release 6 (3GPP R6), of the Internet Engineering Task Force (IETF) Session Initiation Protocol (SIP). In 3GPP Release 7, IMS incorporated the NGN concepts promoted by the European Telecommunications Standard Institute's (ETSI) Telecommunications and Internet Converged Services and Protocols for Advanced Networking (TISPAN) division. Also, the mobile networks support for IMS was extended from GSM/UMTS to CDMA2000 through 3GPP2 working specifications.

A major feature of the IMS concept is to build an encompassing open platform that separates the User Layer (access networks), Transport Layer (core and connectivity networks), Control Layer (signaling network and network management), and Service Layer (applications). The fabric of these layers consists of interconnected functional modules and standard interfaces for multimedia communications and signaling. In the visionary view of IMS, the service platform will be accompanied by development tools that will allow development of applications by third parties using standardized Applications Programming Interfaces (APIs).

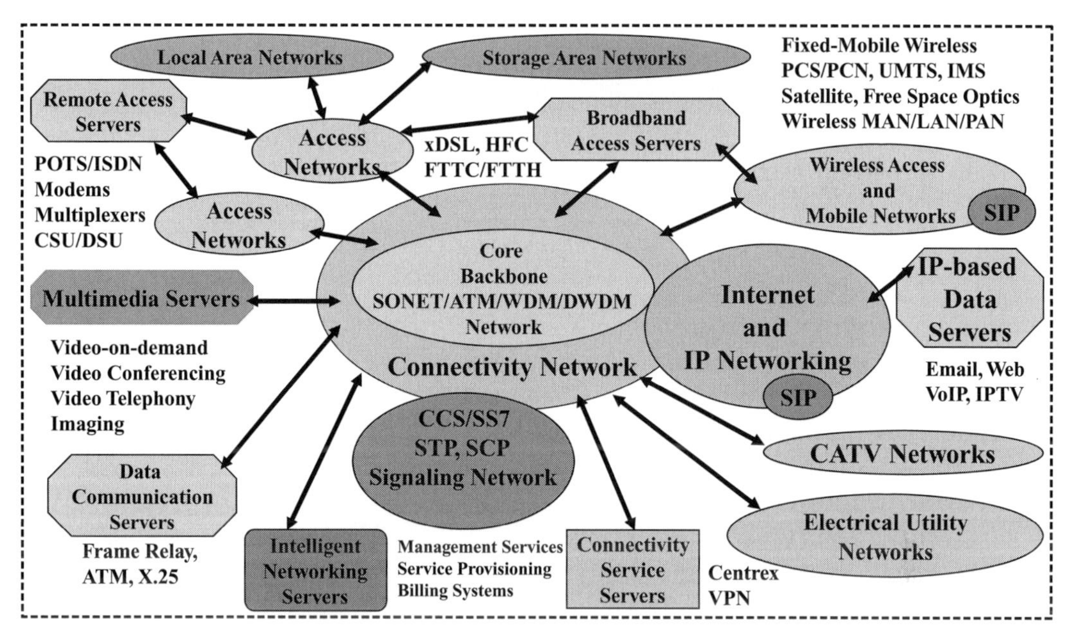

Figure 17.1 Global Telecommunications Network

17.2 The Convergence Path to IMS

To understand the concept of IMS we have to identify the major lines of changes planned in building the IMS architectural framework. Since the goal of IMS is to provide an encompassing framework for wireless and wired networks, we refer again to the complete picture of telecommunications networks, presented in Chapter 1, and reproduced here as Figure 17.1. This global network is a collection of subnetworks that provide data, voice, and video communications, using an array of networking technologies that cover access, connectivity, and core backbone networks.

The diagram includes the core backbone network, a fast growing all IP-based Internet network, and competing solutions offered by the CATV networks and Power Line-based networks. Overall communication is supported by specialized servers that provide specific networking services. The telecommunications world contains the entire family of wireless networks from mobile networks to satellite, to free space optical networks. A simplified view of the telecommunications network is shown in Figure 17.2.

In this diagram, networks are presented as layered transport mechanisms that involve access, connectivity, and core backbone components, with an overlay of signaling networks. Distinct domains, orthogonal to all these networks, are those allocated to the applications/services and network/service management. The IMS framework follows the same approach by creating three separate grand domains: **Transport Layer, Service/Applications Layer, and the IMS Layer**. The IMS Layer is essentially a common multi-functional **control plane**, responsible for convergence across networks through

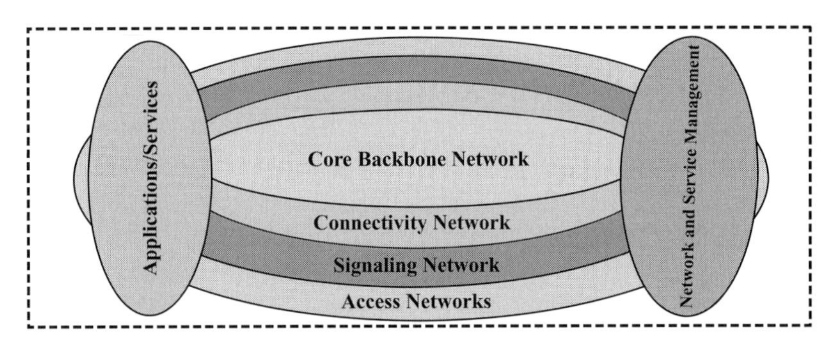

Figure 17.2 A Simplified View of Telecommunications Networks

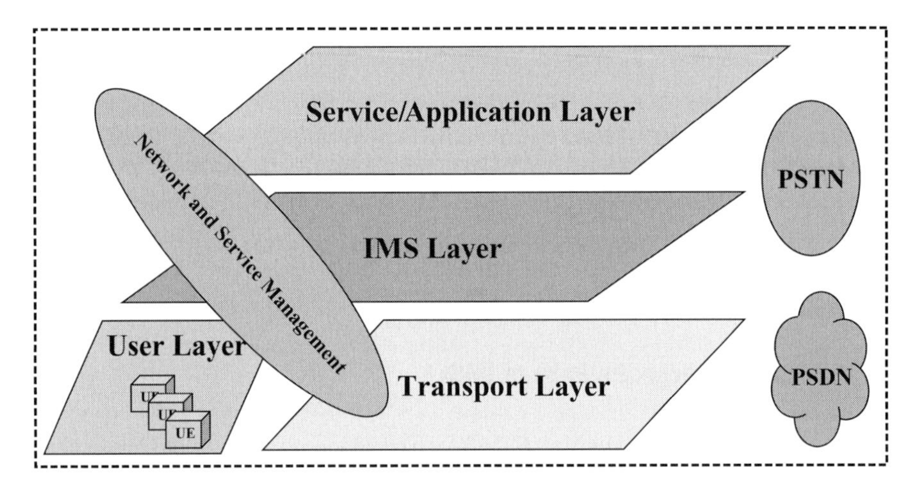

Figure 17.3 A High-level IMS Architectural Framework

signaling and service/applications. A high-level framework for the IMS architecture is shown in Figure 17.3.

The IMS Layer, in the center, resembles a core network. Later we will detail its components and functions. The Access Layer is included within the Transport Layer. Legacy networks, represented here by the Public Switched Telecommunications Network (PSTN) and the Packet Switched Data Network (PSDN), will continue to coexist with the IMS framework. External Network and Service Management tools and applications will be required. Finally, we have to acknowledge a distinct User Layer, where the IMS convergent services are delivered.

The diagram indicates the essential networks where convergence must be achieved. First, at the User Layer is User Equipment (UE) such as mobile phones, dual-mode or multi-mode phones, smart phones, SIP phones, PDAs, PCs, desktops that need to support the signaling protocols which will be used in the IMS. Convergence should also be achieved at the Transport Layer, where there currently are a variety of access methods, access technologies, and transport networks. Signaling should be extended over access and transport networks to the IMS Layer. The IMS Layer allows the services/applications layer to be transparently accessed and used regardless of where calls originate. This is true for calls or data coming from legacy systems, either wired telephone networks or wireless terrestrial mobile networks.

17.3 IMS Networking Overview

IMS, as the long-term solution for the next generation of mobile networks, is the pinnacle of the technical effort that started with the promotion of **Universal Mobile Telecommunications System** (UMTS). All UMTS specifications were developed as part of the 3rd **Generation Partnership Project** (3GPP). UMTS was first released in 1999 (UMTS R99) followed by the Release 2000 (UMTS R00). The first four releases of 3GPP cover definitions of the UMTS standards. UMTS Release 2000 was further split into Releases 4 and 5. Release 4 primarily addressed basic IP connectivity.

3GPP Release 5, frozen in mid 2003, addressed the overall packet-switched domains and introduced, for the first time, the **IP-based Multimedia Subsystem** (IMS) framework. This framework supports VoIP, data, and video over mobile wireless networks. Release 5 IMS [92] was optimized to be used by GSM UMTS networks. Release 5 also introduced the requirements for the QOS needed when providing toll-quality VoIP, and, it mandated use of IPv6.

3GPP Release 6 is the first release that went beyond IMS architectural aspects by adopting IETF's **Session Initiation Protocol** (SIP) as the standardized IMS signaling protocol [93]. It provided the necessary enhancements to the **Call Session Control Function** (CSCF), core component of the IMS Control Layer. In addition, convergence aspects for fixed wireless networks such as WLAN were defined, paving the road for IMS implementation at the User and IMS Layers. These specifications are applicable to a large family of fixed wireless networks that include WiMAX and Bluetooth. Release 6 has also incorporated multi-way conferencing and group management capabilities, necessary for instant messaging and presence services as well as support for the Push to Talk over Cellular (PoC) service. This release has extended support for IPv4, in addition to the mandatory IPv6.

3GPP Release 7 [94] marked incorporation of the **ETSI TISPAN Release 1** specifications into the IMS world. This allows expansion of IMS to the wired world by aligning IMS with broadband access and advanced concepts such as TISPAN NGN and TISPAN QOS. In this way, future interoperability problems can be minimized or avoided. A list of the most important ETSI TISPAN technical standards is provided in Section 17.14 of this chapter.

From the progress made over the last five years, it is clear that IMS development is an ongoing process of designing, changing, adjusting, and enhancing features. This makes it difficult to point to a single IMS system architecture. The difficulty comes from the existence of several versions, each new one adding functional modules and reference points to the older. This will be evident when we present the "full" IMS architecture with all the functional modules in the next sections.

17.4 High-Level IMS Architecture

Following the architectural framework presented in Figure 17.3, we can develop a simplified IMS architectural diagram that contains the essential modules of the IMS Layer and Application/Service Layer. This architecture is presented in Figure 17.4.

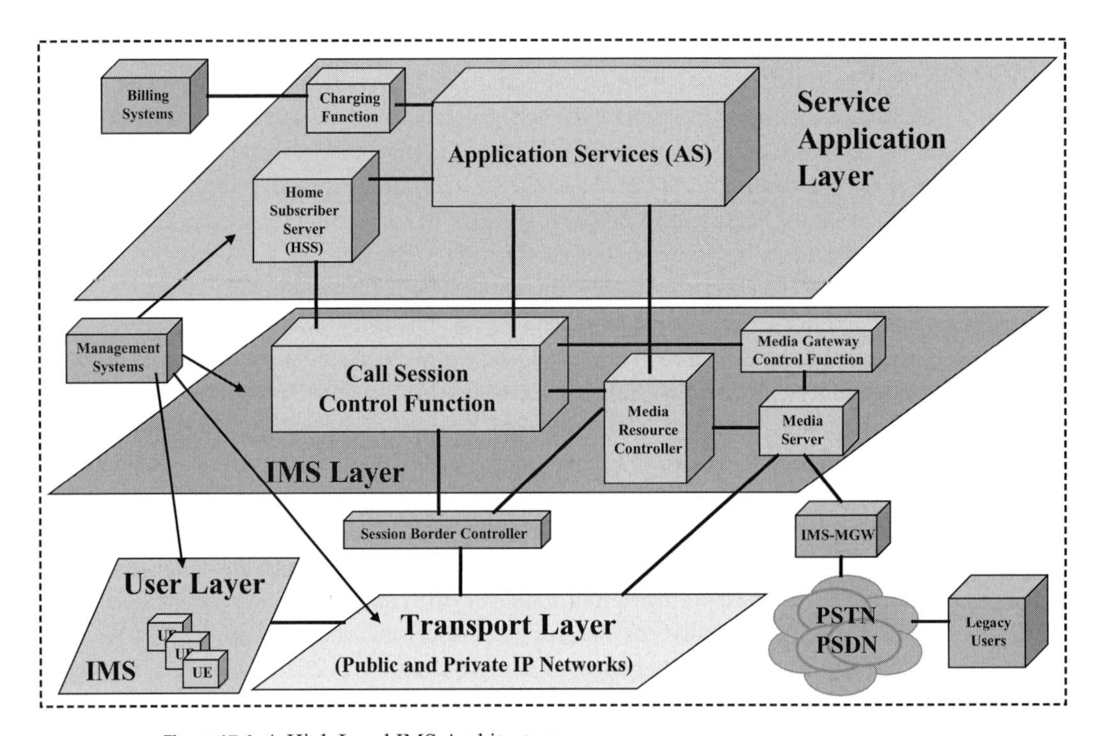

Figure 17.4 A High Level IMS Architecture

The IMS Layer control plane is at the center of the IMS architecture. It provides signaling control of sessions that carry multimedia information. The core IMS Layer component is the **Call Session Control Function** (CSCF) module. The IMS Layer controls the **Applications/Services** provided by the IMS service platform. It also controls resources that are marshaled to provide these services through the **Media Resource Control Function** (MRCF) module. The **Legacy Users** and media information originated in legacy networks such as **Public Switched Telephone Networks** (PSTN) and **Packet Switched Data Networks** (PSDN) are served through **Media Servers** and **IMS Media Gateways** (IMS MGW); the latter controlled through the **Media Gateway Control Function** (MGCF) module.

All operations are triggered by IMS subscribers. The subscribers, through standard **User Equipment** (UE), access the IMS network via the **User Layer** and **Transport Layers**. The Transport Layer is in fact a collection of various public or private IP-based data networks. Subscribers will access the IMS systems in sessions originated from different networks. Routing these sessions within IMS systems will require a **Session Border Control Function** module connecting the Transport Layer with the IMS Layer.

With IMS, all subscriber information, from registration, profiles, location, to service-specific preferences, will be kept in a central repository/directory, hosted by the **Home Subscriber Server** (HSS). This information is used to establish multimedia sessions and deliver IMS services. Access to HSS information should be done in a secure manner, given the wealth of information regarding subscribers and use of services. Security aspects include subscriber authentication and authorization to use the requested services. A **Charging Function** module will be required to provide accounting information to external **Billing Systems** [95]. Additional external **Management Systems** will be required to manage the IMS network and systems components.

17.5 IMS Reference Architecture

Having now a basic understanding of IMS's major functional modules, we can provide a "full" IMS reference architecture, as depicted in Figure 17.5.

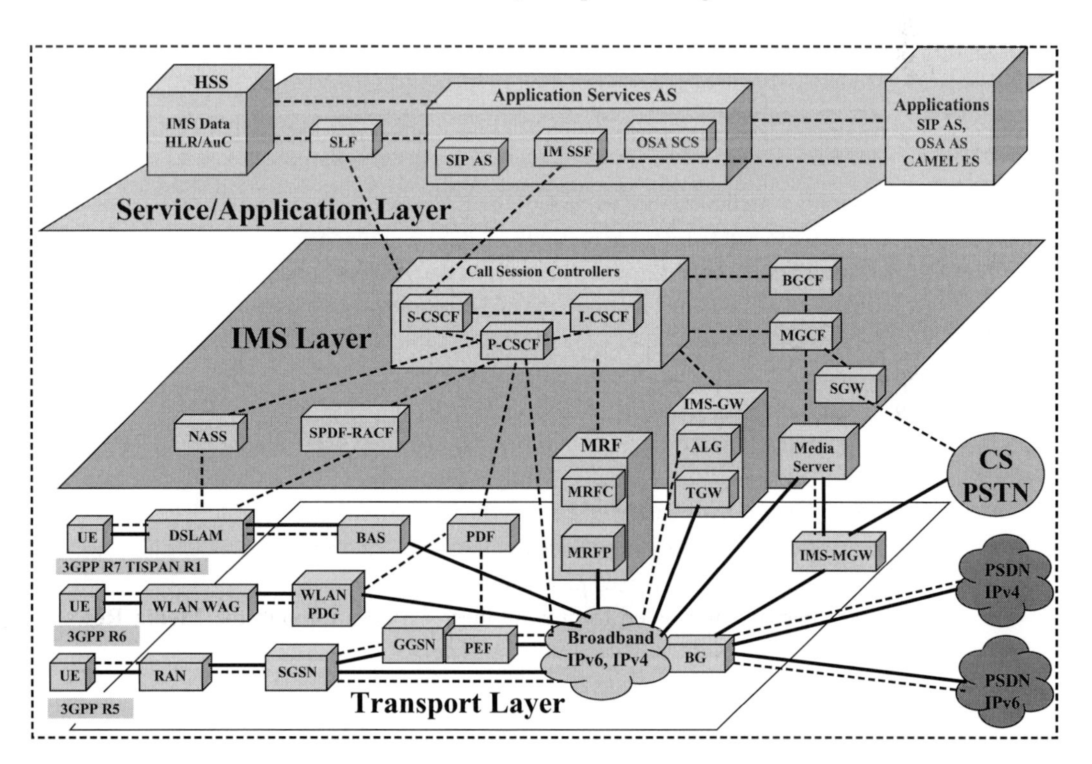

Figure 17.5 IMS Reference Architecture

This diagram, one of the many circulated in the technical literature, follows the architectural layout presented earlier. It places the main components into three major planes: transport, control (IMS core), and applications/services. Within each plane are functional modules consisting of physical (hardware processes) and logical (software processes) components. In practice, how the various functional modules are combined is left to the designers and implementers of IMS systems. These modules or components are linked together to carry signaling information (represented in the figure by broken thin lines), using the IMS **Session Initiation Protocol** (SIP), and user service information (represented by solid, thick lines).

The first requirement of **User Equipment** (mobile phones/handsets, dual-mode handsets in WLANs or laptops connected to broadband DSL lines) is support of the ISM SIP by incorporating a **SIP User Agent** in all endpoint devices. ISM SIP represents an extension of the SIP signaling protocol. The reason is simple. The IMS choice of a signaling protocol is the SIP and the whole handling of SIP is done through SIP servers and SIP proxy servers. IMS SIP is a variation of IETF SIP, specially adapted for IMS. IMS is built as an IP-based infrastructure and may use public or private **broadband data networks** supporting both versions of IP addressing schemes, **IPv4** and **IPv6**.

Calls originated in non-IP-based legacy networks such as Circuit Switched Public Switched Telephone Networks (CS PSTN) that use SS#7 type of signaling and other data protocol stacks, are directed through the **Signaling Gateways** (SGW) and **IMS Media Gateways** (IMS-MGW), then mapped into the IMS supported SIP signaling protocol. User data is carried within the IMS infrastructure by the Real-Time Protocol (RTP).

Given the fact that IMS is not a conduit for media information, there has been some controversy concerning the place and functions supported by the IMS **Media Server**. Some consider media servers as part of the IMS core layer because it is controlled by the **Media Gateway Control Function** (MGCF) module that is part of the IMS control plane. Specifications are limited to the recommendation to use RTP. All negotiations regarding media encoding are left to the end devices and gateways. However, IMS keeps control of media servers and media gateways, routing the RTP streams through media servers in accordance with the service provided.

The **Call Session Control Function** (CSCF) module is a session routing point achieved through three different types of **Call Session Controllers** (CSC). These controllers are SIP proxy servers with slightly different roles. The call is ultimately passed to the Application/Service Layer through the IP Multimedia Service Control (IMSC) interface. More details will be provided in the next section.

- **Proxy CSCF** (P-CSCF): The SIP proxy that provides subscriber access to multimedia services. It is the first stop for any call coming from the user and it front-ends the Serving CSCF;
- **Serving CSCF** (S-CSCF): The primary call controller for registered calls. It maintains the session state for the service requested through its contact with the Application/Service Layer;
- **Interrogating CSCF** (I-CSCF): The SIP proxy responsible for locating and registering roaming subscribers and allocating the proper Serving CSCF.

Home Subscriber Server (HSS): The database server that contains the Home Location Register (HLR) Function for mobile users. This function provides centralized subscriber provisioning, management and authentication/authorization. It keeps track of subscribers that belong to the network and their service capabilities. CSCF consults with the HSS before initiating SIP connections (call setups). HSS, in turn, supplies the CSCF with the aggregated data it needs, provided by multiple sources of information. A **Subscriber Location Function** (SLF) maps user addresses when multiple HSSs are used. Both the HSS and the SLF communicate through the DIAMETER protocol.

Application Services (AS): A collection of IMS-compliant multimedia application servers that support applications such as instant messaging and presence, video conferencing, push to talk, video games, and video content sharing. Services are accessed via a Service Capability Interaction Manager (SCIM) that acts as a broker. The application servers are of two kinds: Front-end Servers that handle incoming SIP calls and Service Servers that provide service features. Depending on the actual service, the AS can operate in the SIP proxy mode, SIP US (user agent) mode, or SIP B2BUA (back-to-back user agent) mode. An AS can be located in the home network or in an external third-party network. If located in the home network, it can query the HSS with the DIAMETER "Sh" interface (for SIP-AS and OSA-SCS) or the Mobile Applications Part (MAP) interface (for IM-SSF).

SIP Application Server (SIP-AS): A native IMS application server. **Open Service Access** (OSA) and **Open Service Access Service Capability Server** (OSA-SCS): An application server that interfaces with OSA Application Servers using **Parlay Group X API**, a set of open service access APIs that allow development of value-added telecommunications Web-based applications on the IMS service platform [96]. These APIs are also part of the 3GPP Release 6/7 specifications along with ETSI TISPAN.

IP Multimedia Service Switching Function (IM-SSF): Interfaces with **Customized Applications for Mobile (Networks) Enhanced Logic** (CAMEL) Application Servers using the **CAMEL Applications Part** (CAP) protocol. More details about CAMEL will be provided in Section 17.15.

Network functionality is provided to applications through a set of **Service Capability Features** (SCF). One special server is the **Voice Call Continuity** (VCC) Server designed to facilitate convergence between fixed wireless networks (WLANs) and cellular mobile networks (GSM/CDMA).

Media Resource Control Function (MRCF): The functional module responsible for controlling resources required to provide services. There is a close interdependence between the **MRCF module**, the **Media Server**, the **Media Gateway** (MGW), and the **Media Gateway Control Function** (MGCF). MGW translates voice or multimedia traffic from the PSTN (TDM-based) into IP packets and vice versa. In some designs, there is a common MGW and Media Server but MGCF still acts as a separate entity. MGW and MGCF modules are used to exchange VoIP and multimedia packets with the PSTN. **Signaling Gateway** (SGW): Used to map IMS SIP-based call control messages into PSTN SS#7 signaling protocol and messages.

IMS **Policy Decision Function** (PDF): A policy-based management server that performs policy decisions regarding QOS and security. PDF checks the HSS database and uses the RFC 2748 Common Open Policy Service (COPS) protocol to control Policy Enforcement Points (PEP) which are embedded in network elements. Initially, this functional module was part of the CSCF module but later it has been detached and made a separate component. More details about COPS are provided in Section 18.5, Chapter 18.

Border Gateway Control Function (BGCF): A functional module analogous to the session border controllers that are used to tie together "VOIP "islands", belonging to diverse carriers. For IMS, BGCF handles only control/signaling messages to other IMS domains.

Breakout Gateway Control Function (BGCF): Using routing functionality based on telephone numbers, BGCF is a functional module realized as a SIP server. It is only used for calls from the IMS to a phone in a circuit-switched network such as the PSTN or to cellular mobile public networks.

Packet Data Gateway (PDG) and WLAN **Wireless Access Gateway** (WAG): They provide secure acces to hundreds of thousands of mobile users that access the IMS network via wired and wireless access points or gateways. They are critical components in fixed-mobile convergence between WLANs and cellular mobile networks.

17.6 Call Session Control Functional Modules

Known collectively as **Call Session Control Function** (CSCF) modules, are three inter-working units in the core of the IMS Layer. These modules perform the role of SIP servers or proxies, and are used to process SIP signaling packets in the IMS network [97].

The **Proxy-CSCF** (P-CSCF) is the first point of contact for an IMS terminal. It is located either in the visited network (in full IMS networks) or in the home network (when the visited network isn't IMS compliant). Some networks might use a Session Border Controller for this function. The terminal will discover its P-CSCF by using the DHCP, or it may be assigned to the terminal in the Packet Data Protocol context (in GPRS). The main P-CSCF features are:

- Is assigned to an IMS terminal during registration, and does not change for the duration of the registration;
- Sits on the path of all signaling messages, and can inspect every message;
- Authenticates the user and establishes an IPsec security association with the IMS terminal; this prevents spoofing and replay attacks and protects the users' privacy. Other nodes trust the P-CSCF, and do not have to authenticate the user again;
- Compresses and decompresses SIP messages to reduce the round-trip delay over slow radio links;
- May include a PDF (Policy Decision Function), which authorizes media plane resources to assure adequate Quality of Service; it is used for policy control and bandwidth management. The PDF can also be a separate function;
- Generates charging records.

An **Interrogating-CSCF** (I-CSCF) is a SIP proxy located at the edge of an administrative domain. Its IP address is published in the DNS of the domain (using NAPTR type of DNS records). This enables remote servers (e.g., a P-CSCF in a visited domain, or a S-CSCF in a foreign domain) to find it, and use it as an entry point for all SIP packets to this domain. The I-CSCF queries the HSS using the DIAMETER Cx and Dx interfaces to retrieve a user's location, and then routes the SIP request to its assigned S-CSCF. Up to Release 6, it can also be used to hide the internal network from the outside world (encrypting part of the SIP message), in which case it is called a THIG (Topology Hiding Interface Gateway). From Release 7 onwards, this function is removed from the I-CSCF and becomes part of the IBCF (Interconnection Border Control Function). The IBCF is used as a gateway to external networks, and provides NAT and Firewall functions.

A **Serving-CSCF** (S-CSCF) is the central node of the signaling plane. It's a SIP server, but performs session control as well. It is always located in the home network. The S-CSCF uses DIAMETER protocol on Cx and Dx interfaces with the HSS to download and upload user profiles. It has no local storage of the user. The main S-CSCF features are:

- Handles SIP registrations, allowing it to bind the user location (e.g. the IP address of the terminal) and the SIP address;
- Sits on the path of all signaling messages, and can inspect each message;
- Decides which application server(s) the SIP message will be forwarded to;
- Provides routing services, typically using ENUM lookups; and
- Enforces the network operator's policies.

17.7 Advantages and Disadvantages of IMS Technology

There are several advantages gained when IMS is implemented:

- Provides a service-based convergence of fixed and mobile wireless networks;
- Is designed to run on public or private IP-based networks using standard protocols and addressing schemes such as IPv6 and IPv4;
- Is conceived as a service oriented architecture and platform that allows development of new applications by third parties using open standard APIs;
- Is applicable to both wireless and wired networks allowing full mobility of users and applications;
- Design is flexible since it is based on functional modules with few restrictions regarding implementation in specialized servers;
- Has the commercial support of major wireless and wired operators. These operators plan and pilot IMS implementations and development of new services.

Other advantages of implementing IMS:

- Separation of the access/transport layer from service/applications layer by introduction of an IMS signaling and control layer;
- Incorporation of advanced NGN features, via ETSI TISPAN, especially those providing multimedia QOS;
- Use of standardized interfaces and protocols (IP, SIP, RTP, RTCP, COPS, DIAMETER, and H.248);
- Single sign-on security, unified messaging, and single billing, regardless of which network originates the service calls;

- By having all the services delivered through one IMS network operator and platform, it becomes a cost effective solution.

There are some disadvantages that come with the implementation of IMS as follows:

- Because it is a very complex system, with numerous functional modules and reference points, there is a need for multiple protocols. It requires to be built by multiple manufacturers/vendors and implies various engineering skills;
- The IMS has become somewhat considerably more complex, and hence more demanding to implement, because of the several new features that were added to extend its initial scope to serve the mobile community;
- Incompletely implemented systems, claimed as “partial IMS” or “IMS-like”, could cause interoperability problems;
- There are no proven business cases to justify implementing IMS on a grand scale, either as total replacement of legacy networks or during gradual transitions;
- There is a legitimate fear of building an IMS with the capability of charging each call session crossing the IP-based network. This could threaten the existence of the “free” Internet.

Significant cost savings can arise from having one network with fewer nodes and lower operating costs. In addition to making convergent user services faster and easier to introduce, common shared resources can increase operational network efficiency. From an investment perspective, it is possible to optimize use of control and media processing resources, reducing the need to replace technologies and lowering the cost of network updates.

17.8 IMS Standard Specifications

IMS was designed by 3GPP as an evolutionary development of the earlier UMTS architecture and specifications. The advantages of this new cellular mobile network technology, based on common signaling, have attracted wired telecommunications operators and even cable operators. Currently, IMS endorsements come from major operators and equipment manufacturers such as Vodaphone, BT, Verizon Wireless, AT&T (Cingular), Alcatel/Lucent, Ericsson, Siemens, Nokia, Nortel Networks, and NEC. IMS technical specifications are included in the following documents:

- **3GPP Release 5**: issued in 2003;
- **3GPP Release 6**: issued in 2004;
- **3GPP Release 7**: issued in 2007.

Additional specifications are provided in IETF Requests for Comments:

- **IETF SIP**: RFC 3261–3265;
- **IETF SDP**: RFC 2327;
- **IETF RTP**: RFC 3550;
- **IETF RTCP**: RFC 3605;

- **IETF COPS**: RFC 2748;
- **IETF DIAMETER**: RFC 3588.

Other standards organizations and activities that have provided contributions to the development and implementation of IMS include:

- **ETSI TISPAN**: (www.etsi.org);
- **3GPP2 IMS Consortium for CDMA2000**: Multi-Media Domain (MMD), (www.3gpp2.org);
- **IETF SIP Working Group**: (www.ietf.org);
- **Alliance for Telecommunications Industry Solutions** (ATIS): (www.atis.org);
- **International Telecommunications Union-Telecommunication Sector**: H.248 for media control, Q1912, signaling SIP/ISUP conversion, (www.itu-t.int);
- **Open Mobile Alliance** (OMA): IOMS service definition, (www.openmobile alliance.org);
- **The Parlay Group**: Consortium to standardize IMS APIs, (www.parlay.org);
- **CableLabs Packet Initiative**: Packet Cable Specs, (www.cablelabs.com).

3GPP Specs

The list below is a small selection from the first three stages of IMS specifications. Note the wide scope of technologies required to architect and implement interoperable IMS systems. These specifications can be found at http://www.3gpp.org/specs/ numbering.htm.

- TS 22.228 Service requirements for the IP multimedia core network subsystem: Stage 1;
- TS 23.228 IMS stage 2;
- TS 23.278 Customized Applications for Mobile network Enhanced Logic (CAMEL) IMS interworking; Stage 2;
- TS 23.979 3GPP enablers for Push-to-Talk over Cellular (PoC) services; Stage 2;
- TS 24.228 Signaling flows for the IMS call control based on SIP and SDP; Stage 3;
- TS 29.278 CAMEL Application Part (CAP) specification for IMS;
- TS 32.260 Telecommunication management; Charging management; IMS charging;
- TS 32.299 Telecommunication management; Charging management; Diameter charging applications;
- TS 32.421 Telecommunication management; Subscriber and equipment trace: Trace concepts and requirements;
- TS 33.978 Security aspects of early IP Multimedia Subsystem (IMS).

IETF Specs

The list below is a selection of the original RFC recommendations, adopted as IMS specifications with some changes. They can be downloaded from http://www.ietf.org.

- **RFC 2327 Session Description Protocol (SDP);**

- **RFC 2748 Common Open Policy Server Protocol (COPS);**
- **RFC 3428 SIP Extension for Instant Messaging;**
- **RFC 3550 Real-time Transport Protocol (RTP);**
- **RFC 3588 DIAMETER Base Protocol;**
- **RFC 3840 Indicating User Agent Capabilities in SIP;**

17.9 IMS Applications

When an IMS architecture is introduced and implemented, it is done with the presumption that it will lead to integration and convergence of wireless networks and the delivery of new multimedia applications. Development of applications by third parties is facilitated by IMS's open service platform design and standardized APIs. Although the current focus is on applications to individual consumers, it is expected that IMS will offer service/applications to enterprises, as well. The IMS range of applications includes:

- **Push to Talk** (Push to talk over Cellular (PoC);
- **Instant messaging**;
- **Conference calls**;
- **Presence information**;
- **Location-based services**;
- **SMS, MMS**;
- **Voice mail**;
- **VoIP**;
- **Voice call features** (caller ID, call waiting, call forwarding);
- **Voice Call Continuity (VCC)** (seamless IP-Circuit Switched handoff);
- **Unified messaging**;
- **Speech-to-text, Text-to-speech**;
- **Announcements**;
- **Peer-to-Peer communications**;
- **Video and content sharing**;
- **Web services**.

17.10 IMS Architecture Reference Points

Over twenty **Reference Points** are defined in IMS, in addition to functional modules. These reference points indicate the endpoints of communication established between functional components along with the type of traffic and protocol supported [98]. The most important reference points and interfaces are represented in Figure 17.6. However, since IMS is still evolving as concept and architectural layout, we may find variations on interpretations of some of these reference points.

The IMS reference points are organized into six different groups, G, M, C, D, S, and U. The "G" group is concentrated in the neighborhood of the GSM GPRS Serving Node

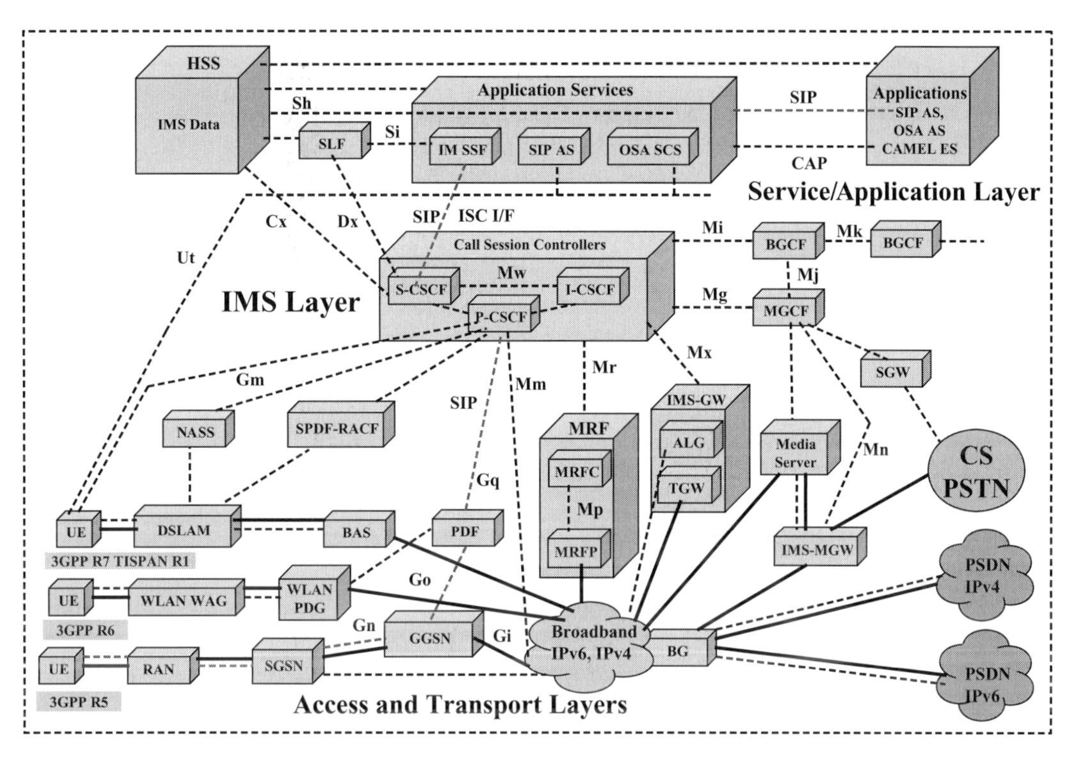

Figure 17.6 IMS Reference Points and Interfaces

(GGSN). The "M" group mainly relates to the IMS Layer, more precisely to the links between call session controller functional modules and other functional modules from the same layer or adjacent layers. In many instances, internal functional modules are not visible and the S-CSCF is considered the named module for all these links. The "C" and "D" groups are organized around accesses to the HSS component from the IMS layer when subscriber information is to be retrieved. The "S" group focus is concerned with access to the same HSS but initiated by functional modules or specialized servers within the Applications/Services layer. The "U" group is envisioned as a means for users to access the IMS functional modules using standardized browsers, and, evidently, supporting HTTP/TCP protocols.

The complexity of IMS can be understood by listing the reference points identified between different architectural components. The reference point descriptions, along with the protocols supported, are given in Table 17.1.

Over ten different protocols are supported in various IMS layers from core control to access to the applications/services layer: SIP/SDP, RTP/RTCP, DNS, GPRS, DIAMETER, Megaco, MGCP, COPS, TCP/UDP/IP, IPSec, DHCP, XCAP, and MAP. However, two protocols dominate the architecture: SIP/SDP and DIAMETER. SIP/SDP is used for signaling. DIAMETER combines signaling with secure access, particularly when accessing the Home Subscriber Server. Two important interfaces are the IP Multimedia Service Control (ISC) Interface which is between the Serving Call Session Control

Table 17.1 IMS Reference Points and Interface Descriptions

Interface Acronym	IMS Entities Involved	Interface Description and Function	Protocol Supported
Gm	UE to P-CSCF	Used for message exchanges between UE and CSCF	SIP
Mw	P-CSCF, I-CSCF, S-CSCF	Message exchanges between CSCF SIP Proxies	SIP
ISC	S-CSCF, I-CSCF, AS	Message exchanges between CSCF and AS	SIP
Cx	I-CSCF, S-CSCF, HSS	Message exchanges between I-CSCF/S-CSCF and HSS	DIAMETER
Dx	I-CSCF, S-CSCF, SLF	Used by I-CSCF/S-CSCF to find the associated HSS in a multi-HSS environment	DIAMETER
Sh	SIP AS, OSA SCS, HSS	Message exchanges between SIP AS/OSA SCS and HSS	DIAMETER
Si	IM-SSF, HSS	Message exchanges between IM-SSF and HSS	MAP
Dh	SIP AS, OSA SCS, SCF, IM-SSF, HSS	Used by AS to find the correct HSS in a multi-HSS environment	DIAMETER
Mm	I-CSCF, S-CSCF, External IP Network	Message exchanges between IMS and external IP network	Not specified
Mg	MGCF to I-CSCF	MGCF converts ISUP signaling to SIP signaling and forwards SIP signaling to I-CSCF	SIP
Mi	S-CSCF to BGCF	Message exchanges between S-CSCF and BGCF	SIP
Mj	BGCF to MGCF	Message exchanges between BGCF and MGCF in the same IMS network	SIP
Mk	BGCF to BGCF	Message exchanges between BGCFs in different IMS networks	SIP
Mr	S-CSCF, MRCF	Message exchanges between S-CSCF and MRCF	SIP
Mp	MRCF, MRFP	Message exchanges between MRCF and MRFP	H.248
Mn	MGCF, IM-MGW	Allows control of user-plane resources	H.248
Ut	UE, AS (SIP AS, OSA SCS, IM-SSF)	Enables UE to manage information related to user's services	HTTP
Go	PDF GGSN	Allows operators to control QOS in a user plane and exchange charging information between IMS and GPRS network	COPS (Release 5) DIAMETER (Release 6)
Gq	P-CSCF, PDF	Allows operators to control QOS in a user plane and exchange charging correlation information	DIAMETER

Function (S-CSCF) module and the IP Multimedia Service Switching Function (IM-SSF). The other, which uses the CAP protocol, is the interface between IM-SSF and the Customized Applications for Mobile Networks Enhanced Logic (CAMEL) Service Environment. More details will be provided in Section 17.15.

IMS architecture does not put constrains on the design of functional modules. Some IMS architectural modules could be aggregated in common hardware or software components. Therefore, some of the reference points or interfaces presented in this diagram might not be exposed to users or operators. The same is true of IMS features such as access or security that might be bundled in various modules, gateways, or servers.

17.11 IMS-based FMC Solutions, Products, and Services

The complexity of the IMS Reference Architecture is a good indication of the challenges presented in achieving fixed-mobile convergence. Many of the functional modules in IMS require special expertise in the areas of hardware components, servers, databases, services, and media conversion. This explains the high number of vendors involved in the implementation of IMS.

Research companies such as **Frost and Sullivan** predict an increase in sales for the IMS market from 2.5 billion dollars in 2006 to 12.5 billion in 2012. iCode Wireless, another analyst, provides a similar forecast. These figures include the softswitch business on the assumption that IMS is a natural evolution of that technology.

Special Applications and SIP Servers

- **Alcatel-Lucent**: 5350 IMS Application Servers, (www.alcatel-lucent.com);
- **BEA Systems-Broadsoft**: BEA WebLogic SIP Server, (www.beasystems.com);
- **Broadsoft**: IMS Application Server Complex, (www.broadsoft.com);
- **Bridgewater Systems**: AAA Service Controller, (www.bridgewater.com);
- **Comverse**: My Call Converged Communications, (www.comverse.com);
- **IBM**: WebSphere Application Server & ParlayX Web Services, (www.ibm.com);
- **Huawei Technologies**: IMS 3.0,Voice Call Continuity AS, (www.huawei.com);
- **Ingate**: SIParator 19, 50/55/65, 90 SIP-based Firewalls, (www.ingate.com);
- **Intertex**: IX67 SIP-enabled Firewall, (www.intertx.com);
- **MIND CTI**: MIND iPhone EX SIP Application Server, (www.mindcti.com);
- **Motorola**: IMS Eco-System SIP Applications Server, (www.motorola.com);
- **Nortel**: Application Server AS 5200, Policy Controller, (www.nortel.com);
- **Polycom**: Proxias Application Server and Application Development Environment, (www.polycom.com);
- **Radvision**: SCOPIA Interactive Video Platform, (www.radvision.com);
- **Redknee**: Policy Decision Rule Server, (www.redknee.com);
- **Ranch Networks**, Asterisk VoIP Security Products RN300, RN20, RN40 and RN41, (www.ranchnetworks.com);
- **Sipera**, IPCS 310 VoIP Security, (www.sipera.com);
- **Sylantro Systems**: Synergy Multiplay Application Server, (www.sylantro.com);

- **Tatara Systems**: Tatara Convergence Server, (www.tatarasystems.com);
- **Telcordia Technologies**: Maestro IMS Portfolio, FMC, (www.telcordia.com);
- **Ubiquity Software**: SIP Application Server, (www.ubiquity.com);
- **ZTE Corporation**: ZIMS Solution Parlay Gateway, (www.zte.com).

Call Session Control Function (CSCF)

- **Alcatel-Lucent**: 5430 Session Resource Broker, (www.alcatel-lucent.com);
- **Ericsson**: Call Session Control Function, (www.ericsson.com);
- **Huawei Technologies**: IMS 3.0 CSCF and PDF, (www.huawei.com);
- **Leapstone Systems (Motorola)**: CCE Service Broker, (www.leapstone.com);
- **HP-Tekelec**: HP-Tekelec Open IMS Solution, TekCore, (www.tekelec.com);
- **Marconi-Ericsson**: IMS Session Controller XCD6000, (www.marconiusa.com);
- **Motorola**: IMS Eco-System Call Session Controller, (www.motorola.com);
- **Nokia Siemens Networks**: CFX-5000 for VoIP and Push over cellular (PoC), (www.nokia-siemens.com);
- **Nortel**: Versatile Service Engine, Call Session Controller, (www.nortel.com);
- **Reef Point Systems**: Universal Convergence Gateway, (www.reefpoint.com);
- **Nortel Networks**: Call Session Controller 1000, (www.nortel.com);
- **Sonus Networks**: SMARRT Solutions, (www.sonusnet.com);
- **ZTE Corporation**: ZIMS Solution P/I/S CSCF, (www.zte.com);
- **Veraz Networks**: User Services Core (UNC) xCSCF, (www.veraznetworks.com).

SIP Session Border Controller (SBC)

- **Acme Packet**: Net-Net SBC, (www.acmepacket.com);
- **Cisco Systems**: SBC on Cisco IOS routers, (ww.cisco.com);
- **Convergence**: Eclipse Access-Edge VoIP SBC, (www.convergence.com);
- **Jasomi Networks (Ditech Networks)**: PeerPoint 100, EdgeWater, BorderWater, (www.ditech.com);
- **HP-Tekelec**: HP-Tekelec Open IMS Solution, BGCF, (www.tekelec.com);
- **Kagoor (Juniper Networks)**: Session and Resource Control Portfolio, C-Series Controllers, (www.juniper.com);
- **NeoTIP SA (Netcentrex)**: IMS SBC and P-CSCF, (www.netcentrex.com);
- **NexTone**: Nextone SBC, (www.nextone.com);
- **Newport Networks**: 1460 SBC, (www.newport-networks.com);
- **Quintum**: Tenor Call relay SBC, (www.quintum.com);
- **Reef Point Systems**: Universal Convergence Gateway, (www.reefpoint.com).

Media Resource Function Controller (MRFC)

- **Alcatel-Lucent**: 5020 Media Gateway Controller, (www.alcatel-lucent.com);
- **Cisco Systems**: Media Gateway Control Function PGW 2200, (www.cisco.com);
- **Huawei Technologies**: IMS 3.0, Media Resource Server, (www.huawei.com);
- **Kagoor (Juniper Networks)**: SRC Policy Engine, SRC DIAMETER Gateway, (www.juniper.com);
- **Nortel**: 1000 and 2000 Media Gateway Controllers, (www.nortel.com);
- **Operax**: Operax Resource Controller 5500, 5700, (www.operax.com);

- **Starent Networks**: ST16 and ST40 Intelligent Mobile Gateway Platform, (www.starentnetworks.com).

Media Gateways (MGW)

- **Alcatel-Lucent**: 7510 and 7515 Media Gateway, (www.alcatel-lucent.com);
- **Cisco Systems**: Media Gateway MGX 8000 Series, (www.cisco.com);
- **Data Connection Ltd.**: Media Gateway, (www.dataconnectionsinc.com);
- **Ericsson**: End-to-End IMS, (www.ericsson.com);
- **Motorola Corporation**: Media Gateway, (www.motorola.com);
- **Nortel**: Converged Media Gateway 15000, (www.nortel.com);
- **Tekelec**: TekMedia, (www.tekelec.com);
- **Veraz Networks**: I-GATE 4000 EDGE MGW, (www.veraznetworks.com).

Home Subscribers Server (HSS)

- **Alcatel-Lucent**: 1430 Unified IMS HSS, (www.alcatel-lucent.com);
- **Bridgewater Systems**: HSS, (www.bridgewater.com);
- **Ericsson**: HSS, (www.ericsson.com);
- **Huawei Technologies**: IMS 3.0, HSS, (www.huawei.com);
- **Motorola**: IMS Eco-System HSS, (www.motorola.com);
- **Nokia Siemens Networks**: CMS-8200 HSS, (www.nokia-siemens.com);
- **Nortel**: Nortel HSS, (www.nortel.com);
- **Verisign**: Real-Time Rating and Payment Billing Services, (www.verisign.com);
- **ZTE Corporation**: ZIMS Solution HSS, (www.zte.com).

IMS Testing

- **Agilent**: J7830A IMS Signaling & Wireless QOS Testing, (www.agilent.com);
- **Spirent Communication**: Spirent PDG Tester, (www.spirent.com);
- **Tektronix**: Tektronix IMS Interoperability Test Suite, (www.tek.com).

The complex IMS design requires that systems be thoroughly tested in the laboratory and in the field. And, since several vendors may be involved in supplying components for a system, overall system and interoperability tests need to be made. IMS products and interoperability demonstrations are performed within two major events: **IMS Forum's**, **Multi-Service Forum's** (MSF) **Global Multi-Service Interoperability** (GMI), and the **Plugfest for Applications and Services**. In 2006, GMI laboratory locations involved were:

- BT Advanced Research and Technology Center, Ipswich, Suffolk, UK;
- KT Technology Lab, Daejon, South Korea;
- NTT Musashino Research and Development Center, Tokyo, Japan;
- Verizon Laboratory, Waltham, MA, USA;
- University of New Hampshire's Interoperability Lab, Durham, NH, USA.

The 2007 plugfest event took place at the University of New Hampshire's Interoperability Lab (UNH-OIL). Vendors involved in the testing were: Acme Packets, Cisco Systems, CommunGate Systems, Empirix, Ericsson, ETRI, Huawei Technologies,

IP Unity, Leapstone Systems, Lucent Technologies, Metaswitch, Mitsubishi Electric Corporation, NEC, NexTone Communications, Nortel Networks, Operax, Samsung, Softfront, Sonus Networks, Spirent Communications, Tekelec, Tektronix, and ZTE Corporation.

Another site where trials and testing takes place is the 3G/IMS Laboratory at the Georgia Institute of Technology, Atlanta. Companies involved there are: AT&T, Lucent Technologies, and Siemens. On a different note, both Siemens and Ericsson, as manufacturers of IMS components and IMS system integrators, are involved in numerous worldwide trials and pilot testing. Spirent Communications (www.spirent.com) developed a product, called Landslide, a comprehensive tool focused on testing Packet Data Gateways, the WLAN access to IMS core network.

Early IMS Implementations and Demonstrations

- **Embarq (Sprint spinoff)** and **NewStep Networks**: SmartConnect (Basic and Plus) FMC services with Starcom Packet PC 6700 IMS handsets;
- **Italtel**: Italtel IMS Solution, (www.italtel.com);
- **NTT DoCoMo**: NEC IMS solutions, Push to Talk Services;
- **AT&T (Cingular)**: VideoShare IPTV Service;
- **China Unicom, Chugwa Telecom, ZTE, Microsoft**;
- **Tekelec, HP, BEA System**: TM Forum Catalyst Project.

17.12 Verizon Wireless IMS Vision

An example of the interest taken in IMS by a major mobile operator is Verizon Wireless. Verizon put together a technology team, formed from its major suppliers to analyze every aspect of IMS architecture and specifications, especially those related to security and implementations [99]. Cisco Systems, Lucent Technologies, Motorola Corporation, Nortel Networks, and Qualcomm were part of this team.

This working team produced a comprehensive analysis and proposed enhancements to IMS called "**Advances to IMS (A-IMS)**". It was presented to the technical community in 2006. It is a practical view of IMS that tries to address the fact that there are two major directions in 3G mobile advancements: WCDMA and CDMA 2000. Also, there are numerous legacy systems that are not SIP-based and some new developments, such as IPTV, are not taking the "all SIP-based road". The proposals are not a repudiation of IMS but they are improvements suggested for adoption by the standards bodies.

The main domains and ideas contained in the A-IMS are:

- Adoption of a "Security Manager" that monitors all security functions from devices to servers and provides not only authentication but selective authorizations and reactions to security threats;
- Adoption of a powerful "Policy Manager", implemented in the PDF module that provides equal treatment of SIP and non-SIP applications;

- Adoption of a "Double IP-addressing" scheme, one for the home address and the other for visiting address, to facilitate quick roaming and soft handovers;
- Adoption of "Multi-tier Service Interaction Management", a service broker that manages features between SIP and non-SIP applications;
- Adoption of an "Application Manager" to authorize access to and management of SIP services;
- Adoption of a "Service Data Manager" to collect subscriber and network control information for charging and billing;
- Adoption of a "Bearer Manager" to allocate resources and monitor bearer traffic for QOS purposes, in line with VoIP services planned for CDMA 2000 upgrade, EV-DO Revision A.

17.13 IMS and Softswitches

This section will analyze the strong resemblances between **IMS** and **softswitch** architectures and functional requirements. Softswitches are key components of telecommunications networks implementing Internet Telephony or Voice over IP (VoIP) technologies for carriers and major enterprises.

At a high level, a softswitch can be described as specialized software running on general purpose computers having three architectural layers: **Application**, **Call Control**, and **Transport/Switching**. This is shown in Figure 17.7.

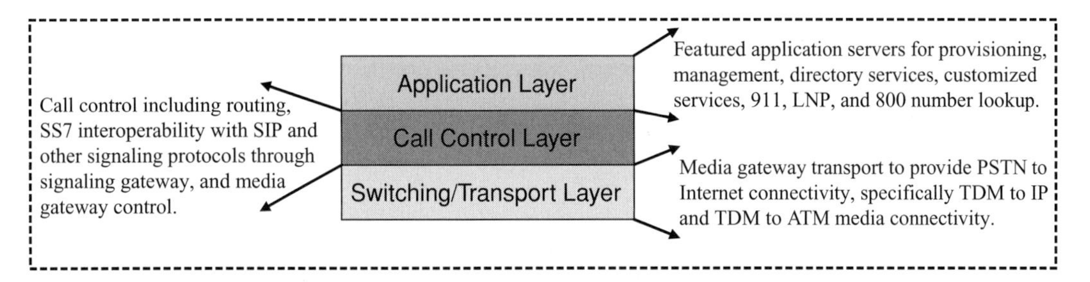

Figure 17.7 Softswitch Layered Architecture

- The **Application Layer** is a collection of standardized applications that support particular services (e.g., analog voice channel, audio/video streams for conferencing, a RTP data stream, or a MP3 audio stream).
- The **Call Control Layer** performs call setup and the correlation of various services to individual data flows. It does this by using a variety of signaling protocols: SIP, H.323, MGCP/Megaco, and SS7 TCAP/ISUP.
- The **Transport/Switching Layer** provides support for **media gateway** and **transport/connectivity functions** to allow convergence of circuit switched and packet switched networks.

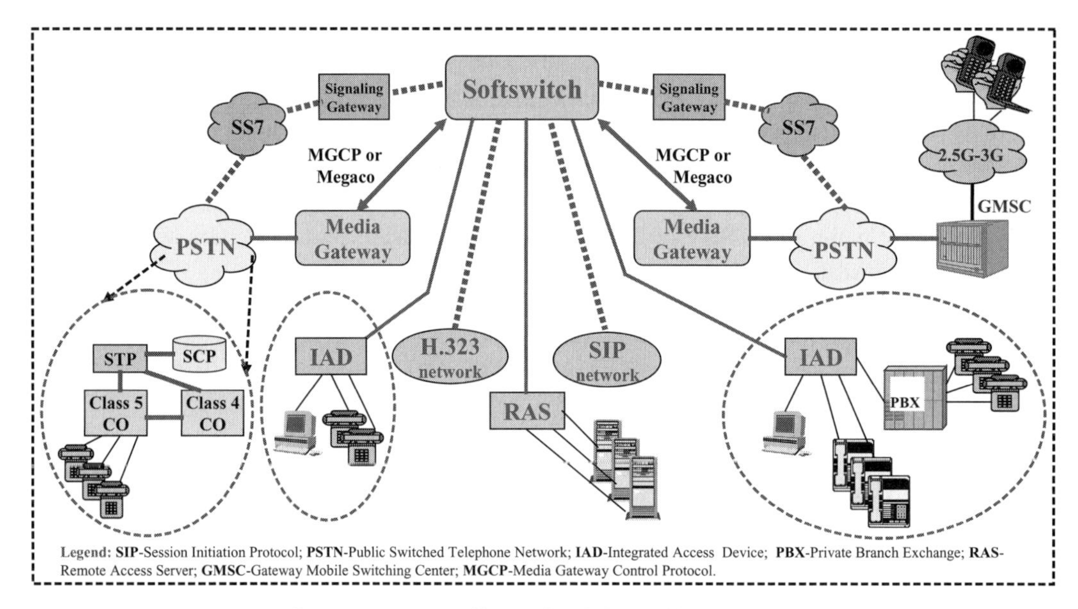

Figure 17.8 Media Gateway Controller (Softswitch) Architecture

Softswitches can be implemented either as standalone hardware/software entities or combined with media gateway functionality.

Softswitch or **Call Agent** is a marketing name for the **Media Gateway Controller** (MGC) architectural component. MGC controls voice and multimedia communications over packet networks at the level of Media Gateways (MGWs) and Signaling Gateways (SGWs). The diagram in Figure 17.8 depicts the position and the role of softswitches in VoIP type of services.

Basically, the softswitch provides signaling call control regardless of the origin of a call (a PSTN central office exchange, an enterprise PBX network with a customer premises gateway, a public cellular mobile network with a Gateway Mobile Switching Center, or a carrier IP network supporting VoIP with either H.323 or SIP-based signaling). One of the softswitch functions is to provide interoperability between signaling protocols such as H.323, SIP, MGCP, SS7 TCAP and ISUP. In some instances, SGW and softswitches can be part of the same physical hardware.

The softswitch does not control the voice or media flow. This means that the actual traffic flow does not pass the softswitch. However, the softswitch controls the media gateways using the Media Gateway Control Protocol (MGCP), Megaco, or SIP, depending on the generation of the softswitch.

The fundamental thrust of the softswitch concept is a functional desegregation of traditional central office switch into components, addressing transport and media gateway functions, those addressing signaling gateway functions, other components addressing call handling and policy management, and the applications servers as the source repository of new services with associated databases. This is the reason that softswitches are logical replacements of the Class 4 and even Class 5 digital switches that dominate the traditional PSTN architecture and services. In other words, to fold softswitches into one digital packet-based network, in line with the Internet.

In addition to its main components, softswitch systems require **Session Border Controllers** (SBC). Two types of SBCs are used: **Access SBCs**, that sit between users/enterprises and service provider IP networks, and **Interconnect SBCs** that sit between service provider point of presence. SBCs handle both call and media flow control, having important security features to protect the systems against unauthorized use of services.

We can see from this analysis that IMS and the softswitch have many features in common, as follows:

- The same layering architecture: transport, control, and application planes;
- The same goal, traffic control via signaling functions;
- Similar components: media gateway controller, signaling gateway, and media gateway;
- Separation of the service/application layer from the control plane;
- Support for an IP-based network and signaling;
- Similar expertise required in design and implementation of softswitches and IMS.

These similarities indicate an important fact, that softswitch systems are precursors of IMS systems and the softswitches can be upgraded to be used in the IMS systems. That explains why IMS as a concept, architecture and set of functions was embraced by all major players from the telecommunications industry, from carriers to manufacturers. A list of softswitch and media gateway vendors is shown below.

Major Softswitch Vendors

- **Cisco Systems**: BTS 10200, (www.cisco.com);
- **Clarent Corp**: NGN Softswitch, (www.clarent.com);
- **Commgates. Ltd**: CSSW4000 softswitch, (www.comgates.com);
- **CommWorks Corp**: Commworks Softswitch, (www.commworks.com);
- **CopperCom**: Local Exchange Softswitch System, (www.coppercom.com);
- **Convergent Net**: ICSX, (www.convergentnet.com);
- **IPAXS**: OmniAXS Gateway Switch, (www.ipaxs.com);
- **IPVerse**: IPVerse Control Switch, (www.ipverse.com);
- **Lucent Technologies**: 5E-Xc, (www.lucent.com);
- **NetCentrex**: CCS Softswitch, (www.netcentrexcom);
- **Nortel Networks**: Succession Comm. Server 2000, (www.nortelnetworks.com);
- **Nuera Corp**: Softswitch, (www.nuera.com);
- **Open Telecomm**: openCallAgent, (www.opentelecommunications.com);
- **Oresis Comm**: ISIS-700, (www.oresis.com);
- **SCS**: ASD Softswitch, (www.strategicinc.com);
- **Snom Technology** AG: Snom 4S SIP Softswitch, (www.snomag.de);
- **Sonus** Net: Insignus Platform, GSX9000 Open, (www.sonusnet.com);
- **SS8**: SS8 Signaling Switch, (www.ss8.com);
- **Sylantro Systems**: Applications Switch, (www.sylantro.com);
- **Taqua**: OCX Taqua Open Compact Exchange, (www.taqua.com);

- **Tekelec**: VXi MGC, (www.tekelec.com);
- **Telcordia Technologies**: Call Agent Class 5 softswitch, (www.telcordia.com);
- **Telica**: Plexus 9000, (www.telica.com);
- **Telos Technologies**: Sonata SE Wireless Softswitch, (www.telostechnology.com);
- **Unisphere** Net: SRX-3000, (www.unispherenetworks.com);
- **VocalData Inc**: VOISS, (www.vocaldata.com);
- **VocalTec Comm**: Architecture Series 3000, (www.vocaltec.com);
- **WestWave Comm.**: Access Switch, (www.westwave.com);

Major VoIP Media Gateway Vendors

- **AddPac Technology**: Multiport (T1/E1, E&M), SIP/H323 Dual Stack Gateways;
- **Adtech**: Gateways compliant with SIP and H.323;
- **Allied Telesyn**: AT-RG200 VoIP Residential GWs;
- **Anatel Communications**: VoIP Gateway;
- **Audio Codes**: MGs and VoIP boards;
- **Avistar**: Multi-Channel IP Gateway;
- **BCM**: SIP Gateways;
- **Bitwise Comm**: SIP Compliant VoIP;
- **Blue Sky Labs**: XIPlite SIP compliant VoIP GW;
- **Cirpack**: Cirpack G16S SIP Gateway;
- **i3 Micro Technology**: Vood CPEs and Vood Residential Gateways;
- **Jasomi Networks**: SIP to SIP VoIP Gateway;
- **Lucid Voice**: IPico RG-400 Voice and Data Gateway;
- **Mediatrics**: Residential Gateway;
- **MIP Telecom**: SIP Low-density Gateways;
- **Motorola**: ComStruct^{TM} Integrated Gateway Platform;
- **Multitech**: 2-to-60 channel MultiVOIP VoIP GWs;
- **Oki**: BV7000 and BV7050 Multi-Channel Internet Voice Gateways;
- **PureData**: SIP Based IP Telephony Gateway;
- **Servonic**: IXI-ITS Gateway ISDN Gateway for SoftPBX;
- **VegaStream**: Vega100 T1 Carrier-class Gateway;
- **VoiceGenie Technologies**: VoiceXML Gateway.

17.14 ETSI TISPAN

Telecommunications and Internet Converged Services and Protocols for Advanced Networking (TISPAN) is a European Telecommunications Standard Institute (ETSI) project, vision, and technical specification developed to support migration from circuit switched networks to packet switched networks for both wired and wireless communications. TISPAN specifications encompass architecture, protocols, services, security, quality of service, and management aspects of this transition, all under a common umbrella called Next Generation Network (NGN) [100]. ETSI is a European

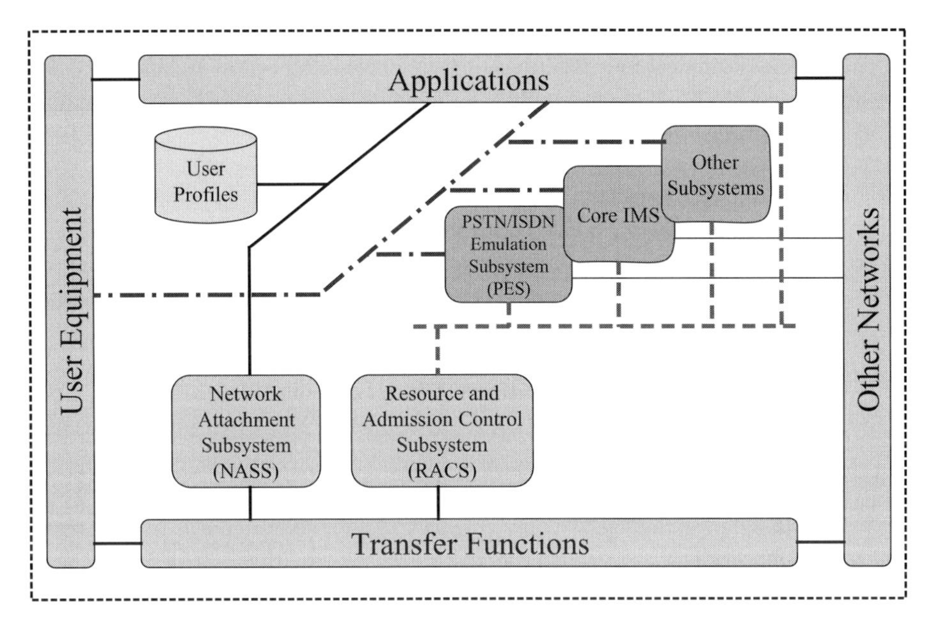

Figure 17.9 ETSI TISPAN High-level Architecture

Commission recognized standard body that was established to promote solutions for the European market.

TISPAN Release 1 was adopted as part of the 3GPP Release 7 IMS specifications. Its goal is to separate the access, transport, and application/service layers by introducing a control plane, the IMS Layer, as presented in previous sections of this chapter. The overall goal of TISPAN NGN is an all IP network, similar to IMS, as the basis for convergence of all types of wired and wireless networks. Included are mobile access across all generations of GSM/GPRS/EDGE/LTE (2G, 2.5G, 3G, B3G) networks and broadband wireless access to networks such as WiMAX. It also includes wired narrowband and broadband access such as ISDN, xDSL, and Packet Cable Modem.

A high-level TISPAN NGN architecture is shown in Figure 17.9. Currently, TISPAN, in collaboration with 3GPP, is working on TISPAN Release 2 specifications. These are expected to be published in 2008.

TISPAN NGN Architecture Release 1 consists of four entities: User Equipment, Network, Applications, and Transfer Functions [100]. The network itself separates the Transport Layer (below) from the Service Layer (above). Two major subsystems provide support for the overall Transport Layer: the Network Attachment Subsystem (NASS) and the Resource Admission Control Subsystem (RACS). NASS provides lower layers adaptation of access networks to be attached to the IMS. RACS is responsible for policy control, including resource reservation and admission control in the access and aggregation networks. Service-oriented platforms provide interworking at the Service Layer: they are the PSTN/ISDN Emulation Subsystem (PES) and the Core IP-based Multimedia Subsystem. The role of these subsystems is to allow users to receive the same benefits from services in NGN as they get in legacy networks.

Some features of the standards are provided below. The full list of ETSI TISPAN NGN Technical Specifications (TS), Technical Reports (TR), and ETSI Standards (ES) standards can be found at the http://www.tech-invite.com/Ti-tispan-standards.html web site.

ETSI TR 180.000: NGN Terminology
This TR provides a common pool of terminology for use within the TISPAN-NGN project. It is aligned with the 3GPP terminology document (3GPP TR 21.905).

ETSI TR 180.001: NGN Release 1-Release Definition
This TR provides a description of the content of NGN Release 1 and defines the release through a catalogue of the Release 1 documentation set. The inter-relationships between Release 1 documents and the features supported are also provided. The document is structured as follows:

- Clause 4 provides an overview of Next Generation Networks (NGN);
- Clause 5 describes NGN Release 1, the Requirements to support Release 1 and outlines the major capabilities that Release 1 will fulfill;
- Clause 6 defines the content and capabilities of NGN Release 1;
- Clause 7 introduces the Release 1 plan documentation set.

01 TISPAN Services
- **ETSI TR 181.001**: Videotelephony over NGN: Stage 1 Service Description;
- **ETSI TR 181.002**: Multimedia Telephony with PSTN/ISDN simulation services;
- **ETSI TR 181.003**: Services Capabilities, Requirements and strategic direction for NGN services.

02 TISPAN Architecture
- **ETSI ES 282.001**: **NGN Functional Architecture Release 1**;
- **ETSI ES 282.002**: PSTN/ISDN Emulation Subsystem (PES)-Functional Architecture;
- **ETSI ES 282 003**: Resource and Admission Control Subsystem (RACS)-Functional Architecture;
- **ETSI ES 282 004**: **NGN Functional Architecture-Network Attachment Subsystem (NASS)**;
- **ETSI TR 182 005**: **Organization of user data**;
- **ETSI TS 182 006**: IP Multimedia Subsystem (IMS)-Stage 2 description;
- **ETSI ES 282 007**: **IP Multimedia Subsystem (IMS)-Functional Architecture**;
- **ETSI TS 182 008**: Presence Service-Architecture and functional description;
- **ETSI TS 182 009**: NGN Architecture to support emergency communications;
- **ETSI ES 282 010**: Protocols for Advanced Networking (TISPAN)-Charging;
- **ETSI TS 182. 012 IMS-based PSTN/ISDN Emulation Subsystem-Functional Architecture**.

03 TISPAN Protocols
04 TISPAN Numbering, Addressing & Routing
05 TISPAN Quality of Service

- **ETSI TS 185 001: NGN QOS framework and requirements.**

06 TISPAN Testing
07 TISPAN Security
08 TISPAN Network Management

- **ETSI TS 188 001**: NGN Management-Operations Support Systems Architecture;
- **ETSI TS 188 003**: OSS Requirements-OSS definition of requirements and priorities for network management specifications for NGN;
- **ETSI TS 188 0041**: NGN Management-OSS Vision;
- **ETSI TR 102.647**: Operations Support Systems Standards Overview and Gap Analysis.

17.15 Customized Applications for Mobile Networks Enhanced Logic (CAMEL)

An important component of the IMS architecture is the **Customized Applications for Mobile Networks Enhanced Logic (CAMEL)** service platform. CAMEL, designed to provide a means to separate the service platform from the switching and transport platforms, is part of the Intelligent Networks (IN) architectural concept. Initially, it was developed for GSM/GPRS/UMTS mobile networks, as ETSI standard 123078 [101] followed by 3GPP TS 22.078 and TS 23.078 [102 and 103]. The main attribute of CAMEL is that it allows development and deployment of value-added services on top of regular mobile services in both circuit switched and packet switched network environments. It has evolved, in a span of 10 years since its inception, and CAMEL4 is being incorporated into the IMS architecture. We do not intend to discuss in greater detail the CAMEL architecture, features, and operations. CAMEL went through several stages of development.

- Phase 1: Defined as part of GSM Release 96 and 97; The GSM Service Control Function (gsmSCF) associated with CAMEL had the ability to bar calls, to modify call parameters, and to monitor call connections and disconnections;
- Phase 2: Defined as part of Release 199x; It enhanced Phase 1 capabilities and added call duration monitoring, call transfer, and multi-party call handling;
- Phase 3: Defined for 3GPP Release 99; It enhanced Phase 2, adding capabilities such as handling announcements, SMS control, and monitoring not-reachability and roaming features;
- Phase 4: Defined as part of IMS Release 5, 2002 [104]; It builds on Phase 3 capabilities. The most important new features are:
  - Control and monitoring of optimal routing;
  - Control over mobile terminating SMS in both circuit and packet switched networks (SMS for pre-paid, SMS charging while roaming, VPN);

- Provision of location information of called subscribers;
- Inclusion of operator determined barring data;
- Inclusion of flexible tone injection (various pre-paid warning tones);
- Call party handling (conference calls, wake-up calls);
- Location information during ongoing calls (subscriber position);
- GPRS any time interrogation (GPRS location and state query);
- Differentiated charges for voice and multimedia calls;
- Enhanced dialed service.

Backward compatibility between phases is provided through the Transaction Capabilities Application Part (TCAP) Application Context (AC) negotiation procedure. The context is specific for each phase. CAMEL4, also known as CAMEL/IMS, went through two stages of implementation: CAMEL/IMS Stage 2, TS 23.278, and CAMEL/IMS Stage 3, TS 29278. An enhanced version of CAMEL4 is provided in 3GPP Release 7.

The most important feature of CAMEL4 (roughly Phase 4) is its incorporation into the IMS architecture with the ability to control IMS sessions. In cases where the CAMEL Service Environment (CSE) is called, the session and the services provided to the caller will be handled by the CAMEL specific Service Control Function (SCF). The relationship between CAMEL and other IMS functional modules is shown in Figure 17.10.

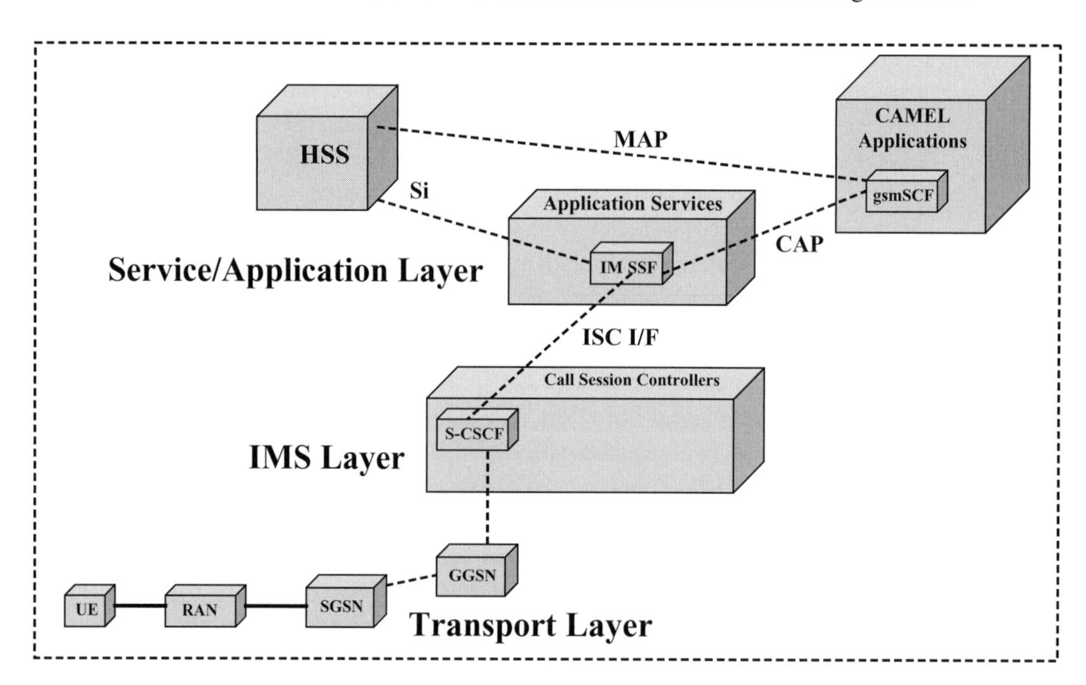

Figure 17.10 CAMEL/IMS Network Architecture

In this simplified IMS diagram, the mobile terminal accesses the core IMS network through the standard Radio Access Network (RAN) and the GPRS Network. The module involved in the call set up, the Serving Call Session Control Function (S-CSCF), is the central node of the IMS Layer signaling plane. This module communicates with the IP Multimedia Service Switching Function (IM-SSF) across the IP Multimedia Service Control (ISC) interface via the SIP protocol. ISC is the interworking module that negotiates services by checking the caller subscription with the Home Subscriber Server (HSS) across the "Si" interface. The same module will request service activation from the CAMEL Application Server (AS), represented by the CAMEL GSM specific Service Control Function (gsmSCF), using the CAMEL Applications Part (CAP) protocol [104]. The ISC interface is detailed in the 3GPPP TS 24.229 specs, the IM-SSF-gsmSCF interface is detailed in the 3GPP TS 29.278 specs, and the HSS-IM-SSF is described in the 3GPP TS 29.002 specs.

Part VI

Fixed-Mobile Convergence Services, Industry Trends, and Implementation Issues

18 QOS in Fixed Wireless Cellular Mobile Convergent Networks

18.1 Fixed-Mobile Convergent Network Management

In Chapters 2 and 3 we provided an extensive introduction to the concepts of network and service management, with no specific reference there to wireless networks or fixed-mobile convergent networks. At this point, we use the layered architecture known as Telecommunications Management Network (TMN) to analyze these concepts. The layered TMN architecture is presented in Figure 18.1.

TMN consists of five logical/functional layers implemented as interconnected management applications. The first three layers, Network Element Layer (NEL), Element Management Layer (EML), and Network Management Layer (NML) deal with network management as a whole and network management sub-divisions such as EML and NEL. The main characteristics of each layer are shown in Figure 18.1.

The next layer, the Service Management Layer (SML), provides management of services, monitors the connectivity between multiple service providers, and deals with specific aspects of service management such as Quality of Service and Service Level Agreements.

Management of convergent networks means management of all those networks that are part of an integrated mobile and fixed wireless network. Network management includes monitoring and controlling the capabilities of all vital network components, including diagnosing faults, modifying configurations, measuring performance, providing billing/charging information, and securing network operations. A high-level depiction of convergent wireless networks management is shown in Figure 18.2.

Convergent network management can be achieved through dedicated network management systems that provide separation of management aspects from the switching and transport aspects. To achieve this separation, call set up control or signaling is needed that provides access, monitoring, and controlling capabilities of the resources that participate in the process of communication. The essential condition of convergence means access to all networks and transparent communications across wireless networks.

Given the fact that convergence is done under the banner of transparent mobility, network management must provide transparent handover between various wireless networks without affecting the expected quality of service. Network management systems can be complex platforms or dedicated management software, known as Operations Support Systems (OSS). These OSSs can be designed according to TMN specifications or to the TeleManagement Forum's TOM specifications that were presented in Chapter 3.

Business Management Layer (BML)

Service Management Layer (SML)

Network Management Layer (NML)

Element Management Layer (EML)

Network Element Layer (NEL)

TMN BML
- Total enterprise view and management policies
- Agreements between TMN service operators
- Strategic planning and executive actions

TMN SML
- Managing service level agreements
- Interfaces with other service providers
- Management services to customers
- Network technology and topology independence

TMN NML
- Management of network view of network elements
- Support for OS-NE interactions across networks
- Provisioning of networks for customer support

TMN EML
- Management of collections of network elements
- Gateway to the network management layer
- Maintain management data about network elements

TMN NEL
- Provides the telecommunications network functions
- Mixture of standard and proprietary interfaces
- Management of individual network elements

Figure 18.1 TMN Layered Architecture

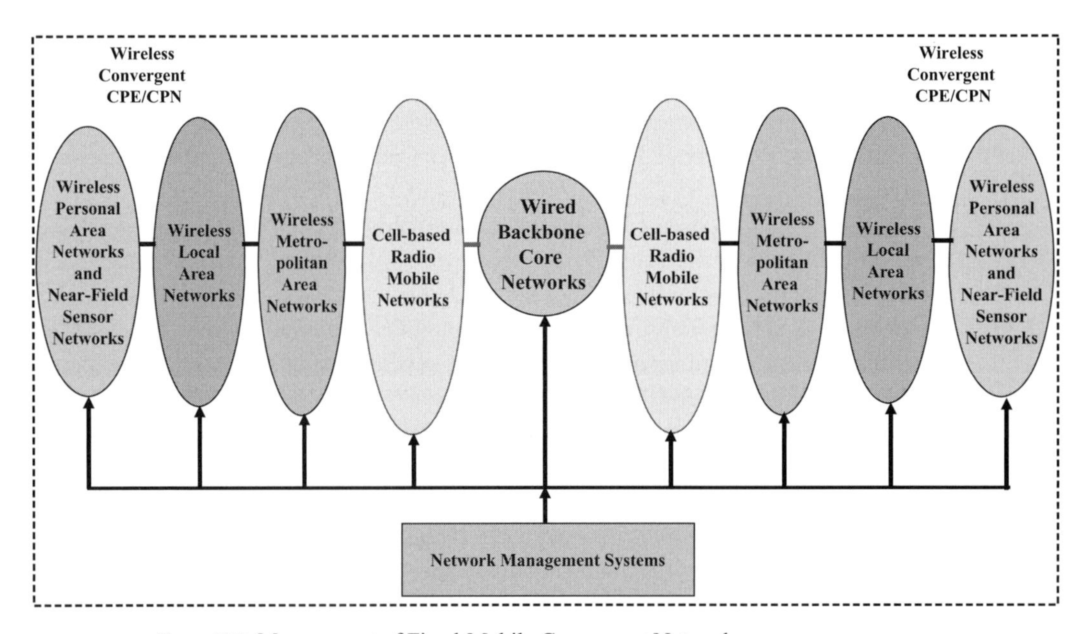

Figure 18.2 Management of Fixed-Mobile Convergent Networks

In the current state of development of Fixed-Mobile Convergence (FMC) networks management, a centralized hierarchical architecture is needed where a manager of managers exchanges information with multiple Element Management Systems (EMSs). Each EMS is responsible for a specific network or network elements that are part of cellular mobile or fixed wireless networks. The hierarchy might have multiple EMS layers. This depends on the size of the networks and the presence of families of individual network elements (base transmission stations, base station controllers) from multiple vendors that have differing standard or proprietary interfaces. Examples of convergent network management will be given in subsequent sections.

18.2 Service Level Management in Wireless Convergent Networks

To address Service Level Management (SLM) in wireless convergent networks we will follow up on the general concept presented in Chapter 3. There, we introduced a general model of service management and the definitions of Quality of Service (QOS), Classes of Service (COS), and Service Level Agreement (SLA). QOS is expressed by metrics such as availability, throughput, response time, delay, jitter, and packet loss. Because transmission requirements differ in some aspects, QOS must be defined specifically for each of voice, data, video, and multimedia communications.

Because convergent networks mean interoperable fixed and mobile networks, QOS must be defined on an end-to-end basis. This idea is presented in Figure 18.3.

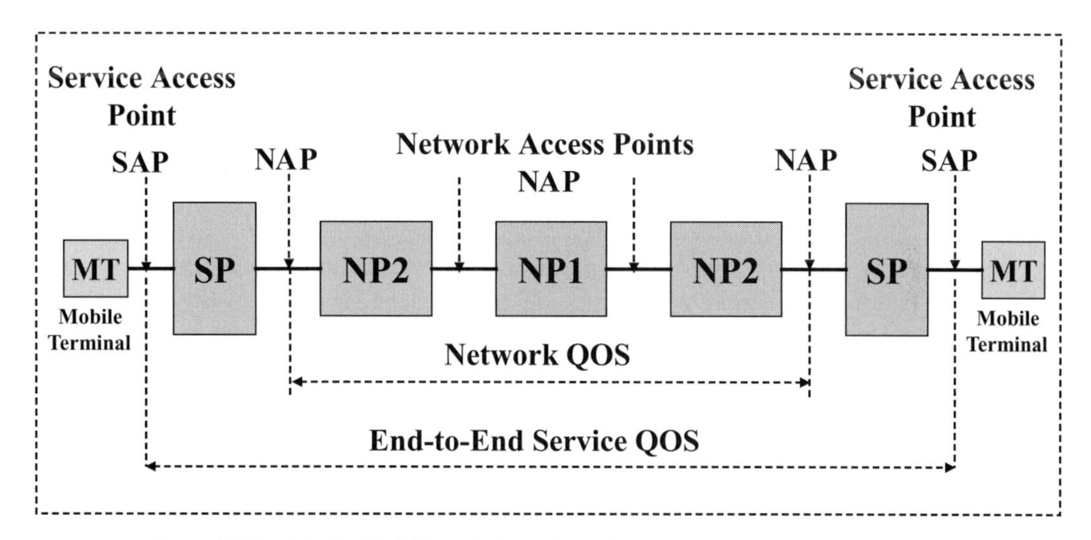

Figure 18.3 End-to-End QOS in Wireless Networks

The overall convergent network consists of multiple networks supported by various Network Providers (NPs). Therefore, we have an **End-to-End Network QOS**, as applied to the concatenation of several networks. One network, say NP1, might be the mobile network and the two symmetric networks flanking it, NP2, fixed wireless networks such as WLAN or WPAN. Network QOS is evaluated based on the QOS metrics between well-defined Network Access Points (NAPs).

Similarly, convergent services are provided by Service Providers (SP). In this case, **End-to-End Service QOS** is defined between two Service Access Points (SAPs). End-to-end service QOS includes the end-to-end network QOS and depends on the service provider's network. The SP network consists of hardware (servers) and software

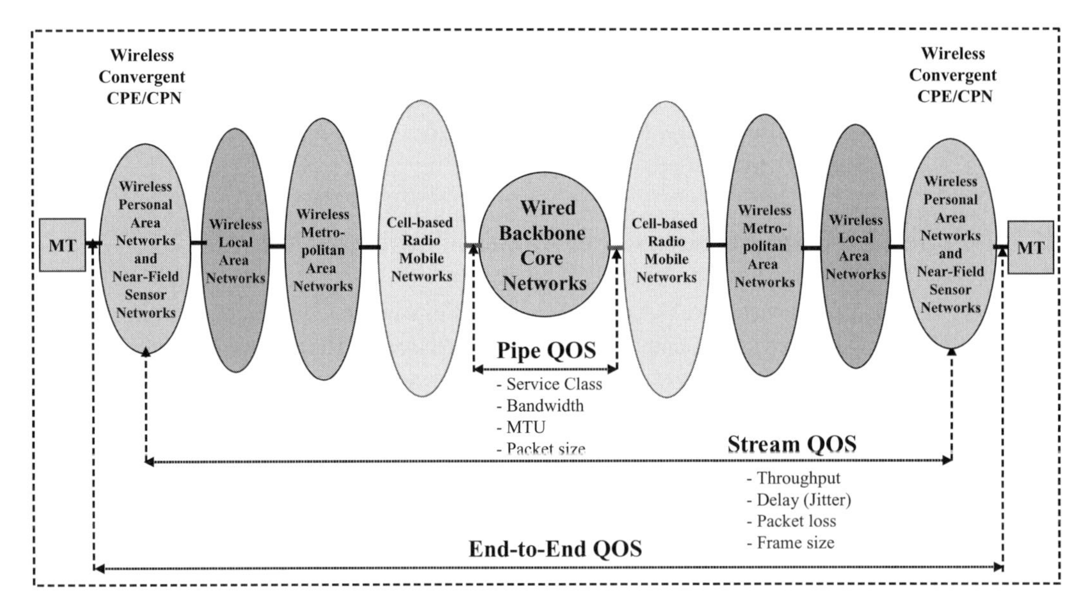

Figure 18.4 QOS in Fixed-Mobile Convergent Networks

(applications) elements that support the delivery of services. In this model, the Mobile Terminal (MT) is not included in End-to-End QOS.

Application of the model to the fixed-mobile convergent network presented in the previous section is shown in Figure 18.4.

Three types of QOS are defined in FMC networks: Pipe QOS, Stream QOS, and End-to-End QOS. **Pipe QOS** is a network QOS, the realm of core backbone network providers. The main characteristics are service class and transmission bandwidth. It also depends on the sizes of the Maximum Transmission Unit (MTU) and transmitted packets. **Stream QOS** is an extended network QOS that covers all the FMC networks. It is characterized by QOS metrics that determine the performance of voice and data communications; namely, throughput, delay, jitter and packet loss. **End-to-End QOS** is defined between Mobile Terminals and includes the stream and pipe QOS.

The implementation of **Service Level Management** for end-to-end FMC network architectures depends on how QOS is supported on an end-to-end basis across different fixed wireless networks. Given the development trends of FMC network architectures, designed as all IP-based networks, QOS mechanisms specified for IP core networks can be applied to FMC networks. Three mechanisms will be analyzed here:

- Integrated and Differentiated Services (Int/Diff Services);
- Multi-Protocol Label Switching (MPLS);
- Policy-based Management (PBM).

A fundamental aspect of FMC service management is integration of these mechanisms with the QOS mechanisms specified in 3GPP/3GPP2 standards for UMA and IMS regarding Radio Access Networks (RAN) and fixed wireless networks such as Wi-Fi WLANs, Bluetooth or ZigBee WPANs, and WiMAX.

18.3 Integrated and Differentiated Services

The Internet was designed as a packet switched data network with no inherent guarantees of services beyond "best effort". However, it is not acceptable to delay sensitive communications such as voice, video, and multimedia services. To provide acceptable QOS over Internet type networks, the Integrated and Differentiated Services are used. These mechanisms are applicable to the IP-based wireless convergent networks described earlier.

Integrated Services (IntServices), as a QOS mechanism, focuses on **individual packet flow** between end systems (applications) using TCP/UDP/IP protocols. There are three components recognized as part of the Integrated Services architecture:

- **Admission Control Unit**;
- **Packet Forwarding Mechanism**;
- **Resource Reservation Protocol (RSVP)** (sets the flow states in the routers).

Admission Control determines if the request to establish a "connection" can be carried out by the network. There are several considerations behind this decision such as traffic load, traffic profile that includes required bandwidth, pricing, and other policy considerations. Admission control is performed in the setup of Resource Reservation Protocol (RSVP) flows between routers.

The **Packet Forwarding Mechanism** consists of several operations or procedures such as packet classification, shaping, scheduling, and buffer management in the routers.

- **Traffic Shaping** ensures that traffic entering at one edge of the network adheres to the specified connection profile. This is necessary in deciding how to allocate system resources. Typically, this mechanism is used when there are bursts of packets in a traffic stream;
- **Packet Classification** allows for different QOS treatments. It uses various fields from the IP header (source/destination IP address, protocol type) and higher layer protocol headers (source/destination port number for TCP and UDP);
- **Priority and Scheduling** is required to satisfy the different QOS needs of different connections. Priority provides different delay treatments for higher priority packets on outbound lines. Scheduling mechanisms ensure that connections obtain their promised share of resources such as processing CPU and link bandwidth. In this way, any spare capacity is distributed in a fair manner;
- **Queuing**, based on different algorithms, provides fair treatment of packets, so misbehaving applications will not punish other applications. The result is that average dropped packets will be evenly distributed across data flows and queues. There are different queuing techniques that can be applied in handling congestion:

- **FIFO** (First In First Out): A queuing method in which packets are transmitted in the order received;
- **CBQ** (Class Based Queuing): A queuing method that classifies packets according to certain criteria and reserves a separate queue for each traffic class;
- **RED** (Random Early Detection/Discard): A queuing principle in which packets are discarded randomly to control congestion before the queue becomes full. It extends the native TCP/IP congestion control mechanism;
- **WRED** (Weighted Random Early Detection/Discard): A variant of RED in which the packet-discarding probability depends on the characteristics of the flows or on the level of importance of individual packets; and
- **WFQ** (Weighted Fair Queuing): A packet scheduling mechanism that allows active streams to grab unused bandwidth dynamically. Packets are queued according to projected final order. Supports bandwidth reservation in both Integrated and Differentiated Services.

The **Resource Reservation Protocol (RSVP)** is a connection-oriented, out-of-band, "host-network-host" Transport Layer protocol. It is used to reserve resources for communication along a path of intermediate network nodes. RSVP acts as a signaling protocol providing backward preparation of resources required to handle the upcoming flows of data and finally informing the source about the results. Scalability and capabilities to signal different QOS needs are issues confronting RSVP. For this protocol there are two types of control messages initiated per each session: **PATH** and **RESV**. There are RSVP implementations solutions in both IPv4 and IPv6.

- **RSVP PATH** messages carry traffic specifications from the sender to the receiver reading each RSVP-capable router.
- **RSVP RESV** messages are sent from the receiver to the sender to confirm that the routers have been prepared to accept the upcoming data flow. This means that the flows will be treated according to the agreed on priority or the priority built into each flow according to the differentiated service mechanism.

Standardization of integrated services is specified in:

- RFC 2211 "Specification of the Controlled-Load Network Element Service";
- RFC 2212 "Specification of Guaranteed Services";
- RFC 2215 "General Characterization Parameters for Integrated Service Network Elements".

Differentiated Services (DiffServices) is a QOS mechanism used to handle aggregate traffic. It is based on the former **Type-of-Service** (TOS) field specified in the IPv4 RFC 791 IETF standard. In the original specifications, the first three bits were described as **precedence bits** to be used as a selector of priority in queued interfaces (largely ignored today). The higher the precedence number, the greater is the chance that the packet will be transmitted before other packets. The use of other bits in the TOS field includes:

- 4 bits as flags with the following functions:

- 1 bit to minimize delay;
- 1 bit to maximize throughput;
- 1 bit to maximize reliability;
- 1 bit to minimize monetary cost;
- 1 bit reserved for future use.

Note: All 4 bits set to "0" implies normal service. Only one of the 4 bits can be turned "ON".

Two standards specify DiffServices, both the work of the Integrated Services Working Group. Initially, the group worked on the Resource Reservation Protocol (RSVP) for Integrated Services. But, after it became clear that the protocol had scalability problems, they embarked on specifying **Per-Hop-Behavior** (PHB) as a way to build differentiated classes of services. Differentiated services standards are specified in:

- RFC 2474 "Definition of the Differentiated Services (DS) Field in IPv4 and IPv6 Headers"; and
- RFC 2475 "An Architecture of Differentiated Services Network".

RFC 2474 and RFC 2475 provide a standard process of forwarding packets between ingress nodes, boundary nodes, and egress nodes, each of them subject to some Service Level Agreements (SLAs) and Traffic Conditioning Agreements (TCAs). These standards do not provide service class guidance (it is outside the scope of the RFCs). In Differentiated Services each hop implements a **Per-Hop-Behavior** adjustment. The Differentiated Services architectural model is shown in Figure 18.5.

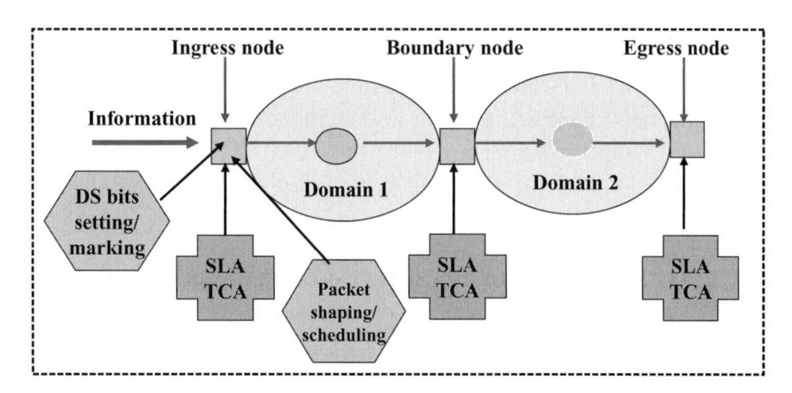

Figure 18.5 Differentiated Services Architectural Model

Technically, PHB is combination of a forwarding mechanism, classification, scheduling, and drop behavior at each hop. RFC 2475 declares explicitly that DiffServ architecture should decouple traffic conditioning and service provisioning from forwarding behavior.

A **PHB class** is a collection of PHBs used when transmitting packets of one application. It means that the service provider can change the PHB within a PHB class (group) but not from one class to another class. A PHB class may belong to a **PHB group**.

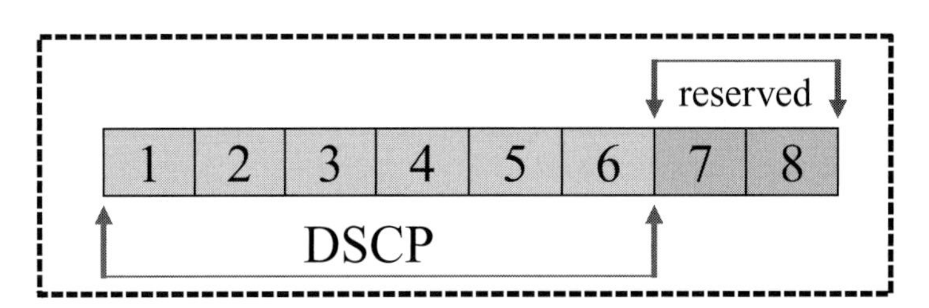

Figure 18.6 Differentiated Services Fields Definition

RFC 2474 replaces the old IPv4 TOS field with a new IPv4/IPv6 Traffic Class definition. In the new layout, the first 6 bits represent the **Differentiated Services Code Point** (DSCP) while the last two bits are reserved for future use, as shown in Figure 18.6.

Although in principle the DSCP field is not restricted, three major categories are considered:

a. **xxxxx0** for standard activities assigned by IANA (allows 32 code points);
b. **xxxx11** for experimental activities and local use (allows 16 code points);
c. **xxxx10** for experimental activities and standard PHB services (allows 16 code points).

Several service models were built based on "standardized" differentiated services. All require interworking between QOS domains where different DSCP code points might be used to specify packet PHBs. One of the proposed model of PHB classes is shown below:

Group A (focus is on minimum loss of packets):
- **AF-1 Assured Forwarding** PHB highest importance level;
- **AF-2 Assured Forwarding** PHB middle importance level;
- **AF-3 Assured Forwarding** PHB lowest importance level;
- **BE "Best Effort"** (default PHB).

Group B (focus is on flexibility and dynamic packet differentiation related to packet loss and bandwidth allocation):
- **DRTP Dynamic Real-time** primary area;
- **DRTS Dynamic Real-time** secondary area;

Group C (focus on delivery with minimum delay):
- **EF-1 Expedite Forwarding** PHB primary area;
- **EF-2 Expedite Forwarding** PHB secondary area.

As indicated in Table 18.1, various applications require different Internet service classes.

Circuit services use the Expedited Forwarding EF-1 PHB class while Best Effort BE, the default PHB class, is adequate for electronic mail. A comparison between IntServices and DiffServices is presented in Table 18.2.

Although there are major differences between IntServices and DiffServices, they can complement each other. Thus, RSVP can be used as an admission control mechanism for DiffServices. Conversely, DiffServices can help RSVP to scale across large networks, a major weakness of Integrated Services.

Table 18.1 Internet Applications and DiffServ Class Attributes

Application	Internet Service Classes	Comments
Circuit Services	Expedited Forwarding EF-1	For leased lines, EF-1 provides the maximum QOS available.
Interactive Video (Video conferencing)	High-bandwidth Expedited Forwarding EF-1 or Assured Forwarding AF-1	The use of AF-1 depends on the compression applied and acceptance level of lost packets. The use of EF-1 is highly dependent on the traffic and it may require jitter buffers and traffic management.
Voice Telephony	EF-1 or AF-1	The same as interactive video.
Broadcast Video	Assured Forwarding AF-1	In this case, the delay is not a significant issue so the use of jitter buffers can smooth the traffic.
IBM SNA over TCP/IP	Assured Forwarding AF-2	IBM SNA traffic requires a reasonable response time and low packet loss. It may require traffic management.
File Transfer	Best Effort BE or Assured Forwarding AF-2	In this case, the delay tolerance depends on the application; packet lost requires retransmission.
Electronic mail	Best Effort BE or Assured Forwarding AF-2	In this case, the delay tolerance depends on the application; packet lost requires retransmission.

Table 18.2 IntServices and DiffServices Comparative Analysis

Service Characteristics	Integrated Services	Differentiated Services
Applicability	Individual flows	Aggregate flows
Type of Service	Deterministic	Relative Assurance
Router's state	Per flow	Per-Hop-Behavior (PHB)
Signaling Protocol	Reservation Protocol (RSVP)	Not required
Scope of Service	Predictable and measurable QOS	Establishment of Class of Services (COS)
Scalability	Limited by the number of flows	Limited by availability of class of service support
Accounting	Based on QOS and flow characterization	Based on class usage
Deployment	Inter-domain with multilateral agreements	Bilateral agreements
Traffic Classification Basis	IP Header fields	TOS/DSCP field in IP header

18.4 Multi-Protocol Label Switching

Multi-Protocol Label Switching (MPLS) is another mechanism that was developed to assure adequate QOS levels for voice, video, and multimedia services in IP-based networks. The following short tutorial will highlight the main aspects of MPLS technology, architecture, and specific protocols. The generic MPLS-based network architecture is shown in Figure 18.7.

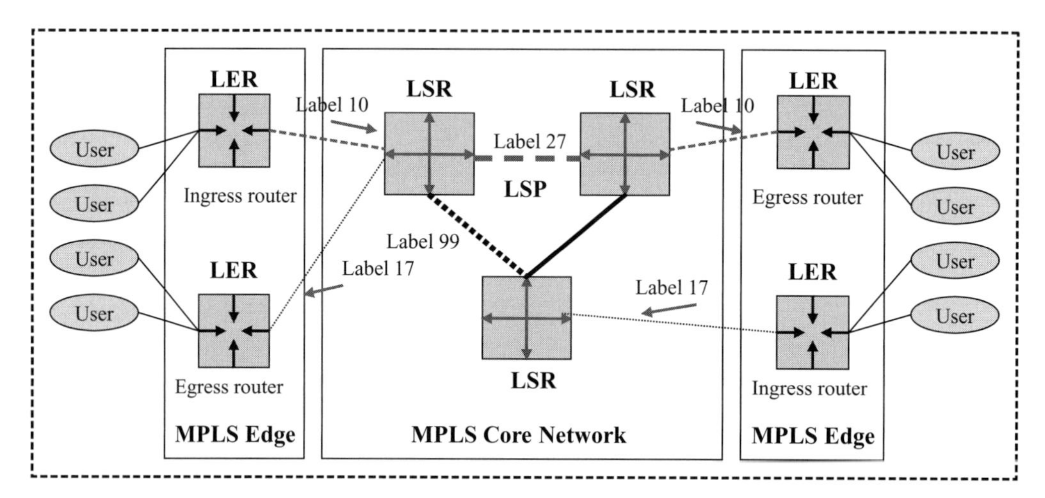

Figure 18.7 Multi-Protocol Label Switching Architecture

Multi-Protocol Label Switching networks consist of **Label Edge Routers** (LER) in the MPLS Edge domain, **Label Switching Routers** (LSR) in the MPLS Core Network domain, and **Label Switching Paths** (LSP). On each physical link, a LSP is represented by a particular label allocated to a flow of packets belonging to a particular application. The labels are distributed using either the specialized **Label Distribution Protocol** (LDP) or extensions of RSVP and the Border Gateway Protocol (BGP).

MPLS was created as a way of forwarding packets in IP-based, Asynchronous Transfer Mode (ATM), or Frame Relay (FR) networks in a simpler and faster manner. However, today, routers are fast enough, so MPLS's role has changed to traffic engineering and support of differentiated services. MPLS is both a packet-forwarding and a path controlling schema and protocol; taking the best of two worlds – the IP packet-based and ATM cell-based networks.

MPLS is an encapsulation protocol, characterized by a new header located between the OSI Layer 2 and Layer 3 headers. The header is essentially a label read by the Label Switching Routers. Based on the label information, packets are switched to the proper port according to a selected Label Switching Path. The MPLS header is shown in Figure 18.8.

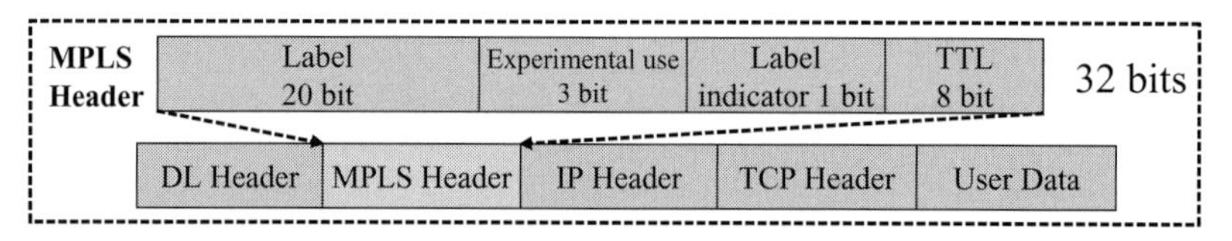

Figure 18.8 MPLS Labeling Header Layout

The **MPLS labeling header** (32 bits) consists of a 20-bit label, a 3-bit experimental field, a one bit label stack indicator, and a Time-To-Live (TTL) field of 8 bits. Each LSR at the ingress point will insert a MPLS header in each packet and the header is removed as the packet leaves the MPLS routing domain.

MPLS incoming labels' content changes as packets advance through the network, replaced by outgoing labels according to the LSR table. This table is simpler and shorter than a routing table. The first label attachment takes place in the ingress Label Edge Router (LER). The router should read only the MPLS header (label) to decide where to switch the packet. This approach is similar to the Label Switching technique, first proposed by Ipsilon Technology, and later adopted by Cisco as a Tag Switching mechanism. The same mechanism was incorporated by the new generation of Layer 3 switches. MPLS can even reside in access devices such as cable modems and DSL modems.

When a MPLS packet traverses an ATM network, the MPLS label header (4 bytes) can be carried in the Virtual Circuit Identifier/Virtual Path Identifier (VCI/VPI) cell header (4 bytes long), as used between ATM switches. In a FR network, the MPLS header is incorporated into the Data Link Connection Identifier (DLCI) field of a FR header.

MPLS uses either a signaling protocol such as RSVP or a dedicated Label Distribution Protocol (LDP). Each LSP can have one or more attributes assigned such as bandwidth, path (normal or dynamically allocated), or a setup priority such as class of service. MPLS improves on existing routing protocols such as OSPF by redirecting traffic to a least traveled path. This is the case of FR and ATM connection-oriented virtual circuit setup. Note that OSPF can congest paths (the routers do not have a long view of packets' final destination).

Multiple MPLS labels can be concatenated, each router examining only one, the label content at the top of the stack. This means packets taking the same path can be assigned the same **Forwarding Equivalence Class** (FEC), a loose logical grouping of traffic with a similar destination. Admission Control is performed in the setup of MPLS paths.

LSR routers use LDP to inform each other of the labels being created among them. LDP is also used between LSRs to learn about their MPLS support and capabilities. In reality, BGP and RSVP have been extended to carry LDP data. Two distribution mechanisms are defined within MPLS: downstream on demand and unsolicited downstream. Some of the labels are retained in a lookup table to provide next-hop information; the others are discarded. LSPs are created or longer paths can be secured by placing multiple labels into a packet. This technique is called explicit routing.

Explicit routing is used when Virtual Private Networks (VPNs) are created over MPLS networks. MPLS can be used as a carrier for VPN traffic, and thus, reduces further the router overhead because encryption and decryption are no longer needed. Use of MPLS will provide a guaranteed path and guaranteed bandwidth.

18.5 Policy-based Management

The third major mechanism used to provide end-to-end QOS in fixed-mobile convergent networks is **Policy-based Management** (PBM). Policy-based Network Management (PBNM) is the encompassing concept of a policy-driven network management that is also applicable to systems and service management. In this context, policies are sets of rules that control user access to applications, configuration information, asset allocations, network, systems, and QOS performance.

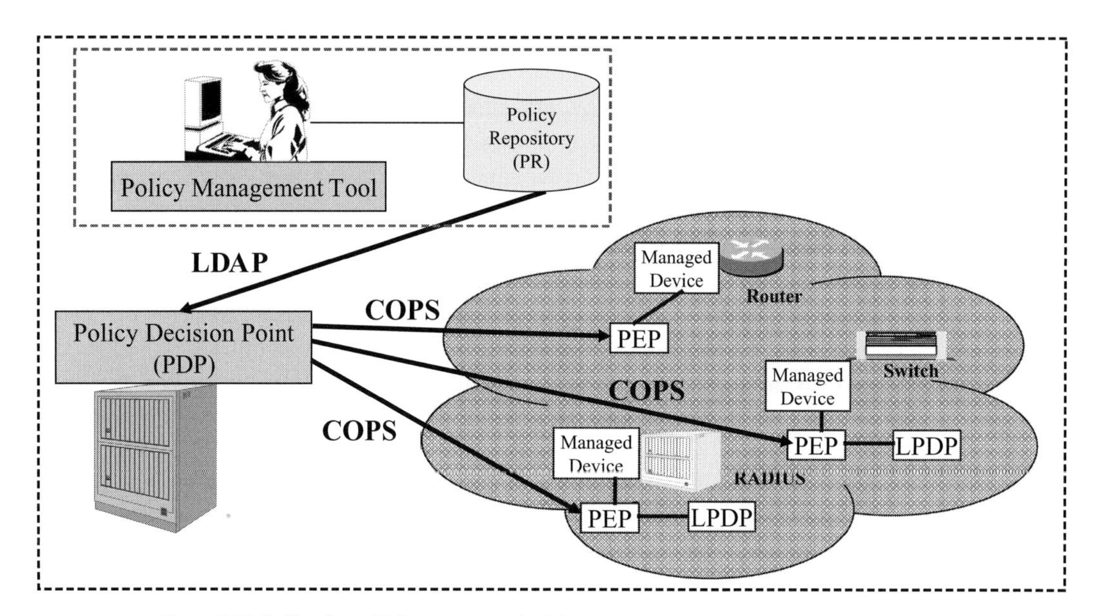

Figure 18.9 Policy-based Management Architecture

Like Integrated and Differentiated Services and Multi-Protocol Label Switching, PMB is primarily focused on time sensitive applications (voice, video, and multimedia services) in IP-based networks. However, PBNM deals with the overall management rather than individual operations such as fault or bandwidth management. The following short tutorial will highlight the main aspects of PBM technology, as an architecture and set of specific protocols. A high level PBM-based network architecture is shown in Figure 18.9.

The Internet Engineering Task Force (IETF) and Distributed Management Task Force (DMTF) have developed a policy-based management framework. It consists of four components: **Policy Management Tool** (PMT), **Policy Repository** (PR), **Policy Decision Point** (PDP), and **Policy Enforcement Point** (PEP). Routers, switches, and Remote Authentication Dial-In User Service (RADIUS) security servers are examples of policy-based managed devices.

The Policy Management Tool and Policy Repository are collocated in this architectural diagram. A **Lightweight Directory Protocol** (LDAP) can be used to access the Policy Decision Point. A special protocol, the **Common Open Policy Service (**COPS), was designed to access any Policy Enforcement Point that might be incorporated into managed devices. A Local PDP (LPDP) can be used to maintain the PEP operation in cases where PDP-PEP links are interrupted. PBNM advantages include:

- **Avoids network over-provisioning**: Providing "unlimited" bandwidth may hide the real latency problems and inefficient use of the existent resources;
- **Allows increased traffic**: The Internet has brought with it new applications such as web browsing, email, instant messaging, and Power Point development that can be monitored, prioritized, and measured;

- **Allows convergence of voice and data**: The Internet has brought latency-sensitive traffic such as voice and video over IP, MP3 music, and games, that can also be monitored;
- **Adds application security**: The security of applications and information can be enhanced by policy-based user restrictions and profiling in support of Virtual Private Networks (VPN), Authentication, Authorization, Accounting (AAA), and Remote Access Dial-In User Services (RADIUS)-based tools;
- **Stops unauthorized applications**: Monitors loading of unauthorized applications that could consume bandwidth and create congestions. Gives high-priority to business-critical applications against unwanted traffic;
- **User satisfaction**: By enforcing minimum service levels for applications, users and departments are assured better availability and higher responsiveness to any problem. Customer-driven QOS management and manual configuration.

A network manager or administrator uses Policy Management Tools (PMT) to define policies that are to be enforced in the network. Policy information generated by the policy management tools is stored in the Policy Repository (PR). For example, information about DiffServices such as Type of Service (ToS) field encoding, Differentiated Services Code Points (DSCP), and Per Hop Behavior (PHB) classes are stored in PRs.

To ensure interoperability across products developed by different vendors, information stored in the PR should use the same Policy Information Base (PIB) information model. The Policy Decision Point (PDP) is responsible for retrieving policy rules stored in the PR, interpreting the policies, and communicating with the PEP. The PEP and PDP may be in a single or different devices.

An example of policy between two routers may state that traffic in the morning has a guaranteed bandwidth of 2,048 Mbps or that between two network access points traffic must be encrypted using Triple Data Encryption Standard (DES). The PR can be accessed using the Lightweight Directory Protocol (LDAP) or the Structured Query Language (SQL) in cases where the PR is a Relational Data Base Management System (RDBMS). The Common Open Policy Service (COPS) or Simple Network Management Protocol (SNMP) can be used between PEP and PDP.

The most used protocol between PDP and PEP is the Common Open Policy Service (COPS) protocol. COPS was standardized by the IETF in the RFC 2748 document. Its main characteristics [105] include:

- It uses a request/reply protocol in a client-server paradigm; PDP is a policy server that supports interactions with multiple PEP clients; PEP sends requests, updates, and deletes to the remote PDP and the PDP returns decisions back to the PEP;
- It runs on top of the TCP/IP stack for reliable exchange of messages between policy-based clients and servers;
- It supports various policy areas identified in the Client-Type field of its header. These areas may include support for DiffServices or RSVP, thereby allowing a flexible PEP configuration;
- It provides message level security for authentication, replay protection, and message integrity; it can use existing protocols for security such as IP Security (IPSEC) or

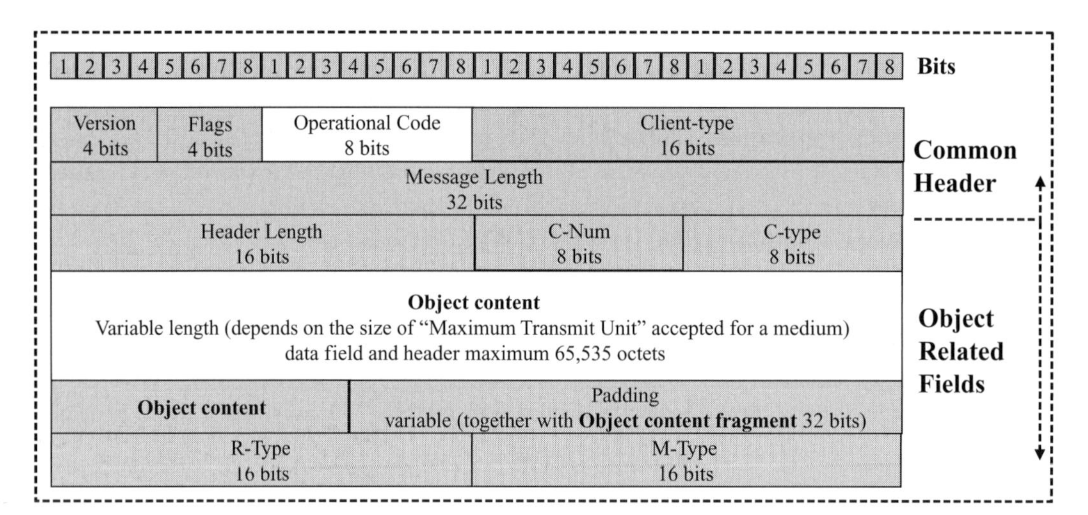

Figure 18.10 COPS Protocol Framing Layout

Transport Layer Security (TLS) to authenticate and secure the channel between the PEP and the PDP.

RFC 2748 provides a detailed layout of the COPS protocol data unit format. It consists of a common header and object related fields that deal with specific policy models. Figure 18.10 shows the layout.

The COPS fields are organized and described as follows:

Common Header

- **Version**: 4 bits, COPS version number; current version is 1;
- **Flags field**: 4 bits, defined flags (only one flag is set to 1);
- **Operational code**: 8 bits; 10 standardized COPS operations;
- **Client type**: 16 bits; identifies the policy client;
- **Message Length**: 32 bits, size of message in octets; includes the standard COPS header and all encapsulated objects.

Object Related Fields

- **Header Length**: 16 bits, describes the number of octets (including the header) that compose the object;
- **C-Num**: 8 bits, identifies the class of information contained in the object;
- **C-Type**: 8 bits, identifies the subtype or version of the information contained in the object;
- **Object Content**: Variable length, there are 12 specific object contents such as: Handle Object, Context Object, Error Object, and Report-Type Object;
- **Padding**: Bits added to assure that the option field is filled to 32 bits;
- **R-Type**: 16 bits, Request Type Flags;
- **M-Type**: 16 bits, Message Type field; Client Specific.

There are two models supported by the COPS protocol: Outsourcing and Provisioning:

- In the **outsourcing model**, **COPS-RSVP** (RFC 2749), each PEP event requires a PDP policy decision. So, when an RSVP router receives an RSVP message, the PEP queries the PDP about pre-established policies and accepts or rejects RSVP reservation requests [106].
- In the **provisioning model**, **COPS-PR** (RFC 3084), policies are installed in PEPs before the events occur and the PEP decides how to treat the coming events. For example, a DiffServ PEP queries the PDP at start-up about the configuration to be installed. Therefore, there is no need to solicit a PDP decision for each DiffServ packet. [107].

COPS can be used to provide dynamic SLA/SLS management in an IP-based Diffserv-oriented network environment by establishing a new COPS extension in the type field (COPS-SLS). COPS-PR will be used in the configuration phase, when PDP tells the clients how to use policies. COPS-RSVP is used to negotiate the SLA-SLS. In this scenario, PEP clients request a new SLS, while the PDP, for example, representing the network bandwidth broker, accepts, rejects, or proposes a new set of SLS. In this capacity, the PDP provides the DiffServ admission control mechanism. IntServ RSVP can be used as the signaling mechanism. A list of policy-based network management products that can also be used for convergent network management would include:

- **Allot Communications**: NetPolicy, (www.allot.com);
- **Aprisma**: Spectrum Policy-based Network Management Suite, (www.aprisma.com);
- **Avaya**: "CajunView", (www.avaya.com);
- **Cisco Systems**: QOS Policy-based Management QPM, (www.cisco.com);
- **Extreme Networks**: ExtremeWare Enterprise Manager, (www.extremenetworks.com);
- **Hewlett Packard**: OpenView Policy Expert, (www.hp.com);
- **Intel Corporation**: NetStructure Policy Manager, (www.intel.com);
- **IP Highway**: Open Policy System, (www.iphighway.com);
- **Lucent Technologies**: RealNet Rules, (www.lucent.com);
- **Nortel Networks**: Optivity Policy Services, (www.nortelnetworks.com);
- **Orchestream**: Enterprise Edition, (www.orchestream.com); and
- **3Com**: Transcend Policy Manager and Policy Server, (www.3com.com).

18.6 QOS in UMA-based Fixed-Mobile Convergent Networks

UMA, as presented in Chapter 15, is a flexible fixed-mobile convergence solution intended primarily to extend GSM/GPRS coverage in the residential (home and public hot-spot) and Small Office/Home Office (SOHO) markets. The UMA specifications, which were transferred to the 3GPP, use Wi-Fi or Bluetooth access combined with fixed backhaul (e.g. based on DSL). This provides convergent GSM/GPRS and WLAN service support with seamless hand-off and roaming within the cellular network.

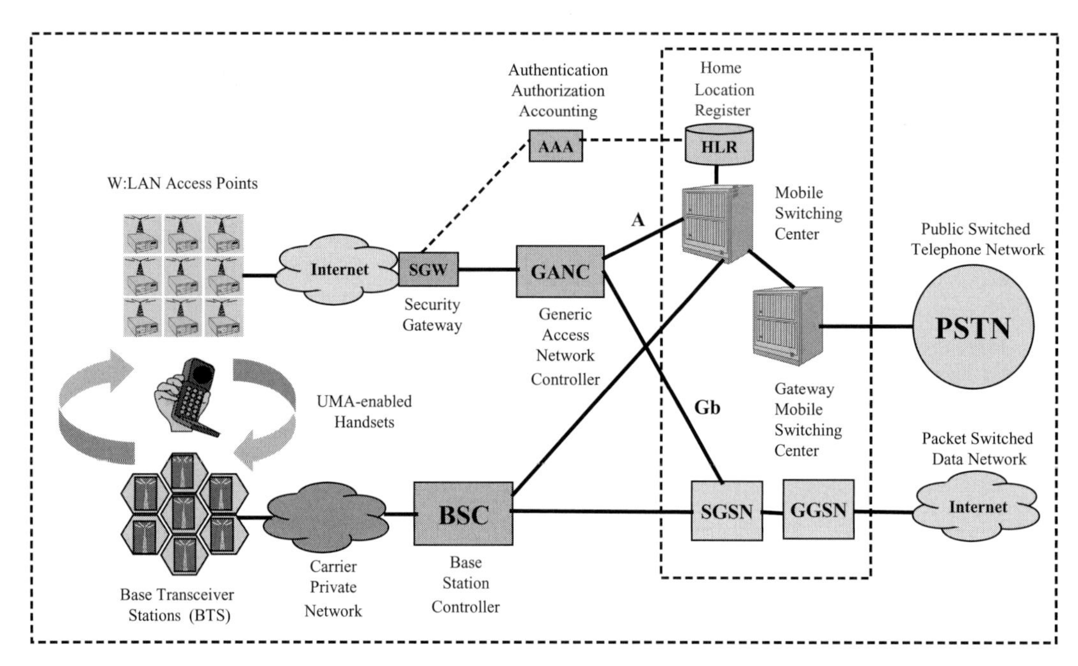

Figure 18.11 UMA/GAN Convergent Architecture

As shown in Figure 18.11, the UMA solution is built around a Generic Access Network Controller (GANC). In this configuration, the GANC mimics the role of a GSM Base Station Controller (BSC). The BSC is connected to the core network using standard interfaces. Voice and data GSM/GPRS services, including Short Message Services (SMS), Multimedia Messaging Services (MMS), Wireless Application Protocol (WAP), and Location-Based Services, are provided in a seamless manner, irrespective of the radio technology used.

Providing QOS in the UMA-based FMC networks will follow the specifications adopted as part of 3GPP network planning described in Chapter 15. UMA QOS depends on the radio transmission technology used, GSM/GPRS or WCDMA. UMA QOS also depends on the fixed wireless access architectures used, WLAN, WPAN, or NFC.

When a user with a dual-mode handset moves within the range of a Wireless LAN (Wi-Fi), the mobile terminal contacts the GANC over the IP access network. Signaling and user data are routed through the GANC using a single IP Security (IPSec) tunnel that provides data integrity and encryption.

3GPP specifies four classes of traffic:

- Conversational class: Traffic requires a guaranteed bit rate with low latency;
- Streaming class: Traffic needs a guaranteed bit rate with minimal latency;
- Interactive class: Corresponds to bursty traffic with various priorities, but with no guaranteed bit rate (in the packet domain only);
- Background class.

Corresponding to the requested service and operator policy, a QOS class will be required over the GSM/GPRS Radio Access Network (RAN). After this request is granted, the

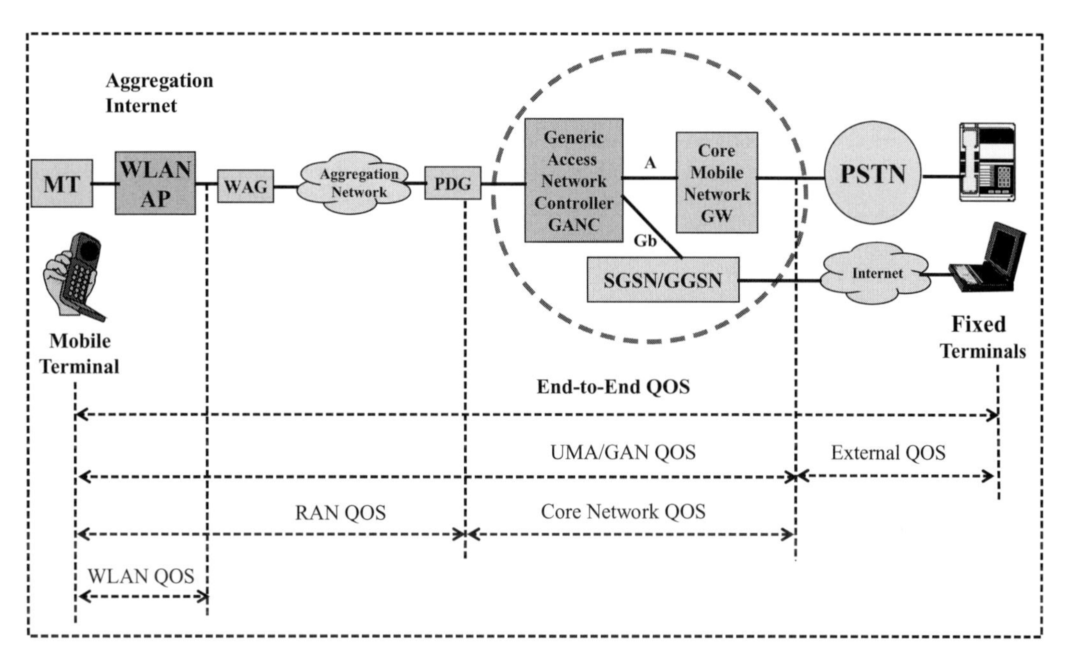

Figure 18.12 QOS in UMA/GAN Convergent Network

GSM/GPRS core network requests the establishment of a suitable bearer to the mobile station. Specifically for UMA/GAN, the GANC determines the appropriate QOS in the RAN based on the traffic class and it then assigns the appropriate Differentiated Services Code Point (DSCP) marking.

The end-to-end QOS model for UMA is shown in Figure 18.12. UMA specifies how the DSCP of an IP packet is to be processed by the Security Gateway (SGW) and the mobile station to ensure adequate QOS along the IPSec tunnel and beyond.

For **downstream packets**: The GSM media gateway (circuit-switched domain) and the SGSN (packet-switched domain) mark IP packets with the appropriate DSCP. The SGW encapsulates the original IP packet into the IPSec tunnel towards the mobile station and builds the IPSec DSCP from the original DSCP. The original DSCP is retrieved from the encapsulated IP packet at the mobile station side.

For **upstream packets**: The mobile station copies the downstream outer IP header DSCP to the upstream outer IP header DSCP to specify the QOS between the mobile station and the SGW. It also copies the downstream inner IP header DSCP to the upstream inner IP header DSCP for QOS beyond the SGW. To guarantee the needed QOS across Ethernet access and aggregation networks, the routing gateway maps the IP DSCP to the appropriate Ethernet QOS class. Additionally, the WLAN access point maps the 802.1p bits to the appropriate 802.11e Wi-Fi multimedia QOS class to provide differentiated channel access. This should result in controllable and predictable transmission times and throughputs, and ultimately QOS.

3GPP standards have specified how 3GPP systems and WLANs can converge. The goal of this convergence is to extend 3GPP packet-based services to the WLAN access environment. Unlike the UMA architecture, which is mainly designed for circuit-switched services, the interworking WLAN architecture is designed for packet-based services. This is done by connecting the WLAN to the same external IP network used by the GPRS access network.

As shown in Figure 18.12, a UMA WLAN is connected to an external IP network via a WLAN Access Gateway (WAG) and a Packet Data Gateway (PDG). The WAG, which is located in the visited network, acts as a firewall, enforcing routing of packets through the PDG and providing charging data. After authentication by the AAA server, an IPSec tunnel is established between the mobile terminal and the PDG.

3GPP has standardized QOS support over interworking WLANs by amending the 802.11 WLAN standards with MAC enhancements for QOS. This work might affect the current IEEE 802.11e standard. Since interworking WLAN services and UMA services are both 3GPP services, and therefore based on the same QOS principles, it is expected that they will use similar mechanisms. User signaling will request and negotiate the service, while network signaling is used to request the appropriate bearer to the access network. This is achieved using the GANC for UMA and the PDG/WAG for the interworking WLAN. Once this has been done, the QOS techniques outlined in the UMA architecture can be used unchanged.

To ensure an adequate performance for real-time and streaming services, broadband fixed and wireless access networks must comply with a number of QOS requirements. The networks must be able to classify subscriber traffic per application type and provide differentiated traffic treatment before aggregating it into limited sets of service classes. Traffic treatment will be performed at network edges. Aggregation and regional networks can provide queuing and scheduling for the aggregate flows.

18.7 QOS in IMS-based Fixed-Mobile Convergent Networks

The IP-based Multimedia Subsystem (IMS) standard convergent architecture, components, and interfaces were analyzed in Chapter 17. IMS is essentially a signaling platform where the paths of the signaling sessions and the user information are different. This is shown in Figure 18.13.

The signaling session flows, using the SIP/SDP protocols, traverse all the networks that provide communications between User Equipments (UEs). The signaling system broker is the IMS Layer, and within the IMS Layer the Common Session Controller Functional (CSCF) modules. The flows of user applications are carried on separate links that do not cross the IMS Layer. They use a combination of RTP and RTCP for normal and real-time media transmissions. IMS facilitates this flow by controlling the media gateways and the media servers, in cases where the architecture includes media servers.

Providing QOS in IMS-based FMC networks depends on QOS specifications for access and transport networks. Access networks can be either wired (ISDN, DSL, Cable Packet) or wireless networks. Among the wireless access networks, we have WLANs/WPANs or complete Universal Terrestrial Radio Access Networks (UTRAN). Following the application flow presented in the previous diagram, we can provide a

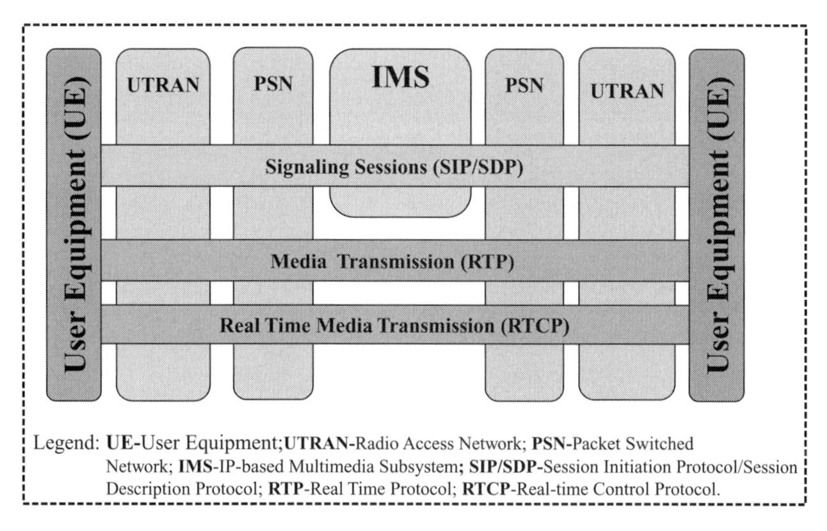

Figure 18.13 Signaling and Application Flows in IMS Networks

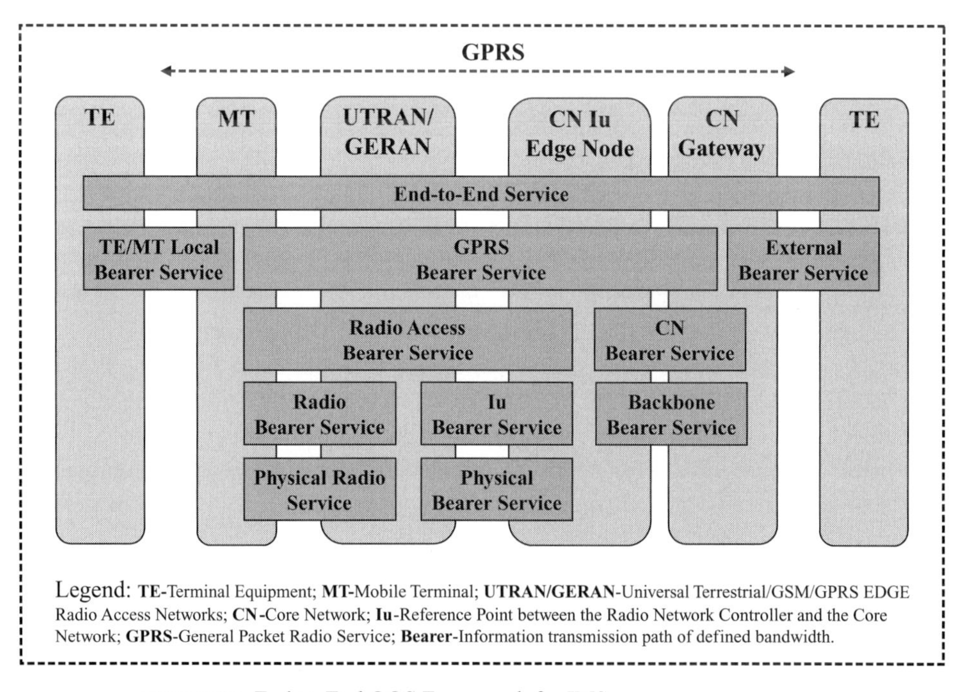

Figure 18.14 End-to-End QOS Framework for IMS

detailed view of the components/networks that have an influence on QOS. Figure 18.14 is a framework for end-to-end QOS for the Packet Switch Domain, including IMS; as provided in the 3GPP Release 5 document [92].

In the 3GPP specs, end-to-end-service is defined to be between Terminal Equipments (TEs) since the framework should be applicable for communications between POTS phones and cell phones. The Mobile Terminal (MT) itself can have an extension into the computing world that can qualify it as combined computer and mobile terminal. In this respect, the end-to-end service and end-to-end QOS may have two components, the TE/MT Local Bearer Service and the External Bearer Service, both outside the realm of wireless mobile networks. This means that IMS QOS is very much dependent on the GPRS service, provided that the IMS implementation is working with a 2G or 2.5G cellular mobile network.

The GPRS QOS, mirroring the GPRS service, consists of two components: the Radio Access network (RAN) QOS and the Core Network (CN) QOS. The core network is defined between a core edge node and the core network gateway (GGSN in GPRS services), the latter providing connectivity to the outside world. The RAN service itself consists of two segments: One segment is the radio access link between the mobile terminal and the UTRAN/GERAN network that includes BTSs and BSCs. The other segment is the interface between UTRAN/GERAN networks and the Core Network, across a reference point, known as the "Iu" interface.

To provide adequate QOS in GPRS services, there is a need to have a controlling agent in the CN Gateway (GGSN), and, optionally, in the terminal equipment. The function of this agent is to monitor QOS metrics and to provide adaptation mapping between QOS parameters of the external network (very likely an Internet IP-based network) and the QOS parameters used in the GPRS domain. One possible solution is the use of Differentiated Services, a mechanism that was presented in Section 18.3 of this chapter along with RSVP Integrated Services.

QOS management can be augmented by using a Policy-based Management (PBM) mechanism, as presented in Section 18.5. In this case, the Policy Enforcement Point (PEP) will be the GGSN module. The role of Policy Decision Function (PDF) will be played by the PDF module from the IMS architecture (see Figure 17.5). The PDF will communicate with the PEP (GGSN) using the standard COPS protocol across the "Go" interface/reference point. This is shown in Figure 17.6 and Table 17.1. The PDF can be incorporated within P-CSCF, one of the Common Session Controllers modules. A simpler layout of QOS in IMS network environment is provided in Figure 18.15.

In this diagram, end-to-end QOS is defined between User Equipment (UE) represented by mobile terminals, the handsets. The relevant QOS, GPRS/EDGE QOS, is defined between MT and the Core Network Gateway (GGSN).

As mentioned earlier, IMS, the standardized solution for SIP-based applications for multiple access network types, is a key component for delivering converged services with telecom-grade quality of service. As part of the convergence, there will be one common user and service management function, a common charging system, and a common identification/authorization system. The 3GPP UMTS QOS TS 23.207 V6.3.0 (2004–2006) document specifies four classes of traffic: conversational, streaming, interactive and background. The characteristics of these classes are presented in Table 18.3. These classes are the same as in the UMA/GAN FMC network architecture.

In this list, voice conversation has the tightest requirements on delays and jitters. Voice communications will have the highest priority and preferential treatment in any schema adopted to control QOS.

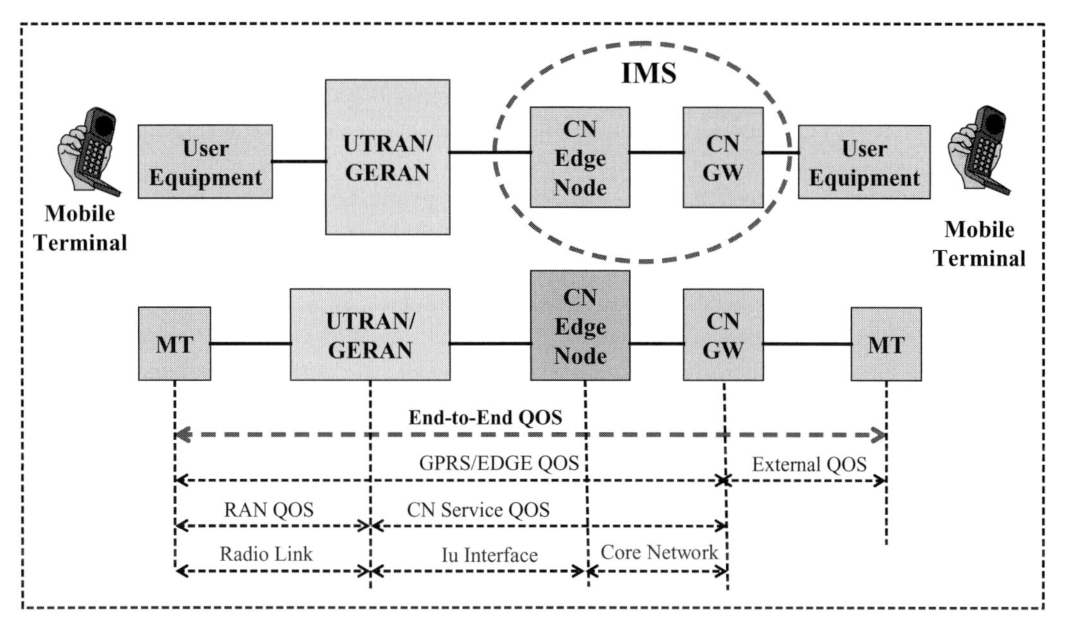

Figure 18.15 QOS in the IMS Network Architecture

Table 18.3 UMTS QOS Traffic Classes

Traffic Class	Basic Features	Additional Features	Example of Applications
Conversational	Preserves time relation (variation) of conversation between end systems	Conversation pattern requires stringent low delays and jitters	Voice conversations
Streaming	Preserves time relation (variation) of streams between end systems	Moderate delays and packet loss are admitted	Video/audio streaming
Interactive	Preserves the payload conveyed between end systems	Requests are followed by response patterns	Web browsing
Background Best Effort	Preserves the payload conveyed between end systems	The destination is not expecting the data conveyed	E-mail

18.8 Open Mobile Alliance and Handset Management

In previous chapters we presented several examples of cellular mobile, fixed wireless, and convergent networks management. There, and in this chapter, we defined the Quality of Service (QOS) and we looked at the management of QOS provided by these networks and convergent networks. In the analysis, the management boundary stopped at the mobile terminal (cell phone, PDA, palmtop, tablet PC) demarcation point and did not include the mobile terminal itself. The Open Mobile Alliance specifically addresses handset management.

The **Open Mobile Alliance** (OMA) was formed in 2002 to facilitate implementation of applications and services on mobile terminals based on service enablers. The service architecture developed by OMA is independent of the radio access technology supported by mobile terminals. It assumes an existent communication link and a transport protocol facility. As of 2007, OMA has over 300 members representing all major mobile phone manufacturers (Nokia, Motorola, Samsung, LG, Siemens, Sony Ericsson, etc.), mobile operators (Vodaphone, Orange, T-Mobile, Telefonica, etc.), and software vendors and content service providers (Microsoft, Symbian, Palm, IBM, Oracle, HP, etc.). OMA acts as an umbrella industry forum [108] because it consolidates the work done by several disjoint forums such as Multimedia Messaging Service (MMS), Wireless Application Protocol (WAP), Location Interoperability, Gaming, Wireless Village, and Mobile Wireless Internet.

OMA enabler specifications are released in two phases. The "Candidate Release" contains an approved OMA-based set of technical specifications that still need to be tested for interoperability. The "Approved Release" reflects OMA enablers that were tested for interoperability and went through the public comment process. Testing is performed in accordance with the OMA Enabler Test Specifications (ETS) and Validation Plan. This includes test cases with expected inputs and outputs, and a methodology of testing and reporting. Many tests take place during OMA Testing Festivals.

OMA Testing Festivals (TestFests) are regular gatherings, organized by OMA, to create an environment where member companies can test their products for interoperability and conformance using various vendor combinations. The ETSs are developed in 15 Technical Working Groups and are ratified by the OMA Technical Plenary. The specifications address various aspects of handset management, applications/services framework, and handset applications interoperability.

The most important OMA specifications are listed below:

- OMA Browsing, Version 2.1, 2.2, 2.3 (2007);
- **OMA Device Management** (OMA DM), Version 1.2 (2006/2007);
- OMA Digital Rights Management (OMA DRM), Version 2.1 (2007);
- OMA Email Notification, Version 1.0 (2006);
- OMA Instant Messaging and Presence Service (OMA IMPS), Version 1.3 (2006);
- OMA Multimedia Messaging Service (OMA MMS), Version 1.3 (2006/2007);
- OMA Push to Talk over Cellular (OMA PoC), Version 1.0, 2007; and
- OMA Presence SIMPLE (OMA SIMPLE), Version 1.0, 2005/2006.

The **OMA Device Management Working Group** has specified mechanisms and protocols that can be used to manage mobile devices including the configuration and monitoring of mobile handsets [109]. Eighty five percent of mobile phone vendors have implemented the OMA DM client specifications. Management includes:

- Setting initial configuration information in mobile devices (provisioning and security);
- Subsequent installation and updates of persistent information in client devices;

- Retrieval of management information from client devices;
- Processing events and alarms generated by client devices.

Configuration, operation, software installation, and application parameters as well as user preferences are the type of management information targeted. Configuration, security, fault, and performance management are areas included in the OMA specifications. The management specifications are worked out in collaboration/liaisons with well established standard organizations (3GPP/3GPP2, TeleManagement Forum, Wi-Fi Alliance) and harmonized with the work done in other OMA working groups.

An example of the magnitude of management services required to support the mobile world is the security aspect. Initially, when mobile terminals come into service, the users must be authenticated, the devices configured, and authorization determined to access particular services. Then, when mobile phones are lost, at a non-trivial rate of about 300,000 per day, all the personal data they contain must be protected from compromise, where possible. Managing this number is itself an issue and specialized products were developed to handle these situations. Examples of vendors and products dedicated to device management are listed below.

- **Bridgewater**: Application Policy Controller, (www.bridgewatersystems.com);
- **InnoPath:** Innopath Mobile Device Management (IMDM) with Firmware Over-The-Air (FOTA), Configuration Management, Diagnostics Management, and Feature Upgrade, (www.innopath.com);
- **MFormation Technology**: Enterprise Manager 5.1, (www.mformation.com);
- **Microsoft Corporation**; Windows Mobile 5.0, (msdn2.microsoft.com);
- **SMobile System**: Virus Guard, (www.smobilesystems.com);
- **Syncronica**: Firmware Update Management Object (FUMO) Syncl/DM Server, (www.synchronica.com); and
- **VeriSign**: Wireless Data Roaming Service, (www.verisign.com).

Handset development will enter a new phase following the recent trends to open up the mobile software platforms to third parties developers. Google's announcement to provide the Android open source mobile software platform, based on the Linux operating system, is in line with this trend. The platform and software development kit will facilitate the promotion of Google's own applications such as Gmail, Google maps, and web-based searching. This platform will compete against well established mobile operating systems provided by Symbian, Microsoft, Apple, Palm, and Research In Motion. To promote this openness, Google, together with 30 major mobile operators and chipset and handset manufacturers such as T-Mobile, Orange, Motorola, HTC, LG, Samsung Electronics, Sprint/Nextel, China Mobile, Telecom Italia, Telefonica, Intel, Texas Instruments, and Qualcomm, has formed, as of November 2007, a new consortium, the **Open Handset Alliance** (OHA) [110]. Ultimately, the success of Google's open platform will depend on the services provided by mobile operators, the easiness of operation, and last but not least the price of handsets.

19 The Economics of Fixed Wireless Cellular Mobile Networks Integration

19.1 Economic Drivers of Fixed-Mobile Convergence

Although the title of this section includes the word "economics", it is not the intent here to provide an economic analysis of fixed-mobile convergence implementations beyond some general considerations. It is not a secret that cost considerations can make or break any technical solution. The convenience of having one mobile phone that works transparently across all wireless networks is very appealing. However there are cost considerations that make such services economically justifiable. Three economics terms are used that need short explanation: Return on Investment (ROI), Compound Annual Growth Rate (CAGR), and Average Return Per Unit (ARPU).

Return on Investment (ROI), or rate of return, is the ratio of money gained relative to the money invested. It is usually calculated on an annual basis. ROI is also used to give a general qualification for a sound investment that brings profit as opposed to losing money. **Compound Annual Growth Rate** (CAGR) is the cumulative annualized growth rate that takes into account the cumulative effect of investment or growth over a period of time; five years for example. **Average Return Per Unit** (ARPU), as used in mobile telephony, indicates the revenue generated per unit, on a monthly basis, as the result of subscription and usage of services. Addition of a new service increases the ARPU.

If we consider a working network infrastructure, the cost of building a Fixed-Mobile Convergent network can be quantified by calculating the value of the following main components:

- Cost of the initial design of the FMC network;
- Cost of additional network infrastructure components;
- Cost of multi-mode convergent terminals;
- Cost of installation and activation;
- Cost of licensed radio spectrum;
- Cost of mobile communications;
- Cost of roaming, i.e., use of other mobile operators;
- Cost of usage of wired network infrastructure; and
- Cost of network provisioning, management, operations, and administration.

Two economic instruments are used in characterizing these costs: capital expenditures and operating expenditures. **Capital Expenditures** (CAPEX or capex) are the totality of costs or values of fixed assets used to build a system or to develop a product. **Operating**

Expenditures (OPEX or opex) represent the cost of running and maintaining a system or a product.

The most important economic drivers of fixed-mobile convergent solutions are:

- Reduced overall cost of communications;
- Reduced response time and higher available bandwidth;
- Availability and continuity of applications and services, irrespective of location;
- Development of new services for customers and overall increased revenue.

In the following sections we provide a sample of statements that indicate development forecasts of fixed wireless and cellular mobile technologies. These forecasts were provided by research organizations, manufacturers, and service providers. The forecasts are based on the analysis of technologies, impact of technology on users and businesses, the state of standardization, market demand, and on the level of current implementations. These estimates, when compared with actual numbers from the past few years, sometimes vary widely, but they are presented here to illustrate the state of current technology compared with projections for the future.

19.2 Projected Mobile Communications Growth

Mobile communications has experienced the fastest growth of any technology, with the highest impact on personal life of individuals and on businesses, in history. It certainly surpassed the impact of personal computing whether in the form of PCs, laptops, pocket PCs, or PDAs. Today, mobile communications provides tremendous computing and communication power to the users, all concentrated in the ubiquitous mobile handsets or cellular phones and their accessories. Thus, the handset becomes not only a voice telephone but also a computer that can access the Internet for emails, an instant messenger, a browser in search of information content, a digital photo camera, a music store and player, a screen to play games, a travel navigator, a personal agenda and scheduler, and much more. The following set of statistics will back the above statements [111].

- The number of worldwide subscribers will surpass 3.1 billion people in 2007 (1.8 billion in 2004, 2.2 billion in 2005, and 2.7 billion in 2006). In 2006 alone, over 1 billion mobile phones were sold (143 million in USA). In 2007, an estimated 1.1 billion new mobile phones will be sold.
- As of 2006, 80% of the world's population had mobile phone coverage. This coverage is expected to increase to 90% by the year 2010;
- As of 2007, over 2.5 billion subscribers will use the GSM/GPRS/EDGE technologies, as opposed to 600 million subscribers using one of the CDMA-based technologies.
- Five major manufacturers have over 70% of the market share: Nokia (36%), Motorola (20%), Samsung (12%), LG Electronics (7%), and Sony Ericsson (5%). Other notable manufacturers are Apple Inc., Kyocera, HTC, Panasonic, RIM, and Siemens;
- The average growth in handset sales of 20% from 2004 to 2006 will see a gradual slowdown in production, because of market saturation. The estimated increases are: 10–15% in 2007, 7–10% in 2008, 5–8% in 2009, and 3–5% in 2010;

- Luxembourg has the highest penetration of mobile handsets in the world (164%). Countries/regions with handset penetrations well above 100% are Hong Kong, Taiwan, Italy, Sweden, and Finland. On average, Europe is close to 100% penetration;
- Currently, the highest growth rate in handsets sold is in Africa, followed by Asia, South America, and Eastern Europe. The largest growth market is India that adds 70–80 million phones every year. India is expected to reach 500 million subscribers by 2010; in China there are over 500 million users;
- 1.6 billion mobile phones will be concentrated in the Asia Pacific region by 2009; (roughly 700 million in China, 500 million in India).

Some estimates of the number of users, and the revenues that will be generated by the availability of multimedia applications and machine-to-machine applications, are offered here:

- The number of gamers on cell phones will reach 134 million by 2009;
- The number of TV mobile subscribers will reach 25 million by 2009 from about 5 million in 2007;
- The number of mobile maps/navigation systems will reach 42 million by 2012;
- The market for social networking applications will reach $500 million in 2009, compared with $47 million in 2006.
- The number of mobile users involved in social networking will reach 174 million by 2010;
- The number of mobile subscribers using Presence and Instant Messaging applications will grow from 12% in 2006 to 625 in 2010;
- By 2012, 45% of cell phone will have touch screens;
- Mobile content and services will generate $150 billion, by 2012;
- The revenue generated by worldwide SMS services will reach $60 billion in 2008;
- The volume of worldwide SMS will reach 3 trillion messages by 2010;
- Mobile broadband W-CDMA connections will reach 40 million users by 2010 with 45% supporting 3.5 G WCDMA/HSDPA standards;
- Qualcomm has developed an embedded global mobile solution for notebooks. Nicknamed Gobi, it consists of software configurable MDM 10000 chipset supporting HSDPA/HSPUPA and CDMA EV-DO Rev. A, a reference design with associated API, and GPS radio. It will be available in 2008;
- Mobile video telephony service revenue (video mail, video calling, and video sharing services) will increase from $1 billion in 2007, to $17 billion by 2012;
- The revenue generated by old handsets recycled for use will reach $3 billion by 2012, with over 100 million handsets involved.

The inStat research company forecasts for 2011 that there will be 3.1 million worldwide installations of Base Transmission Stations, up from the 1.8 million in 2006.

Regarding revenue generated by mobile communications:

- The Average Return Per Unit/User (ARPU) will reach $70 in 2007, an increase from $66 in 2006 and $55 in 2004. The increase is mostly generated by value added data services such as SMS, MMS, instant messaging, and Internet access;

- Deployments of W-CDMA base stations are growing at a Compound Annual Growth Rate (CAGR) of more than 16%;
- Mobile phone chipset businesses alone, in 2006, generated $41 billion out of $250 billion worldwide sales generated by the entire semiconductor industry.

19.3 Projected Development of WLANs and Mesh WLANs

WLAN and WLAN Mesh standards based on IEEE 802.11 are evolving technologies that have undergone numerous improvements and enhancements to achieve higher data rates, to support the addition of voice over WLAN, to extend coverage areas through mesh networking, and to provide dual mode convergent handsets. The following set of statistics show WLAN Wi-Fi development projections [111, 112].

- Worldwide Wi-Fi phone sales topped $535 million in 2006, up 327 per cent from 2005; This can be compared with mobile phone sales of $115.5 billion with a growth rate of only 27% increase from the previous year;
- Shipments of dual-mode Wi-Fi handsets for the next four years, 2007 to 2010, is expected to increase 1300% while the forecast for mobile phones annual increase rate will be a relatively low 26%;
- Symbian OS remains the leading operating system used in smartphones with a penetration of 72%. This is followed by Linux (13%) and Microsoft Windows CE Mobile (6%). The market share of Palm OS, RIM BlackBerry, and OS X from Apple is around 2% each.
- There are over 300,000 hot-spots in the world, and the number is increasing rapidly. However, only a small percentage of hot-spots provide both Wi-Fi voice and data services. An exception is the T-Mobile with over 20,000 worldwide hot-spots (8600 in Germany, and 7400 in the USA);
- In 2008, several airlines will provide in-flight Wi-Fi-based Internet connectivity using satellite services;
- By 2012, 90% of Wi-Fi enabled terminals and WLAN access points will be based on next generation 802.11n chipsets;
- The forecast for dual-mode Wi-Fi cellular mobile handsets is 100 million units by 2009;
- Dual-mode Wi-Fi cellular mobile chipsets market share in 2009 will be 20% of the total Wi-Fi chipsets sold;
- Wi-Fi networking gear is a multibillion dollar business, and in the past four years has increased sales revenue roughly 20% year-after-year. The numbers were $2.5 billion in 2003, and an estimated $4 billion in 2007;
- Shipments of WLAN mesh network components doubled in 2006 and 2007 compared with the previous years. This increase will slow in 2008 because of concerns regarding the economics of the municipal networking business model;
- WLAN vendor revenues are divided almost equally between enterprise WLANs and Wi-Fi for homes or small businesses;

- WLAN market share for enterprises is dominated by Cisco (58%), followed by Symbol Technologies (18%), and Aruba (6.5%) [112];
- WLAN market share for small offices and homes is dominated by Cisco (25%), followed by D-Link and NetGear (together 33%), and Belkin (8.6%) [112].

19.4 Projected Development of Bluetooth and ZigBee Networks

The ubiquitous presence of Bluetooth features in most handsets shipped in 2006 and 2007 is the best indication of the popularity of this technology. For Bluetooth, the major drivers are the hands-free headset and car set devices connected to mobile phones. In addition to cell phones, a myriad of Bluetooth notebooks, PCs, mice, keyboards, speakers, medical devices, gaming units, camcorders, and digital cameras use one of the Bluetooth profiles. Smaller scale developments take place in the ZigBee world that addresses home, building, and industrial automation. The following set of statistics shows Bluetooth and ZigBee development projections [111].

- Over one billion Bluetooth-enabled units were shipped in 2007, supported by hundreds of applications. This is a ten fold increase from 2002; this pace is expected to continue over the next few years, with an attendant increase in the diversity of products and applications;
- The number of ZigBee chipset units will grow to 90–200 million units sold in 2010, compared with 2.5 million units sold in 2005;
- The forecast for 2008 calls for 21.5 million ZigBee-based units sold, half to be used in home and building automation;
- The revenue generated by ZigBee chipsets will grow from $11.2 million in 2005 to over one billion in 2010;
- The global market for wireless sensor network (WSN) systems and services is expected to grow to about $4.6 billion in 2011, up from approximately $500 million in 2006;
- An estimated 10 billion microcontrollers that operate in various automation and controlling products, could be networked by ZigBee technology;
- An estimated 40 million femtocells will be installed by 2012 with over 100 million users in a market that will reach at that time about $1 billion.

19.5 Projected Development of WiMAX Networks

Once the IEEE 802.16e standard was adopted, development of WiMAX technology as an alternate broadband access solution marked a tremendous growth. As of 2007, there were about 250 million broadband access lines already installed. However, the potential of WiMAX will be achieved when, combined with 3G and B3G mobile technologies, it becomes the core of the next generation mobile networks. In 2007, ITU-R designated WiMAX/OFDMA as one of the six radio technologies that support IMT-2000 3G mobile communications. The following set of statistics display the extent of WiMAX development projects and offer some future projections [111].

- The estimated number of 14.5 million WiMAX subscribers in 2007 will increase almost ten fold, to reach 131 million subscribers in 2011;
- WiMAX shipments from Taiwan, a major exporter of WiMAX technology, will reach 50 million units by 2012, generating almost $2 billion in revenue;
- WiMAX capex sales will reach $26 billion by 2011, compared with $3 billion in 2007;
- Five worldwide WiMAX certification labs are expected by the end of 2007;
- WiMAX is a competitive solution for cellular mobile backhauling. The backhaul networking business is estimated to reach $23 billion capex by the end of 2012;
- Sprint/Nextel Xohm cellular mobile and Clearwire WiMAX-based solutions, claimed as B3G/4G implementations, are expected to be offered by mid 2008. The plan is to reach 185 million people in the USA by the end of 2010; in the most recent news, this plan is unraveling as I am writing this chapter;
- Sprint/Nextel will spend $5 billion by 2010 to build up the 4G network and they expect 50 million subscribers will generate a revenue of $2.5 billion a year;
- Japan DoCoMo, together with Acca Networks, will invest $870 million for the deployment of WiMAX-based networks in Japan;
- WiMAX deployment in South Korea (WiBro transitioned to IEEE 802.16e) will involve about $1 billion in investments;
- VSNL, a leading Internet Service Provider in India, will spend about $1 billion to provide WiMAX services to businesses in 120 major metropolitan areas;
- WiMAX, along with Wi-Fi and 3G/B3G, will provide powerful connectivity for future Ultra Mobile Devices (iPhones, PC tablets, Pocket PCs, PDAs); 5 million such devices are expected to be sold by 2012.

19.6 Projected Development of RFID, NFC, and WUSB Networks

Of the fixed wireless technologies and applications, RFID has the greatest range of applicability by its use in tagging and tracking items, goods, and even humans. Similar developments are expected from NFC technology in the realm of financial transactions and electronic commerce as well as from UWB-WUSB for very short range high speed wireless data transfer. The following set of statistics will support these statements [111, 113].

- The estimated RFID global supply chains market of $5 billion in 2007 will increase to $27 billion in the next ten years;
- RFID tag shipments will grow by 100% every year for the foreseeable future;
- By 2010, the value added impact of RFID will increase to 8%, affecting the manufacturing, retail, and transportation economy in Europe, which is estimated to be 62 billion euros at that time;

- In the retail industry, RFID will grow at a CAGR of nearly 57% during the next four years, with revenue exceeding $1.5 billion in 2010;
- In the retail industry, the number of companies adopting RFID technology will increase from 20% in 2006 to 40% in 2010;
- The North American retail RFID market is about 40% of the total global RFID market.
- The Chinese RFID market is forecast to be 82.4% CAGR for 2005–2010, and to reach $2 billion by 2011;
- RFID manufacturing sales generated by German companies will increase from 914 million euros in 2006 to 1.4 billion euros in 2010;
- The overall revenue generated by RFID systems used in airline baggage handling will grow to $27.5 million in 2011, at a CAGR of 18.5%;
- Contactless payment based on NFC technologies will create an estimated global market of $1 billion by 2012, that includes hardware and associated software;
- The North American market for NFC-based converged technology will increase from almost $700 million in 2005 to about $7 billion in 2008;
- The two billion USB devices sold in 2006 are a solid base for WUSB growth in the foreseeable future, estimated at 12% CAGR through 2011.

19.7 Projected Development of UMA-based Convergent Networks

The UMA specifications were developed to provide convergence between GSM cellular mobile networks and IEEE 802.11 a/b/g WLANs, IEEE 802.16 WiMAX, IEEE 802.20 UWB, and IEEE 802.15 Bluetooth WPANs. And, a large number of UMA chipset vendors, dual-mode or multi-mode handset manufacturers, and UMA-based convergent systems integrators have developed products to support this convergence. The following statistics should help to confirm these statements [111].

- UMA-based dual-mode subscriber numbers increased from 400,000 units in 2006 to almost 2 million handsets in 2007;
- According to ABI Research, dual-mode handset production is expected to reach 50 million units in 2009, 100 million in 2010, and 256 million units in 2012;
- More optimistic forecasts come from InStat, predicting 66 million dual-mode handsets by 2009;
- A more reserved forecast comes from Pyramid Research, calling for only 18 million units sold in 2011. In the same span, global UMA revenue will increase from $130 million in 2006 to $9 billion in 2011;
- Cincinnati Bell was the first operator providing a Wi-Fi cellular mobile service, the CB Home Run in June 2007. The service is available in WLAN-based offices, homes, and 300 "Cincinnati Bell ZoomTown WiFi" designated hot-spots;
- T-Mobile is the largest operator in the USA providing GSM cellular mobile and Wi-Fi b/g convergent services, the HotSpot&Home, in more than 7400 hot-spots;

- Several dual-mode handsets including T-Mobile Dash and T-Mobile Wing, Nokia 6086, Samsung t409, and T-Mobile BlackBerry Curve, are offered as part of USA convergent services;
- UMA-based convergent services are provided in Europe by O2 in UK, Orange in France, and T-Mobile International in Germany and other European countries; a total of 27,000 worldwide hot-spots are available;
- Orange's Unik/Unique, the most successful UMA-based service, has over 5 million customers. The service bundles FMC capabilities with broadband services within Livebox home gateways;
- By 2009, 20% of the chipset market for cellular phones will be dual-mode Wi-Fi cellular mobile chipsets.

19.8 Projected Development of IMS-based Convergent Networks

From the description of the IMS architecture and specifications in Chapter 17, it is clear that IMS is still an evolving concept, with only limited implementations of pre-IMS and IMS-like solutions. Therefore, IMS implementations, as presented in 3GPP Release 5, 6, and 7 documents, will take a long time to become a critical mass. This fact is reflected in the scarcity of predictions provided by the research companies, vendors, or mobile service operators who have interest in IMS. The same scarcity characterizes the number of implementations and services. The following set of statistics and news could confirm these statements:

- ABI Research forecasts an increase of IMS related investment from $2.6 billion in 2006 to about $10 billion in 2011. The revenue generated in that interval by IMS-based service/applications will be close to $50 billion. In this account are considered developments intended to support softswitch-based VoIP technology;
- However, a recent study by Infonetics Research shows a decreasing number of service providers planning IMS-based solutions in 2007 and 2008 compared with 2006;
- British Telecom's 21st Century Network (21CN) is the largest scale implementation that operates on the IMS service delivery platform architecture. Although it is an IP-based network, 21CN only loosely follows IMS specifications. The project started in 2006 with the goal of replacing 16 regional BT legacy networks. However, it is expected to take 5 to 10 years to complete;
- Verizon Wireless, the leading vendor of Advances to IMS (A-IMS), a rounded version of IMS that incorporate non-SIP-based applications (VoIP, IPTV, video on demand), does not expect IMS-based services earlier than 2008. Verizon Wireless represents the CDMA world versus the GSM world perspective of IMS;
- Using the Alcatel-Lucent IMS core platform, AT&T started the first IMS-based Video Share service in the first half of 2007;
- Nortel Networks has proposed an IMS-based IPTV solution for standardization to ITU-T. The architecture will support IPTV services such as video-on-demand and broadcast TV over a wired or wireless broadband network;

- China's ZTE Company, as part of the Catalyst Project, has provided a demonstration of the integration of the IMS platform with the TeleManagement Forum's New Generation Operation Systems and Software (NGOSS) framework;
- Nortel tested its IMS platform based on Voice Call Continuity 3GPP standard using Qualcomm's dual-mode chipsets provided by Qualcomm;
- The lack of a standardized service delivery platform is a major impediment in the development of services by third parties;
- It is generally believed that IMS will be successful when mobile networks and terminals became based solely on IP protocol based on Commercial off-the-Shelf (COTS) telecommunication products and platforms.

19.9 Evaluation Criteria for Fixed-Mobile Convergence Solutions

Evaluation of fixed-mobile convergence solutions is an important step in building a business case for creating a network infrastructure with corresponding applications and services. This section will provide a summary of evaluation criteria to be used in the selection of FMC solutions. These criteria can be used in the development of specific Requests for Information (RFIs) or Requests for Proposals (RFPs), and to build evaluation matrices to weigh vendor responses.

Four general areas that can be used in an evaluation are: **technology, applications/features, services/performance, and economics**. Many of the criteria listed below were introduced as we discussed various aspects of FMC solutions and their implementations. Therefore, in most cases we will list them as bullets without a detailed explanation. Criteria falling out of the above four categories will be listed as miscellaneous.

Technology Criteria

- FMC components (cellular mobile, WLAN (Wi-Fi, Mesh), WMAN (WiMAX, FSO) WPAN (Bluetooth, femtocell), sensor networks (ZigBee, RFID, NFC, WUSB), etc.);
- Cellular radio access networks and their generations: GSM, CDMA, 2.5G, 3G, 4G, GPRS/EDGE, HSDPA/HSUPA, CDMA 2000 Releases A/B, iDEN, etc.;
- Multiplexing and modulation scheme for downlink and uplink: OFDM, SC-FDMA, QPSK, 16 QAM, 64 QAM, BPSK, QPSK, 8 PSK, etc.;
- Infrastructure components: BTS towers, BTS/BSC links, femtocell, MSC/GWSC, SGSN/GGSN, WLAN, VoWLAN, WLAN AP, WLAN controller, WiMAX SS, WiMAX BS, Bluetooth master/client, RFID reader/tags, NFC card, etc.;
- Communications protocol infrastructure: TCP/IP and OSI protocol stack, Physical Layer technologies and connectors, MAC sublayer, convergence sub-layers, transport protocols, DIAMETER, COPS, H.248 protocol stacks, etc.;
- Handsets/Cell Phone characteristics: chip sets, antenna types, display and keypad types, form factors, USB, WUSB, Bluetooth, Wi-Fi, RFID, NFC connectivity, MIMO antennas, flash drive storage capability, etc.;
- Handsets classes: standard cell phones, smartphones, dual or multi-mode FMC headsets, BlackBerry, iPhones, etc.;

- Type of convergence supported: Cellular & Wi-Fi, Cellular & Bluetooth, Cellular & WiMAX, Cellular & ZigBee, or Cellular & RFID/NFC;
- Operational spectrum of frequencies: GSM 850/900/1800/1900 MHz, PCS, WiMAX 2.4/3.5/5.8 GHz, Bluetooth, ZigBee, RFID VHF/UHF, WUSB, UWB, licensed, unlicensed;
- Standards supported: 3GPP Release 5, 6, 7, 3GPP2 UMB, IEEE 802.11x, IEEE 802.16 d/e, 802.15x, 802.20, 802.21, EPCglobal G2, proprietary interfaces, etc.;
- Global convergence standards supported: UMA/GAN, IMS, SIP/SDP, OMA;
- Physical coverage: very short proximity, short proximity, home, local, building, campus, metropolitan, wide area, all by distance or radius;
- Transmission characteristics: maximum data rate, average data rate, peak data rate;
- External connectivity: USB 2.0, WUSB, Wi-Fi, Bluetooth, Bluetooth 2.0+EDR, GSM/GPRS/EDGE, WiMAX, etc.;
- Topology and architecture: point-to-point, point-to-multipoint, star, tree, mesh, centralized, hierarchical, distributed, backhaul links, etc.;
- Security characteristics: AAA, encryption type DES, 3 DES, AES, public/private keys;
- Conformance testing: ISO/IEC, ITU-T, IETF, consortium/vendor alliance inspired certification testing, third party conformance/interoperability testing, etc.;
- Storage capability: flash card, micro SD, etc.;
- Battery: built-in, removable, rechargeable lithium-ion, talk time, stand-by time, Internet access over Wi-Fi use time, audio/video playback time, etc;
- Operating system: Symbian, Linux, Windows CE Mobile, Palm OS, RIM BlackBerry OS, Apple OS X, etc.);
- Browser type: WAP Browser, Linux browser, OMA browser, Opera Mobile mini-browser, Microsoft Mobile Explorer, Blazer, Nokia S60, NTT DoCoMo, Apple Safari, user installable browsers, etc.

Applications/Features Criteria

- Types of communications supported: voice, data, video, multimedia;
- Voice applications: Voice, VoIP, voice value-added features, voice conferencing, ringtones, one number across heterogeneous networks, signaling type SIP/SDP, legacy SS#7, connectivity with PSTN, etc.;
- Data: SMS, MMS, voice mail, instant messaging, etc.;
- Video: video streaming, video broadcast, video-on-demand, video phone, mobile IPTV, video conferencing, etc.;
- Push-to-Talk over Cellular;
- PDA features: address book, schedule, calendar, calculator, etc.;
- Navigation kit: built-in GPS, Google Maps, Google Earth, Navteq, etc.
- Music: audio streaming, MP3 download, storage, playback, RF radio, iTunes;
- Built-in digital camera: taking pictures, storing, sending/sharing; slide projection;
- Gaming: game console, on-line gaming, etc.;
- Accessories: hands-free headsets, car sets, etc.;
- APIs supported: Web 2.0 to develop new applications, Parlay Group IMS API, AJAX-Asynchronous Java Script XML.

Service/Performance Criteria

- Voice services: voice, voice over WLAN, voice over Bluetooth, visual voice mail, voice conferencing, etc.;
- Internet access: narrow band, broadband, web browsing, web services;
- Content access: Google, Yahoo, Amazon, MSN, CNN Fox Interactive, iTunes stores, YouTube, My Space, Facebook, etc.;
- Widgets access: weather reports, stock reports;
- Location and presence;
- GPS and navigation maps;
- Performance metrics supported: availability, reliability, MTBF/MTTR, response time, throughput, delay, jitter, packet loss, power consumption, battery life cycle, interference with other sources, environmental conditions, etc.;
- SOA support;
- Mobile to Mobile Communications: file, music, video clips, and photo sharing;
- Cellular mobile and fixed wireless service providers: PSTN, PSDN, etc.;
- Announcements;
- Service provisioning and maintenance;
- Security services: login password, public/private security keys;
- Service class: (8, 9,10 or above);
- QOS class: conversational, interactive, best effort, end-to-end, guaranteed peak rates/throughput, etc.;
- Use of Differentiated Services, MPLS, PBM technologies;
- Service plans: individual, family, home, SOHO, campus, city-wide, business, special, etc.

Economic Criteria

- Number of implementations, number of subscribers;
- Maturity of products, warranty;
- Network infrastructure components costs (CAPEX);
- Installation, testing, and activation costs;
- Multi-mode handsets costs (per unit, volume and plan-based discounts, long term contract, etc.;
- Accessories costs: headsets; external storage, etc.;
- License costs;
- Charging/billing systems: flat monthly subscriptions, planned based usage, roaming charges, pay-as-you-go, prepaid, etc.;
- Estimated ARPU;
- Network operations and maintenance costs (OPEX);
- Estimated ROI;
- Customer service support (help desk);
- Usability of system with other operators: locked, unlocked;
- Partnership with other organizations: chipsets, systems integrators;
- Convergence plans: basic individual, family, enterprise, no additional charge, flat fees, Wi-Fi fees, etc.

Miscellaneous Criteria

- Network Management system/platforms used to manage FMC networks;
- Applications and service management systems;
- Operations management;
- Troubleshooting/simulation tools;
- Scalability, modularity, flexibility;
- Compatibility with legacy systems;
- Future developments.

20 Fixed-Mobile Convergence Implementation: Status, Trends, and Issues

20.1 Benefits of Fixed-Mobile Convergence

Throughout this book we analyzed in great detail the technologies behind fixed wireless cellular mobile networks convergence. In doing this, we used the shortened term Fixed-Mobile Convergence (FMC). However, convergence has multiple facets and a long history of expectations, and sometimes marketing hype. The FMC term has been used since the early attempts to integrate voice and data through the convergence of telecommunication and data communications networks. Later, the term was used for the convergence of digitized services such as ISDN. The development of the new wave of land based wireless communications brought the convergence term back as the on-going process of integration of wireless networks and wired networks. Returning to the narrow definition adopted for this book, fixed wireless and cellular mobile networks convergence, we can summarize the goals and benefits of FMC as follows:

- Communications anytime, anywhere, using any wireless technology;
- Continuous communication and operation regardless of the network used;
- Convenience for the user with a single handset and telephone number;
- Improved cost/performance factors by automatic selection of the network with highest available bandwidth and lowest cost;
- High availability by having multiple networks that can be used for transmission;
- Flexibility in the network design by combining indoor and outdoor, and short range and long range communications;
- Support for fixed, nomadic and highly mobile users;
- Service portability across heterogeneous networks.

The achievement of these goals requires positive answers and solutions in the following areas:

- Economics of FMC with adequate Return on Investment;
- Standardization of transparent handover operations;
- Scalability of convergent architectures which is critical in urban and metropolitan areas;
- Security of operations and integrity of data transmitted;
- Interoperability between different network designs and vendor products;
- Efficient use of both licensed and unlicensed frequency spectrums;

- Reduced interference between networks that use the same spectrum;
- Mobile terminals with low power consumption and long life batteries;
- Service platform architectural design to develop new applications/services;
- Easy access to content of information and social networking;
- Customized solutions and the ability to personalize the use of mobile phones;
- Convergent network management and end-to-end QOS assurance.

20.2 Trends in Fixed and Mobile Wireless Communications

The evolution of fixed wireless cellular mobile networks convergence has two aspects. One aspect of the evolution is the development and implementation of the technologies that are part of the fixed wireless networks: WLANs, WPAN, WMANs, and Near-Field Sensor Networks. Along with this is the development and implementations of cellular mobile networks. Each technology was born out of the need to support communications in local and wider geographical areas. Therefore, development of each technology had its own course driven by specific applications supporting that technology. Another aspect is the evolution of and trends in the convergence between various fixed wireless technologies and cellular mobile networks. A graphical representation of fixed and mobile networks development is shown in Figure 20.1.

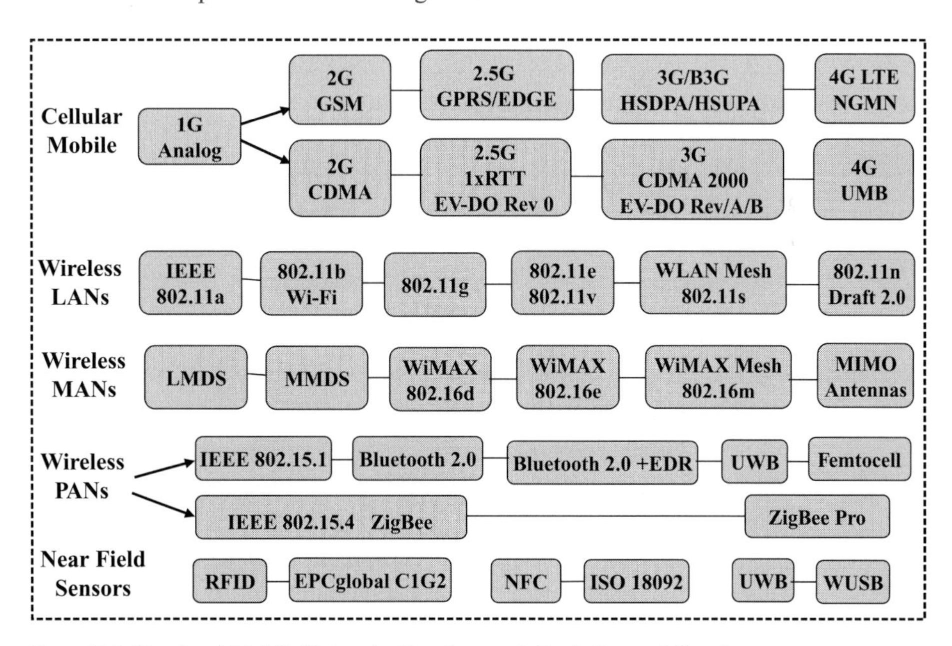

Figure 20.1 Fixed and Mobile Networks Development: Evolution and Trends

In this figure, we recognize individual technologies that exist now and were analyzed in previous chapters of this book. In addition, the figure indicates solutions based only on specifications in draft formats or solutions that did not reach a critical mass market. For example, specifications of **3GPP Long-Term Evolution** (LTE) and **3GPP2 Ultra**

Mobile Broadband (UMB), published in 2007, are expected to go through a long process of adoption. More details about LTE, UMB, and Next Generation Mobile Network (NGMN) will be provided in following sections. In general terms, NGMN is envisioned as the integrated end-to-end IP-based network infrastructure that encompasses all existing and emerging wireless networks [114].

Currently, there is no clear definition of what constitutes a 4G mobile network in the sense of underlying technologies; cellular mobile only or a combination of evolved cellular mobile and WiMAX. However, for the 802.11n-based WLANs, there are several products from major vendors that have been developed from the Draft 2.0 specifications. The concept of femtocell or home-based access points, acting as a micro mobile base transmission station, needs to be embraced by most of the mobile operators. WiMAX definitely made positive steps in its evolution after adoption of the mobile 802.16e amendment. Work is underway to specify and standardize WiMAX mesh configuration and to define the 802.16m specifications. MIMO antennas will help consolidate not only the WiMAX position but also other wireless technologies from cellular mobile to WLANs. There is no clear transition path for the long term evolution of Bluetooth and NFC versus Ultra Wideband (UWB) [115]. Because of incompatible regional and national standards that impede solutions for global chain supply applications there are still some issues in the general adoption of the RFID technology.

In addition to these trends, two developments will have an impact on the future of wireless communications, and, by default, on FMC networking. They are the Mobile-to-Mobile or Peer-to-Peer communications, and Software Defined Radio concepts.

Mobile-to-Mobile (Mo-to-Mo) is related to the larger **Machine-to-Machine** (M2M) concept in that it provides peer-to-peer communications between similar types of embedded wireless devices without involving higher level authorities or core networks. Use of Mo-to-Mo has two major applications. One is peer-to-peer networking, the normal state of communications in WLANs, WPANs, and WiMAX mesh configurations. In this case, WLAN peer access points, WPAN ZigBee nodes, and WiMAX base stations communicate with each other, directly. Standards have been developed to handle the mesh nodes interworking.

The second application uses cell phone mobile networks and services to create M2M networking based on data transmission services such as reading instrumentation and SMS. These relaying types of services are generally controlled by Mobile Virtual Network Operators (MVNO) and use a few, or no, land lines. Home alarm security systems, reading of utility meters, and vehicle tracking with GPS-enabled mobile devices are typical applications of this type. A popular application in Europe is E-Call, where a call giving the exact location and time of that event is made automatically from a disabled car to police and an ambulance service.

In the **Software Defined Radio** (SDR), the concept is quite simple: to program or reconfigure hardware components by downloading specific software. In this way, the same device or component can operate in various environments. By adopting programmable RF front stages that are close to antennas, we can make them work in multiple frequency bands, and with multiple modulation schemes, for both downlink and uplink.

The SDR concept is not quite new, and it has roots in similar applications such as Software Defined Networking (SDN) and Programmable Logic Controller (PLC). SDN was designed and applied by AT&T for its class 4 switches to create software defined Virtual Private Networks (VPNs). PLCs are specialized programmable computers that are used in industrial automation to provide command and control of industrial processes. The programs are downloaded and stored for viability in non-volatile Erasable Programmable Read-Only Memory (EPROM).

SDR can be applied to cell phones working in multiple bands (for example GSM 850, 900, 1800, and 1900 MHz) with multiple mobile technologies/access methods (TDMA, GSM 2.5/3G, and CDMA, TD-SCDMA). It can also be applied to handle a variety of WLANs (802.11a/b/g/n) and RFID (HF/VHF/UHF) technologies, to gateways working with multiple protocol stacks, to encoders/decoders using different algorithms, and to RF modems. SDR can improve the economics of FMC solutions by minimizing the number and complexity of hardware components, and by providing a conduit for control and management. With this idea in mind, the Software Defined Radio Forum was formed, a non-profit organization whose goal was to promote the development and deployment of SDR technologies in wireless systems.

A particular aspect of the SDR is the emerging **Cognitive Radio Technology** based on draft IEEE 802.22 standard. This technology allows unlicensed radio transmitters to operate in the licensed frequency bands where and when the spectrum is not used. The standard covers wireless regional networks to provide broadband wireless access in rural areas by sharing of, but not interfering with, the TV broadcast service spectrum [116].

20.3 Trends in Fixed-Mobile Convergence

Following the analysis of wireless technologies evolution presented in the previous section, we now look at the trends in convergence between various fixed wireless technologies and cellular mobile networks. A graphical representation of fixed and mobile networks convergence development is shown in Figure 20.2.

In this figure we display convergent solutions by pairing one of the fixed wireless technologies/networks with cellular mobile networks. The initial driver was convergence between WLAN Wi-Fi and cellular mobile networks. Convergence solutions are also provided for other technologies: WPAN (Bluetooth, ZigBee), WMAN WiMAX, and Near-Field Sensor Networks (RFID, NFC). The diagram does not show convergent solutions involving multiple fixed wireless networks such as WLAN, WPAN, NFC, and cellular mobile networks, combined.

The same diagram also shows the evolution of two global standardized solutions for convergence across multiple wireless technologies; UMA/GAN and the IMS architecture and specifications. UMA was incorporated into the 3GPP Release 6 specifications and was updated and extended in 3GPP Release 7. Products and implementations from handsets to hot-spots to enterprises are being produced based on these specifications. IMS also evolved from 3GPP Release 5, where the architectural framework was provided, and Release 6 where the SIP protocol was incorporated. Release 7 included the ETSI

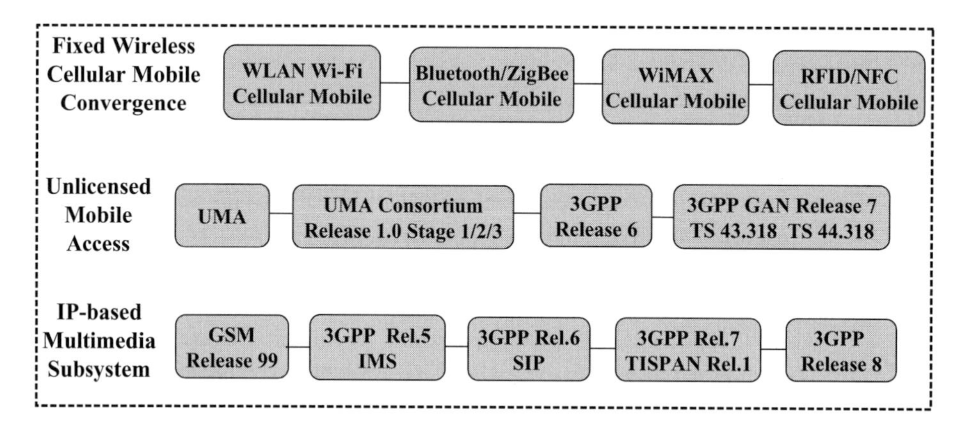

Figure 20.2 Fixed-Mobile Convergence Development: Evolution and Trends

TISPAN Release 1 specifications. IMS products are based on a combination of specifications from Releases 6 and 7. Release 8 will address 3GPP Long-Term Evolution (LTE) and System Architecture Evolution (SAE).

20.4 3GPP Long Term Evolution, 3GPP2 Ultra Mobile Broadband, and NGMN

3GPP Long Term Evolution (LTE) and **3GPP2 Ultra Mobile Broadband** (UMB) are the far-looking developments for the Next Generation of Mobile Networks (NGMN) extending beyond current 3G implementations. LTE follows the line of 3GPP GSM evolution which included HSDPA, HSUPA and W-CDMA. UMB work follows the 3GPP2 CDMA evolution: CDMA 2000 and CDMA 2000 1xEV-DO Revisions A and B. 3GPP LTE development started late 2004 with a completion goal of late 2007. The CDMA Development Group had similar plans for UMB. The intent of this section is to provide the major coordinates of this work in the context of presenting major trends in the development of mobile communications. The high level goals of 3GPP LTE include:

- Highly efficient use of mobile communications channels (high bits/Hz);
- Flexibility in using the allocated spectrum;
- Development of high-speed broadband services;
- Lower power consumption of mobile terminals;
- Open and standard interfaces;
- Lower cost per bit transmitted; and
- Minimum number of options.

The LTE coverage targets the Universal Terrestrial Radio Access Network (UTRAN) architecture. LTE will be based on packet switched networking technologies with full support for conversational class traffic and adequate QOS. LTE will succeed UMTS 3G technologies such as High Speed Downlink Packet Access (HSDPA) and High Speed Uplink Packet Access (HSUPA) mentioned earlier. Technical Report (TR) 25.913 written early in the project's timeframe contains detailed technical requirements in the following areas:

- A downlink peak data rate of 100 Mbps in the 20MHz spectrum allocation;
- An uplink peak data rate of 50 Mbps in the 20 MHz spectrum allocation;
- Transition time between dormant state and full operation less than 100 ms/50 ms;
- A minimum density of 200 users per cell site for a 5 MHz frequency band;
- A downlink average throughput three to four times higher per MHz than HSDPA Release 6;
- An uplink average user throughput two to three times higher per MHz than Enhanced Uplink;
- Mobility across cellular networks at speeds from 120 km/h to 350 km/h;
- Throughput, spectrum efficiency, and mobility maintained for 5 km radius cells and with only slight degradation for 30 km radius cells;
- Operation in spectrum allocations as small as 1.25 MHz as well as sizes of 2.5 MHz, 5 MHz, 10 MHz, 15 MHz, and 20 MHz;
- Compatibility between Enhanced UTRAN (E-UTRAN) and UTRAN/GERAN working on adjacent channels;
- Downlinks will use the Orthogonal Frequency Division Multiplexing (OFDM) while the uplinks use the Single Carrier-Frequency Division Multiple Access (SC-FDMA) access methods. A variety of modulation schemes are supported: QPSK, 16 QAM, and 64 QAM for the downlink and BPSK, QPSK, 8 PSK, and 16 QAM for the uplink; and
- Use Multiple Input Multiple Output (MIMO) antennas.

According to the CDMA Development Group's initial tests, peak data rates of 288 Mbps in ideal conditions and an average throughput of 75 Mbps are possible for UMB with a spectrum allocation of 20 MHz. The design allows mobility of users up to 180 mph, while maintaining a satisfactory level of latency and jitter.

Another development in line with LTE and UMB is the **Next Generation Mobile Network** (NGMN), promoted by a joint organization with the same name formed in 2006. NGMN will establish a roadmap for the development and implementation of future mobile networks, beyond the HSDPA/HSUPA and CDMA 2000 EV-DO Revisions A/B/C architectures and specifications. Seven companies founded NGMN Limited: China Mobile, KPN, NTT DoCoMo, Orange, Sprint Nextel, T-Mobile and Vodafone. AT&T later joined the group in 2007. The NGMN team will act as an advisor and consultant for service providers and mobile equipment manufacturers.

The success of FMC is highly dependent on the level of handset RF integration. Initially, incorporation of GPRS/EDGE and 3G support for both GSM and CDMA technologies required multi-chip solutions. As the complexity of FMC handsets increases, there is a need for multi-band multi-mode, single-chip based handsets design to reduce the total cost of ownership. That involves hardware components such as RF ICs, amplifiers, filters, and switches as well as software components to handle the new content, digital music, mobile TV, streaming audio and video, and gaming [117].

20.5 World Wide Web 2.0 and Service Oriented Architecture

The success of fixed-mobile convergence depends on three major elements. First, the technologies involved and the network infrastructure built based on these technologies must be compatible with the goals of convergence. This book has focused exactly on the technological aspects of FMC. Second, there are applications/services that deal with the software applications that facilitate specific functions. The combination of technology/networks and applications provides the bases for the actual convergent services. These services can be related to the type of communications supported (voice, data, video, or multimedia for example) or to specific functions such as GPS-based location and navigation systems. The third component is the content of information accessed through the convergent network according to the application/services provided.

Using the above points as a guideline, we can identify the intersection of technology, applications/services, and content aspects in the development of a standardized approach for FMC. First, there is the IP-based Multimedia Subsystem (IMS) that represents the technology/network architecture. IMS is, in fact, a service platform that allows separation of the access/transport networking layer from the applications/services layer, based on signaling. The focus of the IMS platform is on network elements. Second, the development of applications and services is tightly linked to the World Wide Web 2.0 and the Service Oriented Architecture [118]. The focus of these two developments is on the Information Technology (IT) integration. The third aspect is the generation, distribution, and delivery of content information. Content can be delivered either by convergent service providers or by specialized providers such as weather forecasts and stock quotes. It can range from dedicated content to a selected group of subscribers, or just simple advertising and promotion. All these developments are part of the same encompassing framework aimed at providing a **Service Delivery Platform** (SDP) [119].

World Wide Web 2.0 (Web 2.0), the new enhanced browser, is, in fact, a collection of web services that facilitate a new generation of collaborative, social networking type of applications. More precisely, these multimedia services are provided over IP-based networks, i.e., support for voice, video, audio, text messaging and other new multimedia types such as Really Simple Syndication (RSS) and mashup for individual users and businesses. Personal applications/services include presence/location, personal blogs, audio/video streaming, exchanging pictures and music files, and self-generated and distributed content (Facebook). Business applications/services add access to enterprise blogs, professional networks, software development tools, enterprise research libraries, and professional training. We must note that Web 2.0 is more a marketing term than a physical network since it has no formal version specification. However, additions that are geared toward extensive collaboration across networks make Web 2.0 into another class of browsing service.

Service Oriented Architecture (SOA) is a high-level paradigm or architecture that supports organization of multiple services that originate from multiple service providers. It is relevant in the context of enterprise businesses that incorporate mobile communications that come with a wide range of multimedia services provided to mobile users. SOA modules can play the role of service provider, service broker, or service requester. They can be implemented using web services, XML, or the Simple Object Access Protocol (SOAP) [120]. SOAP is a protocol used between service provider and service customer as part of the service location and binding process.

20.6 General Issues in Fixed-Mobile Convergence

The convergence of Fixed Wireless and Cellular Mobile networks requires a mechanism to transparently make handovers as calls progress over multiple networks. This raises several issues that need be solved through standardization and agreements between the vendors and operators who will provide fixed-mobile convergence.

- Definitions and criteria for Initial User Access (combined criteria to provide optimal access);
- Level of mobility (fast and transparent intersystem handover for voice, data, and multimedia);
- IP-based interworking between Mobile IPv4, Mobile IPv6, and cellular IP networks;
- Security: Authentication, Authorization, and Accounting security mechanisms;
- Quality of Service: The ability to maintain satisfactory levels of QOS in ever changing environmental and traffic conditions;
- Service Level Agreements: SLA/SLS layout including penalties for QOS violations;
- Reconfiguration of base stations, access points, dual-mode handsets/terminals; reconfiguration is required by changes in radio link budget, external interference, and user speed;
- Use of new technologies: High capacity channels multiple-input multiple-output antennas;
- Extended convergent services: Tri-mode and multi-mode handsets working in multiple WPAN, WLAN, WiMAX and mobile cell-based GSM/CDMA environments of several generations;
- Convergence standards: Harmonized IEEE 802.11x, ETSI HiperLAN2, IEEE 802.16 (WiMAX), IEEE 802.20 (MBWA), and Unlicensed Mobile Access UMA/GAN 3GPP specifications;
- Need for additional components and extension of the current infrastructure to provide continuity of communications in areas not adequately covered by any of the convergent technologies;
- Adequate return on investment based on convergent applications and services.

20.7 Issues in UMA-based Convergence Implementation

Fixed-mobile convergence promoted through UMA/GAN 3GPP Release 6 and Release 7 specifications raises some specific issues:

- UMA early versions **do not fully support the Session Initiation Protocol**, since they use their own signaling down to the mobile station (as opposed to the solutions promoted by the MobileIGNITE Alliance);

- UMA is **not aligned with the Internet Protocol for Multimedia Subsystem (IMS)** architecture and specifications;
- UMA **can not handover between multiple access points**, i.e., **it does not support mesh configurations**, and multi-hop wireless networks;
- UMA **does not mandate short handovers** of 30–50 ms at the Access Points which would give it the ability to compete with wired communications;
- There is **no guarantee of cost savings** using dual mode handsets and fixed-mobile convergence, even though the costs of using IP-based services in WLANs is 85–90% less than that of GSM networks;
- UMA was designed primarily for convergence between fixed wireless WLANs (low cost, high capacity) and mobile wireless networks (high cost, low capacity) at a time when 3G implementations such as High Speed Downlink and Uplink Packet Access (HSDPA and HSUPA) and CDMA 2000 EV-DO were unavailable.

20.8 Issues in IMS-based Convergence Implementation

Fixed-mobile convergence is promoted through IP-based Multimedia Subsystem 3GPP Releases 5, 6 and 7 specifications. It is a most ambitious ongoing project and has as a goal providing integration and convergence of wired and all types of fixed and mobile wireless networks. However, given the scale of the design, there are serious issues when transitioning from concept to reality:

- IMS design, despite its layering and modular, flexible architecture, provides a **centralized** way to track the users and services. This philosophy is in contrast to the concept of a highly distributed network on which the Internet is based.
- IMS is a very **complex** system and is **difficult to implement** because of the numerous functional modules required and the optional or undefined features. And, unfortunately, there is currently no certification testing taking place;
- IMS is a **moving target** that continuously has new features added to it. That threatens homogeneous implementations. Partial IMS implementations run the risk of interoperability problems;
- Given the necessary investments in infrastructure that includes everything from IMS endpoints to core applications servers, there is **no proof of cost savings** using IMS-based fixed-mobile convergence;
- There is a **fear** that IMS's built-in capability of controlling all sessions and Internet-based services might lead to the end of the ubiquitous "free for all" Internet.

20.9 Conclusions

No communications technology that has witnessed such a tremendous growth and exerted such a worldwide global impact in a relatively short period of time as cellular mobile radio. **Fixed wireless and cellular mobile networks convergence** (FMC), the subject of this book, is one technological aspect of wireless communications that happens under our eyes. The ultimate scope of FMC is to provide transparent communications across heterogeneous wireless networks by using **one handset** and **one portable number** that works across any wireless network capable of transmitting digital information.

This book is a comprehensive analysis of the wireless technologies that are being considered in the effort to achieve convergence, and, in addition, we look at the architectural solutions, components, interfaces, and layered protocol stacks that provide the convergent network infrastructure. Particular networks we have examined for convergence are: **GSM/CDMA Cellular mobile networks**, spanned over three generations, **Wireless Local Area Networks** IEEE 802.11x in point-to-point and mesh configurations, **Wireless Personal Area Network** technologies such as **Bluetooth** and **ZigBee**, **Wireless Metropolitan Area Networks** such as IEEE 802.16d/e **WiMAX**, and **Near-Field Sensor Networks** based on **RFID, NFC**, and **Wireless USB**. Manufactured products, applications, services, and real case studies of pair convergent solutions are analyzed along with standardized global solutions for convergence such as the **Unlicensed Mobile Access** (UMA) and **IP-based Multimedia Subsystem** (IMS).

The convergence analysis included a look at the **management capabilities** associated with particular solutions and a look at the **services** provided in conjunction with convergent solutions. Further, we noted the well defined metrics that allow measurement of parameters that quantify **Quality of Service** (QOS). In this respect, technical diagrams, features/components tables, comparative analysis of various solutions, lists of applications, products, vendors, carriers, and service providers are all information that complements architectural solutions to give this book a highly practical use. The book ends with a set of evaluation criteria to qualify fixed-mobile convergent solutions and with the analysis of current state, trends, and issues regarding FMC implementation.

On a final note, we should be aware of the volatility and currency of technical information in this age of rapid technical progress. Therefore, there is a good chance that ideas and figures presented in this book may become obsolete because of new developments, products, features, and new standards that might appear in this ever evolving dynamic picture.

References

1 "Open System Interconnection-OSI Reference Model", ISO 7498 (1978) and ISO 7498/1 (1994).

2 "Federal Communications Commission-FCC, United States Frequency Allocation Spectrum", 2003.

3 Iosif G. Ghetie, "Network and Systems Management: Platforms Analysis and Evaluation", 1997.

4 "Simple Network Management Protocol SNMP Version 1", RFC 1157, 1990.

5 "Common Management Information Protocol CMIP", ISO 9596, 1994.

6 "Telecommunications Management Network-TMN", ITU-T Recommendations M3000 Series, 1995.

7 "Telecommunications Operation Map", TeleManagement Forum, Journal of Network and Systems Management, 2003.

8 "Personal Communications Networks and Services-PCN/PCS", 1900 MHz US Spectrum and US GSM/CDMA 2G Mobile Network Architecture, 1998.

9 "CDMA 2000 High Rate Packet Data Air Specifications", 3rd Generation Partnership Project, 3GPP2, 2004.

10 "3rd Generation Partnership Project-3GPP Technical Specifications TS 05.05", Group GSM/EDGE Frequency Bands and Channel Arrangement, 2003.

11 Rajeev Gupta, "Path to Cellular/WLAN Convergence Crosses UMA", Kineto Wireless. Inc., June 2005.

12 "Mobile Wireless Internet Forum Technical Report", MTR-006, v1.0, 2000.

13 "IP Mobility Support for IP V4", IETF RFC 3344, 2002.

14 "HP OpenView TeMIP Product Family 5 Data Sheet", 2007.

15 "3G Technical Specification (TS) 23.107 V3.3.0 Specification Group Services and System Aspects; QOS Concept and Architecture", Release 1999.

16 "Methods of Subjective Determination of Transmission Quality", ITU-T P.800, (Mean Opinion Score), 1996.

17 Gary Audin and Fiona Lodge, "Call Quality is more than Voice Quality, Business Communications Review, August 2006.

18 "GSM Association, GSM World, GPRS Capability Types and Multislot Classes".

19 "IBM Tivoli Netcool OMNIbus Centralized Management System", 2007.

20 "Wi-Fi Alliance; Wi-Fi Certification Program; Wi-Fi Certified Products", 2007, (www.wi-fi.org).

21 "Wireless LAN Medium Access Control (MAC) and Physical Layer (PHY) Specifications: Enhancement for Higher Throughput", IEEE 802.11n.

22 Paul DeBeasi, "Wireless LANs – a New Battle Begins", Business Communications Review, June 2007.
23 "Local and Metropolitan Area Networks, Part 11: Wireless LAN MAC and PHY Specifications: MAC QOS Enhancements", IEEE Standard, 2005.
24 "Perceptual Evaluation of Speech Quality (PESQ), An Objective Method for End-to-End Speech Quality Assessment of Narrowband Telephone Networks and Speech Codecs", ITU-T P.862 standard, 2002.
25 "Wireless Medium Access Control (MAC) and Physical Layer (PHY) Specifications for Wireless Personal Area Networks (WPANs)", IEEE 802.15.1.
26 "Wireless Medium Access Control (MAC) and Physical Layer (PHY) Specifications for Low Rate Wireless Personal Area Networks (LR-WPANs)", IEEE 802.15.4.
27 "Bluetooth Special Interest Group (SIG)", 1998–2007, (www.bluetooth.com).
28 "Bluetooth Specifications Version 1.2", 2003.
29 "ZigBee Alliance; Wi-Fi Certification Program; Wi-Fi Certified Products" 2007, (www.zigbee.org).
30 "Standard for Broadband over Power Line Hardware", IEEE P1675.
31 "Powerline Communication Equipment: Electromagnetic Compatibility (EMC) Requirements – Testing and Measurement Methods", IEEE P1775.
32 "WiMAX Forum; Certification Program; Certified Products", 2007, (www.wimaxforum.org).
33 "Local and Metropolitan Area Networks Part 16: Air Interface for Fixed Broadband Wireless Access Systems", IEEE Standard 802.16–2004, 2004.
34 "Local and Metropolitan Area Networks Part 16: Air Interface for Fixed and Mobile Broadband Wireless Access System", IEEE Standard 802.16e, 2005.
35 "Mobile WiMAX Part I Technical Overview and Performance Evaluation", 2006, WiMAX Forum.
36 "FDD Enhanced Uplink Overall Description", 3 GGP TS25.309 Version 6.1.0, 2004.
37 Kejie Lu and others, "A Secure and Service-Oriented Network Control Framework for WiMAX Networks", IEEE Communications Magazine, May 2007.
38 Roger Marks, "IEEE 802.16 Wireless MAN Standard: Myths and Facts", Wireless Communications Conference", June 2006.
39 "Mobile WiMAX Part II A Comparative Analysis", 2006, WiMAX Forum.
40 "Federal Communications Commission-FCC, United States Frequency Allocation Spectrum", 2003.
41 "Radio Frequency ID (RFID) Forum", Europe, 2007, (www.rfidforum.org).
42 "The Air Interface for Radio-Frequency Identification (RFID) Devices Operating in the 860 MHz to 960 MHz Industrial, Scientific, and Medical (ISM) Band Used in Item Management Applications", ISO/IEC 18000-6C, 2004.
43 "ISO/IEC 18000 Part 7 Parameters for Air Interface Communications at 433 MHz".
44 "Near Field Communications (NFC) Forum: Operations Specifications", 2007, (www.nfc-forum.org).
45 "Multiband OFDM Alliance (MBOA)", 2007, (www.intel.com).
46 "Wireless USB Specifications Revision 1.0", 2007, (www.usb.org).
47 "Universal Serial Bus Implementors Forum (USB-IF)", (www.usb.org).
48 "Media Independent Handover Services", IEEE Draft 802.21.
49 "College Plans 802.11n WLAN Rollout", John Cox, Network World, 2007.

50 "Toll Quality Voice and High-Capacity Data Over Wireless", Meru Networks Product Sheet, 2007.
51 John Cox, "New York College Test First 802.11n WLAN", Network World, 2007.
52 "World-Wide-Walkie-Talkie: United States Patent 6763226" http://www.freepatentsonline.com/6763226.html
53 "Motorola Canopy and MOTOwi4: HotZone Duo Dual-Radio Meshed Wi-Fi System", Motorola Corporation.
54 "iPhone Technical Specifications", (www.apple.com/iphone/features), 2007.
55 Brad Smith, "iPhone's Impact on AT&T's Network", Wireless Week, September 2007.
56 "HiPath Wireless Driving Value through Open Mobility Solutions", Siemens Communications, 2006.
57 Wei Song and others, "Resource Management for QOS Support in Cellular/WLAN Interworking", IEEE Network, September/October 2005.
58 "BLIP Systems Delivers the World's Largest Health Synchronization Network for Health Care", BLIP Systems News.
59 "Norwegian Center for Telemedicine, Tromso" Conference and Exhibition, University Hospital, News, 2005.
60 Andrew Wheeler, "Commercial Applications of Wireless Sensor Networks Using ZigBee", IEEE Communications Magazine, April 2007.
61 Mikhail Galeev, "Will Bluetooth, ZigBee, and 802.11 All Have a Place in Your Home?", Embedded Systems Design, April 2004.
62 "ZigBee Alliance Document 053474r06 ZigBee Specifications Version 1.0", December 2004.
63 Brad Smith, "ZigBee Generates Power", Wireless Week, July 2007.
64 Axel Sikora, "ZigBee and the ISM Coexistence Issue", University of Cooperative Education, Lorrach, Germany, June 2005.
65 Brad Smith, "A Base Station in Every Home", The concept of Femtocells, Wireless Week, 2007.
66 Shridhar Mubaraq Mishra and others, "Detect and Avoid: An Ultra-Wideband WiMAX Coexistence Mechanism", IEEE Communications Magazine, June 2007.
67 Gangxiang Shen and others, "Fixed Mobile Convergence Architectures for Broadband Access: Integration of EPON and WiMAX", IEEE Communications, 2007.
68 Glen Kramer, "From 1 to 10 Gbps in Five Years, The EPON Story", IEEE 802.3av Task Force, 2007.
69 "WiMAX, Wi-Fi, and Intel Technologies Transform the Ironman Triathlon Experience", Intel Mobile Technology Case Study, 2007.
70 "Hopling Technologies Enhances Metro Mesh in Collaboration with Fujitsu" Hopling Technologies Case Study, 2007.
71 Brad Smith, "For Sale: Wireless Beachfront Property 700 MHz Spectrum Band", Wireless Week, September 2007.
72 "US Frequency Allocation Chart. Radio Spectrum", FCC US Department of Commerce.
73 Steve Louis, "An Introduction to RFID Technology and its Use in the Supply Chain", LARAN RFID, January 2004.
74 "RFID Basics Updated Including Generation 2", Monarch Products and Services, 2006.
75 Swapna Dontharaju and others, "The Unwiding of a Protocol", IEEE Applications and Practices, April 2007.
76 "RFID Solutions for HealthCare Service Providers", Alvin Systems.
77 Steffen Schaefer, "IBM's Secure Trade Lane Solution", RFID im Blick, CEBIT 2007.

78 Dirk Spannaus, "Der Sclussel zur Wirtschaftlichkeit mit RFID", RFID im Blick, CEBIT 2007.
79 "Near Field Communications in the Real World", Innovision Research and Technology. Tutorial.
80 Mary Catherine O'Connor, "Sports Fans Use NFC to Pay and Play", RFID Journal, December 2005.
81 Marco Conti and Silvia Giordano, "Multihop Ad-hoc Networking: The Reality", IEEE Communications Magazine, April 2007.
82 S. Jajashree, C. Siva Ram Murthy, "A Taxonomy of Energy Management Protocols for Ad-hoc Wireless Networks", IEEE Communications Magazine, April 2007.
83 "Overview of 3GPP Release 6; Summary of all Release 6 Features", Version TSG#33, ETSI Mobile Competence Center, 2006.
84 "UMA Consortium, UMA Architecture (Stage 2)", September 2004.
85 "Bluetooth Network Encapsulation Protocol (BNEP) Specification", Bluetooth SIG, 2001.
86 "Nokia 6301 Handset Data Sheet", (www.nokia.com/Nokia_6301) data_sheet, 2007.
87 "SIP: Session Initiation Protocol", IETF RFC 3261, 2002.
88 "Internet Message Format", IETF RFC 2822.
89 Sridhar Ramachadran, "Session Controllers Join H.323 and SIP", Network World, October 2003.
90 "Mobile IGNITE Announces Release v1.0 of Functional Specifications for Fixed Mobile Convergence Handover", 2006.
91 "The E.164 to Uniform Resource Identifiers (URI) Dynamic Delegation Discovery System (DDDS) Application (ENUM), IETF RFC 3761.
92 "Overview of 3GPP Release 5; Summary of all Release 5 Features", 2003.
93 " Technical Specifications Group Core Network and Terminals; IP Multimedia Subsystem; Call Control Protocol based on Session Initiation Protocol (SIP) and Session Description Protocol (SDP), Stage 3", 3GPP, TS 24.229, 2005.
94 "Technical Specification Group GSM/EDGE Radio Access Network Generic Access to the A/Gb Interface", 3GPP TS 43318 Release 7 v3.0 Stage 2, 2007.
95 "Telecommunications Management; Charging Management; IP Multimedia Subsystem (IMS) Charging", 3GPP TS 32240, Release 7, 2006.
96 "Open Service Access (OSA) Applications Programming Interfaces (API), Parlay 5", Parlay Group, 2005.
97 "Technical Specifications Group Core Network; IP Multimedia Subsystem; Session Handling; IM Call Model", 3GPP TS 22.228, Release 5, 2005.
98 Lin Cheng, Frank Toupin, "IMS The Road Towards Convergence", Connect-World Europe, Article 19, 2007.
99 Brad Smith, "Verizon Unveils IMS Vision Team", Wireless Week, August 2006.
100 "Telecommunications and Internet Converged Services and Protocols for Advanced Networking (TISPAN) Next Generation Network (NGN) Functional Architecture Release 1", ETSI (ES 2820010), 2005.
101 "ETSI TS 123078, Version 4.0.0", 2001.
102 "3GPP TS 22.078, CAMEL Service Description, Stage 1", 2004.
103 "3GPP TS 23.078, CAMEL Technical Realization, Stage 2", 2004.
104 "3GPP TS 29.278: 3rd Generation Partnership Project; Technical Specification Group Core Network; Customized Applications for Mobile Network Enhanced Logic (CAMEL) Phase 4 CAMEL Application Part (CAP) Specification for IP Multimedia Subsystems (IMS)".

105 "The COPS (Common Open Policy Service) Protocol", IETF RFC 2748.
106 "COPS Usage for RSVP", IETF RFC 2749.
107 "COPS Usage for Policy Provisioning (COPS-PR)", IETF RFC 2750.
108 "Open Mobile Alliance Principles", 2004.
109 "OMA Device Management Working Group Charter", 2006.
110 "Open Handset Alliance", Home Page, (www.openhandsetalliance.com).
111 "IT Facts All Mobile Usage Statistics from Various Sources: ABI Research, Business Communications Review, InStat, IDC, Gartner, Forrester Research, Giga Research, Infonetics Research, Pyramid Research, NPD Group, Telecom View, Wireless Week, Yankee Group, etc.", 2007, http://www.itfact.biz.
112 Aaron Vance, "Catching up on the World of Wi-Fi", Business Communications Review, March 2007.
113 Marc Bovenschulte and others, "Status and Perspectives of Applications based on RFID on National and International Markets," CEBIT 2007.
114 M. Rubayat Kibria and Abbas Jamalipour, "On Designing Issues of the Next Generation Mobile Network", IEEE Network, January/February 2007.
115 Stratos Chatzikyriakos, "Boosting Bluetooth to 250 Mbps with Ultra Wideband", Wireless Design and Development, December 2007.
116 "Cognitive Radio-based PHY/MAC Air Interface. Wireless Regional Area Networks", IEEE 802.22 Draft 2.0, 2007.
117 Kent Heath, "Clearing the Path to 3G Handset RF Integration", Wireless Design and Development, March 2007.
118 Manuel Wexler, "The Nomadic Consumer: Rethinking Work and Play", Business Communications Review, September 2007.
119 Cristopher J. Pavlovski, "Service Delivery Platforms in Practice", IEEE Communications Magazine, March 2007.
120 Donna Griffin and Dirk Pesch, "A Survey on Web Services in Telecommunications", IEEE Communications Magazine, July 2007.

Note: The author has made every attempt to reference the resources used in this book. Therefore, he apologizes for any possible omissions while expressing his gratitude and thanks to all contributors.

Index